W9-BND-492

IMPLICATIONS OF CONTINENTAL DRIFT TO THE EARTH SCIENCES

VOLUME 2

Implications of Continental Drift to the Earth Sciences

VOLUME 2

NATO Advanced Study Institute, April 1972,
The University of Newcastle upon Tyne

Edited by
D. H. TARLING AND S. K. RUNCORN
School of Physics, The University,
Newcastle upon Tyne,
England

1973
ACADEMIC PRESS · London & New York

ACADEMIC PRESS INC. (LONDON) LTD
24–28 Oval Road
London NW1

U.S. Edition published by
ACADEMIC PRESS INC.
111 Fifth Avenue
New York, New York 10003

Library of Congress Catalog Card Number: 72–7706
ISBN: 0–12–683702–3

PRINTED IN GREAT BRITAIN BY
THE WHITEFRIARS PRESS LTD., LONDON AND TONBRIDGE

Contributors to Volume 2

W. ALVAREZ — Lamont-Doherty Geological Observatory of Columbia University, Palisades, New York 10964, USA.
R. BLACK — Department of Geology, University of Haile Selassie, Addis Ababa, Ethiopia.
M. H. P. BOTT — Department of Geology, University of Durham, Durham, England.
P. J. BUREK — Institut für Geologie-Paläontologie, Universität Tübingen, Sigwartstr, 10, Germany.
K. BURKE — Erindale College, University of Toronto, Mississauga, Ontario, Canada.
W. M. CADY — US Geological Survey, Denver, Colorado 80225, USA.
P. J. CONEY — Department of Geology, Middlebury College, Middlebury, Vermont, USA.
B. W. DARRACOTT — Department of Earth Sciences, University of Leeds, Leeds, England.
P. R. DAWES — Geological Survey of Greenland, Copenhagen, Denmark.
E. R. DEUTSCH — Department of Physics, Memorial University of Newfoundland, St. John's, Newfoundland, Canada.
J. F. DEWEY — State University of New York, Albany, New York 12222, USA.
R. S. DIETZ — NOAA, Atlantic Oceanographic and Meteorological Laboratories, 15 Rickenbacker Causeway, Miami, Florida 33149, USA.
J. D. FAIRHEAD — Geological Survey, Pretoria, South Africa.
H. FAURE — Laboratoire de Géologie Dynamique, Université de Paris, Paris, France.
I. G. GASS — Department of Earth Sciences, The Open University, Wilton, Bletchley, Buckinghamshire, England.
A. G. SMITH — Department of Geology, Sedgwick Museum, Downing Street, Cambridge CB2 3EQ, England.
R. W. GIRDLER — School of Physics, University of Newcastle upon Tyne, Newcastle upon Tyne, England.
A. M. GOODWIN — Department of Geology, University of Toronto, Toronto, Ontario, Canada.
N. S. HAILE — Department of Geology, University of Malaya, Kuala Lumpur, Malaysia.
S. A. HALL — School of Physics, University of Newcastle upon Tyne, Newcastle upon Tyne, England.
W. HAMILTON — US Geological Survey, Denver, Colorado 80225, USA.
J. C. HOLDEN — NOAA, Atlantic Oceanographic and Meteorological Laboratories, 15 Rickenbacker Causeway, Miami, Florida 33149, USA.
P. M. HURLEY — M.I.T., Cambridge, Massachusetts, USA.
P. JAKES — Geological Survey, Hradebri 9, Praha, Czechoslovakia.
G. A. L. JOHNSON — University of Durham, Durham, England.

J. P. Kennett	Graduate School of Oceanography, University of Rhode Island, Kingston, R.I., 02881, USA.
P. E. Kent	The British Petroleum Co. Ltd., Britannic House, Moor Lane, London, England.
L. C. King	University of Natal, Durban, South Africa.
P. S. Mohr	Smithsonian Astrophysical Observatory, 60 Garden Street, Cambridge, Massachusetts, USA.
E. M. Moores	Department of Geology, University of California, Davis, California 95616, USA.
A. E. M. Nairn	Department of Geology, Case Western Reserve University, Cleveland, Ohio 44106, USA.
M. F. Osmaston	The White Cottage, Sendmarsh, Ripley, Woking, Surrey, England.
J. D. A. Piper	Department of Earth Sciences, University of Leeds, Leeds, England.
K. V. Rao	Department of Physics, Memorial University, St. Johns, Newfoundland, Canada.
R. A. Reyment	Paleontologiska Institutet, Uppsala University, Box 558, S75122 Uppsala, Sweden.
S. K. Runcorn	School of Physics, University of Newcastle upon Tyne, Newcastle upon Tyne, England.
R. W. R. Rutland	Department of Geology and Mineralogy, University of Adelaide, Adelaide, South Australia.
R. M. Schackleton	Department of Earth Sciences, University of Leeds, Leeds, England.
N. H. Sleep	Department of Earth Sciences, M.I.T., Cambridge, Massachusetts 02139, USA.
E. Stump	Institute of Polar Studies and Department of Geology, Ohio State University, Columbus, Ohio 43210, USA.
J. Sutton	Department of Geology, Imperial College, London, England.
F. J. Vine	School of Environmental Sciences, University of East Anglia, Norwich, England.
P. R. Vogt	US Naval Oceanographic Office, NRL/CBD, Chesapeake Beach, Maryland 20732, USA.
N. D. Watkins	Graduate School of Oceanography, University of Rhode Island, Kingston, R.I. 02881, USA.
R. A. Wells	Space Sciences Laboratory, University of California, Berkeley, California 94720, USA.
A. J. Whiteman	Department of Geology, University of Bergen, Bergen, Norway.

Introduction

Quantitative geophysical evidence for the reality of continental drift was first obtained from the comparison of palaeomagnetic directions in igneous and sedimentary rocks from different continents. More recently Wegener's concept of continental drift has been beautifully complimented by the hypothesis of sea-floor spreading. Again the palaeomagnetism of the ocean floor has provided quantitative evidence for its occurrence. Thus the older qualitative arguments from the geological record, presented so imaginatively by Alfred Wegener, have been vindicated. In recent years we have seen a marked change in the climate of scientific opinion about the reality of major horizontal movements of parts of the Earth's surface and, from palaeomagnetic and other geophysical studies, the positions of the continents in different geological periods and the evolution of the ocean basins are being determined. It is still not very clear how these movements take place in time and there is still considerable uncertainty about the precise relationships of different parts of the Earth's surface during the geological past.

These developments have, however, essentially ended the long debate about whether or not the classical lines of geological evidence, palaeoclimatic, palaeontological distributions, global tectonic patterns and lithological relationships, support or refute drift. What is now scientifically significant is the study of the geological record in the light of the known horizontal displacements. This is of great potential importance to various other sciences involved in the study of our environment, e.g. biology, global meteorology.

These two volumes are the proceedings of the April 1972 NATO Advanced Study Institute held in the University of Newcastle upon Tyne. They commence mainly with reviews of the objective evidence for the past position of the continents, i.e. the palaeomagnetism of continental and oceanic rocks. The palaeontological evidence is then examined to see how the creation of supercontinents and their fragmentation affects the mobility and rate of evolution of the biota on and around them. This data must also be examined carefully in order to delineate evidence which still appears inconsistent with current views of the past distribution of the continents to see if our present views need modification or whether such discrepancies can yield further information about our planet in the past, for example, the distribution of topographic, predatory or climatic barriers to the migration of terrestrial fauna. Similarly, the movement of continental fragments into different climatic belts obviously has an effect on their prevailing climate, but this movement, particularly the formation or fragmentation of supercontinents, must also have a drastic effect on the climatic belts themselves.

The change in perspective of geological studies is possibly most significant in terms of the relationship between continental movements and economic deposits. Obviously the fragmentation of pre-existing ore formations means that the matching of one half with another allows knowledge gained in one section to be applied directly in another. However, there is increasing evidence that the fragmentation of the continental blocks can give rise directly to the conditions in which ore deposits become concentrated and the structures are formed within which oil and natural gas can accumulate. With this in mind, the relation of various mineral deposits is examined and the processes of rift formation are considered in terms of their effects on geological structures forming along the edges of an opening ocean. Examples of such locations, from the East African rifts through to the Atlantic Ocean, are then examined prior to studies of the evolution of selected mey or neglected areas, mainly during the last 200 million years or so. The fact that present sea-floor data is restricted to the last 150–200 million years does not mean that spreading oceans were not present in earlier times, and the evidence for such past oceans is examined before discussing the relevance of such large scale horizontal movements to the evolution of the continents. An epilogue is included which sets the planetary scene and gives a more light-hearted view of the whole subject.

Clearly, in a five-day conference, the coverage of topics is necessarily incomplete, and this is particularly true for many aspects of sedimentology, geomorphology and stratigraphy. Any discussion of the mechanisms causing the motions were, as far as possible, deliberately excluded—although obviously the understanding of these is becoming increasingly vital to our understanding of the Earth.

We wish to thank the Vice Chancellors of the Universities of Newcastle upon Tyne and Durham and the Lord Mayor of Newcastle upon Tyne for entertaining participants, the Master of University College, Durham, Mr. Slater, for his permission to hold the conference dinner in the Great Hall of Durham Castle, and the Dean of Durham and his colleagues for arranging, as they have so readily done for conferences in previous years, a fascinating tour of the Norman Cathedral.

Grants provided by the NATO Advanced Study scheme enabled many research students to attend the meeting and four travel grants were also made available by the National Science Foundation. Contributions towards Institute costs were received from British Petroleum and Gulf Oil. The administration was undertaken by Mr. W. F. Mavor, Mrs. Marion Turner, Miss Anne Codling, Mrs. Sheena Mavor, and the success of the meeting, to a large extent, was due to them. We also wish to thank our colleagues in the School of Physics and Department of Geology for their help and cooperation. It is intended that any profits from this publication should be used for the financial support of research students in the Earth Sciences, thus furthering the NATO Advanced Study scheme and the theme of the meeting. We would also like to thank the authors for their cooperation in keeping to their restricted time allowance at the meeting and for submitting their manuscripts at the times requested.

D. H. TARLING
S. K. RUNCORN
January, 1973.

Contents

Part 8 Palaeogeographic Implications

Part 9 Ancient Oceans

Part 10 Continental Evolution

Part 11 Epilogue

CONTENTS OF VOLUME 1

6

RIFT MARGINS AND CONTINENTAL EDGE STRUCTURES

6.1

J. D. A. PIPER
Department of Earth Sciences,
University of Leeds,
Leeds, England

The Geology of Constructive Plate Margins

Introduction

The highly successful theory of plate tectonics has explained the geometry of motions of the Earth's crust in terms of the relative movements on the surface of a sphere between rigid plates. The plates are bounded by extensional margins (ridge axes) connected to each other or to compressive margins (trench systems) by conservative margins (transform faults). The conservative margins merge abruptly with the other types of margins in oceanic areas and more gradually in the continental areas (Wellman, 1971). Although plates may comprise either continental or oceanic crust or both, only the oceanic portions can participate in the processes of plate growth and destruction. The oceanic portions grow by the process of sea-floor spreading which involves the continuous generation of new crust within the asthenosphere below the ridge axes. This paper outlines present knowledge of the geological structure and spatial distribution of ridge axes, and examines explanations for the different types.

The Lithosphere and Asthenosphere at Extensional Plate Margins

Many works have emphasized that the geological and geophysical data on sea-floor spreading require the presence of a layer with significant strength and about 100–150 km in thickness, the lithosphere, moving on top of a layer of no effective strength, the asthenosphere. Issacks, Oliver & Sykes (1968) have shown that this concept is consistent with the seismological data. Molnar & Oliver (1969) have examined the propagation of S_n waves in the upper mantle and have found that the uppermost mantle beneath ridge crests is not a part of the rigid lithosphere but only becomes so about 200 km from the axis. Bott (1965) has found a partial melting to be the most acceptable mechanism for the production of the low density mantle below ridge axes and Oxburgh & Turcotte (1968) consider that the upper mantle is undergoing partial melting to produce basalt. Kay, Hubbard & Gast (1969) have concluded that the main chemical characteristics of basalts from ridge axes can be explained in terms of a partial melting of the order of 30% at depths of 15–25 km.

The concept has therefore developed in which plates of lithosphere are thinning greatly near their constructive margins and are being penetrated by rising columns of asthenosphere. It has been argued (Cann, 1970) that the sides of this wedge of asthenosphere must be steep and that the asthenosphere must rise relatively rapidly; if this is not so then the material will cool by conduction and not pass through the basalt melting curve.

The transition between the asthenosphere and the lithosphere would appear to define the zone where the interstitial melt leaves the parent material to become a separate phase, since it is unlikely that it will be controlled by conduction along the whole length of the narrow wedge. The temperature at the lithosphere/asthenosphere contact must be close to the melting point of basalt and the melt will migrate upwards along the interface to accumulate in a magma chamber, or chambers, below the ridge crest. This magma may remain and cool by conduction where it is or it may be intruded into the pre-existing crust if the stress conditions are appropriate.

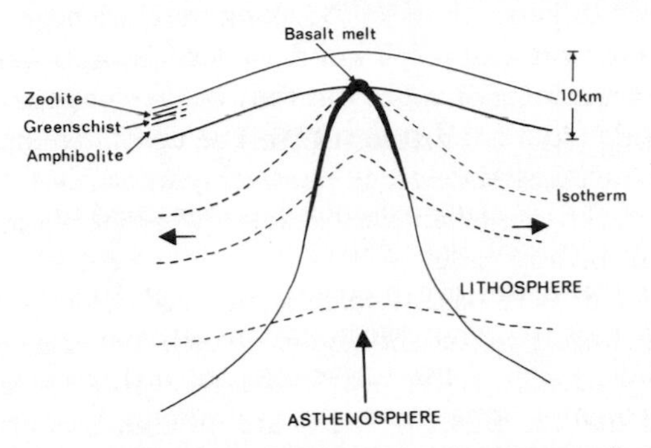

FIGURE 1 Schematic cross section through the crust and upper mantle at a growing plate margin based on the hypothesis that the crust grows by the addition of basaltic derivatives from a rising plume of ultrabasic material. The crust and upper mantle move apart as two rigid plates coupled at a zone of weakness along the axial zone. As this zone fractures basalt is continuously intruded and extruded from the body of magma (after Cann (1970) and Vogt *et al.* (1969)).

This model for the process of crustal generation at constructive plate margins is based on laboratory studies and theoretical considerations, and is generally consistent with knowledge of the upper mantle below ridge axes. The presence of basaltic melt below a ridge axis has received support from studies in Iceland. From magnetic variation studies Hermance & Garland (1968) have inferred the presence of a region of high conductivity below the island. It has not been established whether this layer coincides with the base of the crust, but Hermance & Grillot (1969) have calculated from magnetotelluric data that the temperatures at the base of the crust are close to the melting point of basalt and are similar to those obtained by the downward extrapolation of thermal gradients observed near the surface.

Crustal Structure of Ridge Crests

Extensive dredge hauls have shown that the bulk of the topography along mid-ocean ridge axes is built up by extrusive volcanism. Basalts have generally been erupted as pillow lavas and when eruption takes place in this manner it characteristically builds up piles with slopes of up to 25° (Jones, 1966) and a conical cross section about the eruptive fissure. Later eruptions will be controlled by the earlier topography and tend to fill-in irregularities. From studies of the mid-Atlantic ridge crest at 45° N, Aumento (1967) has found that the composition of dredged samples varies continuously from olivine tholeiites, flooring the median valley, to alkali basalts capping the crestal peaks on the flanks of the valley. All rock types can be produced by fractional melting of a suitable parent, and a typical cycle commences with outpourings of olivine tholeiites on the floor of the median valley. Subsequent eruptions are progressively enriched in alkalis until the former valley floor has become uplifted and capped with alkali basalt to form the flanking mountains during an interval of 1–2 m.y.

Ridge axes, however, vary greatly in their characteristics. The Reykjanes ridge has a smooth crest and no median rift valley, while the ridge continuation north of Iceland has a rougher topography and a median rift although the spreading rate is similar. The width of the crest increases from around 30 km north of Iceland to about 80 km near Jan Mayen (Vogt *et al.*, 1970). A median valley is developed along most of the the mid-Atlantic ridge south of 50° N, along the Carlsberg and the Gorda ridges; it is typically 10–20 km wide and 0·5–3 km deep. On the mid-Atlantic ridge at 45° N faults are absent from the median valley but they begin to appear in the crestal mountains flanking each side (Aumento *et al.*, 1971). The uplifts of the crestal mountain at 22–23° N in the Atlantic takes place along faults dipping at 30–45° inward (van Andel & Bowin, 1968) and in the immediate crestal area the outer blocks are uplifted higher than those nearer the median valley.

In addition to fresh extrusive basalts some sections of the ridge system have yielded metamorphic basalts belonging to the zeolite greenschist and amphibolite facies; sometimes these metamorphic rocks are found at high topographic levels. Cann (1970) has postulated that the first two facies are produced by the metamorphism of the extrusive lava pile since only this phase has access to abundant water. The relatively water-deficient amphibolites might result from the metamorphism of a dyke or other intrusive complex at somewhat greater depths. The seismic layer 2 outcrops at the ridge axes and is known to be composed, at least in its upper levels, of extrusive basalts. The interpretation of crustal layer 3, however, presents more problems since it is not everywhere present. Le Pichon *et al.* (1965) did not locate it closer than 100 km to the mid-Atlantic ridge axis. They have observed an inverse relationship between layers 2 and 3 so that the total crustal thickness remains about the same, and Le Pichon (1969) has also noted that layer 3 generally increases in thickness with distance from the ridge axis. The interpretation of the Troodos Massif of Cyprus as an upthrust portion of ocean crust (Gass, 1968) suggests that the lower parts of the oceanic crust are composed of a mixture of basic plutons and dykes. However from general considerations it seems likely that a gabbro pluton/feeder dyke complex is universally present beneath, or grading into, the extrusive pile, and cannot by itself be the cause of crustal layer 3. Hess (1962) originally proposed that this layer was composed of serpentinized peridotite but Cann (1968) has shown that the thermal gradients, uniform velocities and crustal relationships of this layer are not compatible with the hypothesis. He has argued that it is formed by the conversion of basaltic material to amphibolite and since this commonly takes place when the temperature gradients through the crust are sufficiently high, layer 3 will be restricted to zones of high heat flow.

Earthquakes at constructive plate margins are largely confined to depths of less than 10–20 km (Issacks *et al.*, 1968). They appear to be restricted to within 20 km of the ridge axis which includes the median valley and immediate flanking crestal mountains where they occur. The ridge axes give fault plane solutions with normal components while the earthquakes along the transform faults give exclusively strike-slip solutions. The earthquakes associated with ridge axes normally have magnitudes of less than 5 and only the fracture zones are associated with large earthquakes. This may be a consequence of larger stresses building up under compression than under tension (Stéfansson, 1967) and the concentration of activity along the active fracture zones compared with the ridge segments leads to the conclusion that very little of the work done in moving the plates apart is manifested as brittle fracturing on a seismic scale. The ridge segments are characterized by earthquakes swarms (Sykes, 1967). They are believed to be an expression of volcanic activity here and are confined to the ridge crests; Sykes (1967) has discussed an example of an earthquake located outside of the median valley which however, was not associated with a swarm.

Present knowledge of ridge axes suggests that there is a good correlation between rough topography, the presence of a median valley, earthquake frequency, irregular magnetic stripe anomalies and probably variable petrology (Cann, 1968). These ridges (distinguished here as 'rough') contrast with ridges showing a smooth topography, no median rift valley, infrequent earthquakes, regular stripe anomalies, and with only unmetamorphozed basalts cropping out at the surface (distinguished as 'smooth' ridges). The factors defining these two ridge types do not seem to be related to spreading rate, the magnitude of the heat flow and Bouguer anomalies and the presence or absence of layer 3. Consequently there appear to be two sets of factors controlling the gross structure of constructive plate boundaries: the first set controls the small scale features (up to about 20 km in scale) defining rough and smooth ridges, and the second set controls the rate of growth of the plate (Cann, 1968).

Distribution of Volcanism at Ridge Crests

The distribution of earthquake activity suggests that tectonic activity is confined to a narrow band about the crest of the ridge. Where well developed magnetic stripe anomalies are present they further imply that magmatic activity is confined to the axial region. Strong linearity is shown by the magnetic anomalies over the Reykjanes ridge where dredge hauls (Krause & Schilling, 1969) have revealed fresh basalts only within about 15 km of the ridge axis, and basalts at greater distances are altered, and have acquired a manganese coating. Matthews & Bath (1967) have developed a model for injecting dykes with a normal distribution about the ridge axis, and have produced anomalies with a similar complexity to those observed at 45° N over the mid-Atlantic ridge with a standard deviation of 5 km. This suggests that nearly all the dykes must be injected within 10 km of the ridge axis or within the median valley in this area. Harrison (1968) has refined this model and used the polarity time scale to simulate anomalies over the Reykjanes and East Pacific ridges; he increased the standard deviation of the normal distribution until critical anomalies caused by reversed events of short time duration were suppressed. A standard deviation of less than 3 km can explain the features of the anomalies observed over these ridges and the results imply that 95% of the dykes are injected within 6 km of the axis.

In reality the distribution may be more nearly rectangular with volcanism restricted within a band about the ridge but being fairly constant within that band. The present author has examined the case of the Reykjanes ridge using both normal and rectangular probability distributions. The method followed was essentially that used by

Harrison (1968). Dykes 20 m thick are injected at a frequency appropriate to the spreading rate parallel to the ridge axis and infinite in this direction, the injection sequence is reversed and the magnetization averaged over successive group of 20 dykes for the anomaly calculation. The dykes were assumed to terminate at the mean depth of the sea floor over the Reykjanes ridge and to extend 2 km below this, and the observation height was taken as 0·47 km above sea level so that the profiles are directly comparable with those obtained by Heirtzler *et al.* (1966) and slightly smoothed relative to those obtained by Godby *et al.* (1968) at 0·3 km above sea level. Figure 2 shows the anomalies generated for rectangular distributions with several widths of injection (D) and a normal distribution with several standard deviations (T), and using the polarity time scale of Cox (1968). The magnetization is assumed to be entirely remanent with an intensity of 0·0065 Gauss and to be oriented along the dipole field. On some of the observed profiles over the Reykjanes ridge the flanks of the axial anomaly are modified by an anomaly attributable to the Jaramillo event. On the models this anomaly is practically suppressed at T values of 0·5 km and completely at D values of 2·0 km, suggesting that if this anomaly is real actual distributions must be narrower than this. The next control is obtained from the anomaly due to the Mammoth and Kaena events which can be isolated on a few profiles and where present restricts T values to less than about 2 km and D values to less than about 3 km. The anomaly due to the Olduvai–Gilsa event is universally present and on the models this is nearly lost at D = 3 km and T = 5 km. The studies suggest that if volcanism takes place about the Reykjanes ridge with a normal distribution it must have a maximum standard deviation of between 0·5 and 3 km, and if it is more uniformly distributed and approaches a rectangular distribution it can be spread across widths of 5 km or possibly more before critical event anomalies are eliminated.

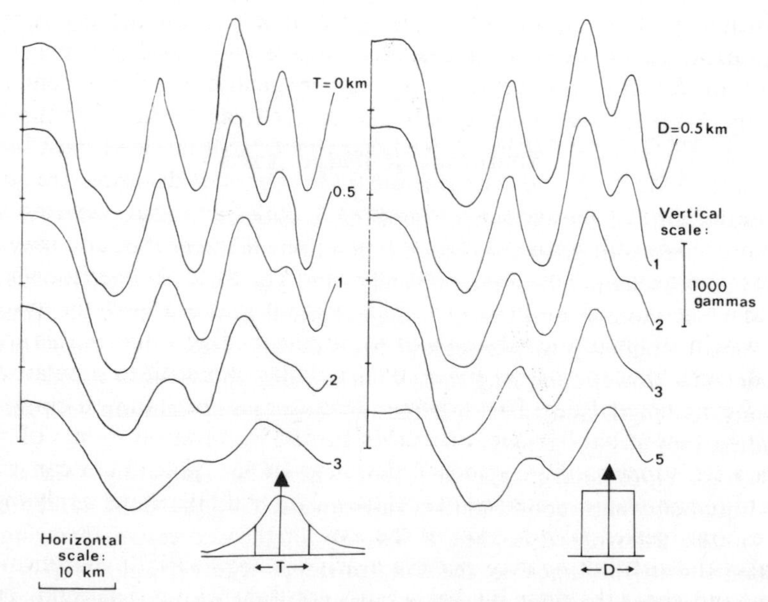

FIGURE 2 Results of simulated dykes model studies for the case of the Reykjanes ridge using rectangular and normal probability distributions. Computations were made on the Imperial College IBM 7094 computer and details of the model are given in the text.

Although these models are artificial they do give limiting conditions and actual dispersions of volcanism about the ridge axes must be smaller because of a variety of complicating effects to be expected in practise such as topography and faulting of the magnetic layer. They do not accommodate a lateral spreading of the extrusive piles but Watkins & Richardson (1971) have shown that if the horizontal extent of the extrusive bodies of the order of, or less than, the depth of the ocean then the sharpness of the anomaly pattern is not diminished. Magnetic models are rather insensitive to the depth chosen for the base of the magnetic bodies, but it is likely that the magnetization of the ocean crust is concentrated in the unmetamorphozed part of the extrusive pillow lava piles (Cann, 1970).

The approach used here to determine the distribution of volcanism about the ridge axis breaks down when the anomalies close along their length, and show such a lack of symmetry that it is impossible to correlate them with magnetic events. The irregular anomalies cannot be attributed to a very great dispersion of volcanism about the ridge axis since they may still be well defined and of high amplitude but merely show no regular pattern. Also irregular magnetic anomalies are associated with rough ridges so it is likely that a faulting of the magnetic layer contributes to this effect.

In Iceland volcanism is distributed over a distance an order or more wider than the Reykjanes ridge, but Hast (1969) has measured horizontal compressive stresses within the crust on the flanks of the neovolcanic zones and directed approximately away from these zones. His nearest station to a neo-volcanic zone (Keflavik) is only 12 km from a field of open dilation fissures. It appears that pure tension exists only at shallow depth and within a restricted zone about the ridge axis. If magnetic activity takes place outside of this area it may not reach the surface to form extrusive piles but will follow a path controlled by the stress distribution in the crust. Piper & Gibson (1972) have suggested that the features of rough ridge axes can be explained at least in part, by a greater distribution of volcanism than that associated with smooth axis. This concept is developed further in the next section.

Tectonics of Ridge Axes

Most explanations of the growth of plates at constructive plate margins have been controlled by the assumption that rough topography is a consequence of slow spreading. This has been seen to be not completely true. The East Pacific Rise has a smooth topography but the combination of a high thermal gradient and fast spreading rate may prevent the build up of stresses which would be released by brittle fracture on a large scale. The slower spreading mid-Atlantic ridge shows both rough and smooth types along its length and their distribution shows no correlation with the variation in spreading rate along the axis.

Vogt *et al.* (1969) have suggested that when the spreading rate is slow the basaltic liquidus cannot continuously maintain itself at the base of the crust. The growth of the lithosphere is then a function of the successive intrusion of ultrabasic diapirs into the crack in the lithosphere at intervals of about 3 m.y. As a consequence of this a wedge-shaped graben may develop, and crustal widening and volcanism continue until the valley has been replaced by high crestal mountains. A new ultrabasic intrusion then displaces the old one and the mechanism of growth is not a steady state one, but is episodic and can explain the variations in petrology between the floor of the median valley and the crestal hills observed by Aumento (1967).

Cann (1968) has postulated that the fluidity of the mantle under fast spreading ridges allows a smaller turning radius than is possible under slow ridges. This might give a narrower zone of tension while a broader turn of more viscous material would distribute tension through a zone that is wider than the basalt producing zone. The crust is then produced at the immediate axis and passes through a zone of normal faulting about 20 km wide with a rift valley developing in the stagnant zone between the divergent flow lines below the crust. This is a steady state hypothesis which makes no attempt to explain why subsidence occurs in the stagnant zone, but unfortunately it makes drastic assumptions on the geometry of flow of the asthenosphere (see next section) and cannot easily be related to the narrow wedge-like shape of the asthenosphere/lithosphere contact at the plate boundary. Osmaston (1971) has examined the problem of the median valley of continental and oceanic rift, but the mechanism he proposes is difficult to envisage and does not accommodate the gravity field observed over exposed portions of the rift system and the dilational features observed within them.

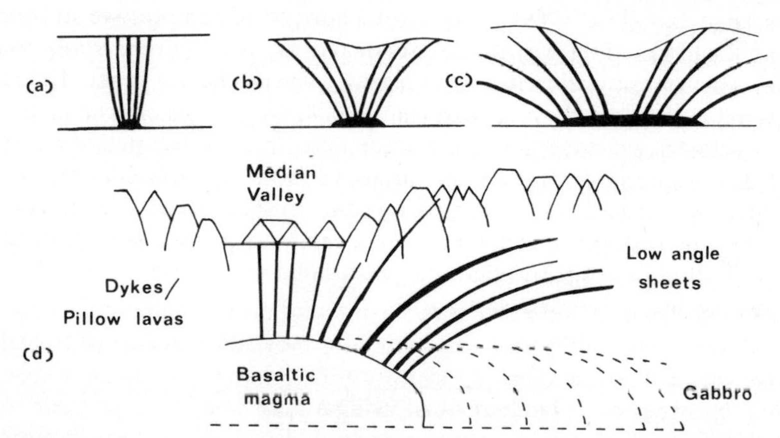

FIGURE 3 Schematic diagram to show how the mode of intrusion into the crust varies as the width of basalt intrusion about the axial zone is increased ((a)–(c)). (d) illustrates how the lateral intrusion of low-angle sheets into the crust may affect the structure of crust by faulting the margin and developing a median rift.

Piper & Gibson (1972) have shown that the way in which a plate margin is constructed will depend on the distribution of stress and basalt intrusion about the plate boundary. For a given plate thickness, if this is narrow the magma chamber of the plate contact will fail and basalt will be intruded vertically upwards in the form of dykes. As the width of the zone of stressing at the base of lithosphere is increased the zone of tension at the surface diminishes and principal stress trajectories are deflected away from the plate surface. More material is then intruded laterally into the plate at low-angle sheets and less material reaches the surface of the plate to feed extrusive volcanics (Fig. 3). This means that the plate margins are incremented not only at the boundary between the plates, but also for some distance on either side. The upper part of the plate will be warped by this addition and a conjugate system of curved stress trajectories will result and develop into normal faults as the crust migrates away from the plate boundary. This mechanism would allow for the development of a median valley and the conjugate fault system could upfault deep crustal levels to high topographic levels. With this model it is not necessary to assume anything about the asthenospheric motion, or that the median valley where it occurs represents a failure of isostasy and is held down relative to the flanking crests. Whether the ridge develops a smooth or rough form is then a function of the width of the zone across which basalt magma accumulates at the base of the crust and thus on the shape of the lithosphere/asthenosphere contact in this vicinity. This mechanism could accompany a non-steady mechanism of the kind proposed by Vogt *et al.* (1969).

Properties of Ridge Axes

The system of segments of ridge axis offset at intervals by transform faults, together with the global pattern of the world rift system, clearly shows that the relationship between plate movement and asthenospheric motion must be a loose one. This is to be expected from the great thickness of the plates and it is only in the ocean basins that we may expect to see evidence of movements of the asthenosphere. Along most of the ridge axes the crustal thicknesses are remarkably uniform (Raitt, 1963) although spreading rates vary greatly, so that whatever may be the motion of the asthenosphere, basalt production must keep pace with the spreading rate. However in some parts of the ocean basins basalt production has been excessive and thick basalt ridges, seamount chains and islands have been produced (Wilson, 1965b). In Vogt's (1972) analysis of sea-floor features south of Iceland he suggested velocities of asthenospheric motion of 20 cm/year along the ridge axis with an unknown but certainly slower velocity in the direction of sea-floor spreading.

Oblique ridge segments offset by transform faults do not represent a stable configuration and Vogt *et al.* (1969) have shown that the most efficient case of crustal growth is for ridge segments to be offset by transform faults at right angles. Ridge systems might thus be expected to acquire this configuration shortly after their initiation. Wilson (1965a) has suggested that the transform fault/ridge systems may represent the form of the original continental fracture in areas such as the Atlantic Ocean. Morgan (1971) has stated that the plates are apparently strong and resistant to major changes since they do not commonly move to new locations. In the Gulf of Aden where continental fracturing and sea-floor spreading are recent (Laughton, 1966) it is possible to demonstrate the permanence of the ridge offsets by restoring the continental fragments to their pre-drift configuration along the transform faults.

However this generalization breaks down in two cases. Firstly a collision of the continental portions of two plates will lead to a readjustment of plate motions, and a change in the poles of rotation of the plates will lead to a reorientation of the offsets and may change the entire character of the sea-floor spreading. Vogt *et al.* (1971) have recognized three episodes of sea-floor spreading in the North Atlantic; the middle one is associated with the extensive development of transform faults while in the latest phase there has been a reorientation of the ridge axis and an elimination of many of the offsets. Secondly along some sections of the ridge system the axis of spreading has moved laterally with time. A major shift in the location of the ridge axis has been suggested in the region north of Iceland (Vogt *et al.*, 1970). Within Iceland the ridge axes are features of limited life which are not offset by transform faults but grow by deformation of the plate boundary; they subsequently decay and the ridge axis shifts laterally (Piper, 1973, and this volume). This may also be the mechanism of crustal growth within the Afar Depression of Ethiopia (Gibson & Tazieff, 1970). Both Iceland and Afar overlie mantle hotspots identified by Morgan (1971) and plate margins of this kind may be restricted to these localities. The length of *en echelon* ridge segments is a function of the length of their offsets (Piper, 1973) but an examination of the ridge segment/transform offset system of the ocean basins shows that no such relationship holds for these parameters here. This relationship for *en echelon* ridge axes suggests that they represent a fracture pattern and they may result from the stressing of lithospheric plates where they are exceptionally thick.

References

Aumento, F., 1967. The mid-Atlantic ridge near 45° N, 2, Basalts from the area of Confederation Peak, *Can. J. Earth Sci.*, **5,** 1–21.
Aumento, F., Loncarevic, B. D. and Ross, D. I., 1971. Hudson geotraverse: geology of the mid-Atlantic ridge at 45° N, *Phil. Trans. Roy. Soc. Lond. A* (in press).
Bott, M. H. P., 1965. Formation of oceanic ridges, *Nature, Lond.*, **207,** 840–3.
Cann, J. R., 1968. Geological processes at mid-ocean ridge crests, *Geophys. J. Roy. astron. Soc.*, **15,** 331–41.
Cann, J. R., 1970. New model for the structure of the ocean crust, *Nature, Lond.*, **226,** 928–30.
Cox, A., 1968. Geomagnetic reversals, *Science*, **163,** 237–45.
Gass, I. G., 1968. Is the Troodos Massif of Cyprus a fragment of Mesozoic ocean floor? *Nature, Lond.*, **220,** 39–42.
Gibson, I. L. and Tazieff, H., 1970. The structure of Afar and the northern part of the Ethiopian Rift, *Phil. Trans. Roy. Soc. Lond. A*, **267,** 331–8.
Godby, E. A., Hood, P. J. and Bower, M. E., 1968. Aeromagnetic profiles across the Reykjanes Ridge southwest of Iceland, *J. Geophys. Res.*, **73,** 7637–49.
Harrison, C. G. A., 1968. Formation of magnetic anomaly pattern by dyke injection, *J. Geophys. Res.*, **73,** 2137–42.
Hast, N., 1969. The state of stress in the upper part of the earth's crust, *Tectonophysics*, **8,** 169–211.
Heirtzler, J. R., Le Pichon, X. and Baron, J. G., 1966. Magnetic anomalies over the Reykjanes ridge, *Deep-Sea Res.*, **13,** 427–43.
Hermance, J. F. and Garland, G. D., 1968. Deep electrical structure under Iceland, *J. Geophys. Res.*, **73,** 3797–800.
Hermance, J. F. and Grillot, L. R., 1969. Correlation of magnetotelluric, seismic and temperature data from southwest Iceland, *J. Geophys. Res.*, **75,** 6582–91.
Hess, H. H., 1962. History of the ocean basins. *In:* Petrological Studies: A Volume in Honor of A. F. Buddington, edited by A. E. J. Engel *et al.*, 599–620. Geol. Soc. Amer.
Issacks, B., Oliver, J. and Sykes, L. R., 1968. Seismology and the new global tectonics, *J. Geophys. Res.*, **73,** 5855–900.
Jones, J. G., 1966. Intraglacial volcanoes of southwest Iceland and their significance in the interpretation of the form of the marine basaltic volcanoes, *Nature, Lond.*, **212,** 586–8.
Kay, R., Hubbard, N. J. and Gast, P. W., 1969. Chemical characteristics and origin of oceanic ridge volcanic rocks, *J. Geophys. Res.*, **75,** 1585–1613.
Krause, D. C. and Schilling, J. G., 1969. Dredged basalt from the Reykjanes Ridge, North Atlantic, *Nature, Lond.*, **224,** 791–3.
Laughton, A. S., 1966. The Gulf of Aden, *Phil. Trans. Roy. Soc. Lond. A*, **259,** 150–171.

Le Pichon, X., 1969. Models and structure of the ocean crust, *Tectonophysics*, **7**, 385–401.
Le Pichon, X., Hontz, R. E., Drake, C. L. and Nafe, J. E., 1965. Crustal structure of the mid-ocean ridges, 1. Seismic refraction measurements, *J. Geophys. Res.*, **70**, 319–40.
Matthews, D. H. and Bath, J., 1967. Formation of magnetic anomaly pattern of mid-Atlantic ridge, *Geophys. J. Roy. astron. Soc.*, **13**, 349–57.
Molnar, P. and Oliver, J., 1969. Lateral variations in alternation in the upper mantle and discontinuities in the lithosphere, *J. Geophys. Res.*, **74**, 2648–52.
Morgan, W. J., 1971. Convection plumes in the lower mantle, *Nature, Lond.*, **230,** 42–43.
Osmaston, M. F., 1971. Genesis of ocean ridge median valleys and continental rift valleys, *Tectonophysics*, **11,** 387–405.
Oxburgh, E. R. and Turcotte, D. L., 1968. Mid-ocean ridges and geothermal distribution during mantle convection, *J. Geophys. Res.*, **73,** 2643–61.
Piper, J. D. A., 1973. The fine structure of *en echelon* ridge axes and crustal deformation at constructive plate boundaries, *Bull. Geol. Soc. Amer* (in press).
Piper, J. D. A. and Gibson, I. L., 1972. Stress control of geological processes at extensional plate margins, *Nature*, **238,** 83–86.
Raitt, R. W., 1963. The crustal rocks. *In:* The Sea, Vol. 3, 85–102. John Wiley and Sons, New York.
Stéfansson, R., 1967. Some problems of seismological studies on the mid-Atlantic ridge. *In:* Iceland and Mid-Ocean Ridges *Soc. Sci. Islandica*, **38** (edited by S. Bjornsson), 190–9.
Sykes, L. R., 1967. Mechanism of earthquakes and nature of faulting on the mid-oceanic ridges, *J. Geophys. Res.*, **72,** 2131–53.
van Andel, T. H. and Bowin, C. O., 1968. Mid-Atlantic ridge between 22° and 23° N latitude and the tectonics of mid-ocean rises, *J. Geophys, Res.*, **73,** 1279–98.
Vogt, P. R., 1972. Asthenosphere motion recorded by the ocean floor south of Iceland, *Earth and Planetary Sci. Lett.*, **13,** 153–60.
Vogt, P. R., Schneider, E. D. and Johnson, G. L., 1969. The crust and upper mantle beneath the sea, *Am. Geophys. Union*, *Geophys. Monogr.*, **13,** 556–617.
Vogt, P. R., Ostenso, N. A. and Johnson, G. L., 1970. Magnetic and bathymetric data bearing on sea-floor spreading north of Iceland, *J. Geophys. Res.*, **75,** 903–20.
Vogt, P. R., Johnson, G. L., Holcombe, T. L., Gilg, J. G. and Avery, O. E., 1971. Episodes of sea-floor spreading recorded by the North Atlantic basement, *Tectonophysics*, **12,** 211–34.
Watkins, N. D. and Richardson, A., 1971. Intrusives, extrusives and linear magnetic anomalies, *Geophys. J. Roy. astron. Soc.*, **23,** 1–13.
Wellman, H. W., 1971. Reference lines, fault classification, transform systems and ocean-floor spreading, *Tectonophysics*, **12,** 199–209.
Wilson, J. T., 1965a. A new class of faults and their bearing on continental drift, *Nature, Lond.*, **207,** 343–7.
Wilson, J. T., 1965b. Submarine fracture zones, aseismic ridges and the ICSU Line: Proposed western margin of the East Pacific Ridge, *Nature, Lond.*, **207,** 907–11.

6.2

J. D. A. PIPER

Department of Earth Sciences,
University of Leeds,
Leeds, England

History and mode of Crustal Evolution in the Icelandic Sector of the mid-Atlantic Ridge

Introduction

The neovolcanic zones of Iceland have been the subject of interest and study for some years (Van Bemmelen & Rutten, 1955; Kjartansson, 1960; Thorarinsson, 1960, 1967). Although understood in a general way, the geology of the Tertiary lava pile was not elucidated in detail until the studies of Walker (1959, 1960, 1963, 1964), who demonstrated that there has been a continuity in the growth of the basalt plateau in Iceland, with the processes taking place along the neovolcanic zone being comparable with those which have built up the lava pile. Bodvarsson & Walker (1964) have examined the significance of the Icelandic geology in terms of crustal spreading and concluded from the information available at that time that it is consistent with a partial crustal spreading. The model deduced from the geology however, does not agree well with the geophysical data, and other authorities (Einarsson, 1965, 1968) have found it impossible to reconcile many facets of Icelandic geology with sea-floor spreading.

In recent years much new information has been forthcoming from explosion seismology (Pálmason, 1970), from earthquake seismology (Sykes, 1965; Stéfansson, 1967; Ward *et al.*, 1969), on the magnetic field over Iceland (Sigurgeirsson, 1967; Serson *et al.*, 1968), and from geodetic studies within the neovolcanic zone (Tryggvason 1968, 1970; Decker & Einarsson, 1971). Geological and palaeomagnetic studies of the lava pile have provided new information on its structure (Gibson, 1966; Piper, 1971), and other studies have demonstrated the phenomena of a growing neovolcanic zone terminating within a plate (Piper, 1973b). This paper presents a brief outline of current knowledge of the geology and geophysics of Iceland. The features peculiar to this portion of ocean crust are discussed together with an interpretation of their possible significance.

Structure of the Lava Pile

The exposed crust in Iceland is composed of a thick pile of predominantly basaltic lavas. They are mainly subaerial and according to Walker (1959) tholeiite and olivine basalts alternate with no apparent rhythm. Central volcanoes occur at intervals within the lava pile, and are defined by acid and intermediate extrusive and intrusive rocks which interdigitate with the basaltic lavas of the remainder of the pile. In eastern Iceland acid and intermediate volcanics make up about 20% of the pile and are characteristically surrounded by zones of heavy alteration (Walker, 1964). Isotope (Moorbath & Walker, 1965) and chemical (Carmichael, 1964) data demonstrate that the acid rocks are probably derived from crystal fractionation from a basaltic parent magma. Gibson (1970) has shown that Carmichael's data is also consistent with the acid and intermediate rocks deriving from partial melting of a basaltic crust. That this is the case for at least some of the silicic rocks is suggested by the high temperature gradients observed near currently active silicic centres (Piper, 1971).

The lavas of the pile are predominantly the products of fissure eruptions and where it is exposed to considerable depths the feeder dykes are found to occur in swarms. These swarms are confined to narrow belts which pass through the central volcanoes and the dykes decrease in number upwards within the pile (Walker, 1960). They have a dip which is approximately at right angles to the dip of the lavas, although this latter parameter is not constant and tends to diminish with altitude so that the lavas are asymptotic to a level a short distance above the present mountain summits (Walker, 1965). As a result of this the succession thins updip. Gibson (1966) has observed units which also thin downdip and he has suggested that the lava pile is also asymptotic in this direction; the observed updip thinning of the bulk of the lava pile is then purely a function of the present depth of erosion of the pile. Gibson (1969) has shown that components of the Reydarfjordur central volcano in eastern Iceland reach their maximum development over the dyke swarm through the silicic centre, and they thin rapidly at right angles and less rapidly along the axis of the swarm. He has used this evidence to suggest that the lava pile grows as lenticular units along dyke swarms focused about the silicic centres. The lenticular nature of the lava pile has been confirmed by palaeomagnetic studies in southwest Iceland (Piper, 1971) which show that the stratigraphy of the pile thickens through the silicic centres. The central volcanoes thus define the regions where basalt eruptions are concentrated and acid and

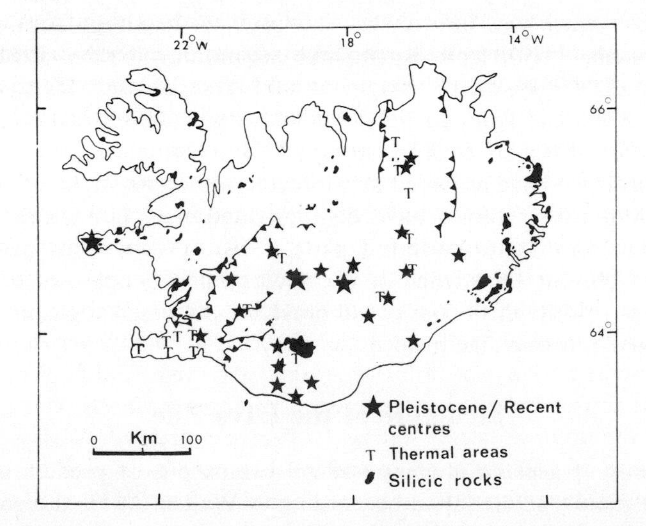

FIGURE 1 Map showing the distribution of acid volcanic rocks in Iceland (based mainly on Walker (1966)) and active central volcanoes and thermal areas (after Saemundsson in Ward *et al.* (1969)).

intermediate volcanism is localized. The silicic centres of the Tertiary lava pile have analogues in the active central volcanoes of the present neovolcanic zone which are also as a rule high temperature areas. They appear to have life spans of the order of 1 m.y. The distribution of active central volcanoes, thermal areas and extinct silicic centres in Iceland is shown in Fig. 1.

The lava pile in Iceland has apparently grown by the more or less continuous addition of basaltic lavas and development of silicic centres (Walker, 1964), and no major stratigraphic breaks have yet been reliably established within the pile. The reliable age dates available at the present time confirm a general younging of the succession towards the neovolcanic zones (Gale *et al.*, 1966; Moorbath *et al.*, 1968). Moorbath *et al.* have dated the oldest rocks in eastern Iceland at about 12·5 m.y. and those in northwestern Iceland at about 16 m.y.

Structure of the Neovolcanic Zones

The neovolcanic zones in Iceland are defined by postglacial, interglacial and intraglacial eruptions. The first two types have either followed fissures to produce elongate subaerial flows and crater lines, or have been confined to a central vent to build up low shield volcanoes. The latter eruptions have produced hyaloclastite ridges or table mountains capped to a variable extent by subaerial lavas. The intraglacial volcanics have been produced during glacial episodes which in Iceland appear to be confined to about the last 3 m.y. (McDougall & Wensink, 1966); they constitute the Moberg formation of Icelandic literature (Kjartansson, 1960).

The trend of the eruptive fissures is closely followed by open (non-eruptive) fissures which cut young lavas along much of the neovolcanic zone. They occur in both dextral and sinistral *en echelon* arrays and result from the continuous dilation that takes place in these areas (Nakamura, 1970). The amount of dilation represented by these fissures has averaged about 0·5 cm/year in the Thingvellir and Myvatn areas (Walker, 1965) in postglacial times. While many of these features are purely dilational others show a vertical displacement with the development of graben structures. Decker & Einarsson (1971) have suggested that extension and subsidence are simultaneous and that igneous injection may result from them, and Walker (1965) has proposed that they represent dykes which have failed to reach the surface.

The lava pile is tilted and uplifted within distances of about 50 km of the neovolcanic zones by 1–2 km. This results in part from uplift along normal faults that are prominent along some flanks of the zones (e.g. Hengill–Thingvellir section), but these are not universally present and it seems that much of the uplift is accommodated by a warping of the lava pile. Mapping of the pile in eastern Iceland (Walker, 1959) shows that movements across normal faults are small and displacements across vertical dilation fissures (represented by dykes) are only rarely observed. Uplift of the lava pile may result from the intrusion of low-angle sheet complexes at depth. These complexes are visible where the crust is deeply exposed in eastern Iceland (Walker, 1964).

The mean rate of lava eruption in Iceland has been estimated by several workers from K-Ar and palaeomagnetic evidence (Table 1). The mean rate is likely to vary considerably along the neovolcanic zone in view of the concentration of activity near silicic centres. On average individual flows are thicker away from the source dyke swarm

TABLE 1

	Source:	Estimated rate:
1.	Dagley *et al.* (1967) Eastern Iceland	1 lava per 28,000 years
2.	Dagley *et al.* (1967) recalculated using age data of Moorbath *et al.* (1968).	1 lava per 18,000 years
3.	McDougall & Wensink (1966) Northeastern Iceland	1 lava per 40,000 years
4.	Moorbath *et al.* (1968) Eastern Iceland	1 metre per 600 years
5.	Piper (1971) Southwest Iceland	1 lava per 27,000 years

probably because they flowed down low lava shields (Gibson, 1969), but this effect is more than compensated for by the increasing number of units near the dyke swarm. Icelandic basaltic lava flows typically have a volume of only about 1% of that in continental plateau basalt areas (Gibson, 1969). The largest known single basaltic eruption has produced the Eiriksjökull intraglacial tablemountain near the northeast end of the western neovolcanic zone and has a volume of 40 km³. The postglacial Thjórsa lava (8000 years B.P.) has a volume of about 15 km³ and the Laki eruption (1783 A.D.) produced 12 km³ of material (Thorarinsson, 1960).

The distribution of postglacial and late intraglacial eruptive fissures and centres is shown in Fig. 2. The localities of postglacial eruptions have been taken from the

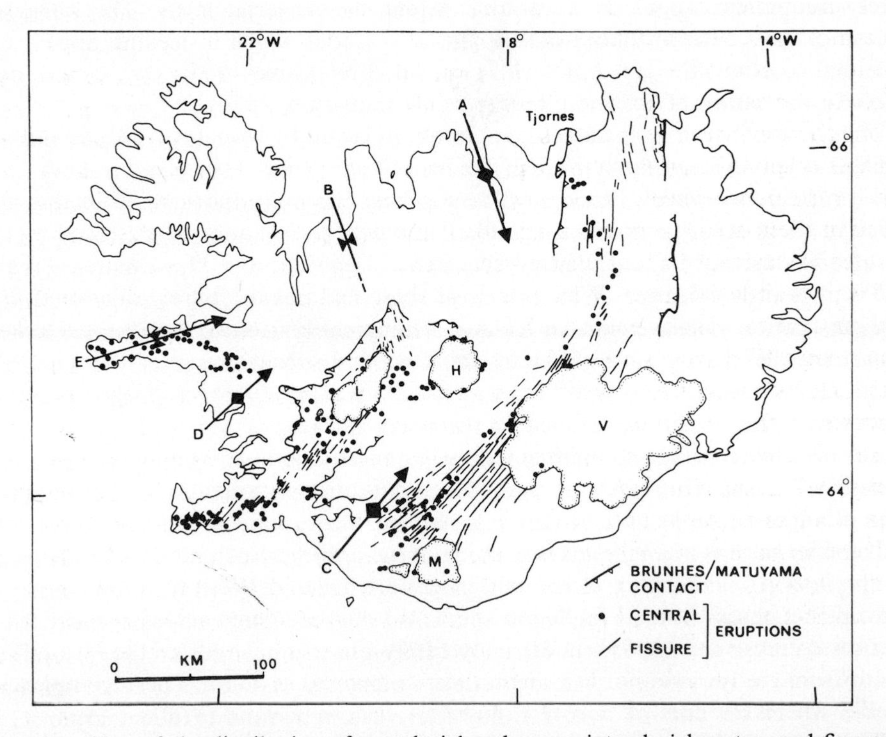

FIGURE 2 Map of the distribution of postglacial and young intraglacial centres and fissures in Iceland. The arrows show the distribution and direction of plunge of anticlines and synclines within the lava pile. The folds of the Eyafjordur anticline (A), the Hunafloi-Langjökull syncline (B), the Hreppar anticline (C), the Bordarfjordur anticline (D) and the Snaefellsnes syncline (E). The icesheets are: Langjökull (L), Hofsjökull (H), Myrdalsjökull (M) and Vatnajökull (V).

Geological Map of Iceland (sheets 2, 3, 5 and 6, Scale 1:250,000) and from the map by Saemundsson (in Ward *et al.*, 1969). The distribution of intraglacial and interglacial eruptions has been taken from the Geological Map of Iceland, Piper (1973b) and unpublished data of the author. Although the distribution is incompletely known in northern Iceland and only mapped at a reconnaissance level in the southern part of the eastern volcanic zone, Fig. 2 gives a fair impression of the distribution of young volcanics which can still be related to an eruptive centre, and probably includes the bulk of the eruptions within about the last 50,000–100,000 years (Piper, 1973b).

There is a marked reduction in both the number and the total volume of young eruptives in the northern part of Iceland. In the south there are two lines of active volcanism with northeast–southwest orientations. The western (Reykjanes-Langjökull) neovolcanic line is defined by both fissure and central eruption and undergoes a sharp change in strike near latitude 64°45′ from north–southwest to north-northeast–south-southeast, and recent volcanism dies out altogether within 30 km of this strike change. The width of this zone reaches its narrowest near Lake Thingvellir (64°10′ N) and widens to nearly 40 km in region of the Langjökull icesheet. The Reykjanes peninsula is a 70 km long east–west line of volcanism, and the open and eruptive fissures show a large scale sinistral *en echelon* pattern implying a left lateral strike slip movement at depth (Tryggvason, 1968). The eastern neovolcanic zones show more gradual change in strike near 64°40′ N from northeast–southwest to near north–south, and the width of the zone of postglacial volcanics decreases from about 70 km in the south to 40 km in the north.

In addition to showing considerable variation in the width of volcanic activity the neovolcanic zones also show varying structural characteristics. Some parts of the zone (e.g. Reykjanes peninsula) show an intermingling of fissure eruptions and small central eruptions. Other parts (e.g. southern half of the eastern neovolcanic zone) show a preponderance of fissure volcanism and centralized eruptions occur only in the silicic centres. The active zone along the southeastern side of the Langjökull is defined exclusively by large central eruptions.

There is little evidence of an east–west structural connection between the eastern and western neovolcanic lines in Iceland. The tectonic features within each zone are nearly parallel across most of the island. On the north east side of the Langjökull there are two small east–west trending eruptive fissures, and a further postglacial fissure with this orientation occurs on the northwest side of Vatnajökull. In addition there have been isolated small intraglacial and postglacial eruptions around the Hofsjökull icesheet but this area is separated from the eastern neovolcanic zone by an area of older tilted plateau basalts (Geological Map of Iceland, sheet 5).

The distribution of earthquakes in the Icelandic region shows no simple relationship to the observed neotectonic features. Figure 3 has been compiled from the earthquake data of Stéfansson (1967, Figs. 3 and 4) and Ward *et al.* (1968); they are based largely on records taken since 1955 and Ward *et al.* (1969) have published further epicentres for microseismic observations. The earthquakes within the neovolcanic zones are mostly small with magnitudes of 5 or less, and there has been no recorded seismic activity within the Snaefellsnes region. Large earthquakes are restricted to two east–west zones (Stéfansson, 1967), the first running for 150 km eastwards from Cape Reykjanes, and the second running along the north coast between the northern termination of the eastern volcanic zone and the Skagi peninsula. Earthquakes from both of these zones

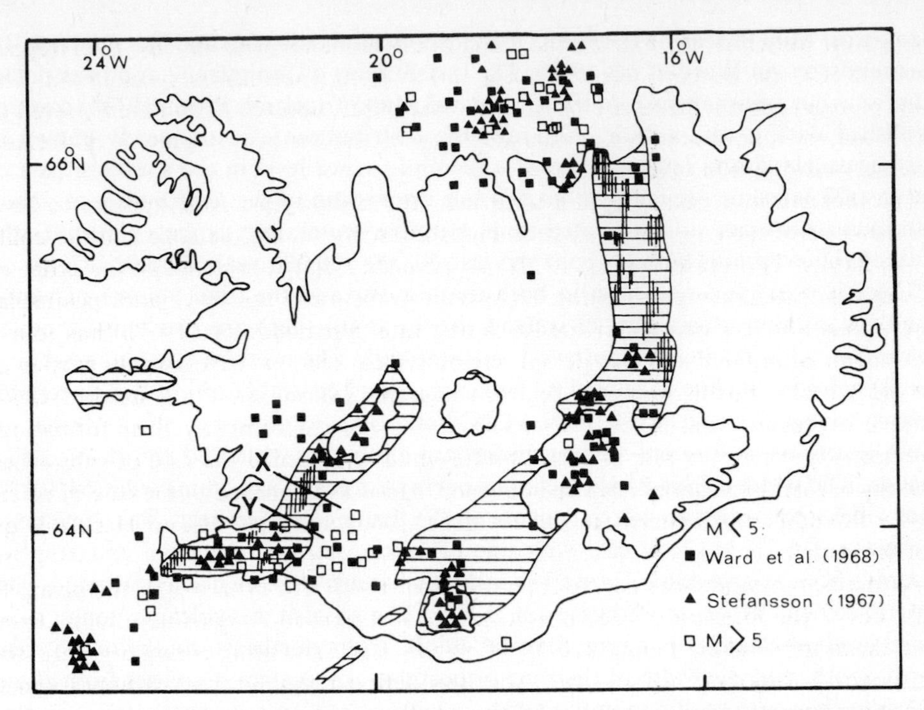

FIGURE 3 Distribution of earthquake activity in Iceland. The distribution of earthquakes with magnitudes 5 or less is shown after to Stéfansson (1966) and Ward *et al.* (1969). The areas of horizontal shading are zones of postglacial volcanism and the areas of vertical shading are fields of open dilational fissures.

have given strike-slip solutions (Sykes, 1965; Ward, 1971), but these fault zones show little surface expression and are not true transform faults. The Tjornes peninsula in northern Iceland (Einarsson 1958) shows a number of structural complexities which might result from proximity to the northern fault zone, it is the one area in Iceland where a considerable sedimentary succession has been built up, the rocks dip northwestwards or away from the neovolcanic zone and the area is cut by northwest–southeast faults. Ward *et al.* (1969) and Ward (1971) have advocated a transform fault along the belt of seismicity in southern Iceland, but there is no surface expression of this fault and it cannot have a transform nature since it does not terminate the Reykjanes–Langjökull neovolcanic zone. It would be a valid explanation if this active zone is in the last stages of extinction (Sigurdsson, 1970) and a transform fault is in the process of establishing itself. If this is not the case then the strike-slip motion must be a consequence of internal deformation of the plate (Piper, 1973a).

History of Growth of the Lava Pile

The lava pile in Iceland shows a small tectonic dip towards the neovolcanic zones along their margins. This dip is less than 10° over most of the pile but is locally steeper in the vicinity of the silicic centres. In certain areas away from the immediate flanks of the neovolcanic zones in the western half of Iceland, the direction of the dips are reversed and they define large low anticlines and synclines. These large scale structures are the

Eyafjordur anticline (A, Fig. 2), the Langjökull–Hunafloi syncline (B), the Hreppar anticline (C), the Borgarfjordur anticline (D) and the Snaefellsnes syncline (E). The folds plunge towards the centre of Iceland close to latitude 65° N and are believed to be of major importance in the interpretation of the history of growth of Iceland (Saemundsson, 1967a; Piper, 1973a.)

The inclination acquired by the lavas near the neovolcanic zone appears to be a consequence of the continuous dilation and subsidence within the zone and the uplift of the flanks. Tryggvason (1968, 1970) has estimated that subsidence of the order of 0·5 cm/year has taken place in the Thingvellir graben and the Reykjanes peninsula, and is thus about a quarter of the spreading rate. Saemundsson (1967b) has found movements along faults in the Hengill region which become progressively greater in the older rocks. Furthermore the central volcanoes seldom show any marked topographic expression and a continuous subsidence must accompany their formation. Occasionally where the rate of extrusion has greatly exceeded the rate of subsidence composite shield volcanoes have been formed. This has happened in the case of silicic centre Breiddalur in eastern Iceland (Walker, 1963) and the volcano Heckla is an active example.

As the neovolcanic lines control the direction of dip, the synclines in the lava pile must define the location of old neovolcanic lines. The shallow syncline running from near the north end of Langjökull to Hunafloi is in structural continuity with the Reykjanes–Langjökull active zone. The line defining the limit of Bruhnes epoch volcanism turns eastwards north of Langjökull (Fig. 1, Piper, 1973b) but the line between here and Skagi was an active volcanic zone along much of its length in the Matuyama epoch (Einarsson, 1962), and this synclinal axis defines a now-extinct volcanic line. The Eyafjordur and Hreppar anticlines have grown by the division of a single volcanic zone and growth of lava pile between them. The two zones have moved apart and the northern half of the western one has ceased to be active.

The magnetic anomalies over Iceland show no symmetry and only poor linearity about the neovolcanic axes (Serson *et al.*, 1968), and they are of little use in reconstructing the history of crustal growth as has been possible over the Reykjanes ridge. However it is possible to draw several conclusions from them. The large amplitude short wavelength anomalies over the neovolcanic areas are spread over a broad area spanning the two neovolcanic zones in the south, while in northern Iceland they are restricted to the single neovolcanic line and are lower in amplitude. Furthermore the positive anomaly over the Reykjanes–Langjökull neovolcanic line is not found north of Langjökull and correlates with the observed termination of the line.

Crustal Structure

Gibson & Piper (1972) have reviewed the geological and geophysical data on the structure of the crust and have concluded from eruption rates, heat flow data and lava pile geometry that the pile is probably confined to the crustal layer 2 (seismic P velocities averaging 5·08 km/sec). The crustal layer 3 (seismic P velocities averaging 6·35 km/sec) is seismically more homogeneous and believed to be entirely intrusive. In order that this model should hold, the percentage of the crust composed of feeder dykes must increase greatly with depth towards the seismic layer 2/layer 3 interface and a near exponential rather than a linear increase is favoured by Gibson & Piper

(1972). The mean thickness of layer 2 from 52 determinations is 2·15 km (Pálmason, 1970) and the depth to the interface with layer 3 varies from 0·4 to 9·9 km. The topography of the interface is apparently very uneven and it tends to be at shallow depths beneath high temperature areas in the neovolcanic zones and beneath silicic centres in the Tertiary lava pile. This correlation may be interpreted in terms of the proximity of dyke or other intrusive complexes in these areas, or in terms of a metamorphic change effected by the high temperatures prevailing in these areas. Pálmason (1970) has suggested that the interface represents the metamorphism of an extrusive basalt crust to amphibolite as Cann (1968) has originally suggested for the oceanic crustal layer 2/layer 3 interface. It is however difficult to account for an oceanic crust built up entirely of extrusive material (Cann, 1970) and the same difficulties apply to the case of Iceland. The base of the lava pile probably coincides with the amphibolite/greenschist facies interface for the reason given by Cann (1970), namely the paucity of water in a complex built up of dykes and other intrusions.

The layer 3/layer 4 interface has only been mapped in a few parts of western Iceland (Pálmason, 1970) where it occurs at depths between 8 and 16 km compared with depths of 5–6 km under the axis of the Reykjanes ridge (Ewing & Ewing, 1959). Layer 4 (seismic P velocities 7·2–7·4 km/sec) is interpreted as anomalous upper mantle and if temperature gradients observed near the surface are used to extrapolate the temperature at this discontinuity, it is found to be in the range 800°–1300°C, or within the melting range of basalts. From earthquake evidence Tryggvason (1961) concluded that the lower boundary of layer 4 lies at a depth of 100–140 km and is about 1000 km wide. In a subsequent paper (1964) he made a revised estimate of 240 km for the depth of this layer. Francis (1969) has subsequently made a more precise study of this layer and he finds that it is confined to a depth of about 250 km and a width of about 300 km about the ridge axis between latitudes 50° to 70° N. Thus both the crust and upper mantle in the Icelandic region have structures which are exceptional for a mid-ocean ridge axis and imply that an exceptional volume of the mantle is subject to partial melting here (Bott, 1965).

Plate Movements within Iceland

The history of crustal growth in Iceland has been complex, with a bifurcation and lateral migration of ridge axes occuring at least twice in the period represented by the growth of the lava pile (12–16 m.y.). Growth has been most active in the southern half of the island where the amount of postglacial volcanism and the current spreading rate is about twice that in the northern half of the island. The greatest concentration of silicic centres is also in the southern half of the island. This zone of excess crustal growth does not terminate against a transform fault and is accommodated by deformation within the plate incorporating western Iceland. This deformation is clearly not responsible for the folds within the lava pile since the dips of the lavas show no correlation with the disequilibrium in spreading, and the evidence outlined in the previous sections implies that they have a different origin. It may, however, be responsible for fracturing within the plate and Saemundsson (1967a) has drawn attention to fault zones with west-northwest–east-southeast trends between the Snaefellsnes area and the termination of the Reykjanes-Langjökull zone. The Pleistocene and Recent craters and vents of the Snaefellsnes region of western

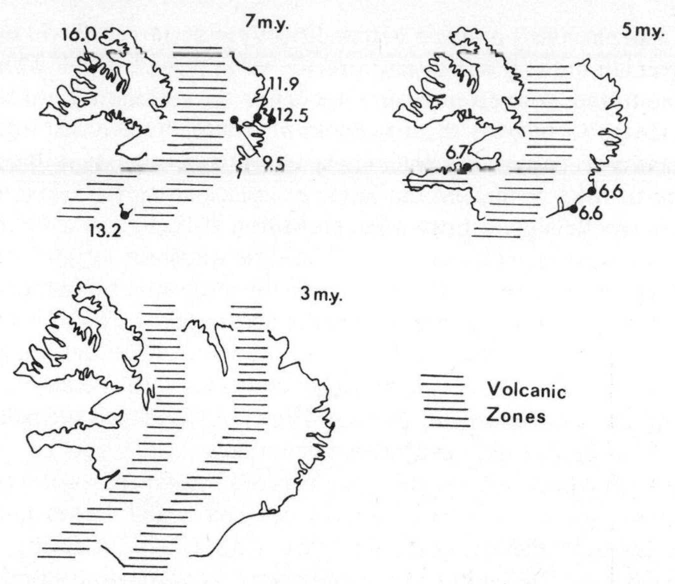

FIGURE 4 Diagrammatic representation of the growth of the crust in Iceland showing the disposition of active neovolcanic zones in relationship to pre-existing crust. This model is based on the disposition of anticlines and synclines within the lava pile and the K-Ar age dates of Moorbath *et al.* (1968) shown by the figures in millions of years. At about 7 m.y. B.P. two neovolcanic zones were active to produce the Borgarfjordur anticline between. The western zone subsequently became extinct and only one neovolcanic zone seems to have been active about 5 m.y. B.P. This later bifurcated to produce the Hreppar and Eyafjordur anticlines. The northern part of the western zone has been extinct since the Matuyama epoch.

Iceland lie uncomformably on the Tertiary plateau basalts and follow three west-northwest–east-southeast trending volcanic and tectonic lines which continue on the sea-floor to the northwest (Sigurdsson, 1970). The belt of faulting associated with this area runs eastwards to within 30 km of the Reykjanes-Langjökull active zone. Saemundsson (1967a) has also recorded two other fault belts with north-northwest–south-southeast and west-southwest–east-northeast trends that continue from the Snaefellsnes fault belt to near the termination of the western neovolcanic zone; some of these faults are very young and open fissures are present. Although it is not certain what interpretation should be placed on these fault systems the prominent Snaefellsnes system has an orientation consistent with a right lateral movement within the western Icelandic plate.

Plate Tectonics of the Icelandic Region

Vogt *et al.* (1970) have distinguished three major spreading epochs in the ocean crust south of Iceland since the separation of Norway and Greenland 60–70 m.y. ago. The first interval of fast spreading lasted until 40 m.y. B.P. and was followed by slower spreading with the development of fracture zones until 18 m.y. B.P.; in the latest episode which has included the formation of the Reykjanes ridge, the spreading rate has been faster and the fracture zones have disappeared. The history of the crust between the Reykjanes ridge and the Mohns ridge is more complicated. The Jan Mayen ridge has a sedimentary structure with flat-lying sediments overlying dipping sediments

(Johnson & Heezen, 1967) and it is believed by Vogt *et al.* (1970) to have been split off from Greenland during a relocation of the ridge axis at about 42 m.y. B.P. The seamount line in the Norwegian basin may define the extinct axis and there is some magnetic evidence to support this (Avery *et al.*, 1968; Vogt *et al.*, 1970, Fig. 4). If this interpretation is correct the phenomena of shifting ridge axes is not confined to the mid-Atlantic ridge in Iceland but has taken place along the whole section of the ridge between the Reykjanes ridge and Jan Mayen (Fig. 5).

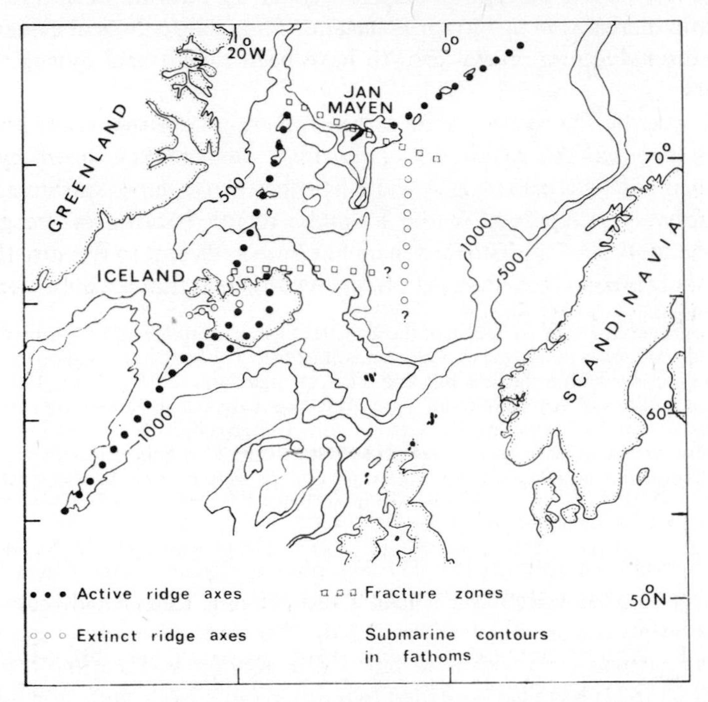

FIGURE 5 Regional setting of Iceland in relation to the active and extinct ridge axes of the North Atlantic.

The complex history of excessive basalt production, shifting ridge axes and complex tectonics of the Icelandic region is accompanied by anomalous mantle conditions. Iceland is one of the mantle 'hotspots' of Wilson (1965) and Morgan (1971) and the explanation of the features of the Icelandic region in terms of asthenospheric motion has been developed by Vogt (1972). The topographic features of the Reykjanes ridge (Talwani *et al.*, 1971) and probably the whole Icelandic region (Fig. 4) cut across the crustal isochrons deduced from other evidence and have been explained by Vogt in terms of the southward migration of an elliptical front in the asthenosphere. This is an attractive hypothesis and may be the explanation for great basalt production, large numbers of silicic centres, and the widespread area of high heat flow in southern Iceland. It may also explain the progressive lengthening of the subaerial ridge axes in Iceland with time (Fig. 4). Bodvarsson & Walker (1964) have estimated the mean rate of postglacial extrusive volcanism in Iceland at 0·025 km^3/yr, and Thorarinsson (1967) has estimated the rate of extrusive volcanism in Iceland between 800 A.D. and

1965 A.D. at 0·040 km^3/yr. Since only about 20–25% of the volcanism is extrusive, the total rate of volcanism is 0·125–0·200 km^3/yr or 0·0002–0·0036 km^3/yr per kilometer of ridge axis, which is 2–3 times that of most submarine portions of the mid-Atlantic ridge. It is unlikely that this rate of volcanism has been maintained during the history of the Greenland–Iceland–Faeroes ridge (Vogt, 1972). The anticlinal structure between the two neovolcanic ones in Iceland was initiated about 3–4 m.y. ago, and a third neovolcanic zone may have been active at this time (Fig. 4). The major topographic rise on the Reykjanes ridge observed by Talwani *et al.* (1971) is about 5 m.y. old and increases in age towards Iceland. Thus the geological evidence suggests that basalt discharge and crustal growth have been exceptional during the past few million years.

Whatever may be the nature of the asthenosphere giving rise to the hotspot under Iceland it is seen that this region does *not* exhibit the tendency shown by most plate margins of being self-perpetuating. Although the plates may have considerable strength, which is probably exceptional in the Icelandic region because of the great crustal thickness, the motion of the asthenosphere has been sufficient to fracture the plate and relocate the ridge axis. This observation further suggests that such motion can be the driving mechanism of the plates.

References

Avery, O. E., Burton, G. D. and Heirtzler, J. R., 1968. An aeromagnetic survey of the Norwegian Sea, *J. Geophys. Res.*, **73,** 4583–4600.

Bemmelen, R. W. van and Rutten, M. G., 1955. Tablemountains of Northern Iceland. Leiden.

Bodvarsson, G. and Walker, G. P. L., 1964. Crustal drift in Iceland, *Geophys. J. Roy. astr. Soc.*, **8,** 285–300.

Bott, M. H. P., 1965. Formation of oceanic ridges, *Nature, Lond.*, **207,** 840–3.

Cann, J. R., 1968. Geological processes at mid-ocean ridge crests, *Geophys. J. Roy. astr. Soc.*, **15,** 331–41.

Cann, J. R. 1970. New model for the structure of the ocean crust, *Nature, Lond.*, **226,** 928–30.

Carmichael, I. S. E., 1964. The petrology of Thingmuli: a Tertiary volcano in eastern Iceland, *J. Petrol.*, **5,** 435–60.

Dagley, P., Wilson, R. L., Ade-Hall, J. M., Walker, G. P. L., Haggerty, S. E., Sigurgeirsson, T., Watkins, N. D., Smith, P. J. and Edwards, J., 1967. Geomagnetic polarity zones for Icelandic lavas, *Nature, Lond.*, **216,** 25–27.

Decker, R. W. and Einarsson, P., 1971. Rifting in Iceland: (abstract), *Trans. Amer. Geophys. Un.*, **52,** 352.

Einarsson, Tr., 1958. A survey of the geology of the area Tjornes–Bardardalur in northern Iceland, including palaeomagnetic studies, *Soc. Sci. Islandica*, **32,** 79pp.

Einarsson, Tr., 1962. Upper Tertiary and Pleistocene rocks in Iceland, *Soc. Sci. Islandica*, **36,** 196pp.

Einarsson, Tr., 1965. Remarks on crustal structure in Iceland, *Geophys. J. Roy. astr. Soc.*, **10,** 283–8.

Einarsson, Tr., 1968. Submarine ridges as an effect of stress fields, *J. Geophys. Res.*, **73,** 7561–76.

Ewing, J. and Ewing, M., 1959. Seismic refraction measurements in the Mediterranean Sea on the mid-Atlantic Ridge and in the Norwegian Sea, *Bull. Geol. Soc. Am.*, **70,** 291–318.

Francis, T. J. G., 1969. Upper mantle structure along the axis of the mid-Atlantic ridge near Iceland, *Geophys. J. Roy. astr. Soc.*, **17,** 507–20.

Gale, N. H., Moorbath, S., Simons, J. and Walker, G. P. L., 1966. K-Ar ages of acid intrusive rocks from Iceland, *Earth and Planetary Sci. Let.*, **1,** 284–8.

Gibson, I. L., 1966. The crustal structure of eastern Iceland, *Geophys. J. Roy. astr. Soc.*, **12,** 99–102.

Gibson, I. L., 1969. A comparative study of the flood basalt volcanism of the Columbia Plateau and eastern Iceland, *Bull. Volcanol.*, **33,** 419–37.

Gibson, I. L., 1970. Origin of some Icelandic pitchstones, *Lithos*, **2,** 343–9.

Gibson, I. L. and Piper, J. D. A., 1972. Structure of the Icelandic basalt plateau and the process of drift, *Phil. Trans. Roy. Soc. Lond. A*, **271,** 141–50.

Johnson, G. L. and Heezen, B. C., 1967. The morphology and evolution of the Norwegian–Greenland Sea, *Deep-Sea Res.*, **14,** 755–71.

Kjartsansson, G., 1960. The Moberg formation. *In:* On the Geology and Geophysics of Iceland, edited by S. Thorarinsson. Guide to Excursion No. A2, Int. Geol. Congress, Session 21, 21–28, Reykjavik.

McDougall, I and Wensink, H., 1966. Palaeomagnetism and geochronology of the Pliocene–Pleistocene lavas in Iceland, *Earth and Planetary Sci. Let.*, **1,** 232–6.

Moorbath, S. and Walker, G. P. L., 1965. Strontium isotope investigation of igneous rocks from Iceland, *Nature, Lond.*, **207,** 837–40.

Moorbath, S., Sigurdsson, H. and Goodwin, R., 1968. K-Ar ages of the oldest exposed rocks in Iceland, *Earth and Planetary Sci. Let.*, **4,** 197–205.

Morgan, W. J., 1971. Convection plumes in the lower mantle, *Nature, Lond.*, **230,** 42–43.

Nakamura, K., 1970. *En echelon* features of Icelandic fissures, *Acta Naturalia Islandica*, **2,** No. 8, 15pp.

Pálmason, G., 1970. Crustal structure of Iceland from explosion seismology, *Sci. Institute, Univ. of Iceland*, 239pp.

Piper, J. D. A., 1971. Ground magnetic studies of crustal growth in Iceland, *Earth and Planetary Sci. Let.*, **12,** 199–207.

Piper, J. D. A., 1973a. The fine structure of *en echelon* ridge axes and crustal deformation at constructive plate boundaries (in press).

Piper, J. D. A., 1973b. Volcanic history and tectonics of the North Langjökull region, central Iceland, *Can. Jour. Earth Sci.* (in press).

Saemundsson, K., 1967a. An outline of the structure of SW Iceland. *In:* Iceland and Mid-Ocean Ridges, *Soc. Sci. Islandica*, **38,** (edited by S. Bjornsson), 151–61.

Saemundsson, K., 1967b. Vulkanismus and Tektonik des Hengill Gebietes in Sudwest-Island, *Acta Naturalia Islandica*, **2,** No. 7, 105pp.

Serson, P. H., Hannaford, W. A. and Haines, G. V., 1968. Magnetic anomalies over Iceland, *Science*, **162,** 355–7.

Sigurdsson, H., 1970. Structural origin and plate tectonics of the Snaefellsnes volcanic zone, western Iceland, *Earth and Planetary Sci. Let.*, **10,** 129–35.

Sigurgeirsson, T., 1967. Aeromagnetic surveys of Iceland and its neighbourhood. *In:* Iceland and Mid-Ocean Ridges, *Soc. Sci. Islandica*, **38** (edited by S. Bjornsson), 91–96.

Stéfansson, R., 1967. Some problems of seismological studies on the Mid-Atlantic Ridge. *In:* Iceland and Mid-Ocean Ridges, *Soc. Sci. Islandica*, **38** (edited by S. Bjornsson), 80–90.

Sykes, L. R., 1965. The seismicity of the Arctic, *Seism. Soc. Amer. Bull.*, **55,** 501–36.

Talwani, M., Windisch, C. and Langseth, M., 1971. A comprehensive geophysical survey of the Reykjanes Ridge, *J. Geophys. Res.*, **76,** 473–7.

Thorarinsson, S., 1960. On the geology and geophysics of Iceland, section 6, 7 and 8. Guide to Excursion No. A2, Int. Geol. Congress, Session 21, 55–74, Reykjavik.

Thorarinsson, S., 1967. Hekla and Katla, *In:* Iceland and Mid-Ocean Ridges, *Soc. Sci. Islandica*, **38** (edited by S. Bjornsson), 190–9.

Tryggvason, E., 1961. Wave velocity in the upper mantle below the Arctic–Atlantic Ocean and northwest Europe, *Ann. Geofis.*, **14,** 379–92.

Tryggvason, E., 1964. Arrival times of P waves and upper mantle structure, *Bull. Seismol. Soc. Amer.*, **54,** 727–36.

Tryggvason, E., 1968. Measurement of surface deformation in Iceland by precision levelling, *J. Geophys. Res.*, **73,** 7039–50.

Tryggvason, E., 1970. Surface deformation and fault displacement associated with an earthquake swarm in Iceland, *J. Geophys. Res.*, **75,** 4407–22.
Vogt, P. R., 1972. Asthenosphere motion recorded by the ocean floor south of Iceland, *Earth and Planetary Sci. Let.*, **13,** 153–60.
Vogt, P. R., Ostenso, N. A. and Johnson, G. L., 1970. Magnetic and bathymetric data bearing on sea-floor spreading north of Iceland, *J. Geophys. Res.*, **75**, 903–20.
Walker, G. P. L., 1959. Geology of the Reydarfjordur area, eastern Iceland, *Quart. J. Geol. Soc. Lond.*, **114,** 367–93.
Walker, G. P. L. 1960. Zeolite zones and dike distribution in relation to the structure of the basalts in eastern Iceland, *J. Geol.*, **68,** 515–28.
Walker, G. P. L. 1963. The Breiddalur central volcano, eastern Iceland, *Quart. J. Geol. Soc. Lond.*, **119,** 29–63.
Walker, G. P. L. 1964. Geological investigations in eastern Iceland, *Bull. Volcanol.*, **27,** 351–63.
Walker, G. P. L., 1965. Evidence of crustal drift from Icelandic geology, *Phil. Trans. Roy. Soc. Lond. A*, **258,** 199–204.
Walker, G. P. L. 1966, Acid volcanic rocks in Iceland, *Bull. Volcanol.*, **29,** 375–406.
Ward, P. L., 1971. New interpretation of the geology of Iceland, *Bull. Geol. Soc. Amer.*, **82,** 2991–3012.
Ward, P. L., Palmason, G. and Drake, C., 1969. Microseismic survey and the mid-Atlantic Ridge in Iceland, *J. Geophys. Res.*, **74,** 665–84.
Wilson, J. T., 1965. Submarine fracture zones, aseismic ridges and the ICSU line: Proposed western margin of the East Pacific Ridge, *Nature, Lond.*, **207,** 907–11.

6.3

M. F. OSMASTON
*The White Cottage, Sendmarsh,
Ripley, Woking, Surrey,
England*

Limited Lithosphere Separation as a main cause of Continental Basins, Continental Growth and Epeirogeny

Introduction

Recent work in the study of large scale horizontal motions of the lithosphere has led to progress in understanding the vertical movements of the ocean floor. This progress serves to emphasize how very little is yet understood about the vertical movements that affect the continental crust. Vertical movements have been of primary importance in the formation of many continental basins and geosynclines and in elevating the plateaux from which the sediments to fill them have often been derived. The geological record confirms the major, and at times even dominant, role that such movements must have played in the evolution of much of the continental crust to its present state. By extending the study of lithosphere separation to environments intermediate between those of ocean ridges and continental rift valleys, the opportunity occurs to make valuable progress with this fundamental problem. The paper concludes with brief examples relevant to the origin of the North Sea basin, the evolution of the northern Appalachians, and the origin of the Ross Sea basin in Antarctica.

Recent Work on Continental Epeirogeny and Growth

The following outline makes no claim to completeness, being intended to draw attention only to those aspects upon which this paper bears.

Compressive folding is directly responsible for only a small proportion of the present elevated land areas of the world, most such areas owing their present elevations to events having no obvious genetic connection with earlier periods of folding (Billings, 1960; Belousov, 1962; King, 1967). It is widely recognized that subsidence mechanisms, other than mere isostatic adjustments due to sedimentary loading, are required to account for observed geosynclinal sediment thicknesses and for the large and long-continued subsidence of many continental basins and shelves (Jeffreys, 1959, p. 336; Belousov, 1962, 1966; Stoneley, 1969). Sub-crustal erosion, phase changes within the crust or underlying mantle, or combinations of these, have been invoked as subsidence mechanisms by various authors (Belousov, 1960, 1966; Gilluly, 1964; Van Bemmelen, 1966; Joyner, 1967; Van de Lindt, 1967; Collette, 1968; Magnitinsky & Kalashnikova,

1970; Sleep, 1971), but such hypotheses do not explain why subsidence should occur so selectively, frequently cutting sharply across structural belts, the physical state of whose underlying lithosphere would seem unlikely to vary so greatly or so sharply along the belt concerned. Also, the existence of continuing and appropriately intricate sub-crustal erosive movements within the continental lithosphere would hardly seem compatible within the increasingly secure framework of studies in plate tectonics.

Bott and his co-workers (Bott, 1964, 1971; Bott & Johnson, 1967; Bott & Dean, 1972) have confined their attention to differential vertical movements between crustal units of differing character, proposing that the stress differences which arise result in the horizontal flow transfer of lower crustal or uppermost mantle material from one unit to the other. Such proposals were not offered as relevant to geosynclinal subsidence, nor, it seems, could they apply to the many basins which interrupt structural belts. Collette (1968) has also pointed out that, as the Moho of basins is shallower near the middle, density considerations would prevent migration of deep crustal material towards the margins.

Many authors have accepted that limited addition to the continental crust occurs by sedimentary outbuilding over oceanic crust, notably at river deltas. It is generally accepted that some additions to the area of the continental crust also occur along lines where oceanic lithosphere turns downward along Benioff zones, but this involves a characteristically intense folding and disruption of the material at the time of addition. While there is good evidence that this process has occurred in many fold belts (e.g. Dewey & Bird, 1970) we shall argue that, in areal terms, limited lithosphere separation with sedimentary and volcanic infilling may be a far more significant means of addition to continents.

Carey (1958) was the first to argue that sedimentary basins might be formed by crustal separation. Osmaston (1969) suggested briefly that the Verkhoyansk geosyncline might have been formed by separation of the Kolyma platform from the Angara and Aldan shields, and McGinnis (1970) proposed that limited separation could have caused some intracontinental basins in North America. Impetus to this approach has come from evidence that the small ocean basins behind island arcs, the crusts of which are intermediate between oceanic and continental in character (Menard, 1967) and exhibit high heat flow (Sclater & Menard, 1967; McKenzie & Sclater, 1968), have been produced by lithosphere separation (Karig, 1970, 1971; Matsuda & Uyeda, 1971; Packham & Falvey, 1971) as orginally suggested by Carey (1958). The possibility that sedimentary fillings of such basins could become folded during continental collision has been accepted in recent discussions of geosynclinal development (Mitchell & Reading, 1969; Bird & Dewey, 1970; Dickinson, 1971).

In view of this interest we discuss here the geophysics and some of the geological consequences of lithosphere separation in an essentially continental environment. The amount of separation envisaged in what follows is greater than that of continental rift valleys (and therefore forms a logical extension of a previous study (Osmaston, 1971)), but is limited in the sense that separation ceases before separation has attained oceanic dimensions. A very important feature of our discussion is the attention given to the effects of heat conducted laterally into the older bounding lithosphere. Throughout this paper we ignore departures from isostasy such as may arise from differences between the time constants of the agent and the response processes. We accept, however, the evidence (James & Steinhart, 1966) that crustal thickness, crustal density, and mantle density, perhaps to at least the depth of the base of the lithosphere, all contribute to the attainment of isostatic balance.

Terminology — Chasmic Faults and the Lithosphere

When Carey (1958) proposed the formation of sphenochasms and rhombochasms by crustal separation he gave no name to the faulted boundaries thus produced. Wilson (1965), in introducing transform faults and their variations, seems to have been preoccupied with continuous separative movements. The term *chasmic fault* is therefore proposed to define any major age discontinuity extending through the lithosphere (Fig. 1). Each such 'fault' might be single or a complex fault zone, depending upon the tectonic details of early separation. Thus the search for fits between continental shelf outlines has as its purpose the delineation of the corresponding pairs of chasmic faults. Similarly, if continental basins are formed by separation, chasmic faults may be a common feature of continental interiors. This is a significant matter for the study of epeirogenesis because, not only will chasmic faults initially divide young and hot lithosphere from that which is older and cooler, but also they will mark boundaries between areas of lithosphere that may differ greatly in composition and physical state and will therefore respond differently to changes in temperature at depth.

Nothing certain is known about motions and horizontal forces in the material beneath the lithosphere. Consequently, even if the seismic evidence used to define the base of the lithosphere (e.g. Kanamori & Press, 1970) is in fact relevant to the long-term creep properties of the material (which is disputable), it is at least possible that substantial thicknesses of sub-lithosphere material move integrally with the lithospheric plates they underlie. Therefore, although we shall use the term lithosphere separation, it should be borne in mind that this may affect material to depths beyond the base of the lithosphere as currently defined.

Crustal Thicknesses

Analysis of crustal seismic refraction measurements (Lee & Taylor, 1966) shows that the oceanic crust is 6–8 km thick and that the continental crust averages 35 km thick, but with wide variations (James & Steinhart, 1966) in which it appears that the crust of Precambrian shields and platforms tends to be at least 40 km thick.

The fact that the oceanic and the continental lithosphere are substantially in isostatic equilibrium suggests that if the mantle parts of both are not very different in composition, as the most recent studies suggest (Sclater & Francheteau, 1970), the emplacement of a sufficient amount of sialic (and some basic) material on top of oceanic crust might produce a suitable initial basis for the development of continental crust. Our concern here is with the means of such emplacement and with its consequences, and not with processes that may have the effect of adding material to the base of the crust. It has been argued (James & Steinhart, 1966; Drake & Nafe, 1968; Kosminskaya & Zverev, 1968; Sheridan, 1969; Osmaston, 1971, p. 397) that, at the base of the crust, material exhibiting V_p in the range 7·2–7·9 km/sec is probably altered mantle (e.g. partially amphibolized or serpentinized (Sheridan, 1969)) and may not be permanent. In particular, such material may be restored to more normal mantle velocity by sufficient orogenesis to cause upward loss of its volatile content. For this reason we shall regard $V_p > 7{\cdot}2$ km/sec as applying to subcrustal material.

Seismic refraction data, especially where these have been tied in with gravity data, show that the crust of major continental basins and continental shelves is markedly thinner than the overall continental average. Crustal thickness studies of small and land-surfaced sedimentary basins are limited both by the refraction line lengths required and by the need to place large explosive charges in bore-holes, so the relevant data are correspondingly under-represented.

Ocean-floor Subsidence

Plate accretionary models of ocean-floor spreading (Langseth *et al.*, 1966; Vogt & Ostenso, 1967; Morgan, 1968; Sclater & Francheteau, 1970; Le Pichon & Langseth, 1970; Sclater *et al.*, 1971; McKenzie & Sclater, 1971) show conclusively that the 3–4 km subsidence of the oceanic lithosphere (Fig. 1) during the first 100 m.y. after its emplacement at ocean ridge crests is due to its increasing density as it loses the extra heat it possessed at the time of emplacement. This density increase involves solid-state phase changes as well as thermal contraction. Estimates of oceanic plate thickness (which, as we have noted, may not be synonymous with lithosphere thickness) based on these studies are about 100 km, but this figure is sensitive to the considerable uncertainty in mantle thermal conductivity and water content and to the condition assigned to the sub-plate material.

Sclater & Francheteau (1970) argue that the continental parts of tectonic plates may be twice as thick as the oceanic parts, and petrogenetic evidence (Ringwood & Lovering, 1970) indicates that more than 300 km may be attained beneath shields. If the present argument is correct, that continents grow by incorporation of young lithosphere, major variations in lithosphere thickness within continents are to be expected.

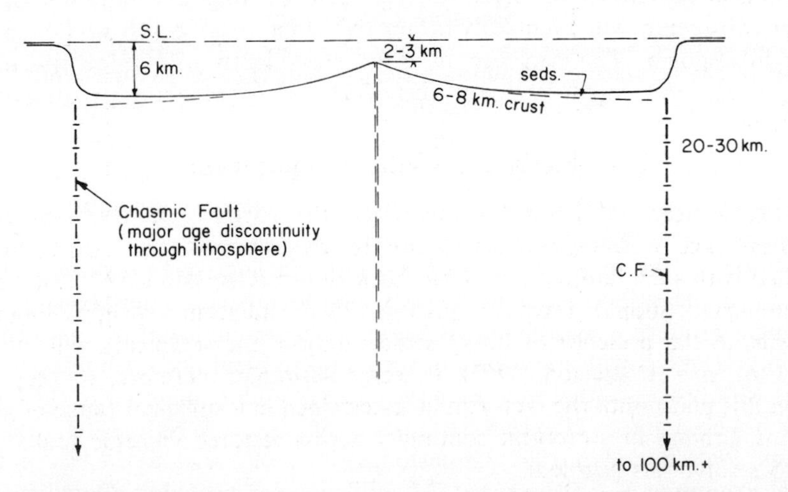

FIGURE 1 Generalized vertical relationships in an ocean of Atlantic type, with wide continental shelves, after symmetrical lithosphere accretion has been proceeding for about 100 m.y. Note the large vertical exaggeration.

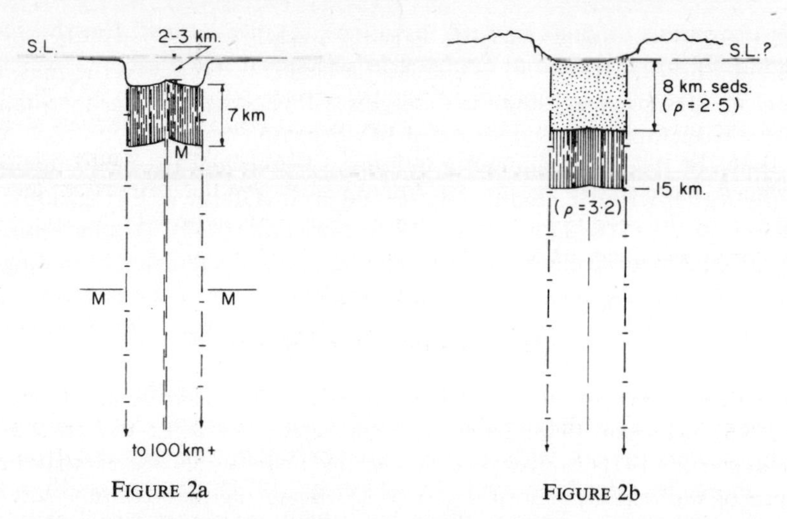

FIGURE 2a

FIGURE 2b

FIGURE 2a Direct result of arbitrarily closing up the ocean shown in Fig. 1 to represent the situation after less than 10 m.y. Oceanic crust, vertical ruling. M is base of crust. No allowance has been made for lateral heat transfer across the chasmic faults (see text).

FIGURE 2b Simplified result of separation similar to Fig. 2a but in a more continental environment providing abundant sediments, and taking into account the marginal upwarping resulting from heat transfer across the chasmic faults. Sediments (stippled) may include much additional volcanic material which would increase the density and thickness of the accumulation.

In succeeding sections of this paper we make the simplifying initial assumption that separation of any lithosphere has the consequences already established for the ocean floor, namely that fresh lithosphere is emplaced with an oceanic crust 7 km thick in a water depth of 2·5 km and subsides 3·5 km in 100 m.y. through loss of heat from its upper surface. It may be noted, however, that separation of lithosphere thicker than that of mid-oceans will draw material from greater depth, giving, in general, an increased heat content for the fresh material, and resulting in a shallower ridge crest, but that subsidence will eventually attain the same water depth as for an initially thinner lithosphere. The decrease of ridge crest depth with ridge spreading rate (Sclater *et al.*, 1971) might be due to this effect if slow spreading produces a thicker lithosphere.

Early Lithosphere Separation

It is not at all clear that major downfaulting of the margins is an inevitable feature of the earliest stage of lithosphere separation, for dykes intrude the crust without being associated with such faulting and it has been shown (Osmaston, 1971) that on ocean ridges important support forces are provided by the adjacent new upwelling material. Moreover, in the presence of heavy sedimentation, the sediments will buttress the original margins (Osmaston, 1969). It seems desirable, therefore, to keep an open mind on this point until the facts can be ascertained in a sufficient range of geological situations. Studies of structural continuity across restored chasmic faults could do much to resolve this problem.

Figure 2a has been obtained by simply closing up Fig. 1 to represent an earlier stage of separation (e.g. 100–200 km). Sedimentation on to the new floor is assumed small, due either to wide shelves or to low erosion rates. It is, however, quite certainly an incorrect picture, because at this early stage the fresh lithosphere will be much hotter than the older lithosphere on either side. Major heat transfer across the chasmic faults will therefore take place and the shelf margins will become upwarped isostatically by the density reductions thus induced at depth beneath them.

An upwarp of this kind is in fact a well documented and persistent feature of the edge of the western North Atlantic continental shelf (Officer & Ewing, 1954; Emery, 1965; Berger *et al.*, 1966; Drake *et al.*, 1968; Sheridan, 1969). In most cases the upwarping affects Upper Cretaceous strata and has caused substantial ponding of later sediments. The loading provided by these would tend to perpetuate this structural relationship even after general subsidence had become re-established. Studies of the eastern shelf of the Atlantic from Lisbon to the Faeroe Isles (Stride *et al.*, 1969) have revealed massive Late Cretaceous–Early Tertiary erosion of the shelf edge, before re-establishment of subsidence. North of the Azores fracture zone the age of this upwarping correlates well with the initiation of separation inferred from oceanic magnetic anomalies (Heirtzler, 1969; Pitman *et al.*, 1971). Further south, a pre-Late Jurassic upwarp east of Florida (Sheridan, 1969) may also correlate with early separation there.

In Fig. 2b we illustrate separation similar in amount to that of Fig. 2a, but in an essentially continental environment. The upwarping of the margins has also been taken into account. It shows the sediment thickness required to replace the sea-water (2·5 km depth assumed) in Fig. 2a, without allowing for any heat-loss-dependent subsidence. A normal erosion rate of 80 m/m.y. (Schumm, 1963; Judson, 1968; Gilluly *et al.*, 1970) from a belt 500 km wide would keep such a trough filled to sea level at a lithosphere separation rate of 0·5 cm/yr. If allowances are made for increased erosion due to uplift, or due to the separation being located in a young mountain belt, and for volcanic contributions to the filling (see below), separation at perhaps ten times this rate could be accommodated without deep-water sedimentation being involved. Thus it appears entirely possible that in small basins the whole of this initial filling could be of continental facies. In larger basins this would also apply to the early-formed marginal areas of the basin.

At this stage in our analysis we have produced a simplified crust some 15 km thick, with the effects of thermal subsidence still to be considered. We may note here, however, that the effect of heavy sedimentation during lithosphere separation would certainly greatly modify the magmatic aspects of the process as compared with normal oceanic crust production. The high cooling rates at ridge crests (Oxburgh & Turcotte, 1969; Osmaston, 1971) would be much reduced by the sedimentary blanket and volcanic activity, producing sills, dykes and perhaps larger intrusions within the sediments, would probably continue for a considerable time after leaving the axial zone. The generation of oceanic-type magnetic anomalies would be most improbable owing to the maintenance of high temperatures for periods much longer than those between geomagnetic reversals. The result would be a crust having high geothermal gradients and probably somewhat thicker (basic intrusions being denser than sediments) than in the simplified model of Fig. 2b.

Post-separative Subsidence

If cooling of the new lithosphere proceeds to the same condition as that attained

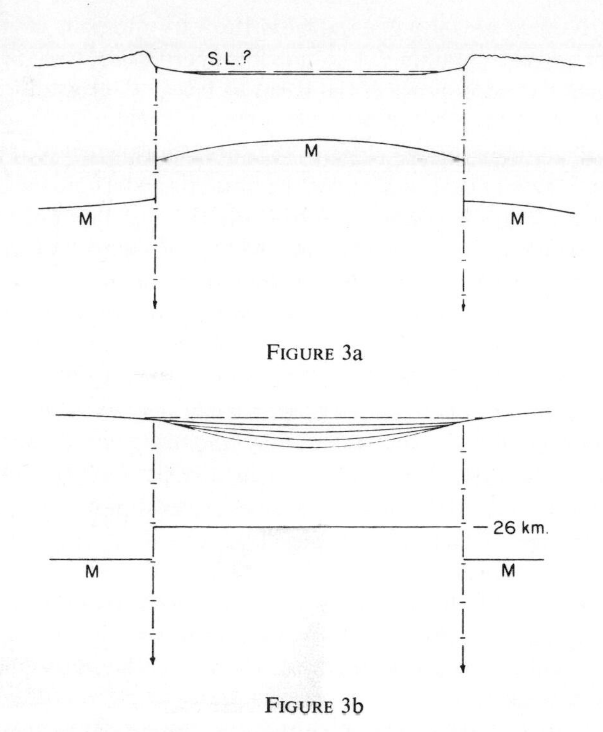

FIGURE 3a

FIGURE 3b

FIGURE 3a Intermediate stage of post-separative subsidence of the basin floor, showing the effect on crustal thickness of accelerated subsidence near the margins, due to lateral heat loss across the chasmic faults.

FIGURE 3b Final stage of basin subsidence, showing the total resulting crustal thickness, on the simplified assumptions discussed in the text. This is probably a minimum value for real conditions. Note the structural sag resulting from the later completion of subsidence near the middle of the basin. Note also the subsidence of and possible marine transgression across the bounding margins of the basin.

beneath the ocean floor after about 100 m.y. (but generally more slowly owing to the thick crustal blanket), the subsidence would permit the accumulation of a further 11 km of sediments, yielding a simplified crust with a total thickness of 26 km (Fig. 3b). In this calculation the density of the mantle was assumed to have risen to 3·33 g/cm^3 during cooling, and the overall mean 'sediment' density to 2·6 g/cm^3, owing to compaction and thermal metamorphism of the lower parts of the accumulation.

We have thus already attained a crustal thickness typical of many basins, particularly if the probable presence of an additional igneous contribution is also taken into account (see preceding section). Some of the details of the basin subsidence stage, however, appear relevant to comparison with actual basins and will now be discussed briefly.

Most continental basins are of dimensions such that the separative phase may have lasted only a few million years at typical average rates of relative plate movement. The opening of the 500 km wide North Sea basin, for example, might have been accomplished as a single movement in under 10 m.y. Therefore, although subsidence due to heat loss begins to affect new lithosphere from the very moment of its emplacement, subsidence will in general continue for so much longer than it took to open the basin that we may regard the lower crust of the basin as being of substantially uniform age across the basin.

Figure 3a shows an intermediate stage in the subsidence process. This draws attention to the effects of lateral heat loss across the bounding chasmic faults. This loss will be most rapid at the margins, decreasing towards the middle of the basin, with the result that the subsidence of the basin crust near the margins will initially be faster, running toward completion sooner than further out in the basin. At this stage, therefore, the basin crust will be thinner near the middle of the basin (Fig. 3a), even if sedimentation has maintained a substantially level upper surface.

Notice that outward transfer of heat across the chasmic faults will involve simultaneous subsidence of the basin and (except insofar as the heat merely maintains previously-elevated temperatures) uplift of the bounding areas. This is one of the most widely evident geological features of basin formation. The chasmic faults will be the main sites of such differential movement, resulting in displacements of a normal fault character extending throughout much of the subsidence life of the basin.

As the heat from beneath the new basin spreads into the older lithosphere on either side, the upwarp, initially of the form shown on the left in Fig. 3a, may be expected to extend to greater distances and take the form shown on the right wherever the lithosphere is especially thick.

Dissipation of the excess heat present in the lithosphere of the new basin will proceed much more slowly as the process nears completion. Subsidence rates will remain greater, however, where the largest amount of undissipated excess heat still remains, namely at the centre of the basin (Fig. 3b), thereby developing the characteristic structural sag which we believe has proved so misleading in previous attempts to understand basin formation. At about this time, a substantial measure of thermal equilibrium across the chasmic faults having been attained, both old and new lithosphere at the basin margins will subside together, yielding a marine transgressive phase in appropriate circumstances.

Subsidence Rates

We now consider briefly the four types of epeirogenic process in which an increase of density results from the extraction of heat. These are (1) thermal contraction, (2) liquid-to-solid phase change, (3) solid-state phase changes, and (4) reactions involving combination of the rock with a volatile phase (usually H_2O) that was already present in its interstices. The first and third are strictly reversible, but the second and fourth may be only partly so, owing to the possibility of the released liquid or volatile component migrating toward the surface on reheating. It should be noted particularly in relation to hydroxylation processes that in the study of epeirogenic movement we are concerned with changes in volume of columns of material extending to great depths, and commonly of considerable horizontal extent, so hypotheses involving introduction of the H_2O from 'elsewhere' are quite inappropriate to the problem, except, perhaps, in the case of the top kilometre or two of the oceanic crust. The horizontal migration of sedimentary pore fluids, familiar to the petroleum industry, and fluid loss to the surface during sediment compaction, do, of course, contribute to observed vertical movements but will not be considered explicitly here.

Epeirogenic processes differ widely in the amount of heat that has to be extracted to produce a given amount of contraction. To evaluate this we define the *thermal sensitivity* (B) of the process as the volume change resulting from unit heat loss. Then, for thermal contraction we have

$$B = \frac{\alpha_V}{\sigma_V} \qquad \text{(i)}$$

where α_V is the volume coefficient of thermal expansion and σ_V is the volumetric heat capacity.

For processes of types (2), (3) and (4) we use a form of the Clausius-Clapeyron thermodynamic relation (Ramberg, 1952, p. 16; Fyfe *et al.*, 1958, p. 115; Turner & Verhoogen, 1960, p. 22) to yield

$$B = \frac{1}{T_e} \cdot \frac{dT_e}{dP} \qquad \text{(ii)}$$

in which dT_e/dP is the slope of the equilibrium boundary between the two states on a pressure-temperature diagram, and T_e is the absolute temperature of transition at the chosen pressure. The first thing to notice about this relation is that, because dT_e/dP is almost invariably positive, the inverse dependence on T_e results in values of B that tend to decrease with increasing depth. Relation (ii) is applicable to multicomponent mineral assemblages provided that the equilibrium slope has been determined in the presence of all participant phases. This proviso is a substantial cause of uncertainty owing to uncertainties in the compositions of the mantle and crust. The relevant characteristics of the process can be changed markedly by the presence or absence of quite small proportions of some constituents.

For this reason the values of B given in Table 1 have been selected from a fuller review of thermal epeirogenic processes being compiled by the present author and which will be published separately in due course. Such a procedure is to some extent justified by the differences in the magnitude of B for the four types of process.

TABLE 1 Some Properties of Thermal Epeirogenic Processes

Process	B 10^{-3} cm³/cal	*Specific Subsidence Rate** m/m.y. per μcal/cm² sec	
		Under water	Under sedimentation to constant level
Thermal contraction	0·032	15	47
Solidification	0·25	120	380
Hydroxylation reactions	0·5	240	760
Solid-state phase changes	4·0	1900	6000
Real ocean floor 0–10 m.y.		30–60	95–195

*Assumed densities: sediments, 2·5; mantle, 3·2; sea-water, 1·03.

The value of **B** given for thermal contraction relates to mantle material and is probably a good average down to 75 km depth, beyond which a decreasing value is to be expected. Within the continental crust the value may be as much as 30% lower but this depends on the composition. The value for solidification is affected by changes in the water content of the melt and therefore upon the degree of melting. The figure quoted has been estimated for mantle material at a depth of 50–100 km and a degree of melting of 10 wt%. The value for hydroxylation reactions relates to depths of 15–20 km, but individual reactions differ from this by factors as large as two. Note also that the low density of steam at low pressures results in a rapid increase in **B** at shallower depths, but at greater depths **B** will in most cases decrease toward zero on approaching the depth limit for stability of the hydroxylated mineral concerned (probably less than 75 km for most of those likely to occur in sufficient abundance to be epeirogenetically significant).

Compositional uncertainties are of particular significance to the value of **B** for solid-state phase changes. Most such changes are complex reactions between many participant minerals and involve solid solution effects, the limits of which are sometimes intimately dependent upon even quite minor constituents of the assemblage. Moreover, reaction rates are commonly extremely slow below about 900°C and extrapolation of laboratory results to lower temperatures (relevant, perhaps, to conditions in the lower part of the continental crust) is fraught with uncertainty. These problems are well illustrated by recent discussions (MacGregor, 1970; O'Hara *et al.*, 1971). The expected presence of H_2O or hydrated minerals in many situations is a further complication. From data currently available it appears that phase changes relevant to the crust and mantle to depths of 150 km may have values of **B** in the range $(0{\cdot}5\text{–}8{\cdot}0) \times 10^{-3}$ cm^3/cal. The value given in Table 1 should therefore be treated with corresponding caution. It is also important to stress that most phase changes require a considerable change of temperature for completion, so their apparently high thermal sensitivity for epeirogenetic purposes is in fact much reduced by the heat which has to be extracted to achieve the required cooling.

We now consider the surface subsidence rate that would occur if heat flow were entirely vertical and entirely attributable to one of these four types of subsidence process. For unit heat flow (10^{-6} cal/cm^2 sec) we then have the *specific subsidence rate* for the process under the stated conditions. In Table 1 the values of this parameter have been calculated for subsidence under sea-water and under constant-level sedimentation.

In an actual case, the measured heat flow at a point on the Earth's surface may be regarded as having two principal components, one of which is continually replaced by radioactive heat generation within the entire underlying column of material, and therefore involves no subsidence.* The other component may be termed the cooling component and involves subsidence due to some combination of the processes listed in Table 1 appropriate to the distributions of temperature and composition within the column concerned. Since it is a characteristic of rock columns extending through a large depth range that, on cooling, they undergo contractions in addition to pure

* This statement would be strictly correct only if α_v did not vary with depth or if the heat loss and its replacement by radioactivity occurred at the same levels. In young lithosphere, the heat loss will generally occur at shallower depth than that at which it is replaced radioactively, the differences in α_v thus resulting in a small subsidence rate, here neglected.

thermal contraction, it follows that after subtraction of the radioactive component the calculated specific subsidence rate (still assuming vertical heat flow) must always exceed that for pure thermal contraction. In Table 1, the approximate figure for the specific subsidence rate of 'real ocean floor' during its first 10 m.y. after emplacement allows an arbitrary $0{\cdot}9 \times 10^{-6}$ cal/cm^2 sec for the radioactive component and is based on subsidence rates, heat flow, topographic profiles and spreading rates taken from the literature (Lee & Uyeda, 1965; Von Herzen, 1967; Vogt & Ostenso, 1967; Sclater & Francheteau, 1970; Sclater *et al.*, 1971), and we may note that it does in fact exceed the pure thermal contraction rate by a factor of two to four.

For our present purpose, however, the significant figure is the one in the right-hand column purporting to represent the specific subsidence rate of young 'ocean floor' if it were subjected to continuous sedimentation to sea-level. It has been derived from that for submarine ocean floor by applying the appropriate isostatic correction factor (3·15) for sediment loading. Implicit in this calculation is the assumption that the same proportionate mixture of epeirogenic processes would occur in both circumstances. This is clearly improbable owing to the effects of sedimentary blanketing and the depression of the crustal base to deeper levels. However, it is at least a first approximation to the maximum specific subsidence rate that is likely to result from heavy sedimentation on to young lithosphere. Ocean floor heat loss rates and subsidence rates fall off rapidly with time but it is conceivable that under sedimentation a heat loss rate through the sediments might be maintained at up to 3×10^{-6} cal/cm^2 sec for the first 10 m.y. Taking the maximum specific subsidence rate (195 m/m.y. per unit heat flow) the total sediment thickness after 10 m.y. would be nearly 14 km, and the temperature at the base of the new crust would exceed 1000°C, if allowance is made for radioactive heat production in the sediments. This is clearly stretching the figures to rather improbable limits. We tentatively conclude, therefore, that the maximum rate of subsidence under sedimentation, sustainable over a 10 m.y. period, and (it must be remembered) on the assumption that all heat loss from the new lithosphere is vertical, is of the order of 400 m/m.y.

Arguments have already been given for the occurrence of substantial heat transfer across chasmic faults. The amount of such heat transfer is difficult to quantify at present but it is clear that it will have most effect on the rate of subsidence of the new lithosphere when the amount of separation is small, with a fairly rapid diminution of the effect when the separation is comparable with or exceeds the thickness of the old lithosphere on either side. Accordingly, in narrow sedimentary troughs formed by lithosphere separation very considerable enhancement of the subsidence rate inferred above is to be expected.

These conclusions appear remarkably consistent with geologically deduced rates of accumulation in geosynclines (Kay, 1955; Sutton, 1969; Ziegler, 1970). Such rates are not necessarily the rates of subsidence except in the few cases where depositional depths at both ends of the sequence have been ascertained. It appears that accumulation rates of the order 150 m/m.y. for 50 m.y. are fairly common; for periods of 20 m.y., rates up to 600 m/m.y. are found, and the highest rate so far recorded is 1400 m/m.y. for 7 m.y. in the Fossa Magna of Japan.

In view of this accord it appears that the subsidence responsible for many geosynclinal accumulations could have been the direct consequence of lithosphere separation. It should be noted that the initial 8 km, or so, of sediments and/or volcanics (Fig. 2b) needed to bring the sedimentation surface up to sea level will accumulate at rates unrelated to thermal epeirogenic subsidence rates, but it seems possible that this deep part of the crust is as yet unexposed in any of the geosynclines of Phanerozoic age, to which data on accumulation rates are at present restricted.

Folding

Compressive folding of the crust of a basin formed by lithosphere separation would thicken the crust, and the associated metamorphism would increase its mean density. Thickening the crust would tend to raise its surface above sea-level, but the increase in density would offset this to some extent. There is thus no obvious reason why basin crust should not ultimately develop into the thick, dense and complexly deformed crust typical of Precambrian shields.

At present, almost nothing is known about the forces either available or required for the folding of continental lithosphere. The absence of any evidence of folding of the oceanic crust implies either (a) that young continental lithosphere is much softer than young oceanic lithosphere or (b) that compressive forces (if any) developed by the ocean ridge/Benioff zone system are small and perhaps independent of those responsible for the compression of wide basins (e.g. the Late Mesozoic folding of the Verkhoyansk geosyncline (Nalivkin, 1960)). We cannot discuss here the genesis of horizontally compressive geotectonic stress fields, but would emphasize that the detailed distribution of folding of any one age in actual fold belts often suggests that the response to such stress fields has been greatly influenced by sharp variations in lithosphere condition such as might have been inherited from preceding separative movements.

At some stage in its cooling the continental lithosphere probably becomes too stiff to be folded by available forces unless some form of thermal 'softening-up' process takes place first. This might be achieved either by separative formation of an adjacent geosynclinal trough or by Benioff zone volcanism.

In other cases the continental basin crust will be folded before its lithosphere becomes too stiff, which will mean that cooling is incomplete. This will have two important consequences. Firstly, if the crust is still thin and the folding involves only minor crustal shortening, this part of the fold belt may never appear above sea level. Secondly, even if the crust is thick enough to result in an elevated mountain range, the continuing cooling of its lithosphere will result in continuing subsidence. It is easily shown that isostatic adjustment for removal of crust by surface erosion would require that 500–1200 m of material (depending on densities assumed) be removed in order to lower the surface by 100 m. The widely-known fact that many mountain ranges have been eroded to near sea-level, but have obviously not suffered the removal of anything like the isostatically requisite amount of overburden, seems good evidence for the continuance of thermal subsidence after folding.

A succession of lithosphere jostling movements, involving alternate separation and approach, could result in a belt of tightly folded crust interpretable as having originally thinly floored a basin of very great width, when in fact the maximum separation between the tectonic plates concerned may at no time have exceeded a fraction of this amount. This might explain how some wide belts of Precambrian rocks possessing rather uniform strike across their entire width may have been formed.

Limited separation, followed by compression, has two other important characteristics. The first is that the alignment of older tectonic trends on opposite margins of the new folded basin may be substantially preserved. The frequent occurrence of such relationships in the Precambrian of Africa does not encourage the view that such fold systems were produced by oceanic closure because of the improbability that this would restore pre-existing relationships so closely (Hurley, this volume, p. 1083; Shackleton, this volume, p. 1091). The second characteristic concerns the probable non-development of a Benioff zone and its corresponding crustal descent line, or suture, owing to the presence of relatively thick continental basin crust, not oceanic crust, between the convergent plate margins. Crustal shortening undoubtedly requires downward disposal of excess subcrustal lithosphere, but the mere occurrence of folding within the basin presumably implies that the mantle part of the basin lithosphere is still soft enough to be extruded downwards as compression proceeds. The difficulty of finding suture lines in some Precambrian fold belts (Burke & Dewey, this volume, p. 1035) may thus be due to their non-occurrence.

Lateral Heat Flush and the Genesis of Granites

Various consequences of the transfer of heat across chasmic faults in simple separative situations have already been mentioned. If, however, the lithosphere break-up is complex, involving complete detachment of 'islands' of old lithosphere, these will appear as inliers within the basin. Because, unlike the main margins of the basin, they are completely detached from an extensive heat sink consisting of old lithosphere, the lithosphere of these inliers will attain higher temperatures, causing their tops to be uplifted even more than the main margins.

It is now well established from theoretical studies of the separation of 100 km thick oceanic lithosphere that, at a depth of 30 km, temperature excesses (relative to those inferred for oceanic lithosphere more than 100 m.y. old) remain in the range 550–200°C during the first 30 m.y., or so, after emplacement. These figures illustrate the character of the heat input to the lower part of any mature continental crust bounding a new basin. The greater thickness of continental lithosphere would, as already mentioned, probably increase the heat content of the young lithospheric material, but, on the other hand, lateral heat losses will cause the excess temperatures near the basin margins to fall away more quickly. Nevertheless it is quite clear that these conditions imply that any crust of mature continental thickness (say 35–45 km) bordering a new basin will experience a major thermal event and that thermal metamorphism, with the rise of granites and pegmatites, may well occur. These effects are particularly to be expected in the case of wholly detached blocks of lithosphere.

It is proposed, therefore, that many post-tectonic granites, and others whose initial mobilization occurred in the absence of compressive stress, may owe their occurrence to the thermal effects of nearby lithosphere separation. This is not to deny that some granitic plutons (e.g. the Tertiary plutons of western Scotland) may result from heat brought in by local volcanism. On the other hand there is now some reason to question whether the undoubted connection between some granites and the presence of a Benioff zone is the whole story, or whether many of these granites are in fact products of the recently recognized tendency (mentioned briefly at the beginning of this paper) for separative movements to occur at the rear of such zones.

Geological Reconstructions

An obvious test of whether continental basins have in fact been formed by lithosphere separation is that the movement proposed in any particular case should not only account for the existence and subsidence of the basin but that reconstructions on this basis should be consistent with basin geometry and with pre-basin geological structures. We now discuss briefly three examples, selected to illustrate different aspects of the arguments given in preceding sections. To justify each fully from all the data available is clearly impossible within the compass of the present paper and no attempt will be made to do so.

(a) *Northeast Scotland and Norway* (Fig. 4). The subsidence of the North Sea basin has been spectacularly proven by the recent search for oil and gas. The base of the

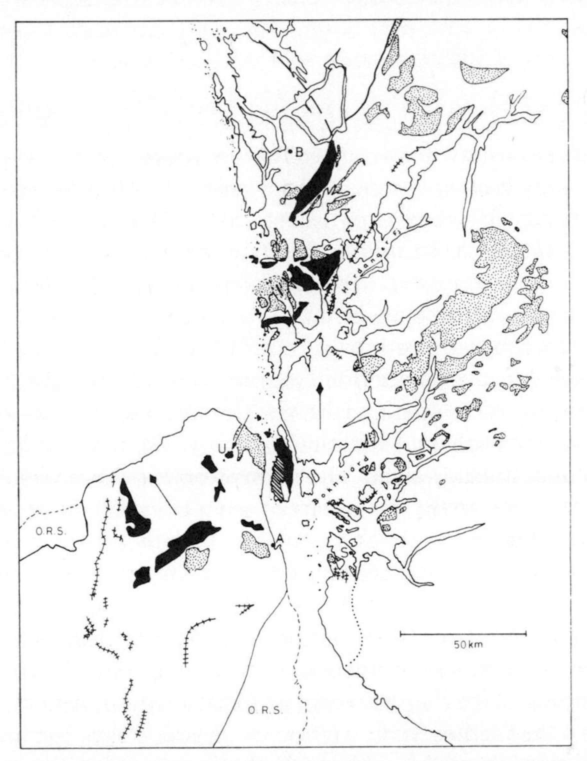

FIGURE 4 Proposed relationship of northeast Scotland to southwest Norway before formation of the North Sea basin. Reconstruction is not drawn to a common projection but from separate maps at 1:1,000,000 scale (Oxford Atlas, 1957 for Scotland; Geological Map of Norway, Holtedahl & Dons, 1960). Black, basic intrusions; stipple, granites; oblique ruling on Karmøy, augen gneiss and breccia. Cross-barred lines are limestone outcrops in Caledonian fold belt. Places marked are: Aberdeen (A), Bergen (B), Stavanger (S), and Utsira island (U). Utsira is shown in its present relationship to Norway, with an arrow suggesting possible displacement. North vectors indicate present north direction on Norway and northeast Scotland. Heavy broken line is outer Bergen arc. Heavy continuous line is northwest limit of Cambro-Silurian 'allochthon' in Norway. Dotted line south of Stavanger shows limit of Cambro-Silurian rocks.

Permian now lies at more than 5 km depth near the middle of the basin but crops out above sea level in northern England (Kent, 1967; Birch, 1969). The uplift, metamorphism and extensive granitic intrusion of the Dalradian rocks of the Scottish central highlands present a major geochronological problem. Recent discussions have been given by Brown & Miller (1969), and by Dewey & Pankhurst (1970). The gabbros in northeast Scotland have been well dated by Rb-Sr whole rock isochron at 486 $\pm$ 17 m.y. (Pankhurst, 1970) and approximates the date of main folding. Geochronological studies in southwestern Norway are still at a very early stage (M. R. Wilson, I. Pringle & I. Bryhni, personal communications, 1972) so the correlation suggested by Fig. 4 cannot yet be discussed.

Dewey & Pankhurst (1970) relate the apparently continuing rise of granites in the Scottish Central Highlands during the period 460 m.y. to 380 m.y. to Benioff zone activity during closure of a proto-Atlantic ocean. The present proposal is that, whether or not there was a Benioff zone there at that time, many of the granites and the uplift responsible for their successive cooling ages were due to lateral heat flush caused by several isolatory separative movements, of which the separation from Norway was one of the last. In Fig. 4 the only Scottish granites shown are those with concordant Rb-Sr and K-Ar ages of 404 m.y. (uncertainties $\pm$ 13 m.y. and $\pm$ 5 m.y. respectively) (Dewey & Pankhurst, 1970, p. 380). All are within 40 km of the Aberdeenshire coast. Similarly, of the Norwegian granites shown in Fig. 4, only those within 40 km of the coast are mapped (Holtedahl & Dons, 1960) as post-tectonic. It is suggested, therefore, that separation from Norway started shortly before 404 m.y. ago. The breccia on Karmøy might be related to this event. Even the 15–20 km of erosion of the Scottish unit, envisaged by Dewey & Pankhurst (1970, p. 383), could have been accomplished by 2–3 km of thermal uplift, as is typical of present-day rift-valley upwarps, accompanied by isostatic adjustment as erosion proceeded.

The details of the relative positions shown in Fig. 4 were decided quite arbitrarily, although with a view to the eventual incorporation of the Scottish Southern Uplands in a suitable position. Offshore marine geology for the Norwegian coast was not available to the writer but could obviously improve control if inliers of a dislocated character can be identified as such. The relative rotation shown (33°) is larger than, but in the same sense as, that (21°) obtained from the palaeomagnetic results for the Arrochar igneous complex in Scotland (dated at 418 m.y.) (Briden, 1970) and the Late Ludlow/Early Downtonian Ringerike sandstone of the Oslo region (Storetvedt *et al.*, 1968).

(*b*) *Basins in the northern Appalachians* (Fig. 5). The somewhat unconventional nature of this map is explained in the caption. Its main purpose is to suggest the formation of a complex of basins, notably the Gulf of Maine and the New Brunswick Basin and parts of the Gulf of St. Lawrence, by a general east-northeasterly movement relative to the Canadian shield, and probably derived from the motion of a major lithospheric unit originally forming the southeastern margin. Such a map obviously omits a number of small areas in which old rocks are known to underlie the cover but these are unlikely greatly to affect the general picture.

The first major problem is to infer when the movements might have occurred. Black (1964) on limited palaeomagnetic evidence, inferred a 30° anticlockwise rotation of Newfoundland at the end of Devonian time. This, and the extensive Carboniferous sedimentation in the basin areas, sets the minimum age of movement. A maximum age

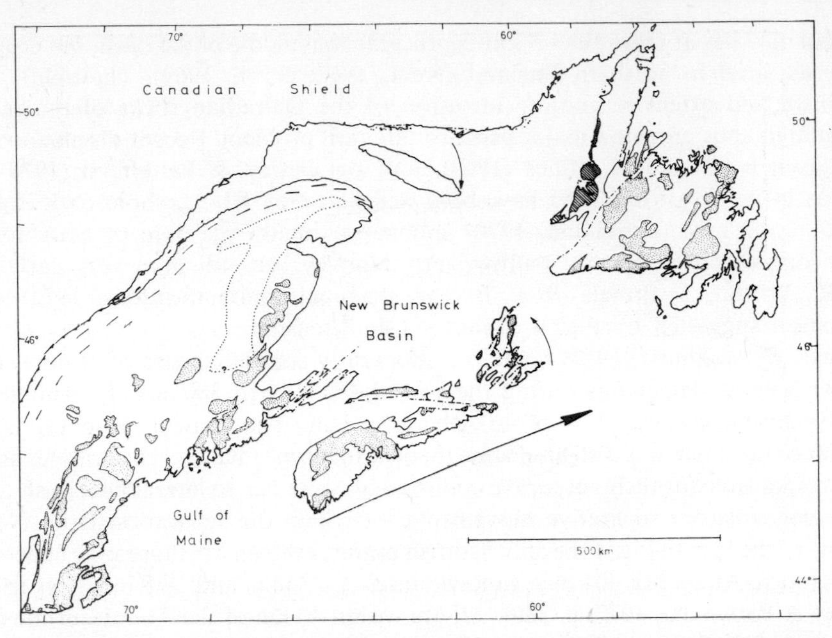

FIGURE 5 Greatly simplified geological map of the Northern Appalachian–Newfoundland region. Areas affected by post-Lower Devonian subsidence have been excluded from the outlines. Acadian granites and granodiorites, stippled. Principal steep or vertical faults, dash-dot lines. Dashes show Taconian (mid-Ordovician) tectonic trends along Appalachian front. Oblique ruling indicates Humber Arm Taconian allochthon in Newfoundland. Data taken from Tectonic Map of North America (King, 1969) and replotted on a Bonne's projection of the area (Debenham, 1962, p. 82) to minimize distortion. Approximate outline of the Matapedia basin, interpreted by Bird & Dewey (1970, p. 1049) as an inter-arc oceanic basin of post-Taconian age, is shown dotted and discussed in the text. The large arrow shows suggested total (Devonian?) motion of a tectonic plate initially attached to the entire southeastern margin of the region. Small arrows show inferred subsididary motions.

is suggested by Fig. 6 in which the mafic, ultramafic and clastic Taconian allochthon of western Newfoundland (Kay, 1969; Williams, 1971) was not in fact transported from the east but is the endwise overflow of the Gaspé Peninsula Taconian fold belt (which also contains ultramafics). Figure 6 shows a rotation of $22\frac{1}{2}°$ for the Long Range Peninsula of Newfoundland, but Belle Isle, shown in black in its present position relative to the peninsula, then overlaps the mainland, implying either that Belle Isle was displaced during the movements or that the peninsula plus Belle Isle should be rotated rather more than shown in Fig. 6.

If Fig. 6 is accepted in principle it is at once obvious from Fig. 5 that the western boundary of the V-shaped New Brunswick basin also needs to be rotated clockwise if it is to accommodate at this time the rest of Newfoundland (assuming any necessary slippage along the Longe Range fault system traversing the island). Closure of the Matapedia basin, which appears to be of immediately post-Taconian age (Bird & Dewey, 1970; McKerrow & Ziegler, 1971), could make such a fit possible, however.

Accordingly, as a working hypothesis for closer investigation it is suggested that the movement of Newfoundland took place in two stages. The first, of immediately post-Taconian (i.e. Upper Ordovician) age, opened the Matapedia basin, thus rotating Newfoundland anticlockwise and detaching its western coast from the end of the

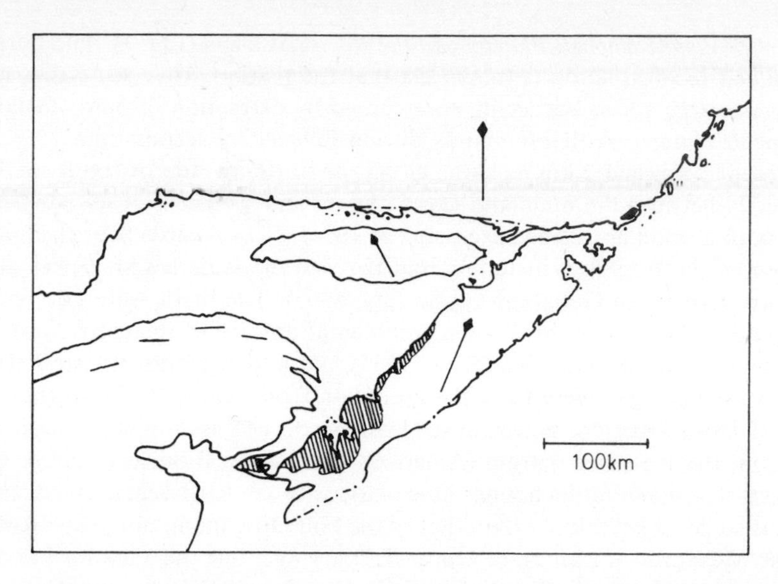

FIGURE 6 Proposed reconstruction of part of Fig. 5 for Mid-Ordovician time, showing relationship between the western Taconian allochthon of Long Range Peninsula and the end of the Gaspé Taconian fold belt. North vectors show present north directions on units involved. Position of Anticosti Island is not closely controlled but that shown would result in basin formation only to the north of the island, with movement on the south side being along a curved fault. The position of Belle Isle (black, superimposed on mainland) is discussed in the text.

Gaspé peninsula. The second, of possibly Lower to Middle Devonian age, may have been the main movement which extracted Newfoundland from the New Brunswick basin, and probably also produced the Gulf of Maine. If this is correct, then proto-Atlantic closure (Wilson, 1966; Dewey, 1969) must have been completed in or by Lower Devonian time because of the difficulty, that would otherwise arise, of producing identical movements on opposite sides of an ocean basin. The continuance of granite dates far into the Devonian could then be the result of lateral heat flush. In this connection it is interesting to note the marginal locations of many of the Acadian granites.

Finally, we return to the matter, mentioned on pp. 643–4, of the low subcrustal seismic velocity (7·2–7·8 km/sec) found beneath much of the basin areas discussed here. Our inference that the crust of these basins has not been subject to folding and metamorphism since formation gives weight to the idea that the presence of volatiles, either combined or free interstitially, may cause the low velocity. The fact that such velocity does not usually occur beneath deep oceans, however (Drake & Nafe, 1968; Maynard, 1970) where folding and metamorphism are also absent, but is found near the axes of ocean ridges, suggests that temperature is the significant factor. Retention of the low velocity beneath basins and continental margins would then be due to the higher sub-crustal temperatures resulting from the greater depth to which the material is depressed. Accordingly, it is suggested that the low velocity may be due to the presence of small amounts of intergranular volatiles. The work of Anderson & Spetzler (1970) on the effects of intergranular films suggests that the amounts required might be volumetrically quite trivial. Nevertheless, if this inference is correct, low-temperature hydroxylation processes may play a rather less important part in basin subsidence than has been suggested.

(*c*) *New Zealand/Campbell Plateau and the Ross Sea Basin* (Fig. 7). The purpose of this simplified reconstruction is to suggest that the present Ross Sea basin, and the mountain upwarps which border it, were caused by extraction of New Zealand and the Campbell Plateau, probably mainly during Lower Cretaceous time.

The present 1 km and 2 km isobaths along the northern side of the Ross Sea run fairly directly between the mainland capes (Adare and Colbeck) at its northern corners, but with a small northward excursion at 180–185° E towards Scott Island. Most of the floor of the Ross Sea, including that beneath the Ross Ice Shelf, lies at about 500 m depth (American Geographical Society, 1965). The main Antarctic feature of the proposed withdrawal is the 2–4 km continental upwarp of the entire western and southern margins of the Ross Sea (King, 1965), tilting peneplains of Late Palaeozoic and Mid-Mesozoic ages away from the mountain front (Gair, 1967), so that sub-ice altitudes of 1·0–0·5 km (corrected for ice loading) do not usually occur until at least 300 km from the Ross Sea margin (American Geographical Society, 1965). Gravity data suggest that major faults bound these margins of the Ross Sea, and that the Ross Sea crust is some 10 km thinner than that of the bounding mountain areas (Robinson, 1964). The McMurdo volcanics, of Upper Tertiary age, line the Victoria Land coast, abutted by New Zealand's South Island in Fig. 7, and Wright (1966) has given

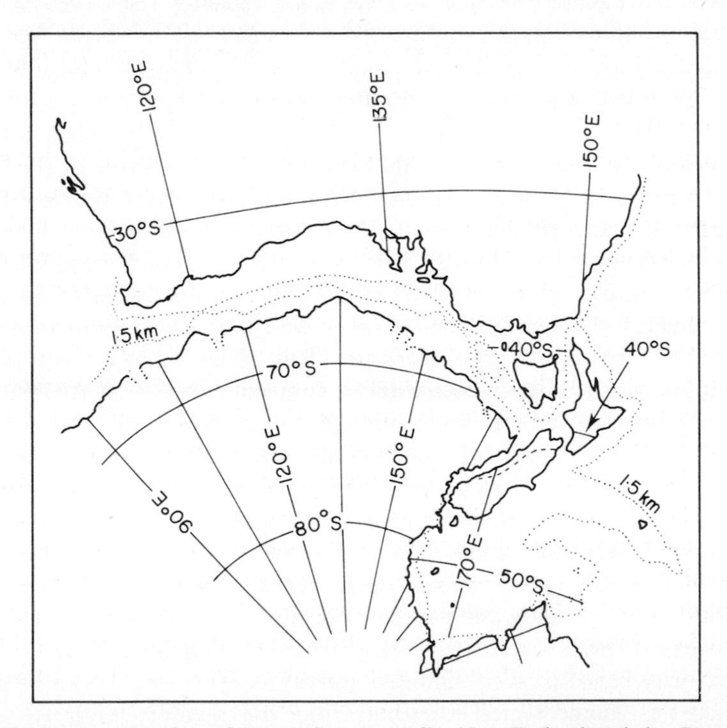

FIGURE 7 Crude reconstruction of Antarctica, Australia, New Zealand and the Campbell Plateau to suggest genesis of the present Ross Sea basin. Drawn from a reconstruction on a 34 cm diameter globe using thin spherically-deformed plastic shells. 1·5 km isobaths taken from Tectonic Map of Australia (1960), American Geographical Society Map of Antarctica (1965) and from Brodie (1964) and Brodie & Dawson (1965). Eastern margin of Ross Sea has poor topographic expression, except at northern corner. The margin shown is the +200 m ice surface contour. Bathymetry on the west side of New Zealand, including the Lord Howe Rise, has been omitted (see text). Dashed line marks the Alpine Fault in South Island, New Zealand.

important correlations between the Late Precambrian and Palaeozoic rocks of South Island and those of Victoria Land. However, both he, and Griffiths & Varne (1972), offered reconstructions in which the southern edge of the Campbell Plateau adjoined the shelf edge of the Ross Sea. We believe that such reconstructions depict only an intermediate stage in the separative evolution of the area because they have been based on the premise that lithosphere separation can only create ocean floor. Accordingly, such reconstructions have been quite unable to place the pre-Carboniferous (Caledonian) structures of South Island, after allowance for Alpine Fault movement, in reasonable structural continuity with those which run through Victoria Land, western Tasmania and southeastern Australia. Figure 7 is attractive in this respect, however, but is achieved at the price of total omission of the Lord Howe Rise, whose crust (like that of the present Ross Sea) is therefore inferred to post-date this reconstruction.

It is proposed that the first extractive movements occurred in Mid-Jurassic time, with the onset of the Rangitaka Orogeny in New Zealand (Fleming, 1970). A computer fit of the Australian and Antarctic continental shelves, slightly closer than Fig. 7 shows, was dated at 45 m.y. by McKenzie & Sclater (1971) so any Mid-Jurassic separative troughs between them would have had about 100 m.y. to fill before final separation got under way. The Rangitaka Orogeny deformed New Zealand by dragging the northern part clockwise with respect to a fixed (?) southern end, and was accompanied by vigorous uplift, particularly on the west. It is inferred that this movement initiated a Late Jurassic trough to the west of New Zealand, and eventually left enough room between New Zealand and Tasmania/Victoria Land for deposition of what is now the southern part of the Lord Howe Rise. The same movements may also have drawn Tasmania southeastwards to form the basins in the Bass Strait. The many horst and basin structures of New Zealand itself, developed during this time and later, appear ideally susceptible to interpretation in terms of limited separation but cannot be discussed here. For this reason all forms of reconstruction within New Zealand, including restoration of the Alpine Fault movement, have been ignored in Fig. 7. Extraction of the Campbell Plateau from the Ross Sea must have been complete by about 80 m.y. B.P., when there begins to be evidence of ocean-floor spreading in the Tasman Sea and south of the Campbell Plateau (Griffiths & Varne, 1972).

Perhaps the most intriguing implication of Fig. 7 is the fact that the Campbell Plateau is so different from Antarctica in its geology (Brodie, 1964), its crustal thickness of 17–23 km (Adams, 1962), and its apparent epeirogenic response to the proposed separation. The islands of the Campbell plateau all lie near its edges, suggesting that they may be inliers based on older crust detached from the Antarctic boundaries and/or that marginal upwarping of this lithosphere unit is responsible. Both are certainly true for South Island of New Zealand. On the other hand, along the southern margin of the Ross Sea, Precambrian and Early Ordovician (Ross Orogeny) folds of metagreywackes and phyllites with north and northwest axes terminate abruptly at the coast (Grindley *et al.*, 1964). There is abundant evidence, however, from the rocks of the eastern side of New Zealand and from the Campbell Plateau and Chatham Rise, that this area was a major basin in Late Palaeozoic and Mesozoic times. Whatever separated from this area to produce this Mesozoic basin may exhibit the continuation of the ancient fold belt of the southern Ross Sea margin. Thus, if Fig. 7 is correct in its implication, the present site of much of the Ross Sea has been the site of at least two major separative events during Phanerozoic time. It is suggested that chasmic faults remain sites of weakness for long periods owing to the protracted differential vertical movement on them which we have shown may persist for upwards of 100 m.y. The important conclusion drawn from this is that, owing to the repeated occurrence of chasmic faulting during geological time, the reconstruction of pairs of chasmic faults of a particular age should not necessarily result in the matching across them of geological structures of appreciably earlier age.

Conclusions and Discussion

We have shown that limited lithosphere separation, in circumstances that ensure an abundant sediment supply, will produce crust having characteristics which compare closely with known features of continental basins (not lying on shields), continental shelves and many geosynclines. In particular, the total crustal thickness, duration of subsidence, and the early (geosynclinal) subsidence rate appear able to attain observed values without difficulty. It is also clear that heavy sedimentation would result in production of a lower crust very different from ocean ridge crust and that the magnetic anomalies so characteristic of ocean-floor spreading would not be produced.

This result rests solely on consideration of the thermal and other consequences of lithosphere separation and new lithosphere emplacement, on the lines now well established by studies of the cooling and subsidence of the oceanic lithosphere. No attention has had to be given to the separative agency except that which is implicit in the assumption of limited separation, i.e. that separation ceases when it has attained values representative of actual continental basins. Extinctions of formerly active ocean ridges are already known, so we have not added to existing constraints upon theories concerning lithosphere separative agencies.

The presence of chasmic faults (Figs. 1–3) marking the boundary between new and old lithosphere is important. The resulting lateral transfer of heat across chasmic faults at depth is a uniquely effective way of accounting for closely juxtaposed vertical movements of opposite sign. Such heat transfer accelerates the subsidence of narrow troughs and the marginal parts of wide basins, whose later subsidence is thus confined to the middle area. Surface expression of chasmic faults will continue as movements of normal fault character throughout much of the subsidence life of the basin. Moreover, because chasmic faults separate lithosphere having different histories and compositions, they may retain almost indefinitely a sensitivity to thermal rejuvenation, even after the adjacent basin crust has been folded. Rejuvenation of ‘ancient lines of weakness’ is a well-known geological phenomenon and may indicate the presence of ancient chasmic faults.

Lateral heat flush into the lithosphere bounding a new basin will vary with the prior condition of that lithosphere. The comparatively minor upwarping of the Atlantic shelf margins and their early reversion to subsidence would be consistent with these margins being fairly young lithosphere themselves. Our discussions of the origin of the North Sea and of the New Brunswick and Gulf of Maine basins suggest that much of the North Atlantic continental shelves north of the Azores may be an elongate basin system formed by a Mid-Palaeozoic limited separation of Europe from North America, immediately following closure of the proto-Atlantic. However the vigorous Jurassic upwarping of Jameson Land in southeast Greenland (Haller, 1970) suggests that initial opening of the Atlantic south of the Azores about 180 m.y. ago

(Francheteau, Vol. I, p. 195) was accompanied by important separative movements further north but that these troughs became added to the shelf area by being filled with sediments.

Where lateral heat flush enters old lithosphere possessing a thick crust, there is evidence not only that the upwarping can be as much as 3 km (largely maintained by isostasy in the face of heavy erosion) but that the heat can cause granites to rise within the crust. Such features, accompanied by the rapid subsidence of troughs and the possibility of flood or other volcanism, are commonly and, we suggest, appropriately regarded as orogenic in character. However, it is obviously essential for the progress of plate tectonics that the effects of lithosphere separation should be clearly distinguished from those of plate convergence, so the terms *separative orogeny* and *convergent orogeny* are proposed for this purpose. In practice the correct attribution of observed structures may be rather less simple because, for example, differential epeirogenic movements can probably give rise to low-angle reverse faulting and apparent thrusting (Sanford, 1959; Osmaston, 1971).

Folding of the continental crust of basins of separative origin has been discussed at some length on pp. 652–3. Such crust is presumably more prone to folding by virtue of its younger and hotter lithosphere. Identification of such areas might help in elucidating the force distribution responsible for horizontal tectonic movements.

In conclusion, there is much to support the proposition that limited lithosphere separation has long been a major factor in producing continental basins and adding to the area of the continental crust. The matter clearly merits further study. It is appropriate, however, to remind ourselves of the central issue of the origin of epeirogenic movements. All such movements—even downward ones to provide more space for sediments—expend energy, and sources for this have to be found. Freshly emplaced lithosphere possesses adequately large amounts of thermal, chemical, and potential energy, and is a source wholly compatible with plate tectonics. Its apparent ability to distribute some of this energy to distances of several hundred kilometres into the surrounding lithosphere suggests that, if the long thermal time constants involved can be successfully taken into account, more detailed epeirogenic problems, such as the post (?) Mesozoic epeirogenetic differentiation and eastward tilting of crustal units within the British Isles, may ultimately become explicable in terms of distant separative orogeny. The large horizontal extent of individual epeirogenic movements and their frequently sharp geographical boundaries already support the suggestion made on page 643 and by Osmaston (1971), that much material currently assigned to the asthenosphere may in fact move integrally with tectonic plates and thus contribute to the scale of epeirogeny.

Acknowledgements

The possibility that lithosphere separation might be relevant to the formation not only of ocean basins but also of continental basins was first pointed out and brought to my notice by S. W. Carey's famous paper (Carey, 1958). In that the development of that hypothesis into the form presented here has been financed entirely from personal resources I am particularly grateful to Professor J. Sutton and to many other former colleagues at Imperial College, London, for their continuing interest and for many hours of stimulating discussion. Professor M. Blackman gave helpful advice on thermal expansion, Professor W. S. Fyfe kindly reminded me of the broad utility of the Clapeyron equation, and I thank Dr. S. Moorbath for advice on geochronology. Finally, I would thank especially Drs. W. S. McKerrow, H. G. Reading and A. Richardson for their critical advice and enthusastic encouragement over a long period.

References

Adams, R. D. 1962. Thickness of the Earth's crust beneath the Campbell Plateau, *N.Z. J. Geol. Geophys.*, **5**, 74–85.

American Geographical Society, 1965. Map of Antarctica, 1:5,000,000.

Anderson, D. L. and Spetzler, H., 1970. Partial melting and the low-velocity zone, *Phys. Earth Planet. Interiors*, **4**, 62–64.

Belousov, V. V., 1960. Development of the Earth and tectonogenesis, *J. Geophys. Res.*, **65**, 4127–46.

Belousov, V. V., 1962. Basic Problems in Geotectonics, edited by J. C. Maxwell, transl. by P. T. Broneer. McGraw-Hill, New York. 816pp.

Belousov, V. V., 1966. Modern concepts of the structure and development of the Earth's crust and the upper mantle of continents, *Q. J. Geol. Soc. Lond.*, **122**, 293–314.

Berger, J., Cok, A. E., Blanchard, J. E. and Keen, M. J., 1966. Morphological and geophysical studies on the eastern seaboard of Canada: the Nova Scotian Shelf, *Roy. Soc. Can. Spec. Pub.*, **9**, 102–13.

Billings, M. P., 1960. Diastrophism and mountain building, *Geol. Soc. Amer. Bull.*, **71**, 363–98.

Birch, R. L., 1969. The search for gas in the North Sea, *Geology*, **1**, 31–37.

Bird, J. M. and Dewey, J. F., 1970. Lithosphere plate: continental margin tectonics and the evolution of the Appalachian orogen, *Geol. Soc. Amer. Bull.*, **81**, 1031–60.

Black, R. F., 1964. Palaeomagnetic support of the theory of rotation of the western part of the island of Newfoundland, *Nature, Lond.*, **202**, 945–8.

Bott, M. H. P., 1964. Formation of sedimentary basins by ductile flow of isostatic origin in the upper mantle, *Nature, Lond.*, **201**, 1082–4.

Bott, M. H. P. 1971. Evolution of young continental margins and formation of shelf basins, *Tectonophysics*, **11**, 319–27.

Bott, M. H. P. and Dean, D. S. 1972. Stress systems at young continental margins, *Nature Phys. Sci.*, **235**, 23–25.

Bott, M. H. P. and Johnson, G. A. L., 1967. The controlling mechanism of Carboniferous cyclic sedimentation, *Q. J. Geol. Soc. Lond.*, **122**, 421–41.

Bott, M. H. P. and Watts, A. B., 1970. Deep structure of the continental margin adjacent to the British Isles. *In*: The Geology of the East Atlantic Continental Margin, edited by F. M. Delaney, Vol. 2, pp. 93–109. ISCU/SCOR Symposium, Cambridge 1970. H.M.S.O., London.

Briden, J. C., 1970. Palaeomagnetic results from the Arrochar and Garrabal Hill–Glen Fyne igneous complexes, Scotland, *Geophys, J. R. astr. Soc.*, **21**, 457–70.

Brodie, J. W., 1964. Bathymetry of the New Zealand region. New Zealand Oc. Inst. Mem. 11. N.Z.D.S.I.R. Bull, 161.

Brodie, J. W. and Dawson, E. W., 1965. Morphology of North Macquarie Ridge, *Nature, Lond.*, **207**, 844–5.

Brown, P. E. and Miller, J. A., 1969. Some aspects of palaeozoic geochronology of British Isles. *In*: North Atlantic: Geology and Continental Drift, edited by M. Kay. *Amer. Assoc. Pet. Geol. Mem.*, **12**, 363–74.

Bunce, E. T., Crampin, S., Hersey, J. B. and Hill, M. N., 1964. Seismic refraction observations on the continental boundary west of Britain, *J. Geophys. Res.*, **69**, 3853–63.

Carey, S. W., 1958. The tectonic approach to continental drift. *In*: Continental Drift: a Symposium, 1956, pp. 177–355. University of Tasmania, Hobart.

Collette, B. J., 1968. On the subsidence of the North Sea area. *In*: Geology of Shelf Seas. edited by D. T. Donovan. Oliver and Boyd, Edinburgh. 15pp.

Collette, B. J., Lagaay, R. A., Ritsema, A. R. and Schouten, J. A., 1970. Seismic investigations in the North Sea, 3 to 7, *Geophys. J. R. astr. Soc.*, **19,** 183–99.

Dainty, A. M., Keen, C. E., Keen, M. J. and Blanchard, J. E., 1966. Review of geophysical evidence on crust and upper-mantle structure on the eastern seaboard of Canada. *In*: The Earth Beneath the Continents, edited by J. S. Steinhart and T. J. Smith. *Amer. Geophys. Un. Geophys. Monogr.*, **10,** 349–69.

Debenham, F. (editor), 1962. The Reader's Digest Great World Atlas. Reader's Digest Association, London. 179pp.

Dewey, J. F., 1969. Evolution of the Appalachian/Caledonian orogen, *Nature, Lond.*, **222,** 124–9.

Dewey, J. F. and Bird, J. M., 1970. Mountain belts and the new global tectonics, *J. Geophys. Res.*, **75,** 2625–47.

Dewey, J. F. and Pankhurst, R. J., 1970. The evolution of the Scottish Caledonides in relation to their isotopic age pattern, *Trans. Roy. Soc. Edinb.*, **68,** 361–89.

Dickinson, W. R., 1971. Plate tectonic models of geosynclines, *Earth Planet. Sci. Let.*, **10,** 165–74.

Drake, C. L. and Nafe, J. E., 1968. The transition from ocean to continent from seismic refraction data. *In*: The Crust and Upper Mantle of the Pacific Area, edited by L. Knopoff, C. L. Drake and P. J. Hart. *Amer. Geophys. Un. Geophys. Monogr.*, **12,** 174–86.

Drake, C. L., Ewing, J. I. and Stockard, H., 1968. The continental margin of the eastern United States, *Can. J. Earth Sci.*, **5,** 993–1010.

Emery, K. O., 1965. Geology of the continental margin off eastern United States. *In*: Submarine Geology and Geophysics, edited by W. F. Whittard and R. Bradshaw, pp. 1–20. Butterworths, London.

Fleming, C. A., 1970. The Mesozoic of New Zealand: chapters in the history of the circum-Pacific Mobile Belt, *Q. J. Geol. Soc. Lond.*, **125,** 125–70.

Fyfe, W. S., Turner, F. J. and Verhoogen, J., 1958. Metamorphic reactions and metamorphic facies, *Geol. Soc. Amer. Mem.*, **73.**

Gair, H. S., 1967. The geology from the Upper Rennick Glacier to the coast, northern Victoria Land, Antarctica, *N.Z. J. Geol. Geophys.* **10,** 309–44.

Gilluly, J., 1964. Atlantic sediments, erosion rates and the evolution of the continental shelf: some speculations, *Geol. Soc. Amer. Bull.*, **75,** 483–92.

Gilluly, J., Reed, J. C. Jr. and Cady, W. M., 1970. Sedimentary volumes and their significance, *Geol. Soc. Am. Bull.*, **81,** 353–76.

Griffiths, J. R. and Varne, R., 1972. Evolution of the Tasman Sea, Macquarie Ridge and Alpine Fault, *Nature Phys. Sci.*, **235,** 83–86.

Grindley, G. W., McGregor, V. R. and Walcott, R. I., 1964. Outline of the geology of the Nimrod–Beardmore–Axel Heiberg Glaciers region, Ross Dependency. *In*: Antarctic Geology, edited by R. J. Adie, pp. 207–18. North-Holland, Amsterdam.

Haller, J., 1970. Tectonic map of East Greenland, 1:500,000, *Medd. om Grønland*, **171** (5), 286pp.

Heirtzler, J. R., 1969. Geomagnetic studies in the Atlantic Ocean. *In*: The Earth's Crust and Upper Mantle, edited by P. J. Hart. *Amer. Geophys. Un. Geophys. Monogr.*, **13,** 430–6.

Holder, A. P. and Bott, M. H. P., 1971. Crustal structure in the vicinity of Southwest England, *Geophys. J. R. astr. Soc.*, **23,** 465–89.

Holtedahl, O. and Dons, J. A. (compilers), 1960. Geological map of Norway, 1:1,000,000, Norges Geologiske Undersøkelse, Oslo.

James, D. E. and Steinhart, J. S., 1966. Structure beneath continents: a critical review of explosion studies, 1960–1965. *In*: The Earth beneath the Continents, edited by J. S. Steinhart and T. J. Smith. *Amer. Geophys. Un. Geophys. Monogr.*, **10,** 293–333.

Jeffreys, H., 1959. The Earth. 4th Edition. Cambridge University Press, 420pp.

Joyner, W. B., 1967. Basalt-eclogite transition as a cause for subsidence and uplift, *J. Geophys. Res.*, **72,** 4977–98.

Judson, S., 1968. Erosion of the land, or what's happening to our continents? *Amer. Sci.*, **56,** 356–74.

Kanamori, H. and Press, F., 1970. How thick is the lithosphere? *Nature, Lond.*, **226,** 330–1.

Karig, D. E., 1970. Ridges and basins of the Tonga-Kermadec island arc system, *J. Geophys. Res.*, **75,** 239–54.

Karig, D. E., 1971. Structural history of the Mariana island arc system, *Geol. Soc. Amer. Bull.*, **82,** 323–44.

Kay, M., 1955. Sediments and subsidence through time, *Geol. Soc. Amer. Spec. Pap.*, **62,** 665–84.

Kay, M., 1969. Thrust shuts and gravity slides of western Newfoundland. *In*: North Atlantic—Geology and Continental Drift, edited by M. Kay. *Amer. Assoc. Pet. Geol. Mem.* **12,** 665–9.

Kent, P. E., 1967. Progress of exploration in North Sea, *Amer. Assoc. Pet. Geol. Bull.*, **51,** 731–41.

King, L. C., 1965. Since du Toit: Geological relationships between South Africa and Antarctica. Alex. L. du Toit Mem. Lect. 9. Geol. Soc. S. Africa. Annexure to Vol. 68, 32pp.

King, L. C., 1967. Morphology of the Earth. 2nd Ed. Oliver and Boyd, Edinburgh, 726pp.

King, P. B., 1969. Tectonic map of North America, 1:5,000,000, Dept. of the Interior, U.S. Geological Survey, Washington, D.C.

Kosminskaya, I. P. and Zverev, S. M., 1968. Deep seismic soundings in the transition zones from continents to oceans. *In*: The Crust and Upper Mantle of the Pacific area, edited by L. Knopoff, C. L. Drake and P. J. Hart. *Amer. Geophys. Un. Geophys. Monogr.*, **12,** 122–30.

Langseth, M. G., Le Pichon, X. and Ewing, M., 1966. Crustal structure of the mid-ocean ridges. 5. Heat flow through the Atlantic Ocean floor and convection currents, *J. Geophys. Res.*, **71,** 5321–55.

Lee, W. H. K. and Taylor, P. T., 1966. Global analysis of seismic refraction measurements, *Geophys. J. R. astr. Soc.*, **11,** 389–413.

Lee, W. H. K. and Uyeda, S., 1965. Review of heat flow data. *In:* Terrestrial Heat Flow, edited by W. H. K. Lee. *Amer. Geophys. Un. Geophys. Mongr.*, **8,** 87–190.

Le Pichon, X. and Langseth, M. G. Jr., 1970. Heat flow from the mid-ocean ridges and sea-floor spreading, *Tectonophysics*, **8,** 319–44.

MacGregor, I. D., 1970. The effect of CaO, Cr_2O_3, Fe_2O_3 and $A1_2O_3$ on the stability of spinel and garnet peridotites, *Phys. Earth Planet. Interiors*, **3,** 372–7.

Magnitinsky, V. A. and Kalashnikova, I. V. 1970. Problems of phase transitions in the upper mantle and its connection with the earth's crustal structure, *J. Geophys. Res.*, **75,** 877–85.

Matsuda, T. and Uyeda, S., 1971. On the Pacific-type orogeny and its model—extension of the paired belts concept and possible origin of marginal seas, *Tectonophysics*, **11,** 5–27.

Maynard, G. L., 1970. Crustal layer of seismic velocity 6·9 to 7·6 kilometers per second under the deep oceans, *Science*, **168,** 120–1.

McGinnis, L. D., 1970. Tectonics and the gravity field in the continental interior, *J. Geophys. Res.*, **75,** 317–31.

McKenzie, D. P. and Sclater, J. G., 1968. Heat flow inside the island arcs of the northwestern Pacific, *J. Geophys. Res.*, **73,** 3173–9.

McKenzie, D. P. and Sclater, J. G., 1971. The evolution of the Indian Ocean since the Late Cretaceous, *Geophys. J. R. astr. Soc.*, **24,** 437–528.

McKerrow, W. S. and Ziegler, A. M., 1971. The Lower Silurian palaeogeography of New Brunswick and adjacent areas, *J. Geol.*, **79,** 635–46.

Menard, H. W., 1967. Transitional types of crust under small ocean basins, *J. Geophys Res.*, **72,** 3061–73.

Mitchell, A. H. and Reading, H. G., 1969. Continental margins, geosynclines and ocean-floor spreading, *J. Geol.*, **77,** 629–46.

Morgan, W. J., 1968. Rises, trenches, great faults, and crustal blocks, *J. Geophys. Res.*, **73,** 1959–82.

Nalivkin, D. V., 1960. The Geology of the USSR, transl. by S. I. Tomkeieff. Pergamon, Oxford.

Officer, C. B. and Ewing, M., 1954. Geophysical investigations in the emerged and submerged Atlantic coastal plain, 7, Continental shelf, continental slope, and continental rise south of Nova Scotia, *Geol. Soc. Am. Bull.*, **65,** 653–70.

O'Hara, M. J., Richardson, S. W. and Wilson, G., 1971. Garnet-peridotite stability and occurrence in crust and mantle, *Contr. Mineral. and Petrol.*, **32,** 48–68.

Osmaston, M. F., 1969. Discussion on paper by R. Stoneley 'Sedimentary thicknesses in orogenic belts.' *In:* Time and Place in Orogeny, edited by P. E. Kent, G. E. Satterthwaite, and A. M. Spencer, *Geol. Soc. Lond.*, Spec. Pub. No. 3, 306–7.

Osmaston, M. F., 1971. Genesis of ocean ridge median valleys and continental rift valleys, *Tectonophysics*, **11,** 387–405.

Oxburgh, E. R. and Turcotte, D. L., 1969. Increased estimate for heat flow at oceanic ridges, *Nature, Lond.* **223,** 1354–5.

Packham, G. H. and Falvey, D. A., 1971. An hypothesis for the formation of marginal seas in the western Pacific, *Tectonophysics*, **11,** 79–109.

Pankhurst, R. J., 1970. The geochronology of the basic igneous complexes, *Scot. J. Geol.*, **6,** 83–107.

Pitman, W. C., III, Talwani, M. and Heirtzler, J. R., 1971. Age of the North Atlantic Ocean from magnetic anomalies, *Earth Planet. Sci. Letts.*, **11,** 195–200.

Ramberg, H., 1952. The Origin of Metamorphic and Metasomatic Rocks. Univ. Chicago Press, 317pp.

Rezanov, J. A. and Chamo, S. S., 1969. Reasons for absence of a granitic layer in basins of the South Caspian and Black Sea type, *Can. J. Earth Sci.*, **6,** 671–8.

Ringwood, A. E. and Lovering, J. F., 1970. Significance of pyroxeneilmenite intergrowths among kimberlite xenoliths, *Earth Planet. Sci. Letts.*, **7,** 371–5.

Robinson, E. S., 1964. Regional geology and crustal thickness in the Transantarctic Mountains and adjacent ice-covered areas in Antarctica. (Abstract), *Geol. Soc. Amer. Spec. Pap.*, **76,** 316.

Sanford, A. R., 1959. Analytical and experimental study of simple geological structures, *Geol. Soc. Am. Bull.*, **70,** 19–52.

Schumm, S. A., 1963. The disparity between present rates of denudation and orogeny. U.S. Geol. Surv. Prof. Pap. 454-H, 13pp.

Sclater, J. G., Anderson, R. N. and Bell, M. L., 1971. Elevation of ridges and evolution of the Central Eastern Pacific, *J. Geophys. Res.*, **76,** 7888–915.

Sclater, J. G. and Francheteau, J., 1970. The implications of terrestrial heat flow observations on current tectonic and geochemical models of the crust and upper mantle of the Earth, *Geophys. J. R. astr. Soc.*, **20,** 509–42.

Sclater, J. G. and Menard, H. W., 1967. Topography and heat flow of the Fiji Plateau, *Nature, Lond.*, **216,** 991–3.

Sheridan, R. E., 1969. Subsidence of continental margins, *Tectonophysics*, **7,** 219–29.

Sheridan, R. E. and Drake, C. L., 1968. Seaward extension of the Canadian Appalachians, *Can. J. Earth Sci.*, **5,** 337–73.

Sleep, N. H., 1971. Thermal effects of the formation of Atlantic continental margins by continental break-up, *Geophys. J. R. astr. Soc.*, **24,** 325–350.

Stoneley, R., 1969. Sedimentary thicknesses in orogenic belts. *In:* Time and Place in Orogeny, edited by P. E. Kent, G. E. Satterthwaite and A. M. Spencer, *Geol. Soc. Lond.*, Spec. Pub. 3, pp. 215–38.

Storetvedt, K. M., Halvorsen, E. and Gjellestad, G., 1968. Thermal analysis of the natural remanent magnetism of some Upper Silurian red sandstones in the Oslo region, *Tectonophysics*, **5,** 413–26.

Stride, A. H., Curray, J. R., Moore, D. G. and Belderson, R. H., 1969. Marine geology of the Atlantic continental margin of Europe, *Phil. Trans. Roy. Soc. Lond. A.*, **264,** 31–75.

Sutton, J., 1969. Rates of change within orogenic belts. *In:* Time and Place in Orogeny, edited by P. E. Kent, G. E. Satterthwaite and A. M. Spencer, *Geol. Soc. Lond.* Spec. Pub. 3., pp. 239–50.

Tectonic Map of Australia, 1960. 1st. edition. 1:2,534,400. Bur. Min. Resources, Geology and Geophysics, Canberra.

Turner, F. J. and Verhoogen, J., 1960. Igneous and Metamorphic Petrology. 2nd Edition. McGraw-Hill, New York. 694pp.

Van Bemmelen, R. W., 1966. On mega-undations: a new model for the Earth's evolution, *Tectonophysics*, **3,** 83–127.

Van de Lindt, W. J., 1967. Movement of the Mohorovicic discontinuity under isostatic conditions, *J. Geophys. Res.*, **72,** 1289–97.

Vogt, P. R. and Ostenso, N. E., 1967. Steady state crustal spreading, *Nature*, *Lond.*, **215,** 810–17.

Von Herzen, R. P., 1967. Surface heat flow and some implications. *In:* The Earth's Mantle. edited by T. F. Gaskell, pp. 197–230. Academic Press, London.

Williams, H., 1971. Mafic-ultramafic complexes in western Newfoundland Appalachians and the evidence for their transportation: a review and interim report, *Geol. Ass. Canada. Proc.*, **24,** No. 1, 9–25.

Wilson, J. T., 1965. A new class of faults and their bearing on continental drift, *Nature*, *Lond.* **207,** 343–7.

Wilson, J. T., 1966. Did the Atlantic close and then reopen? *Nature*, *Lond.*, **211,** 676–81.

Wright, J. B., 1966. Convection and continental drift in the Southwest Pacific, *Tectonophysics*, **3,** 69–81.

Ziegler, A. M., 1970. Geosynclinal development of the British Isles during the Silurian period. *J. Geol.*, **78,** 445–79.

6.4

M. H. P. BOTT
Department of Geology,
University of Durham,
Durham, England

Shelf Subsidence in relation to the Evolution of Young Continental Margins

Introduction

Continental margins of Atlantic type form when a continent splits and the two fragments start to drift apart as new oceanic lithosphere forms between them by sea-floor spreading. Although young margins of this type mark the juxtaposition of continental and oceanic crust, they do not form plate boundaries. In this respect they differ from Pacific type margins. Atlantic type margins have a lifespan set by the time period needed for the ocean to expand to its maximum size, after which the margins may fracture to become plate boundaries such as occur around the Pacific Ocean.

We are concerned in this paper with the causes of the tectonic development of Atlantic type margins. These do not exhibit the strong seismicity of Pacific type margins where oceanic lithosphere is recycled into the mantle. They are associated with a characteristic type of 'quiet' tectonic activity involving substantial subsidence of the adjacent continental region including formation of deep sedimentary basins. This type of tectonic activity appears to continue over the lifespan of the Atlantic type margin. There is convincing evidence that this type of subsidence starts at about the time of the split (Bott, 1971).

Shelf subsidence takes a variety of forms as briefly reviewed by Bott (1971). It may occur as a gentle oceanward sagging of the shelf masked by an oceanward thickening wedge of sediments on the shelf, slope and rise. Subsidence also occurs as relatively localized troughs and basins which may be fault controlled. Such basins may be aligned parallel or sub-parallel to the margin and may be separated from the slope by a basement ridge (Burk, 1968). Around the British Isles, basin subsidence on the shelf extends several hundred kilometres into the adjacent continental region, and includes the Irish Sea basins and the North Sea basin. This paper suggests that the same underlying mechanism is responsible for shelf sagging and for basin subsidence in the vicinity of a margin.

Seismic and gravity observations show that the transition between continental and oceanic crust at an Atlantic type margin occurs over a horizontal distance of about 50–100 km, mainly beneath the slope, but partly also beneath the shelf and rise (Worzel, 1965). Where seismic measurements of adequate precision have been made, there is also evidence for thinning of the crust beneath sedimentary basins such as the North Sea basin (Collette, 1968) and the Hatton–Rockall basin (Matthews & Smith, 1971). Gravity observations show that margins of Atlantic type are in approximate isostatic equilibrium, and that basins such as the North Sea basin are in equilibrium. Thus subsidence of the shelf and of these sedimentary basins appears to have been the isostatic response to crustal thinning. Understanding the mechanism which causes crustal thinning becomes the main key to understanding the mechanism of subsidence.

Several mechanisms have been suggested to explain the crustal thinning. Mechanisms involving subcrustal erosion (Gilluly, 1964) are difficulty to reconcile with present knowledge of the structure of the lithosphere. Supracrustal erosion (Hsu, 1965; Sleep, 1971) can normally account for a limited amount of subsidence following the initial split. Migration of a phase transition boundary may also cause subsidence. While these factors must be taken into account, the most significant mechanism of crustal thinning is probably the outflow of continental crustal material towards the sub-oceanic upper mantle by hot creep occurring below a depth of about 10 km, with the uppermost 10 km responding by faulting.

The development of an Atlantic type margin may be divided into two main stages. Stage 1 is the period of 50–100 m.y. following the initial split when the thermal effects of the split affect the development of the margin. Stage 2 is the succeeding period when tectonic activity can no longer be attributed to effects of the initial split.

Thermal Effects of the Initial Split

A newly formed Atlantic type margin may be uplifted at the time of the split for one or both of the following reasons. First, heating of the continental lithosphere may cause its thermal expansion and consequent uplift. Second, occurrence of low density asthenosphere beneath the newly developed ocean ridge accompanied by thinning of the lithosphere may cause isostatic uplift. After the initial uplift, the continental margin recovers progressively towards its pre-split elevation as the lithosphere cools and the ocean ridge migrates away from the margin. If crustal thinning occurs contemporaneously, then isostatic subsidence of the shelf will occur (Sleep, 1971).

The simplest mechanism for heating the margin is by thermal conduction of heat from the newly formed hot oceanic lithosphere. To estimate the extent of lateral penetration, we assume unidimensional heat flow perpendicular to the margin. The time constant for cooling of the oceanic lithosphere as it spreads laterally from the ridge is known to be about 50 m.y. (McKenzie, 1967). An upper limit to the penetration of the heat can be obtained by solving the unidimensional heat conduction equation in a semi-infinite solid with initial temperature zero and fixed surface temperature over a 50 m.y. period (Carslaw & Jaeger, 1959). The effective penetration is between $x = \sqrt{kt}$ and $x = 2\sqrt{kt}$. Taking diffusivity $k = 0{\cdot}006\ \text{cm}^2/\text{s}$ and $t = 50$ m.y., the penetration is between 30 and 60 km. This is inadequate to explain observed shelf subsidence, but the raised temperatures may accentuate creep of the continental crust at the time of the split and just after.

More extensive heating of the margin may occur as a result of dyke intrusion into the adjacent continental lithosphere at the time of the split. A simple model of this type of heating was studied numerically by Sleep (1971), who showed that the time constant of recovery towards the normal geothermal gradient is 50 m.y., equivalent to that of the cooling of the oceanic lithosphere.

The extent of this thermal effect depends on the heat added to the continental lithosphere. Let us assume the average temperature of the continental lithosphere prior to the split was 600°C. A maximum realistic addition of magma is, say, 15%. Allowing for heat of fusion, this will raise the mean temperature of the continental lithosphere by about 200°C. If we assume that the bulk expansion is entirely relieved by vertical movement, and that the coefficient of bulk expansion is $3 \times 10^{-5}\,°C^{-1}$, then the heating will cause an uplift of 600 m. Thus 600 m is probably the extreme maximum value of uplift which can be caused by this process. The amount of subsidence which can follow depends on the extent of crustal thinning.

Sleep (1971) attributed subsidence of Atlantic continental shelves to thermal recovery of the underlying lithosphere following initial uplift and crustal thinning at the time of the break-up. His numerical model is based on crustal thinning by sub-aerial erosion.

It is doubtful whether this model is adequate to account for observed sediment thicknesses. Taking the density of the sediments to be ρ_s and that of the mantle to be ρ_m, then the maximum possible thickness of sediments corresponding to a zero time constant of erosion for initial elevation h is given by

$$S = \frac{h\rho_m}{\rho_m - \rho_c}.$$

Putting ρ_m=3·3 g/cm^3, ρ_s=2·2 g/cm^3 gives S=3h. If h=600 m, then the maximum possible sediment thickness is less than 2 km, and for a realistic time constant of erosion the possible sediment thickness would be less than 1 km. This difficulty can be avoided if crustal thinning occurs mainly by seaward creep, when the initial elevation is immaterial to the thinning process. A larger initial uplift may also occur if thermal expansion of the lithosphere is accompanied by a mass deficiency in the underlying asthenosphere.

A serious objection to Sleep's hypothesis is that it cannot explain local subsidence of the type occurring on Rockall Plateau. Hatton Basin lies between Hatton and Rockall Banks. It has been investigated geophysically (Matthews & Smith, 1971) and by drilling by the *Glomar Challenger* (DSDP holes 116 and 117). Hatton and Rockall banks have subsided less than 1 km since the Palaeocene, but Hatton basin has subsided much farther (about 2·7 km). Differential subsidence of this type cannot be explained by Sleep's hypothesis although it is a common feature of the shelf. It is also doubtful whether the predicted exponential decay of subsidence is observed in practice. Rather, slow subsidence appears to be accompanied by rapid subsidence over short periods of time. The observational curves presented by Sleep are more consistent with this view.

Undoubtedly the mechanism suggested by Sleep has some relevance to the development of Atlantic type margins. Within the framework of the creep hypothesis, it explains the relatively rapid subsidence in the early stages and the occurrence of unconformities.

Stress Distribution Associated with Atlantic Type Margins

Local variations in the stress pattern of the lithosphere are associated with Atlantic type margins (1) because of differential loading associated with the contrasting density-depth distributions characteristic of oceanic and continental regions, and (2) because the change in the distribution of elastic moduli with depth across the margin causes local modification of any existing stress distribution in the lithospheric plate containing the margin. Stresses of thermal origin are probably of lesser importance. The two main types of stress distribution have been analysed by Bott & Dean (1972) using the finite element method of stress analysis assuming elastic deformation.

The *differential loading effect* arises as follows. Beneath oceans, about 5 km of sea-water (1·03 g/cm³) is underlain by about 6 km of crust (about 2·9 g/cm³ average) which overlies the mantle (3·3 g/cm³). Beneath continents about 35 km of crust (2·85 g/cm³ average) is underlain by mantle. At 5 km depth the vertical pressure is greater beneath continents than oceans because of the difference in density between crustal rocks and sea-water. This additional load on the continental crust is balanced isostatically by an equal and opposite excess upthrust caused by the low density root of continental crust contrasting with denser mantle material beneath the oceans at the same depth. This causes the vertical principal pressure in the continental crust to be greater than at equivalent depths beneath the oceans, thus favouring the occurrence of normal faulting in the continental crust relative to the oceanic crust. A simple analytical model of differential loading illustrating this principle is given in the Appendix.

In order to make a more rigorous study of the stresses arising from differential loading, Bott & Dean (1972) carried out a finite element analysis. In preparation for this, it was necessary to remove the density-depth distribution beneath the oceans from the whole model, leaving an excess load formed by the upper 5 km of the continental crust, and an opposite upthrust caused by the continental root. It was assumed that the continental and oceanic crusts possessed the same density. The result is shown in Fig. 1. The outcome is a closely similar stress pattern to that indicated by the simple analytical model of the Appendix. The principal stresses are nearly horizontal and vertical except immediately beneath the margin, and the maximum principal compression is vertical beneath the continental region. The maximum shearing stress of about 370 bar occurs at about 10 km depth within the continental crust. Beneath the ocean, the maximum principal compression is horizontal, but the stress difference is much less than beneath the continent. The result demonstrates why continental regions are more prone to normal faulting than oceanic regions.

A surprising feature of both the analytical model (Appendix) and the finite element model (Fig. 1) is that the stress differences within the continental crust do not decrease

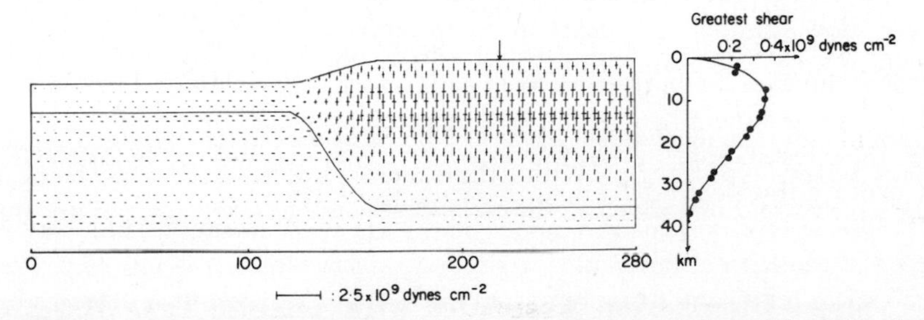

FIGURE 1 Magnitude and direction of principal stresses (compressions) derived by finite element analysis for a model of differential loading and upthrust across a continental margin. The intermediate principal stress is everywhere perpendicular to the model. The variation of greatest shearing stress with depth below the position of the arrow is shown. After Bott & Dean (1972).

away from the margin. This is a rigorous consequence of our assumption that the lithosphere is an elastic body. If, however, the shearing stress in the interior of the continental region can be relieved by non-elastic processes, then the stress differences would be expected to decrease away from the margin towards the relaxed region.

The *redistribution of a regional stress* caused by a margin has also been studied by finite element analysis (Bott & Dean, 1972). Body forces were neglected, and the margin was subjected to a regional tension (or compression) of 500 bar. It was found that the lateral variation in elastic moduli across the margin caused the stress difference to be increased at shallow depth beneath the slope.

Both above types of stress distribution are superimposed in the vicinity of the margin. The stress system caused by differential loading will only change very slowly with time. On the other hand, the regional stress in the lithospheric plate produced by the plate driving mechanisms (and otherwise) would be expected to vary with time. When the lithospheric plate is subjected to regional tension, then normal faulting would be expected to occur most readily beneath the shelf and slope. Under a regional compression, initiation of thrust faulting would be expected to occur most readily at the foot of the slope. Tectonic activity at Atlantic type margins may therefore be the non-elastic response of the lithosphere to shearing stress originating in these two ways.

Rheological Structure of the Lithosphere

The response of the margin to a stress system which exceeds the elastic limit depends on the mechanism of non-elastic deformation. The lithosphere is usually treated as a single rheological layer which deforms by elastic bending or by fracture. This model is adequate for some isostatic purposes and for plate tectonics as so far developed. However, experiments by Griggs *et al.* (1960) show that the model is an oversimplification. Rocks near the Earth's surface are relatively strong and yield by fracture. Both increase in pressure and increase in temperature cause brittle rocks to become ductile. Griggs *et al.* (1960) showed that most rocks deform by flow when the elastic limit is exceeded at 5 kbar pressure and 500°C temperature, corresponding to a depth of about 20 km. On this basis the lithosphere can be subdivided into two main rheological subdivisions: (1) an upper brittle layer about 10–20 km thick, and (2) a lower ductile layer at least 40 km thick.

Support for this subdivision of the lithosphere comes from the study of focal depths of microearthquakes along the San Andreas fault in Central California made by Eaton, Lee & Pakiser (1970). They used dense networks of portable seismometers to locate earthquake foci precisely on several stretches of the San Andreas and related faults. The microearthquakes are 'limited to focal depths of 15 km or less and very few are deeper than 12 km'. As the lithospheric plates sliding past each other must be at least 50 km thick, these results suggest that the upper 10–15 km may deform by brittle fracture but that beneath this depth deformation occurs by creep.

The brittle layer is probably about 10–15 km thick normally, but may be thinner where the geothermal gradient is high, and thicker where it is low (e.g. Precambrian shields). It is relatively strong but when the strength is exceeded it deforms by faulting, jointing or dyke intrusion.

The ductile layer is relatively strong in comparison with the asthenosphere, but deforms by flow when the strength is exceeded. The compressive strength probably decreases with depth within the layer because of the geothermal gradient outweighing the influence of increasing confining pressure. As the temperature within the ductile layer is probably greater than 0·5 T_m, where T_m is the absolute melting temperature, non-elastic deformation may occur by hot creep. The experiments made on rocks by Misra & Murrell (1965) support the view that below about 15–20 km rocks may deform appreciably by transitional or steady state creep, provided a finite creep strength is exceeded.

This subdivision of the lithosphere is of importance in understanding epeirogenic tectonics of continental regions, and in particular of the continental shelf and margin bordering Atlantic type oceans.

Mechanism for Shelf Subsidence

Shelf subsidence can be explained once a mechanism for substantial crustal thinning near Atlantic type margins has been found. Sub-aerial erosion is inadequate and sub-crustal erosion by convection currents is unrealistic. Crustal thinning by phase transformation of gabbro in the lower crust to eclogite appears to be inconsistent with the results of experimental petrology (Ringwood & Green, 1964, 1966). The remaining possibility is that the crust thins by ductile flow of low density continental crustal material towards the margin, as suggested by Bott (1971).

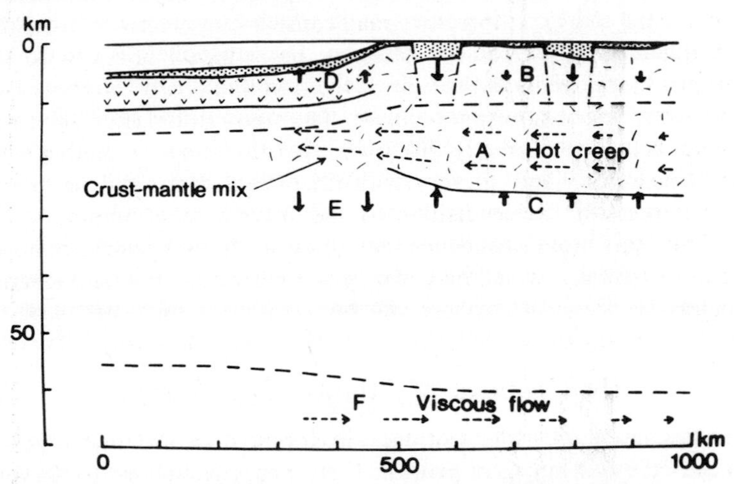

FIGURE 2 The hypothesis for the development of aseismic margins of Atlantic type by oceanward creep of lower continental crustal material accompanied by wedge subsidence in the brittle layer (base marked by dashed line at 10–15 km depth). After Bott (1971).

It was shown above that the stress system in the continental crust beneath the shelf and slope favours stretching of the crust perpendicular to the margin. The brittle and ductile layers respond differently to this stress field. The brittle layer may extend by normal faulting or by dyke intrusion. Such normal faulting is a well-observed feature of the shelf, but the resulting extension is quite inadequate to account for crustal thinning of the required amount (Matthews & Smith, 1971; Bott, 1972). The underlying ductile layer extends by hot creep. In the interior of a continental region, the brittle and ductile layers would be expected to suffer equivalent extension. At a margin of

Atlantic type the situation is different. Much greater extension of the ductile layer can occur because the continental material can flow out into the topmost mantle beneath the oceanic crust with progressive loss of gravitational energy. This flow will cause a progressive broadening of the transition between oceanic and continental crust. The large strain affecting the ductile layer is accommodated by isostatic subsidence of the shelf and uplift of the rise. By this process a very substantial thinning of the crust beneath the shelf can take place without comparable extension of the overlying brittle layer. The hypothesis is illustrated in Fig. 2.

In the absence of normal faulting in the brittle layer, thinning of the crust will lead to gentle seaward sagging of the shelf. If normal faulting occurs in the brittle layer, wedge subsidence affecting the brittle layer will produce local fault bounded subsidence of the type required to explain shelf basins. Thus the mechanism is capable of accounting for the variety of observed types of shelf subsidence. Uplift at the time of the split followed by thermal recovery with a time constant of about 50 m.y. may also occur, and major subsidence may occur at this stage. But the mechanism is not restricted to the stages immediately following the break-up, and can explain long continued subsidence of the shelf.

The subsidence rate may vary with time either because of changes in crustal temperature or because of variation of the stress system. It was shown above that the temperatures in the continental crust adjacent to the new oceanic lithosphere are raised for a period of about 100 m.y. after the break-up. Hot creep is a thermally activated process and a small rise in temperature may cause a large increase in creep rate for a fixed stress difference. Thus crustal thinning immediately adjacent to the margin probably occurs most rapidly at the time of the split and shortly afterwards. The other factor is the stress in the lithospheric plate. The stress differences due to differential loading are unlikely to vary greatly with time, but the stresses associated with the plate driving mechanism are likely to vary with time. Subsidence will be favoured when horizontal compression perpendicular to the margin is minimum, or tension is maximum. Thus maximum subsidence may occur at times when the plate stresses are fluctuating most strongly, which may take place when the relative movement of the plates changes.

Development of Atlantic Type Margins

It has been suggested that the tectonic development of Atlantic type continental margins is caused by two types of process. First, processes related to the initial break-up and the formation of a new ocean ridge between the fragments cause vertical uplift at the time of the split followed by subsidence with a time constant of the order of 50 m.y. Second, anomalous stress distributions associated with the margin may cause long-continuing tectonic activity which only ceases when the margin becomes a plate boundary. Both processes are active during the early stages of development, but the second process persists beyond this stage. Of importance in understanding the tectonics of young margins is the subdivision of the lithosphere into an upper brittle layer about 10–20 km thick overlying a lower ductile layer, a concept which is supported by experimental evidence on rock deformation and by the distribution of earthquake foci beneath the San Andreas fault belt.

The stages of development may be summarized as follows: (1) At the time of the split an anomalously low density upper mantle (part asthenosphere, part lithosphere) develops beneath the new ocean ridge, and the continental lithosphere may be heated by thermal conduction from the new oceanic material and by intrusion of hot magma. This may cause substantial uplift of the newly formed margins and their hinterland. As the ocean ridge migrates away from the margin and the continental lithosphere cools, the margin subsides. Crustal thinning by creep may develop strongly at this stage because of the relatively high temperatures and fluctuating stresses. Some crustal thinning may also occur by erosion at this stage. The maximum rate of subsidence probably occurs shortly after the split. (2) Subsequently, the tectonic activity is mainly the result of stresses associated with the margin. Subsidence of the shelf would be expected to continue, although not as rapidly as during stage one. Subsidence is probably mainly controlled by the stress distribution in the lithospheric plate originating from the plate driving mechanism. (3) The life of an Atlantic type margin is terminated when the ocean can expand no further. The plate becomes subjected to strong compression and thrust faulting is initiated at the foot of the slope as the first stage in the formation of a new Benioff zone when the margin becomes a plate boundary of Pacific type.

Appendix

Analytical Model of Differential Loading Across an Atlantic Margin

Figure 3 below shows a floating two-dimensional elastic block of rectangular cross section and density ρ floating in a fluid substratum of density ρ'. This represents a simplified model of continental crust floating in denser mantle material. The thickness of the block is H and its elevation above the fluid surface is h where $\rho'/\rho = H/(H-h)$. The block is stressed by hydrostatic boundary forces and by the gravitational body force. The stress distribution within the block is obtained by finding an Airy stress function χ which satisfies the biharmonic equation $\nabla^4\chi = 0$ within the block and gives the specified boundary stresses on the surface. Using the formulation of Hafner (1951) which is most appropriate to the Earth, the stresses within the block and on the surface are given by

$$\sigma_{xx} = \frac{\partial^2\chi}{\partial y^2}, \qquad \sigma_{yy} = \frac{\partial^2\chi}{\partial x^2} - \rho g y, \qquad \tau_{xy} = -\frac{\partial^2\chi}{\partial x \partial y}.$$

It is convenient to split the block into two separate regions as shown with stress functions χ_1 and χ_2. It is easily seen that the solution to the problem is as follows:

For $y \leqslant h$ $\quad \chi_1 = 0,$

$$\sigma_{xx} = 0, \qquad \sigma_{yy} = -\rho g y, \qquad \tau_{xy} = 0.$$

For $y \geqslant h$ $\quad \chi_2 = \dfrac{\rho g h}{H-h}\left(\tfrac{1}{6}y^3 - \tfrac{1}{2}hy^2\right),$

$$\sigma_{xx} = \frac{-\rho g h}{H-h}(y-h), \qquad \sigma_{yy} = -\rho g y, \qquad \tau_{xy} = 0.$$

The stress difference increases linearly from zero at $y=0$ to a maximum at $y=h$ and then decreases linearly to zero at $y = H$. The principal pressures are horizontal and vertical, and the maximum compression is vertical everywhere within the block, favouring normal faulting. The stress difference within the block is not restricted to the vicinity of the 'margin' but extends uniformly through the whole length of the block as it is independent of x.

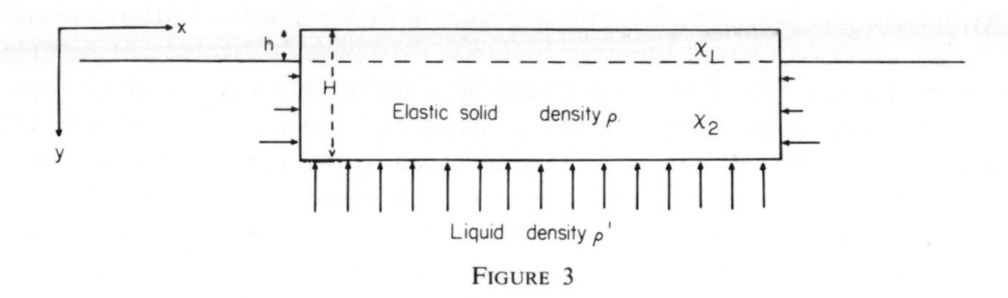

FIGURE 3

References

Bott, M. H. P., 1971. Evolution of young continental margins and formation of shelf basins, *Tectonophysics*, **11**, 319–27.

Bott, M. H. P., 1972. Subsidence of Rockall plateau and of the continental shelf, *Geophys. J. R. astr. Soc.*, **27**, 235–6.

Bott, M. H. P. and Dean, D. S., 1972. Stress systems at young continental margins, *Nature Physical Science*, **235**, 23–25.

Burk, C. A., 1968. Buried ridges within continental margins, *Trans. N.Y. Acad. Sci., Ser. 2*, **30**, 397–409.

Carslaw, H. S. and Jaeger, J. C., 1959. Conduction of Heat in Solids, 2nd edition. Oxford University Press, 510 pp.

Collette, B. J., 1968. On the subsidence of the North Sea area. *In:* Geology of Shelf Seas, edited by D. T. Donovan, pp. 15–30. Oliver and Boyd, Edinburgh and London.

Eaton, J. P., Lee, W. H. K. and Pakiser, L. C., 1970. Use of microearthquakes in the study of the mechanics of earthquake generation along the San Andreas fault in Central California, *Tectonophysics*, **9**, 259–82.

Gilluly, J., 1964. Atlantic sediments, erosion rates, and the evolution of the continental shelf: some speculations, *Bull. geol. Soc. Am.*, **75**, 483–92.

Griggs, D. T., Turner, F. J. and Heard, H. C., 1960. Deformation of rocks at 500° to 800°C, *Mem. geol. Soc. Am.*, **79**, 39–104.

Hafner, W., 1951. Stress distributions and faulting, *Bull. geol. Soc. Am.*, **62**, 373–98.

Hsu, K. J., 1965. Isostasy, crustal thinning, mantle changes, and the disappearance of ancient land masses, *Am. J. Sci.*, **263**, 97–109.

McKenzie, D., 1967. Some remarks on heat flow and gravity anomalies, *J. geophys. Res.*, **72**, 6261–73.

Matthews, D. H. and Smith, S. G., 1971. The sinking of Rockall plateau, *Geophys. J. R. astr. Soc.*, **23**, 491–8.

Misra, A. K. and Murrell, S. A. F., 1965. An experimental study of the effect of temperature and stress on the creep of rocks, *Geophys. J. R. astr. Soc.*, **9**, 509–35.

Ringwood, A. E. and Green, D. H., 1964. Experimental investigations bearing on the nature of the Mohorovičić discontinuity, *Nature, Lond.*, **201**, 566–7.

Ringwood, A. E. and Green, D. H., 1966. Petrological nature of the stable continental crust, *Geophys. Monogr.*, **10**, 611–19.

Sleep, N. H., 1971. Thermal effects of the formation of Atlantic continental margins by continental break-up, *Geophys. J. R. astr. Soc.*, **24**, 325–50.

Worzel, J. L., 1965. Deep structure of coastal margins and mid-oceanic ridges. *In:* Submarine Geology and Geophysics—Colston Papers No. 17, edited by W. F. Whittard and R. Bradshaw, pp. 335–61. Butterworths, London.

6.5

N. H. SLEEP
Massachusetts Institute of Technology,
Cambridge, Massachusetts 02139

Crustal Thinning on Atlantic Continental Margins: Evidence From Older Margins

Introduction and Theory

The simplest and most easily analyzed means of thinning the crust along continental margins is subareal erosion of a region uplifted by thermal expansion. Rapid erosion can remove several times the initial uplift, since vertical crustal movements are magnified by the effect of isostatic compensation.

The length of time required for erosion and thermal contraction to reduce elevation to sea level is about 50 m.y. for a denudation constant of 0·1 km removed by erosion per kilometre of elevation per million years (Sleep, 1971). The amount of material removed by erosion and the length of uplift is sensitive to the value of the denudation constant (Foucher & Le Pichon, 1972).

The period of uplift would appear as a major unconformity in the geologic record.

The effectiveness of erosion is determined by local factors such as climate and rock type. The climate during a period of uplift can be determined only by sophisticated sedimentological methods applied to adjacent areas of sediment deposition. For the Atlantic coast of the United States the relevant sediments are deeply buried on the continental rise.

Poorly understood subsurface processes, including necking due to tension (Artem'yev & Artushkov, 1969), subcrustal erosion by currents (Gilluly, 1955, Heiskanen & Vening Meinesz, 1958), loading by dense intrusions, oceanization (Beloussov, e.g. 1967), and gravitational spreading of the less dense continental crust over the more dense oceanic lithosphere (Gilluly, 1964; Bott 1971; Bott & Dean, 1972; Bott, this volume p. 667) may reduce the buoyancy of the continental crust. Other than for gravitational spreading, thermal contraction of the lithosphere must control the rate of subsidence during continental shelf sedimentation, since the processes can only operate efficiently during the first few million years after continental break-up or other times when the sub-moho temperature approaches the solidus.

Gravitational spreading, although probably more efficient with a hot lithosphere, has been proposed as the principal mechanism of subsidence or Atlantic continental margins (Bott, 1971). This spreading would obey a horizontal diffusion equation since the flow of a viscous fluid is proportional to the hydraulic gradient. Features with pronounced local maxima of subsidence such as mid-continent basins cannot originate by this mechanism since diffusive flow can never create a local minimum in an originally

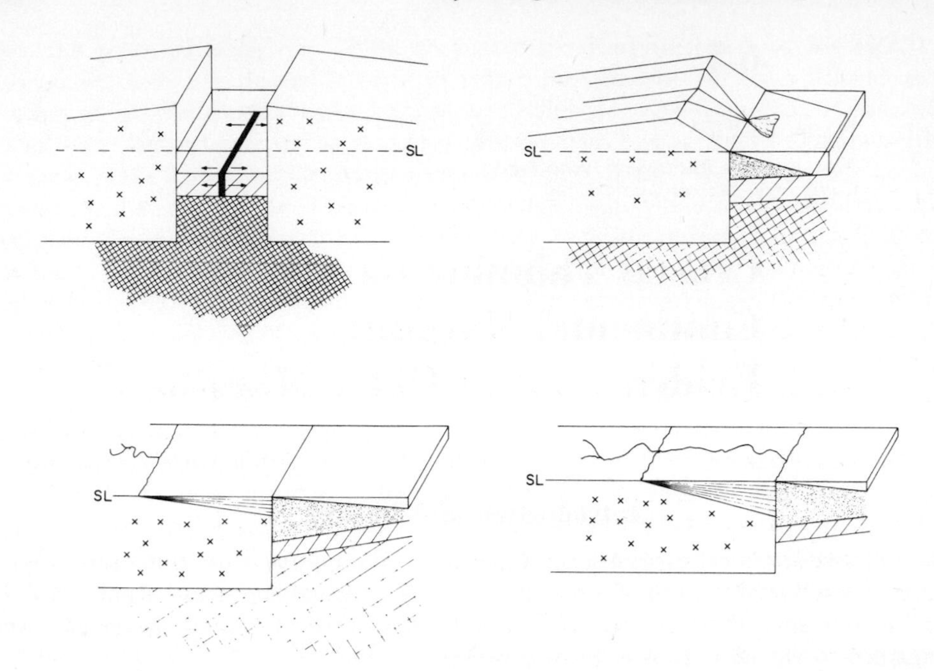

FIGURE 1 Schematic diagram of the evolution proposed for Atlantic continental margins. The density of the hatching indicates the amount of excess temperature present. Top left: A mid-ocean ridge begins spreading beneath a pre-existing continent. Extensive heating of the lithosphere at this time causes the continental margin to be uplifted. The buoyancy of the crust may be reduced by subsurface processes associated with the rifting. The rapid subsidence during rifting is not due to thermal contraction but to the emplacement of more dense oceanic crust. Top right: Subareal erosion thins the continental crust as the lithosphere cools. Bottom left: The continental margin continues to subside as sediments accumulate on the continental shelf. Bottom right: The subsidence ceases after the lithosphere has cooled. The continental margin will remain stable until another process intervenes.

flat area unless a sink of material (i.e. crust) occurs at the centre of the minimum. Sediment accumulation on the oceanic crust could drag down the immediately adjacent continental shelf (Dietz, 1963; Walcott, 1972). Gravity anomalies, however, are incompatible with this being a significant factor in continental shelf subsidence (Sleep, 1971, Bott 1971).

If Dietz' theory were correct, large (100 mgal peak to peak) free air gravity anomalies would be negative on the shelf and positive on the continental (not mid-ocean) rise. Atlantic continental margin gravity anomalies are primarily an edge effect between two isostatic regions (e.g. Worzel, 1968). Any regional isostatic compensation has the sign of the shelf dragging down the rise, since the shelf anomaly is slightly positive for a pure edge effect (Emery *et al.*, 1970).

Dietz' theory is also incompatible with the distribution of sediments in space and time. Progade sediments did not begin to accumulate on the Atlantic continental rise of the United States until Claiborne time (49 m.y.) (Emery *et al.*, 1970); however, the subsidence rate was more rapid before Claiborne time than after it (Sleep, 1971).

The distribution of sediment on older margins and the surface geology of younger margins can be used as data to obtain information on the processes which reduce the buoyancy of continental crust.

The extent of subareal erosion and faulting after continental break-up can be determined if a suitable sequence of marker beds exists beneath the shelf sediments. This cannot be done for the Atlantic coast of the United States since the basement there consists of igneous and metamorphic rocks. A subsurface process must have thinned the crust before rapid Pennsylvanian deposition began in Kansas, since a thinly bedded sequence of immediately older sediments continues beneath the basin (Sleep, 1971). An unconformity below the St Peter Sandstone precedes rapid Ordovician through Devonian deposition in the Michigan and Appalachian basins (e.g. Dott & Murray, 1964). It is unlikely that as much material was removed during the period of the unconformity as was later deposited since extensive Lower Ordovician and Cambrian sediments occur beneath the deposits in these basins. Some minor faulting and folding occurred during the period of unconformity (Sloss, e.g. 1963; McGinnis, 1970).

Regional isostatic effects would not be expected if gravitational spreading of the continental crust were the cause of continental shelf subsidence, since a weak lithosphere is necessary for gravitational spreading to be significant.

If thermal contraction of the lithosphere is the cause of continental shelf subsidence, each increment of contraction will appear as a load where it occurs. If more contraction occurs at point A than point B, the subsidence at point A will drag down point B and point B will buoy up point A by an equal amount. Point A would have a positive isostatic anomaly and point B a negative one. This effect can be observed in gravity anomalies of mid-continent basins (McGinnis, 1970).

The additional dragging down of point B by point A will decrease with time, since the subsidence rate can be considered to decrease exponentially with time at each point (Sleep, 1971). If the regional isostatic stress is relaxed with time, the subsidence at point A will be increased by the rebound and the subsidence at point B decreased. If the rate of rebound at point B exceeds the rate of subsidence due to current local contraction and coupling with concurrent subsidence at point A, erosion rather than deposition

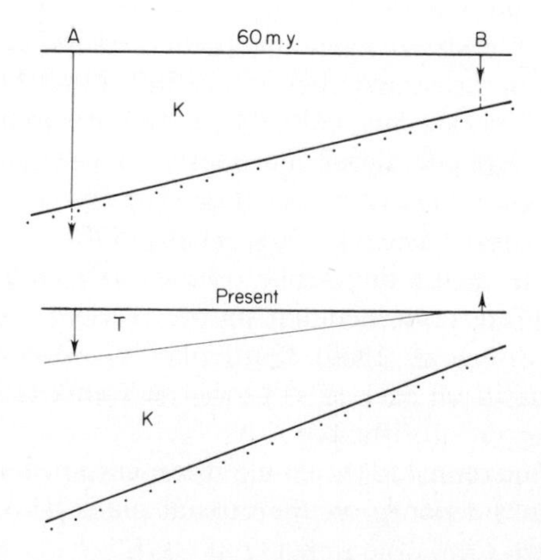

FIGURE 2 Schematic diagrams show the effect of region isostasy on subsidence. During the Cretaceous period (above) more subsidence occurred at point A than point B. Point A is buoyed up and point B dragged down the amount indicated by the dashed lines. During the Tertiary period less subsidence occurred. The regional isostatic stresses built up during the Cretaceous relax causing uplift at point B and additional subsidence at point A. Note that Cretaceous beds outcrop at point B.

will occur at point B (see Fig. 2). Dip-slip faults with the down-throw side toward the basin could form by this mechanism.

The effects of the relaxation of regional isostatic stress would be visible only if the time constant for stress relation was comparable to the time constant of subsidence and not much greater than the age of the margin.

Data

Observation of the continental margins of the Red Sea and the Atlantic coast of the United States are used as data in this paper.

The initial rifting in the Red Sea was probably in the Oligocene or Lower Miocene (20 to 30 m.y.) (Heybroek, 1965; Brown, 1970). Rapid spreading probably began about 10 m.y. ago. (Laughton *et al.*, 1970).

Neogene uplift of about 1 km on both sides of the Red Sea is so obvious that certain geologists (e.g. Van Bemmelen, 1966; Whiteman, 1968) have considered the 'Afro-Arabian' swell to be a primary tectonic feature in the area. The uplift can be shown to have occurred from Late Oligocene through Recent times by studies of wadi capture patterns and the distribution of some fluvial erratics (Brown, 1970).

Much of this uplift may be due to thermal expansion, since high heat flow is observed at the coast (Girdler, 1970) and extensive volcanism occurred inland from Oligocene through Recent times (Brown, 1970). There are many dilatational faults with little slip in any direction (Brown, 1970).

For climatic reasons subareal erosion has not been very effective in reducing the thickness of the crust in the area surrounding the Red Sea. Although some erosion has occurred, large flat pre-rift peneplains have not been affected by erosion since uplift. Subareal erosion is most intense on the northeast end of the Red Sea (Whiteman, 1968). Erosion was more intense during the Oligocene and pluvial Pleistocene periods than at present (Brown, 1970).

The extent of crustal thinning by subsurface processes is difficult to determine since the nature of the crust is not known at enough points. Seismic refraction measurements have indicated that oceanic crust underlies thick evaporites for at least 60 km from the ridge axis (Davies & Tramontini, 1970). Gravity interpretations constrained by drilling records of 4 or more kilometres of evaporites near shore (Heybroek, 1965; see Girdler, 1970 for more records) have indicated that the oceanic crust may extend much closer to shore (Tramontini & Davies, 1969; Allan, 1970).

Pre-rift reconstructions based on satellite photos and land geology indicate that about 50 km of the Red Sea is continental in nature (Abdel-Gawad, 1970). A tighter fit was obtained by McKenzie *et al.* (1970). Girdler (1970) shows necked crust extending about 50 km into the basin on each side. Lowell & Genik (1972) show necked continental crust underlying most of the basin.

The oceanic crust cannot come to the base of the escarpments bounding the Red Sea basin, since pre-rift rocks outcrop on the coastal plain (Brown *et al.*, 1962, 1963; Brown & Jackson, 1969).

Possible subcrustal thinning appears to be limited to the coastal plain and the immediately adjacent sea-floor, a region usually less than 75 km wide. Although dikes are common on the plateaux they do not appear to be abundant enough on the Arabian geologic maps or Brown's photos to significantly affect the buoyancy of the crust. Some block faulting has occurred on the coastal plane (Brown, 1970). Whiteman (1968) believes that monoclinal flexure rather than block faulting is the cause of the escarpments and that the escarpments have retreated to their present position by erosion.

A sequence of shallow water, Cretaceous and Tertiary sediments, which thickens seaward on the coastal plain of the eastern United States, is suitable for determining the history of subsidence in that region.

The effects predicted for regional isostatic compensation and stress release are obvious upon viewing the geologic map of this area. Older sediments are generally exposed near the fall line (Fig. 3) and other local minima of subsidence such as the Cape Fear arch and the Ocala uplift. The distribution of sediment is not explained by erosion of the topmost beds after an eustatic drop in sea-level since the youngest beds are thin or absent at local minima of Cretaceous subsidence independent of the absolute thickness of sediment present. For example, the depth to the base of the Woodbine in wells 3–8–64 and 3–2–25 is both about 1·2 km but the Neogene is absent in the first well on the Ocala uplift and 0·4 km thick at the second well at Cape May, New Jersey (Maher, 1965).

Gravitational spreading of the continental crust could not have produced the distribution of sediment on the east coast of the United States. Once the crust at a point of maximum crustal thickness has begun to spread it cannot rebound. Large, 400 km wide areas such as the Blake plateau would not have subsided uniformly if the

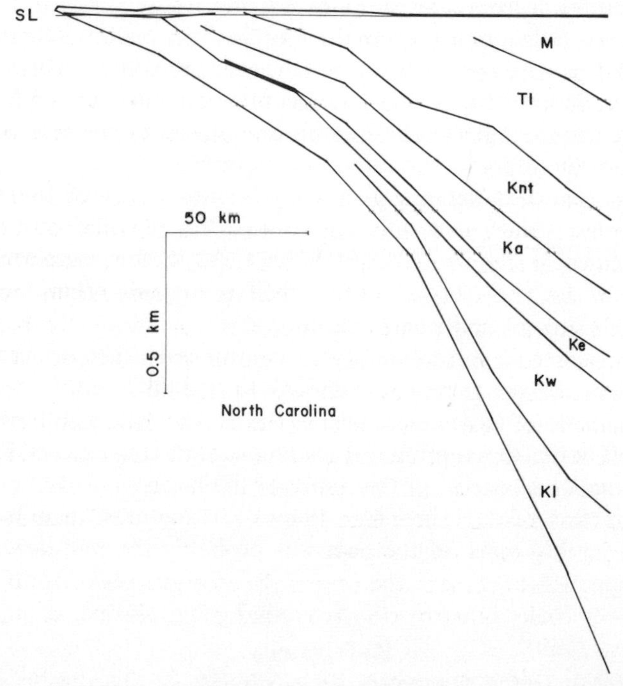

FIGURE 3 Cross section of the sediments on the continental shelf of the eastern United States at Cape Hatteras, North Carolina was determined by drilling (Maher, 1965). Progressively older beds outcrop towards the fall line. Regional isostatic compensation may cause that affect. Standard USGS symbols are used.

process of subsidence was governed by a horizontal diffusion equation in crustal thickness.

If the formation of the Atlantic continental margin of the United States was assumed to accompany the Late Triassic volcanism in that region, the nature of the processes which thinned the crust can be examined. Subareal erosion had time to operate since continental shelf sedimentation did not begin until over 50 m.y. after break-up. No markers suitable for determining the extent of Jurassic erosion exist in the basement below the shelf. Deep drilling into the oldest sediments on the continental rise would give a measure of the Jurassic erosion. No Jurassic sediments occur on land near the region of interest.

Triassic dikes and grabens are common on both sides of the fall line. The location of the most intense rifting which formed grabens does not correspond to local regions of maximum subsidence.

Discussion

Thinning or loading of the crust beneath Atlantic continental shelves probably occurred before the deposition of shelf sediments. Subareal erosion along the margins of the Red Sea and the Jurassic coast of the United States could have occurred if climatic conditions were favourable, since extensive uplift along the margins of the Red Sea and the Jurassic unconformity along the Atlantic Coast of the United States followed break-up. The climate along the Red Sea was unfavourable to erosion. The Jurassic climate in the eastern United States is unknown.

Significant sub-crustal thinning along the Red Sea is probably limited to low lying areas of the coastal plain and possibly some of the near shore areas. This affected region is much narrower, 75 km, than the Atlantic continental shelf of the United States, 250 km. If a wide continental shelf basin ever develops along the Red Sea, significant subareal erosion of the marginal uplift will be required.

The pattern of sediment deposition on the Atlantic coast of the United States suggests that the lithosphere had sufficient strength for regional isostatic stresses to build up during times of rapid subsidence. When these stresses were relieved after the subsidence rate waned the margins of the shelf basin and other local minima of subsidence became sites of uplift and erosion.

Significant gravitational spreading of the continental crust is precluded by the existence of regional isostatic stress. It is difficult to explain by gravitational spreading the uniform subsidence of large areas, such as the Blake plateau and the non-uniform width of the shelf basin. Gravitational spreading cannot cause closed basins unless a sink for continental crust occurs at the centre of the basin.

Thermal contraction of the lithosphere seems to be the most tenable cause of shelf subsidence. The initial source of the heat was probably the mid-ocean ridge which formed the margin.

References

Abdel-Gowad, M., 1970. Interpretation of satellite photographs of the Red Sea and Gulf of Aden, *Phil. Trans. Roy. Soc. Lond.*, **267**A, 23–36.

Allan, T., 1970. Magnetic and gravity fields over the Red Sea, *Phil. Trans. Roy. Soc. Lond.*, **267**A, 153–80.

Artem'yev, M. and Artyushkov, Ye., 1969. Origin of rift basins, *International Geol. Rev.*, English trans., **11**, 582.

Beloussov, V., 1967. Some problems concerning the Earth's crust and upper mantle evolution, *Geotectonics*, English trans., **1**, 1–6.

Bott, M. H. P., 1971. Evolution of young continental margins and formation of shelf basins, *Tectonophysics*, **11**, 319–27.

Bott, M. H. P. This volume.

Bott, M. H. P. and Dean, D., 1972. Stress systems at young continental margins, *Nature Phys. Sci.*, **235**, 23–25.

Brown, G., 1970. Eastern margin of the Red Sea and the coastal structures in Saudi Arabia, *Phil. Trans. Roy. Soc. Lond.*, **267**A, 75–87.

Brown, G. and Jackson, R., 1969. Geology of the Asir Quadrangle, Kingdom of Saudi Arabia, Misc. Geol. Inv. Map 1–217 A, U.S. Geol. Survey.

Brown, G., Jackson, R., Bogue, R. and Elsberg, E., 1963. Geology of the Northern Hijaz Quadrangle, Kingdom of Saudi Arabia, Misc. Geol. Invest Map 1–204A. U.S. Geol. Survey.

Brown, G., Jackson, R., Bogue, R. and Maclean, W., 1962. Geology of the Southern Hijaz Quadrangle, Kingdom of Saudi Arabia, U.S. Geol. Survey Misc. Geol. Invest Map 1–210 A.

Davies, D. and Tramontini, C., 1970. The deep structure of the Red Sea, *Phil. Trans. Roy. Soc. Lond.*, **267**A, 181–9.

Dietz, R., 1963. Collapsing continental rises: an actualistic concept of geosynclines and mountain building, *J. Geology*, **71**, 314–33.

Dott, R. and Murray, G. (eds.), 1964. Geologic cross section of Palaeozoic rocks, central Mississippi to northern Michigan, Am. Assoc. Petroleum Geologists, Tulsa, 32 pp.

Emery, K., Uchupi, E., Phillips, J., Bowin, C., Bunce, E. and Knot, S., 1970. Continental rise off eastern North America, *Am. Assoc. Petroleum Geologists Bull.*, **54**, 44–108.

Foucher, J. and Le Pichon, X., 1972. Comments on 'Thermal effects of the formation of Atlantic continental margins by continental break-up', *J. Geophys. J. Roy. Astr. Soc.* (in press.

Gilluly, J., 1955. Geologic contrasts between continents and ocean basins. *In:* Crust of the Earth, a Symposium, edited by A. Poldervaart, *Geol. Soc. Am. Spec. Paper*, **62**, 7–18.

Gilluly, J., 1964. Atlantic sediments, erosion rates, and the evolution of the continental shelf; some speculations, *Geol. Soc. Am. Bull.*, **75**, 483–92.

Heiskanen, W. and Vening Meinesz, F., 1958. The Earth and its Gravity Field, 470 pp. McGraw-Hill, New York.

Heybroek, F., 1965. The Red Sea Miocene evaporite basin. *In:* Salt Basins around Africa, Institute of Petroleum, London, pp. 17–40.

Kay, M., 1951. North American geosynclines, *Geol. Soc. Am. Memoir*, **48**, 143 pp.

Laughton, A., Whitemarsh, R. and Jones, M., 1970. The evolution of the Gulf of Aden, *Phil. Trans. Roy. Soc. Lond.*, **267**A, 227–66.

Lowell, J. and Genik, G., 1972. Sea-floor spreading and structural evolution of the southern Red Sea, *Amer. Assoc. Petroleum Geol. Bull.*, **56**, 247–59.

Maher, J., 1965. Correlation of subsurface Mesozoic and Cenozoic Rock along the Atlantic Coast, *Am. Assoc. Petroleum Geologists* (Cross Sec. Pub. 3), 18 pp.

McGinnis, L., 1970. Tectonics and the gravity field of the continental interior, *J. Geophys. Res.*, **75**, 317–31.

McKenzie, D., Molnar, P. and Davies, D., 1970. Plate tectonics of the Red Sea and East Africa, *Nature*, **226**, 243–5.

Sleep, N., 1971. Thermal effects of the formation of Atlantic continental margins by continental break-up, *Geophys. J. Roy. Astr. Soc.*, **24**, 325–50.

Sloss, L., 1963. Sequences in the cratonic interior of North America, *Geol. Soc. Am. Bull.*, **74**, 93–114.

Tramontini, C. and Davies, D., 1969. A seismic refraction survey in the Red Sea. *Geophys. J. Roy. Astr. Soc.*, **17**, 225–41.

Van Bemmelen, R., 1966. On Mega-undulations: A new model for the earth's evolution, *Tectonophysics*, **3**, 83–127.

Walcott, R., 1972. Gravity, flexure, and the growth of sedimentary basins at a continental edge, *Bull. Geol. Soc. Amer.*, **83**, 1845–8.

Whiteman, A., 1968. Formation of the Red Sea depression, *Geol. Mag.*, **105** (5), 231–46.
Worzel, J., 1968. Advances in marine geophysical research of continental margins, *Can. J. Earth Sci.*, **5**, 963–83.

6.6

P. R. VOGT
US Naval Oceanographic Office,
Washington, D.C. 20390,
USA

Early Events in the Opening of the North Atlantic

Introduction

The eastern continental margin of North America has been extensively explored geophysically (Hersey *et al.*, 1959; Drake *et al.*, 1968; Taylor *et al.*, 1968; Emery *et al.*, 1970). It has by now become commonly accepted that this is one of the oldest of surviving rifted continental margins, having once been attached to North Africa in a position that is still the object of refinement. Recent continental reassemblies put forth by several authors (Bullard *et al.*, 1965; Dietz *et al.*, 1970; Le Pichon & Fox, 1971) generally do not differ by more than 100 km or so among themselves. Estimates of the time when sea-floor spreading began ranged widely until deep drilling dated the crust near the smooth-rough magnetic boundary (Taylor *et al.*, 1968) at about 155 m.y. ago (Ewing *et al.*, 1970) and westward extrapolation using this age and younger dates on the west flank of the mid-Atlantic Ridge (Peterson *et al.*, 1970) have caused estimates of initial spreading to converge on about 180 m.y. Hallam (1971) reviewed different geological models describing the modification of continental crust in the zone of rifting.

Igneous activity and normal faulting, perhaps labour pains preceding the first spreading, occurred in several spasms, at least back to the 'Alleghenian' events about 240–250 m.y. (Armstrong & Besancon, 1970). There followed the normal faulting and volcanism of the Newark Series (Sanders, 1963), dating from about 200–230 m.y. (Armstrong & Besancon, 1970), and finally the numerous basaltic dikes (de Boer, 1967) injected about 190–200 m.y. (Armstrong & Besancon, 1970). Similar dikes charted on the African side, in Liberia (Behrendt & Wotorson, 1970), appear to be coeval (White & Leo, in press).

It is by now clear that east coast igneous activity did not cease when spreading began; plutonism and perhaps volcanism continued in New England and southern Canada until the Middle Cretaceous, with some suggestion to the writer of peaked activity 170–160 m.y. and 120–110 m.y. ago (Larochelle, 1962; Faul *et al.*, 1963; Christopher, 1969; Armstrong & Besancon, 1970; Zartman *et al.*, 1967; Armstrong & Stump, 1971; Foland *et al.*, 1971). Coney (1971), following Morgan (1971), suggests this magmatic activity was caused by a mantle hot spot or mantle plume, over which the Americas plate moved in a general northwest direction to spawn the igneous trail leading from southern Quebec via the New England seamount chain (Walczak, 1963; Uchupi *et al.*, 1970) to the Corner Rise and thence to the Azores, the present site of this plume.

Even the Cretaceous did not see an end to igneous activity along the supposedly passive rifted margin of eastern North America: Fullagar & Bottino (1969) have discovered Tertiary intrusives in Virginia; and volcanic materials, possibly from distant sources, appear in coastal plain sediments of Eocene age (Gibson & Towes, 1971).

In this paper the continental margin of eastern North America is re-examined, giving particular attention to the problem of reconstructing the continents and identifying the processes that have modified their margins and thereby created mismatch errors. No new data are offered here; rather we draw on the wealth of primarily magnetic data, acquired by the US Naval Oceanographic Office in this area over the last decade, to examine the problems of initial break-up. These magnetic data include the well known east coast aeromagnetic survey (Taylor *et al.*, 1968), shipborne surveys of the magnetic smooth zone (Einwich & Vogt, 1971), the Late Jurassic spreading lineations called the Keathley sequence (Vogt *et al.*, 1971a), and an aeromagnetic survey of the northern Keathley sequence (Vogt *et al.*, 1970a). Syntheses of the spreading evolution in the North Atlantic appear in Vogt *et al.* (1971b) and Pitman & Talwani (1972).

Previous Reconstructions

The most obvious way to reassemble rifted continents is to make their ancient edges match, but the unanswered question 'Where, exactly, are the edges?' inspires the search for constraints other than the bathymetric fit. In the Africa–North America fit, the geometry of reconstruction, although rendered rather flexible by the lack of major irregularities in the shelf edges, nevertheless remains the major constraint used by all investigators. Bullard *et al.* (1965) produced a minimum-error fit at the 500 fm isobath, this depth generally giving the smallest mismatch for all reconstructed continents. The procedure was to estimate an initial fit, then allow the computer to search for an adjustment that minimized the misfit in a least-square sense. Dietz *et al.* (1970) derived a similar but 'looser' fit, using the 1000 fm isobath instead. They argue that the Bahama platform and Blake Plateau represent subsided oceanic crust capped by a carbonate platform. It is not clear why they then presumably still included the Blake–Bahama area in their computerized continental fit, since this would degrade the optimized fit elsewhere. Dietz *et al.* believe that their fit is probably on the loose side, but point out that even in this loose fit, the overlap in the Cape Hatteras area encloses a drill hole that bottomed in crystalline basement. Hence, if the fit is correct, progradation or other post-drift accretion on the African side must be called on to explain the overlap.

Le Pichon & Fox (1971) propose a reconstruction based not only on a match of 500 fm isobaths, but also on the constraint that three major lineaments in the western Atlantic (from north to south, a proposed Kelvin fracture zone related to the seamount chain; a Cape Fear fracture zone related to the Cape Fear arch and the northwest trending Blake–Bahama outer ridge; and a Bahama fracture zone related to the Bahama island chain and the Bahama escarpment) line up with three lineaments on the African side (Canary and Cape Verde 'fracture zones' related to those island groups and a Guinea 'fracture zone,' the south facing scarp of the presumably continental Guinea plateau). Upon closing the ocean, the authors found that the Bahama and Guinea fracture zones as well as the Newfoundland fracture zone (the south edge of the Grand Banks and the submarine Newfoundland Ridge) are segments of small circles about a best-fitting spreading pole at 58·3°3° N, 21·8° W. The Cape Fear lineament lines up with the Cape Verde lineament, whereas the Kelvin and Canary lineaments miss each other by about 150 km and are not quite parallel. One difficulty with the Le Pichon-Fox reconstruction is the large overlap of North Africa and the Grand Banks; but, as the authors point out, this may merely reflect post-drift tectonic activity in northern North Africa.

In this paper the Newfoundland, Bahama–Guinea, and possibly the Blake–Bahama lineaments are accepted as old transform faults, but the overall trend of the Kelvin, Cape Verde and Canary Islands is considered to be the result of plate motion over mantle plumes (Morgan, 1971). (Later we shall examine the possibility that volcanic centres may preferentially develop on intersections of transform and normal fault trends.)

In summary, the reconstructions of Bullard *et al.* (1965) and Dietz *et al.* (1970) are primarily geometric fits of continental edges, whereas Le Pichon & Fox (1971) invoked the additional constraint that related structures on both sides line up in the reconstruction.

Rotated Spreading Isochrons

One of two additional constraints proposed in this paper is the rotation of sea-floor spreading isochrons over the continental edge. These isochrons are defined by magnetic lineations caused by geomagnetic reversals and sea-floor spreading. In general, the configuration of the spreading axis continues to resemble the shape of the rifted margin long after first rifting. For example, the present mid-Atlantic Ridge axis, representing the zero age isochron, still 'remembers' the overall shape of the initial break; south of the Azores the break occurred over 120 m.y. ago. However, the similarity is likely to be greater the older the isochron. Crustal isochrons have the great advantage that they are unaffected by such processes as progradation, volcanism or subsidence that might alter the initial edge of a continent. Thus, by comparing the present continental edge with an old isochron (its exact age is not so important to this method), one could hope to determine which of the two rifted margins should be blamed for particular overlaps and underlaps that result upon continental reconstruction. It should be stressed that any comparison of fits must emphasize *relative* underlaps and overlaps. Since the actual continental edge has not been defined to a precision of less than several tens of kilometres, any fit can be allowed to swing either way about its rotation pole by that amount.

Only in the western Atlantic have pre-Cenozoic isochrons been adequately charted. The oldest magnetic feature is the smooth-rough boundary (Pitman & Talwani, 1972); but because it is not clear that this is an isochron, and because it is actually a transition zone some tens of kilometres wide, it is not used at present. Next come the Keathley anomalies (Vogt *et al.*, 1970a, 1971a), of which J-20 (about 150 m.y.) and J-4 are best defined. This eastern edge of the Keathley anomalies has not been dated by deep-sea drilling; its age was initially estimated at 140 m.y.b.p. (Vogt *et al.*, 1970a, 1971a); more recently Larson & Pitman (1972) verified that the Keathley sequence also exists in the western Pacific. By applying the basement ages recovered from deep drilling in the Pacific, they propose to terminate the Keathley sequence at 110 m.y. It seems reasonable to estimate the age of J-4 as Lower Cretaceous until deep drilling has been accomplished. Using the trial and error method of Pitman & Talawani (1972), we found that when J-4 is rotated westward 3·7° about a pole at 60° N, 10° W, it overlies J-20 generally within present uncertainties of anomaly definition (Fig. 1). Rotating J-4 westward by 13·4° about a pole at 55° N, 25° W (Fig. 2) produces a

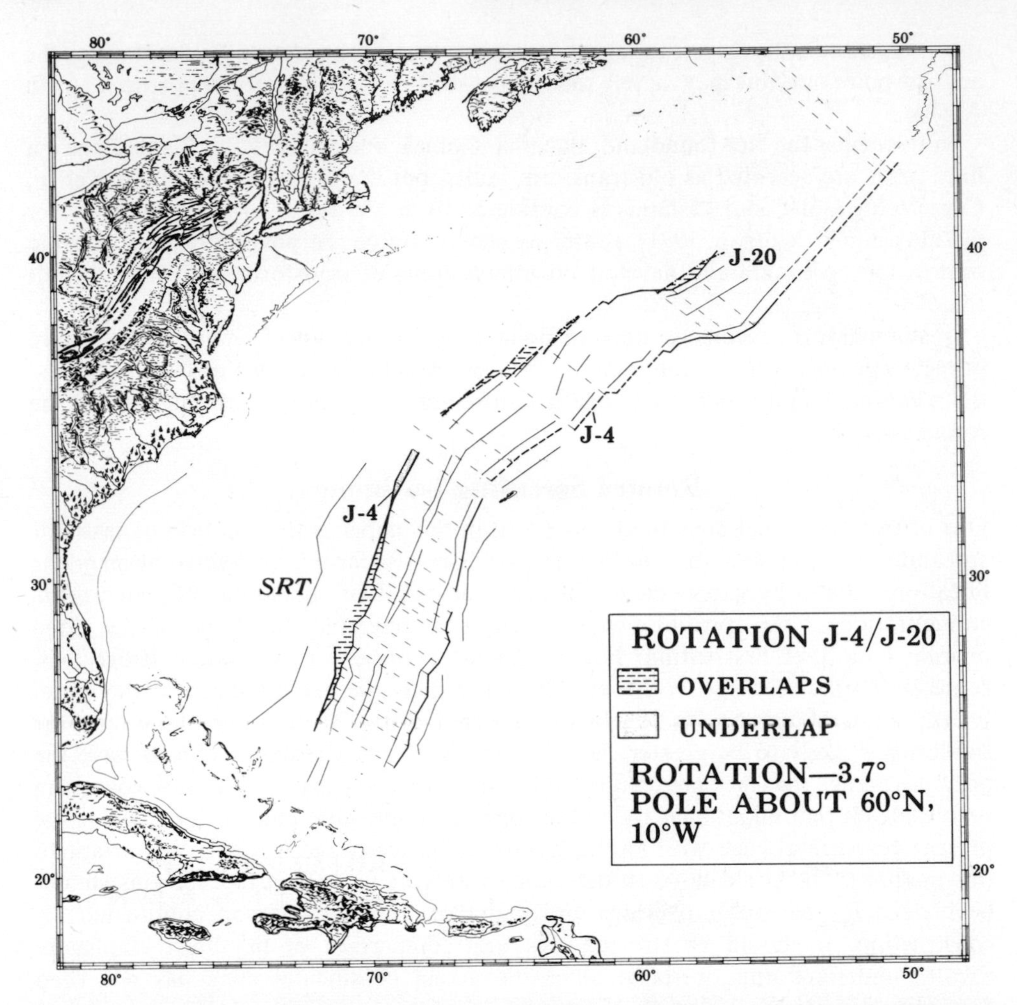

FIGURE 1 When the position of anomaly J-4 is rotated 3·7° about a pole at 60° N, 10° W, it compares closely with the position of anomaly J-20. Straight continuous lines are magnetic anomalies of the Keathley sequence (Vogt *et al.*, 1971a), and SRT denotes the magnetic smooth-rough boundary. Estimated or interpolated anomaly positions denoted by dashed lines; anomalies J-4 and J-20 are shown by heavier lines; dash-dot lines are portions of small circles, showing direction of displacement. Overlap is shown by horizontal dash pattern, and underlaps are dotted.

reasonably good match to the 1000 fm isobath and at the same time keeps the flow lines closely parallel to the Bahama and Newfoundland fracture zones of Le Pichon & Fox (1971). This rotation resembles the 13° about 58·3° N, 21·8° W (the 'pole of early opening') of those authors. The similarity in J4-margin and J4–J20 pole positions suggests that the spreading pole remained relatively fixed over this early period in Atlantic history.

When the overlaps and underlaps of the J4-to-margin fit are compared with the Africa–North America fits (Fig. 3), it is apparent that in the area of the Blake Plateau, rotated J-4 resembles the African edge (Guinea Plateau) in geometry, thereby producing a similar overlap as the reconstructions of Bullard *et al.* (1965), Dietz *et al.* (1970) and Le Pichon & Fox (1971). This further demonstrates what Dietz *et al.*

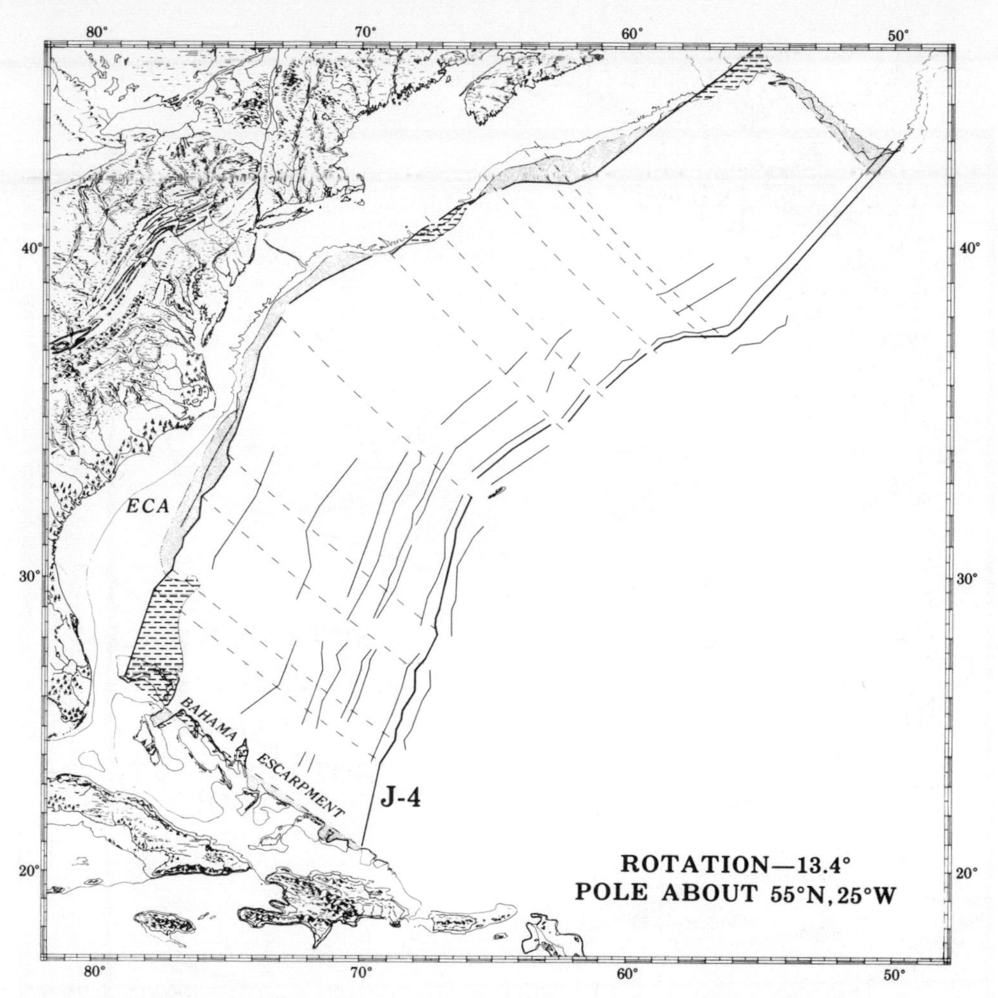

FIGURE 2 When anomaly J-4 is rotated 13·4° about a pole at 55°N, 25°W, it closely follows the 1000 fm isobath. Where J-4 is most accurately known, it is shown by heavier line. ECA denotes crest of east coast magnetic anomaly (Fig. 5). Other conventions as in Fig. 1.

suggested—that the Blake Plateau is an excrescence produced on the North American side subsequent to drift.

North of the Blake Plateau J-4 produces an extensive region of underlap. Although approximately of constant width, the underlap is somewhat greater south of Cape Hatteras, where the underlap is smallest. This relationship is similar to that of the three continental reassemblies, especially that of Le Pichon & Fox (1971). South of Cape Hatteras rotated J-4 resembles the straight North American more than the somewhat indented African margin placed next to it by Le Pichon & Fox. Hence we suggest some subsidence has occurred on the African side. The Bullard and Dietz fits produce large overlaps near Cape Hatteras; if their fits are right (we consider this less likely), the close similarity of J-4 to the American edge suggests that the overlap is produced by sediment prograding or volcanism on the African side.

Proceeding northward up the margin, we next encounter an area of overlap east to

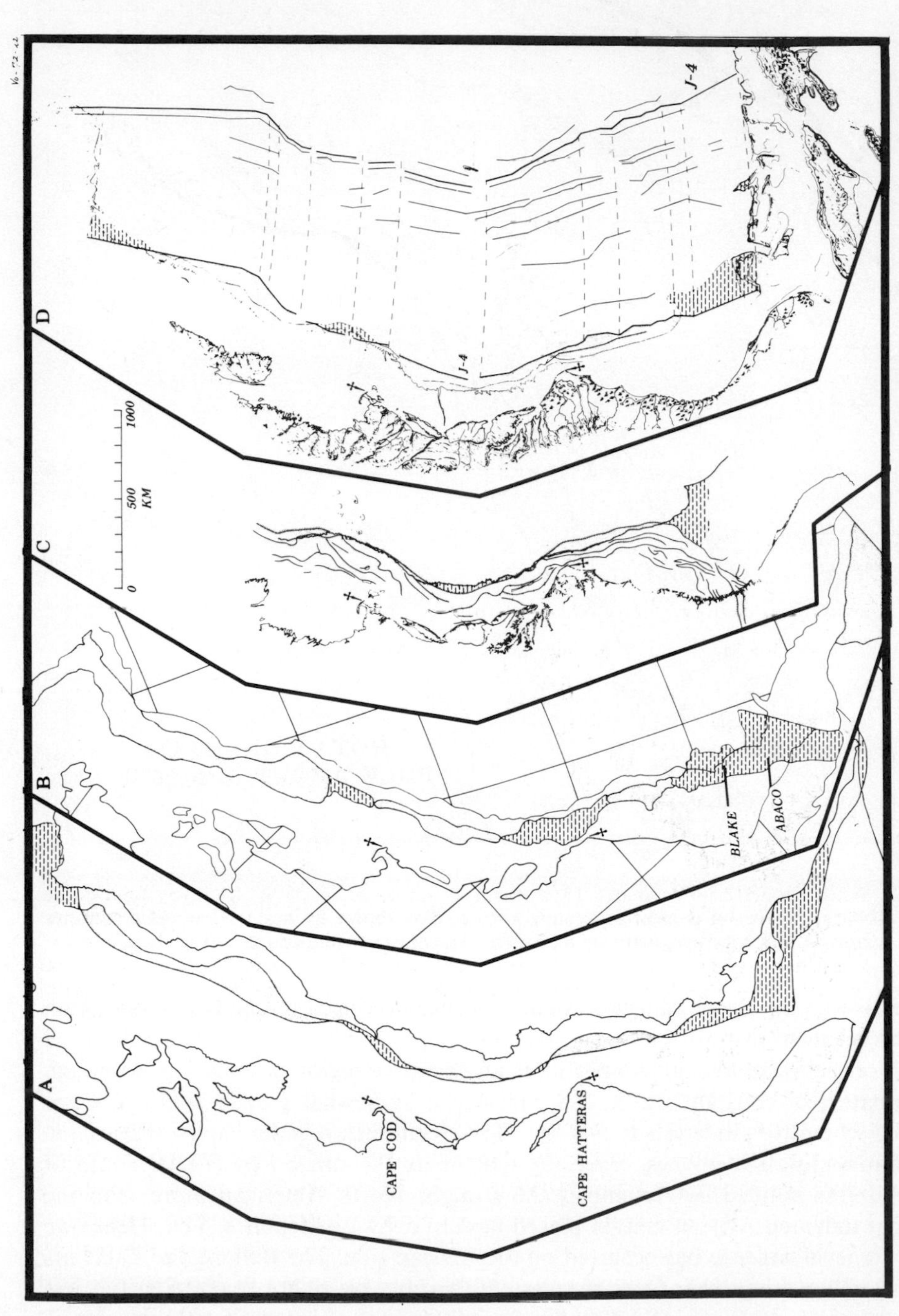

FIGURE 3 Comparison of (B) Bullard *et al.* (1965) and (A) Le Pichon & Fox (1971) Africa–North America reconstructions, using the 500 fm isobaths; also shown (D) is anomaly J-4 rotated over 1000 fm isobath (from Fig. 2), and relation between source of East Coast anomaly and 1000 fm isobath (C) (see Fig. 4). See previous figures for conventions.

FIGURE 4 Magnetic source configuration (constructed from Taylor *et al.*, 1968) and depth to pre-Triassic basement in km (after Emery *et al.*, 1970). Continuous strike and dip lines show edges of magnetic sources in the area of East Coast anomaly. Eastern edge of source for the latter anomaly is alternately inside (horizontal dash pattern) and outside (dotted) the 1000 fm isobath. Labelled contours show approximate magnetic source depths in kilometres. Streamline pattern in New Jersey–Maryland area shows proposed direction of sediment transport and shelf outbuilding associated with Salisbury Embayment.

southeast of Cape Cod (Fig. 3). Similar overlaps appear in all reconstructions. Thus, J-4 resembles the African edge rather the American one, and we are led again to look for a post-drift excrescence to North America. A probable process is not hard to find—the New England (or Kelvin) seamounts intersect North America there; volcanism and sediments piled on the volcanics may have added to North America as the margin moved northwest across a stationary mantle plume. The timing of this event will be considered later. North of the Cape Cod overlap there is an extensive region of underlap that also resembles the one in the continental reconstructions (Fig. 3). Again, if our premise is correct that J-4 reflects the shape of the initial break, it must have been subsidence on the American side that causes the mismatch. We do not wholeheartedly endorse this result, both because subsidence is harder to understand than outbuilding, and because post-drift tectonic modification of the African edge is not unlikely. Part of the mismatch could be explained if the eastern Canary Islands are underlain by continental crust (Rothe & Schminke, 1968); however, seismic refraction and gravity data do not support this view (Bosshard & Macfarlane, 1970).

Finally, at the northern end of the Jurassic Atlantic, rotated J-4 produces a large overlap whereas all three reconstructions produce underlaps. We explain this as follows: Accretion by sediment prograding in the area of the Laurentian Fan filled the embayment from an arc of 230° and modified the American side, while tectonic events —right lateral transcurrent displacements of the margin near Morocco—modified the African side.

The East Coast Magnetic Anomaly

A prominent magnetic anomaly follows the continental margin of North America from the Blake Plateau to well north of the New England seamounts. Taylor *et al.* (1968), publishing a detailed aeromagnetic survey over the east coast of the United States, suggested that the source is a felsitic intrusive defining the Palaeozoic boundary of North America. Here we extend this idea to the proposition that the eastern edge of this source, which tends to follow the 1000 fm isobath (Fig. 4), defines the true, original edge of North America north of the Blake Plateau. To the extent that the anomaly is an edge effect between two magnetically dissimilar plates (Keen, 1969), there is no question but that the anomaly marks the original edge. If the source was an extensive but linear intrusive or complex of intrusives—its age is immaterial—within which the first successful crack developed, parts of this intrusive should survive on the African margin. Rona *et al.'s* (1970) discovery of a West African margin anomaly suggests that this is indeed the case. (We do not inquire here why North America and Africa parted along a crack through this intrusive.) The edges of magnetic sources in the area of the East Coast anomaly were estimated by comparing the residual anomaly chart of Taylor *et al.* (1968) with model profiles computed to fit measured residual profiles, assuming induced magnetization and using approximately the same source depths and magnetization intensities. Source depths along the East Coast anomaly were estimated by Vacquier's method—viz. the source depth is equated to the horizontal distance of constant magnetic gradient. This magnetic source depth is shown contoured in Fig. 4, together with seismically determined depth to the pre-Triassic basement.

Proceeding on the assumption that the eastern edge of the source of the East Coast anomaly defines the original edge of North America, we may compare the various continental reconstructions and the rotated J-4 isochron with the fit of the source edge to the 1000 fm isobath (Fig. 3).

The source edge remains mostly northwest of the Blake Plateau, and all other reconstructions produce overlaps. However, pre-Jurassic rocks are known south of the East Coast anomaly in Florida (Taylor *et al.*, 1968) and the westward curve of the anomaly is not shared by the reconstructed Guinea Plateau. For these reasons it is clear that the East Coast anomaly cannot mark the edge of North America in the area of the Blake Plateau.

North of the Blake Plateau, however, the source edge remains within about 50 km of the 1000 fm isobath. Between the plateau and a point just north of Cape Hatteras, the edge remains seaward of the 1000 fm isobath, then swings inside to a point southeast of New York. This is similar to the rotated J-4 and Le Pichon–Fox reconstructions if these are imagined rotated slightly further on to North America—a perfectly allowable procedure, as stated earlier. There would then be slight overlaps between Cape Hatteras and New York, and slight underlaps between Cape Hatteras and the Blake Plateau. Thus, the edge of the East Coast anomaly source resembles the edge of Africa, rather than the edge of North America; if our hypothesis is correct, progradation has added to North America between Cape Hatteras and New York, while subsidence or erosion occurred between Cape Hatteras and the Blake Plateau. The progradation might be due to sediments supplied to the shelf edge by way of the Salisbury Embayment, as shown schematically in Figs. 3 and 4. Progradation should be associated with subsidence. Indeed, magnetic source depths, although exceeding depth to pre-Triassic basement (Fig. 4), are relatively greater in the region of proposed outbuilding than they are between Cape Hatteras and the Blake Plateau.

In the area where the New England seamounts intersect the edge of North America, continental reconstructions and rotated J-4 produce relative overlaps, while north and south of the seamounts, underlaps occur (Fig. 3). Relatively at least, the edge of the East Coast anomaly source bears a corresponding relation to the 1000 fm isobath: near the seamounts it is on the isobath; to the south it lies about 20 km to the southeast; and north of the seamounts it lies some 40 km seaward (Figs. 3 and 4). Near the seamounts it is undefined for a short distance, perhaps because igneous activity associated with the New England seamounts produced magnetic sources that tend to cancel out the East Coast anomaly.

In summary, the proposition that the outer edge of the source of the East Coast anomaly defines the true edge of North America leads to the same conclusion as the proposition that anomaly J-4 still resembles the original break in shape: With minor exceptions, it has been the shape of North America, not Africa, that was subsequently altered.

The Blake Spur Anomaly

Pitman & Talwani (1972) showed that sea-floor spreading in the North Atlantic has been quite symmetrical since the time of the smooth-rough magnetic boundary, which appears to be at least approximately an isochron. However, the African magnetic smooth zone is of the order 330 km wide, whereas the western smooth zone is some 200 km wider (Rona *et al.*, 1970; Pitman & Talwani, 1972). One could postulate early asymmetrical spreading, or claim, as Drake *et al.* (1968) said about the magnetic smooth zones, that an early, much older basin existed here. We consider both explanations unlikely. Spreading axes are known to jump to new locations, however, and a very early eastward displacement of the mid-Atlantic spreading axis could well explain the wider western magnetic smooth zone. Two more or less well-documented axis jumps occurred about 60 m.y., when the Nansen rift developed along the Siberian continental margin (Pitman & Talwani, 1972) and about 10–20 m.y., when Kolbeinsey Ridge, north of Iceland, developed along the east edge of Greenland (Vogt *et al.*, 1970b). In both cases the new rifts seemed to prefer continental edges, as if these were zones of relative weakness.

We postulate that early in the history of the Atlantic rift, perhaps about 175 m.y., a similar new rift developed along the proto-African continental margin (Fig. 5). Our reasoning for this is based on the enigmatic Blake Spur anomaly, the only prominent magnetic anomaly in the western smooth zone (Taylor *et al.*, 1968). The existence of this relatively high-amplitude (150 gammas vs. less than 50 for most of the smooth zone) feature is troubling for any smooth zone hypothesis except for the constant-polarity theory (Emery *et al.*, 1970) still advocated by some (Larson & Pitman, 1972). The apparent absence of the Blake Spur anomaly on the African side suggests that

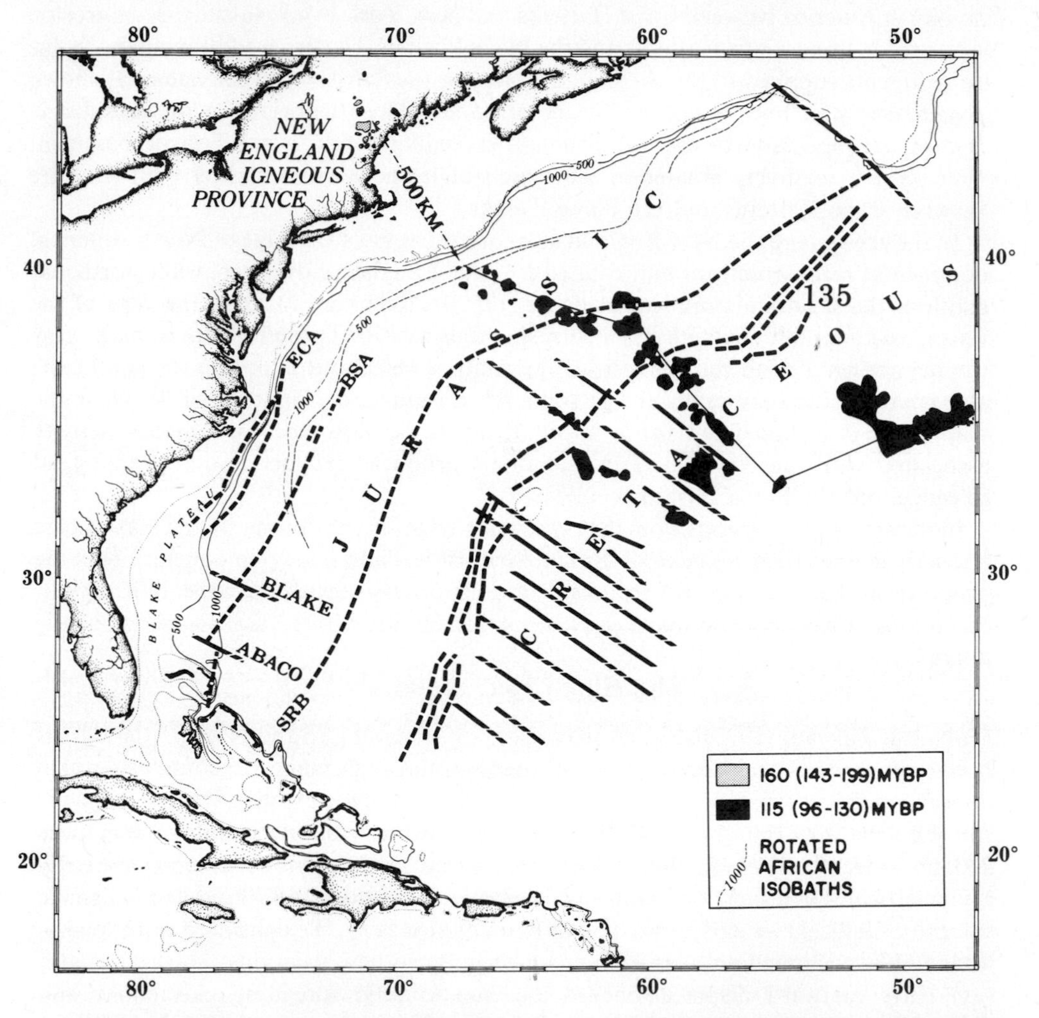

FIGURE 5 New England seamounts and associated basement highs (black), high amplitude magnetic anomalies of Bermuda discontinuity (dashed), Cretaceous and Jurassic fracture zones, all from Vogt *et al.* (1971b). Blake Spur anomaly (BSA) and East Coast anomaly (ECA) from Taylor *et al.* (1968) and Einwich (personal communication). SRB is smooth-rough magnetic boundary after Pitman & Talwani (1972). Dashed thin lines labelled 500 and 1000 are African 500 and 1000 fm isobaths, rotated as described in text. Mesozoic igneous bodies in New England and southern Canada are indicated in simplified fashion, after Foland *et al.* (1971), Armstrong & Stump (1971), Christopher (1969), Zartman *et al.* (1967), Larochelle (1962), and Faul *et al.* (1963). Segmented straight lines through New England Seamount igneous trail shows possible trajectory of Americas plate over mantle.

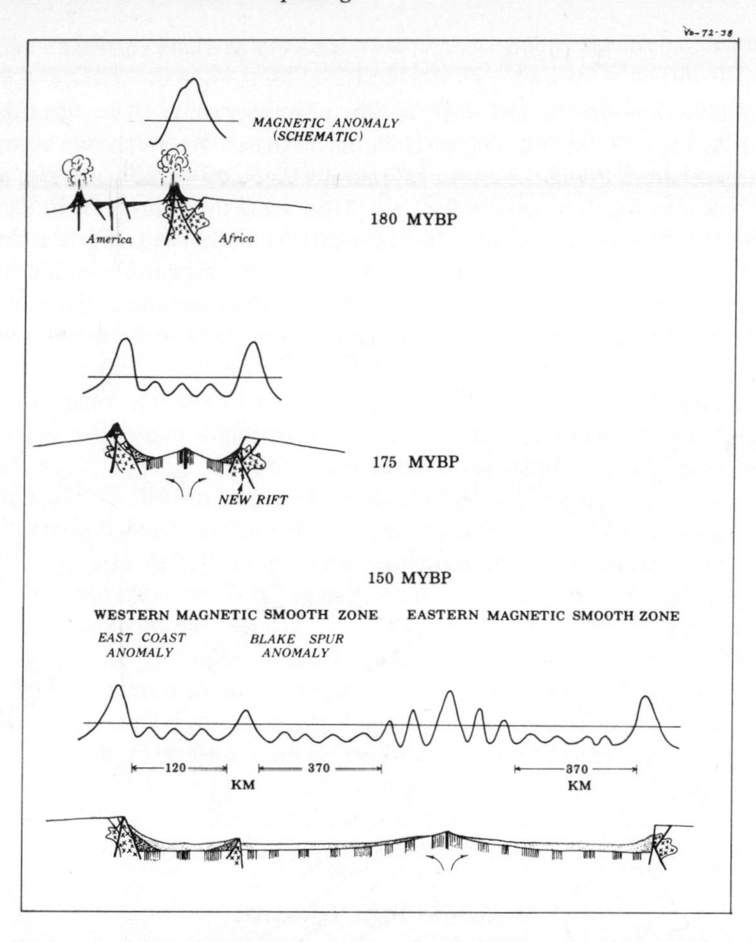

FIGURE 6 Schematic early evolution of the North Atlantic rift, involving an eastward shift of spreading axis about 175 m.y. to account for unequal widths of magnetic smooth zones. East Coast, Blake Spur, and African shelf anomaly (Rona *et al.*, 1970) may reflect three fragments of a once continuous igneous intrusive. The early 'Hatteras rift' contains several small but linear magnetic anomalies, perhaps due to geomagnetic reversals.

this anomaly was created after the crust—we consider this unlikely—or that only sea-floor east of this anomaly occurs on the African side. We propose that the Blake Spur anomaly is a third surviving sliver of a once intact intrusive (Fig. 6). A new rift developing along the early African margin separated this sliver, with rather symmetrical spreading ever since.

At least two requirements must be met if the above hypothesis is correct: First, the Blake Spur anomaly should resemble the edge of the African continent in outline; second, the 'inner smooth zone,' landward of the Blake Spur anomaly, should equal the difference between the width of African and American magnetic smooth zones. Comparison between the edge of Africa, shown on the reconstructions of Fig. 3, and the Blake Spur anomaly (Fig. 5) shows that both are rather straight, northeast-trending lineaments. Two small offsets in the Guinea Plateau shown by Bullard *et al.*'s (1965) rendition of the Guinea Plateau, may, if real, even correspond to two small offsets, of comparable magnitude and position, in the Blake Spur anomaly (Figs. 3 and 5). The latter were recognized by Taylor *et al.* (1968), who called them the Blake and Abaco faults.

The second requirement is also met, within the uncertainty of existing data. We mapped the African 500 fm and 1000 fm isobaths on to the American side according to the following scheme: Using the data of Pitman & Talwani (1972), we first approximated both smooth-rough transitions by lines. Measuring the distance from the African transition to the African 500 fm and 1000 fm isobaths along the Le Pichon–Fox early spreading direction, we then measured off these distances westward from the American smooth-rough transition, again along the Le Pichon–Fox spreading direction. Points on the eastern transition were connected to their counterparts in the west by the use of the post-155 m.y. sea-floor spreading flow lines of Pitman & Talwani (1972). The rotated African isobaths lie about 120 km east of the corresponding American isobaths, approximately at the Blake Spur anomaly. The agreement in position is good, considering the several uncertainties involved in the rotation.

The only adequate test for the hypothesis of Fig. 6 is to drill on the Blake Spur anomaly; short of that, seismic refraction may or may not detect such a small block of presumably granitic material. The refraction sections of Hersey *et al.* (1959) do not indicate any sialic block, but this is an inconclusive result, considering the scarcity of data and the size of the source. Conceivably, of course, the seismic velocity in the source does not differ significantly from that of the surrounding oceanic crust. However, as Taylor *et al.* (1968) pointed out, the crustal profile of Hersey *et al.* (1959) does seem anomalous where it crosses the Blake Spur anomaly. The oceanic layer (V_p = 6·3 to 6·9 km/sec), generally several kilometres thick on nearby stations, thins to a few hundred metres, while a layer with velocity 5·2 to 5·7 km/sec thickens from 1 or 2 to 3 km over the anomaly. Anomalous low velocity mantle (V_p = 7·2 to 7·4 km/sec) rises almost to the −10 km level, its highest level in this region.

Mesozoic Dike Swarms

The relationship between the shape of the first truly oceanic rift and the continental rifting and igneous activity that apparently preceded it is a fundamental unsolved problem. In an effort to evaluate the evidence, the relevant Late Triassic east coast formations (de Boer, 1967) and the only available detailed African data (Behrendt & Wotorson, 1970) has been compiled and is shown (Fig. 7) on the reconstruction of Le Pichon & Fox (1971). The transform and isochron grids have been continued into the Appalachians. This was not done to prove that significant plate motion occurred there about the later spreading pole (although a slight amount of net extension could have occurred). Instead, we ask whether the later transform and ridge trends (Figs. 2 and 3) could have been inherited from lines of weakness that existed in the continental plate or were created by events leading up to first sea-floor spreading.

The Newark Series—Triassic grabens associated with basaltic sills and flows (Sanders 1963) dated at about 220–230 m.y. (Armstrong & Besancon, 1970)—are the earlier of the two events shown in Fig. 7. These features follow the Appalachian structures and also tend to parallel the later oceanic rift directions. There are no orthogonal structures that might be interpreted as ancestral transform breaks. A seismically determined depression in the basement below the shelf (Figs. 4 and 7) could be a buried rift valley similar but more continuous than the Newark Series. This buried valley also tends to parallel the later oceanic rift trend.

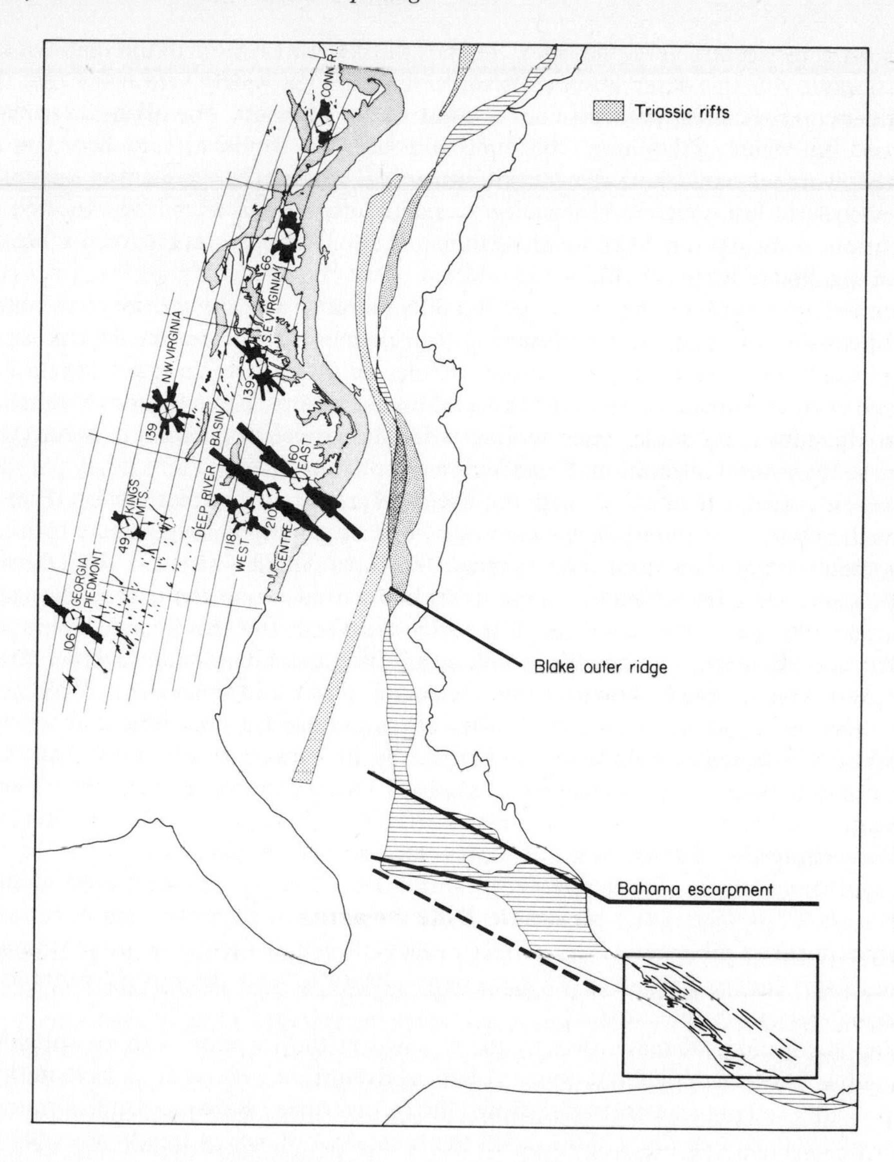

FIGURE 7 Mesozoic dikes and Triassic rifts (de Boer, 1967), and, superposed, a grid of rotated 'ridge' and 'fracture zone' trends generated around our pole of early opening (Fig. 2). Continental reconstruction and possible fractures after Le Pichon & Fox (1971); Liberian dikes and faults simplified after Behrendt & Wotorson (1970). Possible buried Triassic rift, inferred from basement depth chart of Emery *et al.* (1970), incises continental shelf. If there were a post-rift indentation in Guinea Plateau, east–west fracture trend, along south edge of plateau may be spurious.

As sea-floor spreading seems to have begun about 180 m.y. (Ewing *et al.*, 1970), the basaltic dike swarms, found not only in the eastern US (de Boer, 1967) but also in Liberia (Behrendt & Wotorson, 1970), must have been intruded in association with the first spreading. The greatest concentration of isotope dates on those dikes lies between 190 and 200 m.y. (Armstrong & Besancon, 1970), and the few Liberian dikes thus far dated appear to be of the same age (White & Leo, in press). The American dikes tend to crosscut earlier structures (de Boer, 1967). If the dikes are related to the shape of the first oceanic rift, this observation casts some doubt on the widely held belief that the continental break always reflects lines of weakness in the crust. The dikes are usually vertical, but wherever they aren't, the dip is invariably toward the east (de Boer, 1967), in other words towards the axis of future spreading. This makes some sense: the uplift of continental lithosphere and injection of mantle derivatives would be expected to result in a fan-shaped pattern of dikes radiating into the upper crust from a region below the line of incipient rifting.

The relation between the trends of the dike patterns and the trends of the later oceanic rift is interesting but puzzling (Fig. 7). North of the Chesapeake Bay the dikes tend to parallel the future rift; orthogonal trends, indicating incipient transform faults, are not evident. Some dikes with the 'rift' strike exist as far south as western Virginia, a state where the dikes display considerable variation in trend but also begin to turn to a more southeasterly orientation. From Virginia south the dikes are amazingly parallel, but make an angle of 20° to 30° with our average Jurassic transform direction (Fig. 2). Considering the uncertainty of the earliest pole of oceanic spreading, it may turn out that these dikes reflect ancestral, incipient transform trends. Most of the Liberian dikes were amazingly parallel to those in the southern states at the time of emplacement (Fig. 7). This gives some idea of the wide areal extent of this igneous event. We predict the existence of a continuous dike swarm between Liberia and Georgia. Deep burial by later sediments obscures the intervening portion of this swarm. Parallel to this dike swarm lineament are first order physiographic features: the Blake Outer Ridge and Bahama Islands, both interpreted by Le Pichon & Fox (1971) as early transform faults, and the continental edge of Africa from Liberia to the Guinea Plateau.

To complicate matters, there is apparently a second structural trend in the reconstructed Atlantic. In the reference frame of Fig. 7, this trend is east–west to east-southeast. Representing this second trend are the northern edge of South America, a lesser number of dikes and faults in the continental shelf of Liberia, and the southern edge of the Guinea Plateau. This trend more closely follows the transform trend of Jurassic opening (Figs. 2 and 3).

The significance of these two trends is unclear: they cannot both be spreading directions between Africa and North America. Even if the dike swarms have nothing to do with the later transform directions, the reconstructed Bahama–Guinea fracture zone of Le Pichon & Fox (1971) shows the same discordance in trends. Possibly the east trending scarp at the south edge of the Guinea Plateau is a spurious trend resulting from an arcuate post-drift embayment south of the plateau (Fig. 7).

The New England Seamount Chain

A prominent line of extinct volcanoes—the New England, or Kelvin seamounts—crosses the Mesozoic western Atlantic (Fig. 7). The significance and timing of this volcanism may be important to understand how the Atlantic rift came into being, and how the edge of North America came to be subsequently modified.

Drake *et al.* (1968) postulated that the seamounts are related to a major crustal fault, the seaward continuation of a Palaeozoic structure interrupting the Appalachians near 40° N. Uchupi *et al.* (1970) accepted this 'Kelvin fault' and hypothesized that spreading rates were discontinuous across it. More recent magnetic surveys (Vogt *et al.*, 1970a,

1971b) show that these views are untenable. The late Jurassic to early Cretaceous Keathley sequence continues through the seamount chain with only a very minor offset (Fig. 8). The detailed bathymetry of the seamounts (Walczak, 1963) suggests volcanic centres came up along a grid of northwest and southwest oriented fractures, possibly the ridge-normal and transform fault fabric bequeathed to the crust when it was manufactured at the spreading axis. This 'transform' fracture trend, more than the overall trend of the seamount chain, approximates the trend of small circles about the Jurassic pole of opening (Le Pichon & Fox, 1971; Fig. 2). Many other seamount chains suggest that location of volcanic centres was controlled by fractures (e.g., the Canaries, Bosshard & Macfarlane, 1970).

A better explanation for the existence of the seamount chain is the hot spot or plume hypothesis recently elaborated by Morgan (1971). According to this view the New England seamounts are part of a longer igneous trail leading from the White Mountain Mesozoic intrusive complex to the Azores, via the Corner Rise (Coney, 1971). The trail is purported to record the motion of the Americas plate over a mantle plume that is either stationary or moves only slowly.

The existence of a Kelvin hot spot seems plausible enough; we do not agree with the suggestion of Morgan (1971) and Coney (1971) that this plume influenced continental break-up and then maintained itself on the spreading axis. The New England seamounts are surely no older than Middle or Later Cretaceous, i.e. substantially younger than the crust on which they stand. The evidence is all indirect, but we feel it is rather convincing:

(1) Oldest dredged sediments and magnetic pole positions computed from magnetic anomalies are Cretaceous. The relevant data are reviewed by Taylor *et al.* (1968) and Uchupi *et al.* (1970) and will not be repeated here.

(2) Hot spots near spreading axes tend to produce massive aseismic ridges such as the Iceland–Faeroe and Walvis–Rio Grande ridges. Hot spots far from ridge crests, such as Hawaii, seem to produce lines of isolated volcanoes. The Kelvin chain resembles this latter category.

(3) The White Mountain plutonism consists of an older core area, with most radiometric dates in the range 150 to 199 m.y., surrounded by a loose scattering of intrusives many of which date from 100–120 m.y. (Armstrong & Stump, 1971; Christopher, 1969; Foland *et al.*, 1971). (The 216 m.y. Agamenticus plutonism may be an earlier event.) It thus appears that the Kelvin plume was active under New England at about the time that rifting began. However, the Americas plate still had to move 500 km northwest to bring the plume to the westernmost of the Kelvin seamounts. If the Americas plate moved over the mantle during the Cretaceous at similar rates as it did during the Later Tertiary, the results of Morgan (1971) suggest that several tens of million years had to elapse. Magma chambers trapped in the lithosphere and carried with it must be called upon to explain the later, 100–120 m.y. plutonism in New England and southern Canada (Fig. 6).

(4) The Kelvin seamounts are much too tall to have been created at the spreading axis, unless the Jurassic mid-Atlantic Ridge crest was much deeper than the 2·6 km typical of the present mid-Atlantic Ridge (Sclater *et al.*, 1971; Fig. 9). This follows because seamounts rising above sea level are planed by erosion much faster than they can sink. Thus a seamount formed near a spreading axis cannot presently rise much further above the surrounding basement than the depth of the ridge crest where it

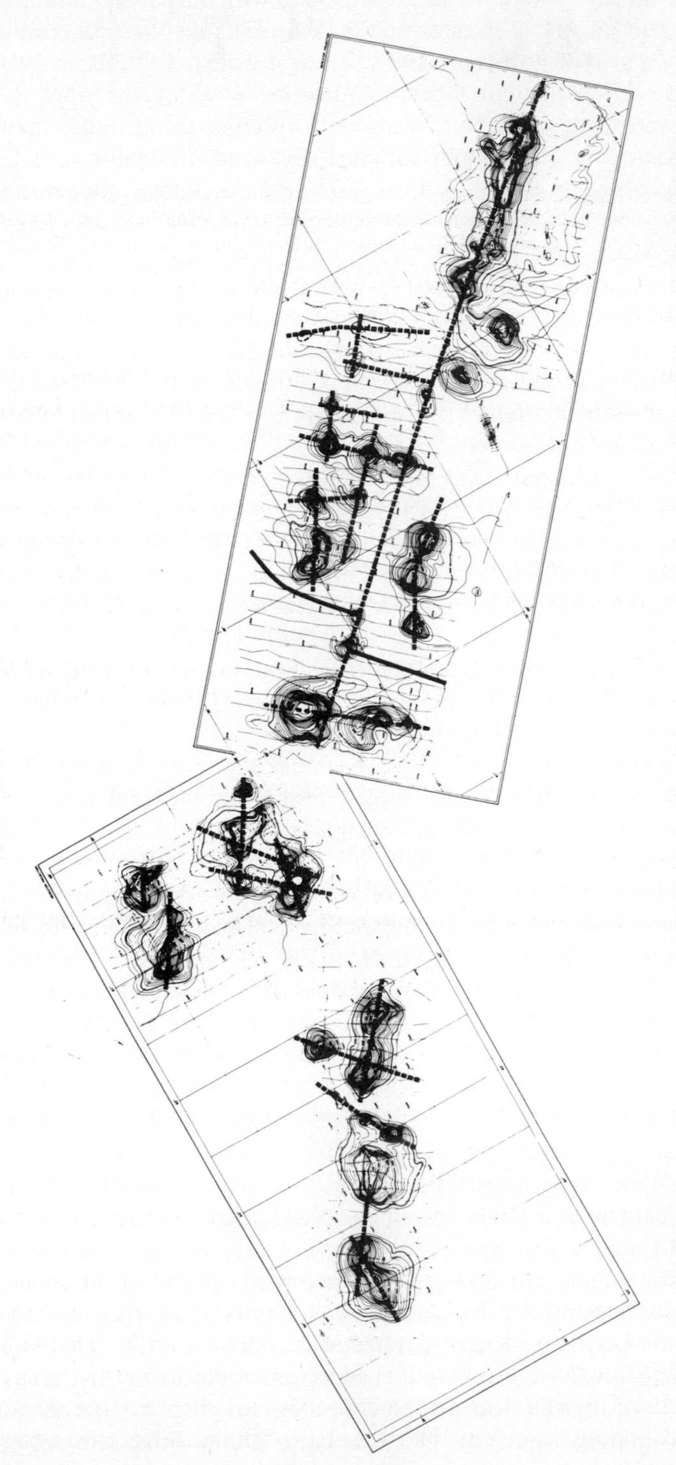

FIGURE 8 Detailed bathymetry of New England seamounts (Walczak, 1963) suggests that location of individual volcanic centres was controlled by pre-existing grid of northwest and northeast oriented fractures (dashed lines), perhaps reflecting transform and normal fault grain created at the early ridge axis. Anomaly J-4 (heavy solid line) appears slightly offset. There is no evidence of the major fault postulated by Drake *et al.* (1968) and Uchupi *et al.* (1970).

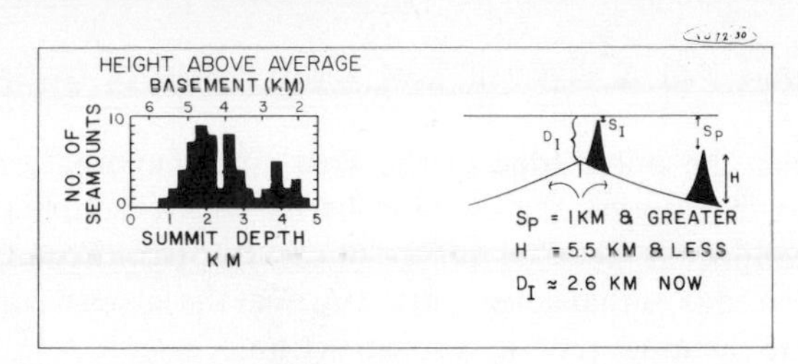

FIGURE 9 Frequency distribution of summit depths for New England seamounts, taken from bathymetric chart of Uchupi *et al.* (1970), and elevation of summits above average regional oceanic basement depth, taken to be 6·5 km (from Emery *et al.*, 1970). Diagram on right shows that if seamounts formed on early ridge axis, assumed to be 2·6 km below sea level as today (D_I), then present height above basement (H) must be 2·6 km or less. Since H ranges up to 5·5 km, basement must have been several km deeper than ridge crest when seamounts were extruded.

formed, unless vulcanism continues long after the seamount is submerged, which is unlikely. The Kelvin seamounts are up to 5·5 km higher than the nearby basement, hence they must have erupted in at least 5·5 km water. The lack of guyots (with one exception) indicates that the volcanoes did not reach sea level at all, hence basement depths could have exceeded 5·5 km. Oceanic basement of 50 to 100 m.y. age is presently about that deep, hence we conclude that the Kelvin seamounts were extruded some 50 or 100 m.y. after the sea-floor had formed.

The one apparently truncated volcano—Bear Seamount—has a top only 1·1 km below sea level (Uchupi *et al.*, 1970). Considering normal ridge subsidence due to thermal contraction of the lithosphere, compounded by sediment loading along the continental slope (Sleep, 1970), it is inconceivable that Bear Seamount was active 180 m.y. ago and then sank a mere 1·1 km. The seismic data compiled by Emery *et al.* (1970) suggest that this is as much as an order of magnitude too low. If, however, Bear Seamount was formed in the later Cretaceous, a subsequent subsidence of 1·1 km is much more plausible.

In summary, we propose the Kelvin Seamounts were formed in middle to later Cretaceous times, long after the sea floor under them. Thus the Kelvin plume cannot be looked upon as a cause of the North Atlantic rift. As the continental edge passed over the plume, volcanic seamounts began to be formed. The continental accretion, suggested in earlier sections to have occurred where the seamounts intersect the shelf, may simply be more volcanics and intrusives, buried by prograded shelf sediments. Below the plume volcanics there should be older Cretaceous and Jurassic sediments. The overall trend of the seamount chain reflects plate motion over the plume but on a more local scale volcanoes may have followed a pre-existing spreading fabric.

Conclusion

The detailed geophysical surveying carried out in recent years in the western North Atlantic has provided a new set of constraints on the Africa–North America reconstruction. The Le Pichon–Fox reassembly or the relatively slight variation we are presently developing, seems to satisfy the available constraints. An early eastward hop of the rift axis must have closely paralleled the previous edge of the rift, otherwise the continents could not be so well reconstructed. A reconstruction fitting Africa against the Blake Spur anomaly is being developed to show the Atlantic at the end of this early rift, for which we propose the name Hatteras rift. Deep drilling is necessary to verify the hypothesis that a sliver of East Coast intrusive causes the Blake Spur anomaly.

If, as we propose, the outer edge of the East Coast anomaly source defines the original edge of North America, the use of isobaths, which include subsequent modifications, can be avoided. A detailed survey of any African 'East Coast anomaly' is needed—the western edge of that source, fitted against the eastern edge of the American one, should provide the most precise reconstruction.

It appears that the rotated J-4 isochron, as well as the outer edge of the East Coast magnetic source (north of the Blake Plateau), resemble the edge of Africa generally better than they do the edge of North America. Thus, the American edge has been most modified, and the following processes may have operated:

(1) The Blake Plateau is an excrescence, a carbonate platform subsequently built on oceanic crust (Dietz *et al.*, 1970). The plateau is bounded on the east by the Blake Spur anomaly; thus the platform was built on sea-floor of the early Hatteras extinct rift, perhaps because it was shallower or had been filled by sediments ponded behind the basement ridge associated with the Blake Spur Anomaly (Fig. 6). A mantle plume, such as Iceland, could also have existed here, creating sea-floor shallow enough to form a base for carbonate sedimentation (Dietz, personal communication, 1972).

(2) Between the Blake Plateau and Cape Hatteras, the edge of North America has subsided. Alternatively current erosion by the Gulf Stream, which flows vigorously in the area, may have removed sediments from the shelf and slope, or at least prevented their buildup as the edge of North America subsided due to thermal contraction and regional loading. Additional subsidence could have affected the corresponding edge of Africa.

(3) Between Cape Hatteras and the New England seamounts, slight prograding associated with the Salisbury Embayment built the American shelf outward a few tens of kilometres. This accretion was associated with extra subsidence, as reflected by exceptional source depths along the East Coast anomaly.

(4) Near the New England (Kelvin) seamounts a larger amount of accretion may reflect Cretaceous hot spot volcanism, intrusion, and subsequently controlled sedimentary outbuilding.

(5) Just north of the New England seamounts a substantial underlap suggests either American subsidence and erosion, or African subsidence, erosion, or tectonism. The latter is most likely.

(6) At the northern limit of the early rift, the Laurentian channel and Laurentian cone suggests outbuilding of North America and post-drift tectonic modifications of Africa. The exact combination of these two processes remains uncertain.

The position of the earliest spreading pole is still unclear. Until offsets in smooth zone isochrons—or seismic surveys of transform fractures—pinpoint the pole, it is impossible to say whether the Mesozoic dike swarms are ancestral transform directions. The parallelism between these dikes in the southern US and the Blake Outer Ridge, the western Bahamas, and the edge of Africa between Liberia and the Guinea Plateau is strongly suggestive, however (Fig. 7).

The basic problem concerning the New England seamounts is their age. A systematic drilling and dredging programme is required. In the meantime, the evidence favours formation by a mantle plume, but one that erupted and remained under the interior of the Americas plate, not close to the mid-Atlantic Ridge, at least while the Kelvin seamounts were extruded.

One significant conclusion of this study has thus far remained implicit, viz. that the original continental edge is 'sharp' and can be defined to a resolution of a few tens of kilometers, or perhaps better, if sufficient data are available. This means that mismatches in continental reconstruction (e.g., Bullard *et al.*, 1965) imply post-drift modifications instead of stretching and deformation associated with the rifting process. This does not mean that the rifted edge of a continent has no transition zone: The continental crust is undoubtedly thinned, perhaps by subaerial erosion (Vogt & Ostenso, 1967; Sleep, 1971), and injected with basic intrusions prior to the first true sea-floor spreading. From that time on, we feel, faulting and igneous modification of the margin become generally minimal (unless the margin moves over a plume); thermal subsidence and sedimentation effects dominate the subsequent history of the margins.

Acknowledgements

The author acknowledges the technical help of A. Einwich; B. Grosvenor, C. Fruik, L. S. Edwards, and B. Trott prepared illustrations and text; and F. Moore gave editorial assistance. Discussions with P. T. Taylor were helpful.

References

Armstrong, R. L. and Besancon, J., 1970. A Triassic time-scale dilemma: K-Ar dating of Upper Triassic mafic igneous rocks, eastern USA and Canada and post Upper Triassic plutons, western Idaho, USA, *Elogae Geol. Helvetiae*, **63**, 15–28.

Armstrong, R. L. and Stump, E., 1971. Additional K-Ar dates, White Mountain magma series, New England, *Am. J. Sci.*, **270**, 331–3.

Behrendt, J. C. and Wotorson, C. S., 1970. Aeromagnetic and gravity investigations of the coastal area and continental shelf of Liberia, West Africa, and their relation to continental drift, *Geol. Soc. Am. Bull.*, **81**, 3563–74.

Bosshard, E. and Macfarlane, D. J., 1970. Crustal structure of the western Canary Islands from seismic refraction data, *J. Geophys. Res.*, **75**, 4901–18.

Bullard, E., Everett, J. E. and Smith, A. G., 1965. The fit of the continents around the Atlantic. Symposium on Continental Drift, *Roy. Soc. London Phil. Trans.* **258**A.

Christopher, P. A., 1969. Fission track ages of younger intrusions in southern Maine, *Bull. Geol. Soc. Am.*, **80**, 1809–14.

Coney, P. J., 1971. Cordilleran tectonic transitions and motions of the North American plate, *Nature*, **223**, 462–5.

De Boer, J., 1967. Palaeomagnetic-tectonic study of Mesozoic dike swarms in the Appalachians, *J. Geophys. Res.*, **72**, 2237–50.

Dietz, R. S., Holden, J. C. and Sproll, W. P., 1970. Geotectonic evolution and subsidence of Bahama platform, *Bull. Geol. Soc. Amer.*, **81**, 1915–28.

Drake, C. L., Ewing, J. I. and Stockard, H., 1968. The continental margin of the eastern United States, *Can. Jour. Earth Sci.*, **5**, 993–1010.

Einwich, A. and Vogt, P. R., 1971. Continued studies of the magnetic smooth zone in the western North Atlantic, *Trans. Am. Geophys. Union*, **52**, 195.

Emery, K. O., Uchupi, E., Phillips, J. D., Bowin, C. O., Bunce, E. T. and Knott, S. T., 1970. Continental Rise off eastern North America, *Bull. Am. Petrol. Geol.*, **54**, 44–108.

Ewing, J., Hollister, C., Hathaway, J., Paulus, F., Lancelot, T., Habib, D., Poag, C. W., Luterbacher, H. P., Worstell, P. and Wilcoxon, J. A., 1970. Deep sea drilling project: Leg II, *Geotimes*, **15**, 141–6.

Faul, H., Stern, T. W., Thomas, H. H. and Elmore, P. L. D., 1963. Ages of intrusions and metamorphism in the northern Appalachians, *Am. J. Sci.*, **261**, 1–19.

Foland, K. A., Quinn, A. W. and Giletti, B. J., 1971. K-Ar and Rb-Sr Jurassic and Cretaceous ages for intrusion of the White Mountain magma series, northern New England, *Am. J. Sci.*, **270**, 321.

Fullagar, P. D. and Bottino, M. L., 1969. Tertiary felsite intrusions in the valley and ridge province, Birginia, *Bull. Geol. Soc. Am.*, **80**, 1853–8.

Gibson, T. G. and Towes, K. M., 1971. Eocene volcanism and the origin of Horizon A, *Science*, **172**, 152–4.

Hallam, A., 1971. Mesozoic geology and the opening of the North Atlantic, *The Journal of Geology*, 129–57.

Hersey, J. B., Bunce, E. T., Wyrick, R. F. and Dietz, F. T., 1959. Geophysical investigation of the continental margin between Cape Henry, Virginia, and Jacksonville, Florida, *Bull. Geol. Soc. Am.*, **70**, 437–66.

Keen, M. J., 1969. Possible edge effect to explain magnetic anomalies off the eastern seaboard of the U.S., *Nature*, **222,** 72–4.

King, P. B., 1969. Tectonic map of North America, U.S. Geological Survey.

Larochelle, A., 1962. Palaeomagnetism of the Monteregian Hills, southeastern Quebec, *Bull. Geol. Sur. Can.*, **79**, 1–43.

Larson, R. L. and Pitman, W. C., 1972. World-wide correlation of Mesozoic magnetic anomalies, and its implications, *Bull. Geol. Soc. Am.*, **83,** 3645–62.

Le Pichon, X. and Fox, P. J., 1971. Marginal offsets, fracture zones, and the early opening of the North Atlantic, *J. Geophys. Res.*, **76**, 6294– 308.

Morgan, W. J., 1971. Convection plumes in the lower mantle, *Nature*, **230**, 42–3.

Peterson, M. N. A., Edgar, N. T., Cita, M., Gardner, S., Jr., Goll, R., Nigrini, C. and von der Borch, C., 1970. Initial Reports of the Deep Sea Drilling Project, Vol. 2, pp. 413–72, U.S. Govt. Printing Office, Washington, D.C.

Pitman, W. and Talwani, M., 1972. Sea-floor spreading in the North Atlantic, *Bull. Geol. Soc. Am.*, **83**, 619–45.

Rona, P. A., Brakl, J. and Heirtzler, J. R., 1970. Magnetic anomalies in the northeast Atlantic between the Canary and Cape Verde Islands, *J. Geophys. Res.*, **75**, 7412–20.

Rothe, P. and Schminke, U., 1968. Contrasting origins of the eastern and western islands of the Canaries Archipelago, *Nature*, **218,** 1152–4.

Sanders, J. S., 1963. Late Triassic tectonic history of the northeastern United States, *Am. J. Sci.*, **261**, 501–23.

Sclater, J. G., Anderson, R. N. and Bell, M. L., 1971. Elevation of ridges and evolution of the central eastern Pacific, *J. Geophys. Res.*, **76**, 7888–915.

Sleep, N. H., 1971. Thermal effects of the formation of Atlantic continental margins by continental break-up, *Geophys. J. R. astr. Soc.*, **24**, 325–30.

Taylor, P. T., Zietz, I. and Dennis, L. S., 1968. Geologic implications of areomagnetic data for the eastern continental margin of the United States, *Geophysics*, **33**, 755.

Vogt, P. R. and Ostenso, N. A., 1967. Steady-state crustal spreading, *Nature*, **215,** 810–17.

Vogt, P. R., Anderson, C. N. and Bracey, D. R., 1971a. Mesozoic magnetic anomalies, seafloor spreading, and geomagnetic reversals in the southwestern North Atlantic, *J. Geophys. Res.*, **76**, 4796–823.

Vogt, P. R., Johnson, G. L., Holcombe, T. L., Gilg, J. G. and Avery, O. E., 1971b. Episodes of sea-floor spreading recorded by the North Atlantic basement, *Tectonophysics*, **12**, 211–34.

Vogt, P. R., Lorentzen, G. R. and Dennis, L. S., 1970a. An aeromagnetic survey of the Keathley magnetic anomaly sequence between 34° N and 40° N in the western North Atlantic, *Trans. AGU*, **51**, 274.

Vogt, P. R., Ostenso, N. A. and Johnson, G. L., 1970b. Bathymetric and magnetic data bearing on sea-floor spreading north of Iceland, *J. Geophys. Res.*, **75**, 903.

Uchupi, E., Phillips, J. D. and Prada, K. E., 1970. Origin and structure of New England seamount chain, *Deep-Sea Res.*, **17**, 483–94.

Walczak, J. E., 1963. A marine magnetic survey of the New England seamount chain, *US Naval Oceanographic Office. Tech. Rep.*, **159**, 37.

White, R. W. and Leo, G. W., Geological reconnaissance in western Liberia, Liberian Geol. Survey. Spec. Paper 1 (in press).

Zartman, R. E., Brock, M. R., Heyl, A. V. and Thomas, H. H., 1967. K-Ar and Rb-Sr ages of some alkalic intrusive rocks from central and eastern United States, *Am. J. Sci.*, **265,** 848–70

6.7

P. J. CONEY
Department of Geology
Middlebury College
Middlebury, Vermont, USA

Non-Collision Tectogenesis in Western North America

Introduction

Plate tectonics has revolutionized interpretations of mountain system evolution by providing suitable kinematics to explain complex petro-tectonic associations exposed in these belts. Gross mountain-system architecture is now more comprehensible (Dewey & Bird, 1970; Coney, 1970) and most sedimentary, igneous, and metamorphic assemblages are reasonably well identified in the unifying hypothesis (Dickinson, 1970, 1971a, 1971b; Hamilton, 1969a, 1969b; Miyashiro, 1971).

Belts of folds and thrusts in the world's mountain chains have not been so clearly explained, although the gross architecture of collision systems, such as the Himalaya, yield readily to an imagery of intercontinental collision and resultant massive telescoping as one continental edge abortively attempts to underthrust the other (Gansser, 1966). Non-collision systems, however, such as the North American Cordillera, have yielded less easily, and debate over mechanisms due to regional compression versus vertical uplift and resulting 'gravity tectonics' continues (see, for example, abstracts for symposium on Foreland Fold and Thrust belt of the North America Cordillera, 24th Annual Meeting, Calgary, Alberta, Geol. Soc. America, 1971). I suggest cordilleran systems are equally 'compressive' and develop as complex strain patterns on leading edges of actively driving 'continental' plates as they override zones of subduction dipping beneath them. Furthermore, I suggest that timing and intensity of compressive strain in these belts are governed by kinematics of the driving plate itself and are little affected by motions of the subducted 'oceanic' plate.

The North American Cordillera and the Opening of the North Atlantic

The North American Cordillera extends over a distance of 8000 km from the Bering Sea through Alaska, western Canada, western United States, and Mexico and attains a maximum width in western United States of nearly 2000 km. Extensions of the belt can be tracked over an additional 2000 km southeastward through Central America and eastward through the Greater and Lesser Antilles connecting the system with the Andes of western South America.

The Cordillera, no doubt, had its origins in a period extending from later Precambrian to Late Palaeozoic when a 'Pacific margin' was formed during periods of rifting away of former extensions (Burchfiel & Davis, 1972), activation of off-shore arcs, and closure of marginal seas (Silberling, in press). Most of present strain and complex plutonic-volcanic patterns, however, have developed since later Palaeozoic time. It can be safely said that the North American Cordillera is mainly a Mesozoic–Cenozoic feature developed on the leading edge of the North America plate as it rafted westward during the opening of the North Atlantic and Arctic Oceans (Coney, 1971a, 1971b, 1972).

Post-Palaeozoic geologic history in western North America evolved in three major phases. From Permian to Middle Jurassic time arc volcanism, minor plutonism, and associated arc tectonics formed a narrow belt on a 'Pacific' margin. Middle Jurassic to Early Tertiary time was the main period of Cordilleran tectogenesis. During this period most of the major plutons were emplaced, there was widespread volcanism, and tectonic telescoping occurred over a belt eventually reaching nearly 2000 km in width. Middle and Late Tertiary tectonics response has been highly differentiated with variable and complex events from region to region which contrast sharply with earlier phases. The three main periods of tectonic evolution, and even subdivisions within the periods, correlate well with events in adjacent oceans and are inescapably linked with plate motions as recorded on sea-floors (Fig. 1).

Late Palaeozoic–Middle Jurassic

Final closure of a proto-Atlantic Ocean apparently occurred in Late Carboniferous–Early Permian time as North America collided with Africa and South America to produce the southern Appalachian, Allegheny and Ouachita orogenies and south-western extensions into Mexico and Central America. Wrenching of the North America plate at initial contacts no doubt relayed through the craton to produce Wichita–Ancestral Rockies deformation.

For the next several tens of million years North America plate motion must have been reduced as plate regimes slowly reorganized. During this time the far-western margin was activated by narrow arc-trench systems and back-arc marginal seas. Arc assemblage rocks of this age can be tracked along a narrow belt through the Sierra Nevada and Klamath Mountains and Central Oregon into western Canada and Alaska (Dott, 1961; Dickinson, 1962). Deep-sea small ocean basin sediments were deposited behind this arc in central western United States. This marginal sea collapsed in Early Triassic time and the Golconda thrust emplaced deep-water sedimentary contents over shelf rocks to the east (Burchfiel & Davis, 1972; Silberling, in press). West of the marginal arc a trench subducting 'Pacific' oceanic lithosphere is inferred. The palaeogeography was perhaps not unlike present plate regimes in the western Pacific Ocean where narrow arcs fring marginal seas east of 'quiet' cratonic margins (Eardley, 1962, p. 89).

Middle Jurassic–Early Tertiary

Sometime between Late Triassic and Middle Jurassic plate regimes reorganized and Africa–South America began separation from North America–Eurasia. Rifting and diabase igneous activity in Triassic grabens between about 225 m.y. and 210 m.y. may herald the separation in eastern North America (Armstrong & Besancon, 1970). The Atlantic opened up south of the Azores, and apparently through the Caribbean, and east of the Azores into Tethys. The separation presumably started slowly, but by Middle Jurassic North America northwestward motion was under way. The response

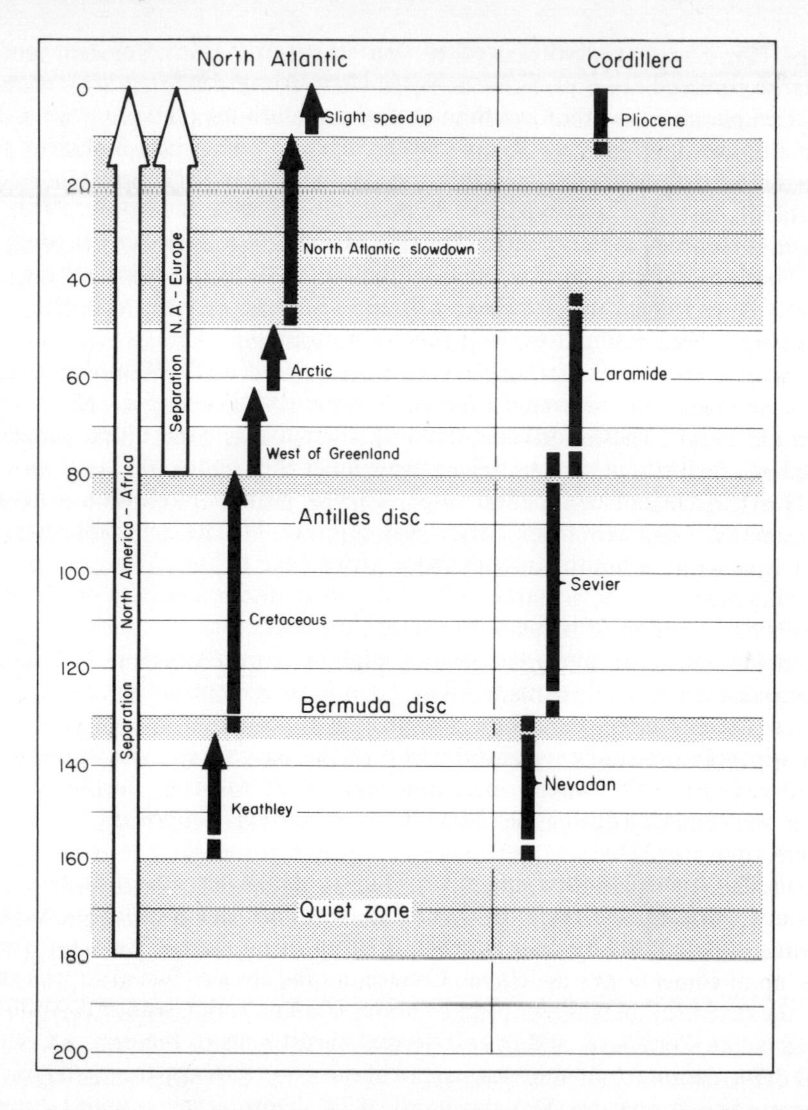

FIGURE 1 Comparison of major tectonic events in the North American Cordillera and North Atlantic Ocean. Heavy black arrows indicating major phases of spreading in the North Atlantic and Arctic Oceans after Vogt *et al.* (1971) and Pitman & Talwani (1972). Main periods of tectogenesis in the Cordillera taken from Armstrong (1968).

on the leading edge of this new North America motion was immediate, initiating the main period of tectonogenesis in the Cordillera.

Activation of the leading edge of the North America plate from Middle Jurassic time can be tracked from Mexico to northeastern Siberia. In the western Sierra Nevada and Klamath Mountains intense west-vergent deformation, metamorphism, plutonism, and emplacement of ultramafics (Nevadan orogeny) took place (Clark, 1964; Irwin, 1966; Davis, 1966; Lanphere *et al.*, 1968). Back-arc thermal pulses in a compressional stress field may have produced infrastructure and initial east-vergent thrusting which lifted the Cordilleran geanticline for the first time (Armstrong, 1968; Stanley *et al.*,

1971, p. 17). The first detrital wedges shed eastwards from this belt on to the Cordilleran foreland are Late Jurassic in age (Armstrong, 1968).

Major emplacement of the Canadian coastal batholith may have started in Middle to Late Jurassic time (Monger & Hutchison, 1971). Similarly, the first clear pulse of detritus shed eastward from the eastern crystalline belt of the Canadian Rockies is Late Jurassic, just as it is in western United States (Price & Mountjoy, 1970). Major plutonism is recorded in the Aleutian Peninsula in Middle to Late Jurassic time (Burk, 1965). The Verkhoyansk and related mountain systems in northeastern Siberia apparently were formed in Late Jurassic time as the leading edge of North America collided with Siberia (Hamilton, 1971; Churkin, in press).

All of these events in western North America correlate with the Keathley sequence of magnetic anomalies in the western North Atlantic dated at about 165 to 135 m.y. (Vogt *et al.*, 1971). These northeast-trending anomalies are the oldest sequence yet identified and lie just southeast of the magnetic quiet zone about 700 km from the base of the North American continental slope. They appear to have been generated as North America separated from Africa just prior to a plate reorganization which initiated separation of South America from Africa near 135 m.y. This reorganization in spreading near 135 m.y. is marked by the Bermuda discontinuity (Vogt *et al.*, 1971) and may mark a period of reduction in spreading rates.

Widespread activation along the leading edge of North America continued from latest Jurassic–earliest Cretaceous to about 80 m.y. In western United States Franciscan trench assemblage on the Pacific margin was deposited and nearly continuously deformed in high-pressure environment west of the Sierran arc and in the Klamath Mountains (Ernst, 1970; Suppe & Armstrong, 1972). Most of the Sierra Nevada batholith was emplaced during the same time span, strongly supporting the concept of paired metamorphic belts.

East-vergent, low-angle thrusting and folding in the Sevier belt of eastern Nevada and western Utah extends up to 1200 km east of the arc-trench complex to the west (Armstrong, 1968). The thrusting may have evolved from at least Late Jurassic time, but overlap of some thrusts by Upper Cretaceous and lowest Tertiary rocks suggests most of the deformation was complete by about 80 m.y. In the eastern crystalline belt of the Canadian Cordillera, and in east-vergent thrust belts to the east, a Columbian phase of deformation from latest Jurassic to about 80 m.y. is apparently distinct from later Laramide deformation (Douglas *et al.*, 1970). Thus, a belt of latest Jurassic to Late Cretaceous Sevier–Columbian deformation can be tracked from northwestern Canada to southwestern United States over a distance of nearly 5000 km.

Emplacement of granite in the coastal batholith of Canada apparently continued from latest Jurassic to Late Cretaceous (Hutchison, 1970). Southern extensions of the belt track into southwestern British Columbia and northwestern Washington as the Cascade Range, where a major period of Cretaceous deformation apparently terminated by about 80 m.y. (Misch, 1966; McTaggart, 1970).

At about 80 m.y. a major reorganization of plate regime took place in the North Atlantic Ocean marked by the Antilles discontinuity (Le Pichon & Fox, 1971). Rifting broke north from the Azores separating North America from Greenland and Europe in Labrador Sea (Laughton, 1971; Le Pichon *et al.*, 1971). The transition seems to mark initiation of a change in North America plate motion from northwesterly away from Africa to westerly and southwest by westerly away from Europe (Pitman &

Talwani, 1972). This reorganization is clearly signalled in the North American Cordillera by initiation of Laramide orogeny from the Greater Antilles to Alaska. After initial rifting in Labrador Sea the spreading also broke east of Greenland at about 65 m.y. and extended north into the Arctic on the Reykjanes and Nansen ridges (Pitman & Talwani, 1972).

In western United States a distinctive belt of Laramide compressive deformation extends from Montana to southern New Mexico well to the east of Nevadan–Sevier belts (Eardley, 1962, pp. 295–301; Armstrong, 1968). The deformation gripped the Cordilleran foreland in a distinct Laramide vise and extended in large part over terrain which prior to 80 m.y. had been cratonic shelf. The most intense deformation in thrust-bounded block uplifts of the Wyoming, Colorado, and New Mexico Rockies (Fig. 2) took place between 65 m.y. and about 50 m.y., but initiation of structural relief began earlier in Late Cretaceous (Berg, 1962; Baltz, 1965; Keefer, 1965). Laramide timing is suggested on the far Pacific margin as well by west-vergent thrust emplacement of Great Valley rocks over Franciscan rocks in the California Coast Ranges and Klamath Mountains (Page, 1966), and post-80 m.y. pre-Middle Eocene deformation in the Cascade Mountains (Miller & Misch, 1963).

In western Canada Laramide thrusting forms a narrow belt along the eastern part of the Cordillera from southern British Columbia and western Alberta north to Alaska (Douglas *et al.*, 1970). Plutonism was widespread, particularly in the coastal batholith. In southern Alaska the Aleutian Peninsula–Alaska Range plutonic arc was reactivated near 80 m.y. and south-vergent telescoping of arc-trench gap and trench rocks occurred around the Gulf of Alaska and Bering Sea (Plafker, 1960; Reed & Lanphere, 1969; Moore, 1971).

In the Sierra Madre Oriental of Mexico, quiet shelf conditions in the Mexican geosyncline terminated near 80 m.y. There followed deposition of flysch wedges, scattered plutonism and widespread compressional thin-skinned, northeast-vergent folding and thrusting (de Cserna, 1971). Laramide deformation can be tracked into Central America and the Caribbean. The Greater Antilles appear to have been a Laramide arc-trench system subducting Atlantic sea-floor from the northeast from Late Cretaceous to about 40 m.y. (Butterlin, 1956).

The most astonishing fact about the Laramide orogeny (Fig. 2) is that it started at 80 m.y. and virtually ceased between about 50 m.y. and 40 m.y. from Alaska to the Caribbean throughout the entire Cordillera. Overlapping continental sediments placing upper brackets on folding and thrusting are well documented in southern British Columbia, the Cascade Mountains, Klamath Mountains, in the California Coast Ranges, throughout the Rockies of western United States, in Mexico, Central America, and the Caribbean, (Coney, 1972). Furthermore, major plutonism in the Canadian coastal batholith ceased near 40 m.y. This significant tectonic fact correlates with a reduction in spreading rates in the North Atlantic after about 50 m.y. (Vogt *et al.*, 1970; Pitman & Talwani, 1972; Williams & McKenzie, 1971). This presumably caused a slowing of North America plate motion and quieted tectonism along the leading edge in the Cordillera. Another observation is that in many places throughout the Cordillera the most intense activity of Laramide deformation initiated near 65 m.y. This is apparently the time when spreading broke north into the Arctic east of Greenland, placing a spreading ridge along the entire northeastern and eastern margin of the North America plate from which it drove southwesterly about a pole in northeastern Siberia.

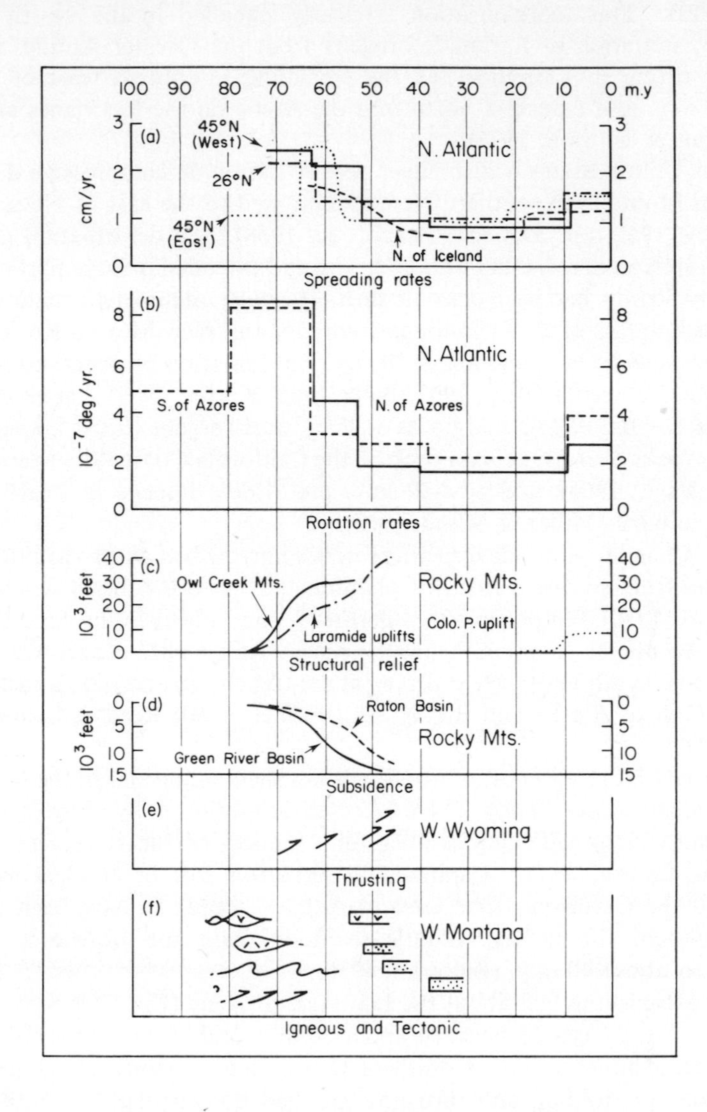

FIGURE 2 Correlation of tectonic timing during Laramide tectogenesis in western North America with spreading rates in the North Atlantic and Arctic Oceans. (a) Spreading rates in the North Atlantic and Arctic Oceans. 45° N (west) = spreading half-rates west of spreading ridge (Pitman & Talwani, 1972). 45° N (east) = spreading half-rates east of spreading ridge (Williams & McKenzie, 1971). 26° N = spreading half-rates west of ridge (Pitman & Talwani, 1972). North of Iceland spreading rates from Vogt *et al.* (1970). (b) Rotation rates in 10^{-7}: degrees/year north and south of the Azores from Pitman & Talwani (1972). (c) Cumulative structural relief produced during Laramide tectogenesis as a function of time in thousands of feet. Solid curve is relief produced between Laramide Owl Creek thrust uplift and Wind River basin (Keefer, 1965) in Wyoming. Dash-dot curve composite summary of Laramide thrust-bound uplifts in Wyoming and Colorado (Berg, 1962). Uplift of southern margin of Colorado plateau after McKee & Anderson (1971). (d) Cumulative subsidence of Laramide basins. Green River basin after Berg (1963); Raton basin after Baltz (1965). (e) Times of thrusting in Laramide Wyoming overthrust belt after Armstrong & Oriel (1965). (f) Times of volcanism (v pattern), plutonism (dash pattern), folding (wave pattern), and thrusting (barbs) in western Montana near Boulder batholith after Robinson *et al.* (1968). Dotted patterns are selected upper brackets from overlapping sediments.

Furthermore, much of the most intense activity during Laramide orogeny seems to have taken place when spreading rates, and presumably North America plate motion, were most rapid. Vogt *et al.* (1971) have shown that basement roughness in the North Atlantic was influenced by spreading rates in that roughness correlates with a period of slower spreading between Early Eocene (about 50 m.y.) and Middle Miocene (about 18 m.y.). This suggests there may be an inverse 'roughness' relationship on leading edges in that tectonic compressional deformation and uplift correlate with faster spreading rates just as roughness of basement topography on trailing edges correlates with slower rates. The threshold in each case may be between 2 and 4 times 10^{-7} degrees/year angular rotation rate.

Middle Tertiary to Present

After 40 to 50 m.y. North America moved slowly southwest by west toward the East Pacific rise. Eventual contact between the Cordillera and the rise along two growing transforms in Middle Tertiary time explain much of post-Laramide tectonics in western North America (Atwater, 1970). Northwesterly motion of the Pacific plate in contact with a much more slowly westward moving North America plate has produced right-slip and associated compressive tectonics in California along the San Andreas transform. A vast outburst of ignimbritic volcanism between about 40 and 20 m.y. from Idaho south through the Great Basin and Rocky Mountains and into western Mexico (Lipman *et al.*, 1971) apparently resulted from back-arc diapirs off dying subducted slabs overrun during rapid North America plate motion from 80 m.y. to 50 m.y. and slower North America plate motion after 50 m.y. Once the ridge was extinguished and transform faulting began between 30 m.y. and 20 m.y. more rapid Pacific plate motion apparently began taking part of North America with it over a broad transform zone between the Rio Grande rift in New Mexico and the Pacific northwest to produce Basin and Range faulting and associated basaltic volcanism (Christiansen & Lipman, in press), perhaps helped by one or more 'hot spots' such as what may be below Yellowstone (Morgan, 1971). Continued slow subduction of the Farallon plate has produced arc volcanics and tectonics in the Cascade Mountains. The Pacific plate transforms northwestward along the Queen Charlotte fault off western Canada and is subducted beneath the Aleutian–Alaska Range and Gulf of Alaska provinces to produce obvious subduction and arc tectonics there (Plafker, 1960; Stoneley, 1967). Generally westward motion of both North and South America has been slightly convergent past a near stationary Caribbean plate and explains most tectonic features on the margins of the Caribbean plate, such as oblique right-slip in northern Venezuela (Bell, in press), compressive left-slip and possible oroclining in Chiapas and Guatemala, volcano-tectonic features on the western margin of the Caribbean plate in Central America (Dengo *et al.*, 1970), disruption of the Greater Antilles and Central America along the Bartlett trough–Motagua fault zones (Hess & Maxwell, 1953), Lesser Antilles post-Early Tertiary arc volcanism and tectonics (Butterlin, 1956), and present Caribbean seismicity (Molnar & Sykes, 1969).

Since Middle Miocene time the North American Cordillera has experienced a distinct and most puzzling massive uplift, rejuvenation of drainage, and exhumation of structural relief, particularly in eastern portions. Uplift of the southern margin of the Colorado Plateau, for example, is well dated between 10 and 6 m.y. (McKee & Anderson, 1971). Post-Miocene intensification of tectonics in California is also recorded

(Crowell, 1971). How much of this Pliocene 'flare-up' in tectonism in the west is due to the slight speed-up in North Atlantic spreading rates and how much is due to changes in 'Pacific' plate regimes is as yet indeterminate. Possibly both are at work, Atlantic events controlling uplift and rejuvenation in eastern portions and Pacific events controlling western rejuvenation. This Late Tertiary tectonic rejuvenation is apparently global and relates, no doubt, to widespread changes in sea-floor spreading (Dott, 1969). Similarly, the 80 m.y. and 50 m.y. changes in sea-floor spreading are important in other oceans (Vine, this volume). This is to be expected since at any given time the global pattern of spreading and subduction should sum to zero.

Conclusions

I have proposed (Fig. 3) (Coney, 1971a, 1971b, 1972) that northwesterly motion of the North America plate away from Africa between 180 and 80 m.y. brackets Nevadan and Sevier–Columbia deformation and more southwesterly to westerly motion of North America away from Europe between 80 and 40 m.y. brackets Laramide deformation. A slowing of North America plate motion after 50 m.y. subdued Laramide deformation. Initiation of rapid separation between 180 m.y. and 165 m.y. is well recorded in Nevadan orogeny. Sevier–Columbian deformation may correlate with a period of spreading after the plate reorganization which initiated separation of South America from Africa after 135 m.y. The well marked tectonic transition near 80 m.y. which ended Sevier–Columbian phases correlates with initiation of separation of North America from Europe. The most intense periods of Laramide deformation between 80 m.y. and 40 m.y. apparently correlate with faster spreading rates in the North Atlantic and Arctic Oceans. The near extinction of Laramide deformation seems to correlate with a slowing of spreading rates in the North Atlantic (Fig. 1).

Using the Azores 'hot spot' as a reference frame for tracking North America plate motion, I assume the Azores hot spot produced the White Mountain magma series in New England after 180 m.y. and generated the Kelvin seamount chain as North America plate drifted northwesterly over the hot spot between 180 and 80 m.y. (Morgan, 1971; Coney, 1971a, 1971b, 1972). North America (Fig. 4(b)) must have driven up to about 1800 km northwestward. Arc-trench assemblage rocks are well known throughout the Cordillera in this age bracket and it is inescapable that a series of east-dipping subduction zones existed on the leading edge of North America throughout the entire motion. Between 80 m.y. and 40 m.y. North America plate drove another 1400 km southwest and west (Fig. 4(c)). Again, east-dipping subduction rimmed the leading edge. In other words, North America plate overrode up to 3200 km of 'Pacific' sea-floor by virtue of its own motion alone. It also, obviously, overrode whatever excess was generated off the East Pacific rise or other 'Pacific' spreading centres. Furthermore, up to 8 times 10^6 km^3 of sediment eroded off the Cordillera since 180 m.y. and deposited on Pacific sea-floor to the west is missing. North America apparently overrode this mass as well which probably accounts for the enormous volume of plutonic and volcanic products which so characterize the North American Cordillera (Gilluly, 1969).

The assumed westward motion of North America plate over east-dipping subduction zones between 165 m.y. and Middle Tertiary presents a paradox since the implication is that the trench on the leading edge must have moved, or sequentially

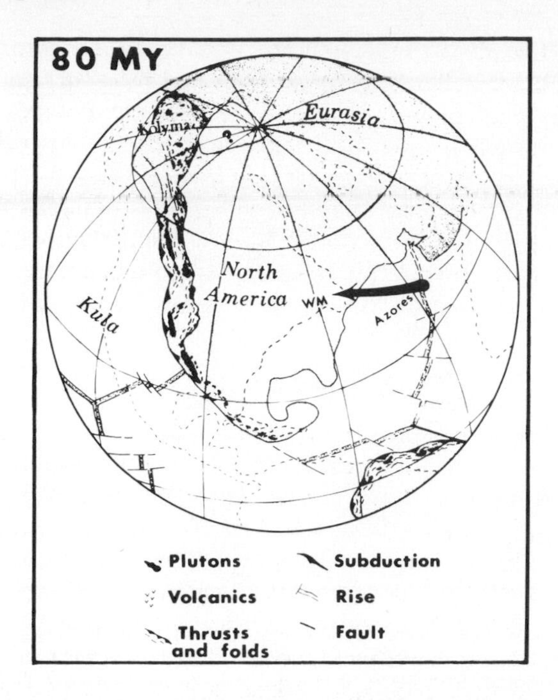

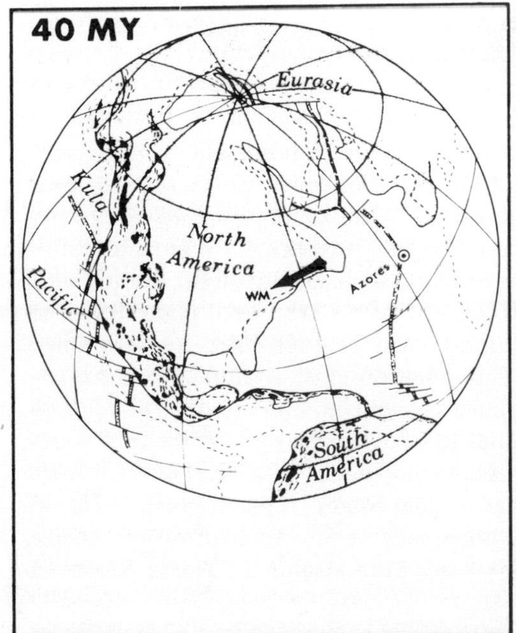

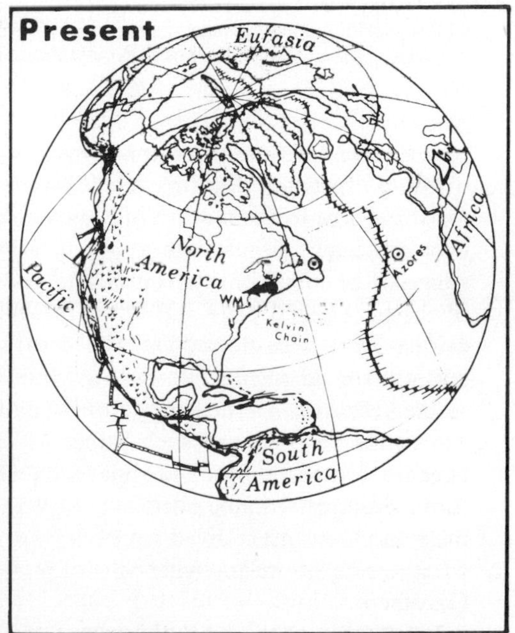

FIGURE 3 Plate configurations at 80 m.y. (a), configurations at 40 m.y. (b), and configurations at present (c). Light dashed line in (a) and (b) is present position of continental outlines. Heavy black arrow in each figure is assumed absolute motion vector for North America plate from 180 to 80 m.y. (a), 80 to 40 m.y. (b), 40 to 0 m.y. (c). Position of White Mountain magma series (WM) is at tip of arrow in each figure and is assumed to have lain over the Azores at 180 m.y. (Morgan, 1971). Tectonic features are generalized and represent structures that developed during Nevadan–Sevier–Columbian deformation (a), 80 to 40 m.y. during Laramide deformation (b), and 40 m.y. to present (c).

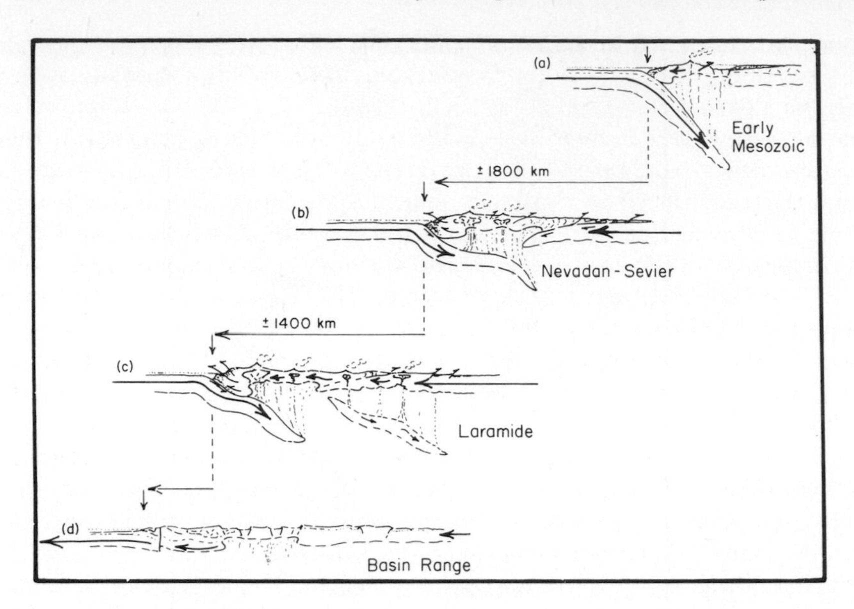

FIGURE 4 Diagrammatic lithosphere plate interactions in western North America during Late Palaeozoic–Early Mesozoic (a), during Nevadan–Sevier tectogenesis from Late Jurassic to 80 m.y. (b), during Laramide tectogenesis from 80 to 40 m.y. (c), and during Basin and Range deformation from 40 to 0 m.y. (d). Arrow in (b) and (c) indicates amount of absolute motion of North America plate while overriding Pacific plates during both phases of deformation. Detached lithosphere slab in (c) after proposed double Benioff Zone of Lipman *et al.* (1971).

jumped, west with that motion. This process must involve bending and displacement of the down-going 'Pacific' slab (Elsasser, 1971, p. 1106) which would seem to be a resistive force to the motion of the overriding plate. The lack of observable deformation in oceanic lithosphere about to descend beneath advancing continental plates suggests the effects of this 'retrograde' resistive action are taken up on the leading edge of the continental plate and in the down-going part of the subducted plate. It is also evident that once descended, the oceanic plate must be overridden and entrained beneath the advancing continental plate. This situation must eventually become unstable, resulting in a shearing off of the entrained slab and initiation of new subduction. Thus may be explained such things as initial low-angle dips seen on Benioff zones beneath South America and apparently double Benioff zones (Fig. 4(c)) inferred from Early Tertiary volcanic chemistry in western United States (Lipman *et al.*, 1971). In these mechanisms, coupled with back-arc diapirs, may lie the explanation for massive compressive failure and wide belts of plutonism and volcanism in the North American Cordillera.

It has been proposed by Morgan (1971, in press) that mantle convection produces 'hot spots' such as the Azores, Iceland, and Hawaii, and drives the plates. Vogt *et al.* (1971) have suggested that amounts of volcanic discharge from 'hot spots' vary directly with spreading rates. There is perhaps no reason why mantle flow should want to drive all plates at similar rates at any given time, but if we assume, other things being equal, that plates *would* move at similar rates, then the 'retrograde' resisted motion of continental plates overriding oceanic plates might be slower than unresisted motion of oceanic plates moving freely into subduction zones (Elsasser, 1971). Certainly rates for North and South America (1 to 4 cm/year) are up to 50% less than rates given, the Pacific plate (4 to 8 cm/year).

During both Sevier–Columbian and Laramide deformation detachment fold and thrust belts developed on the Cordilleran foreland. They were obviously controlled by favourable stratigraphy and most belts seem to have formed in miogeoclinal terrane just east of what had been the shelf-break and the edge of pre-Mesozoic continental crust. In the eastern Canadian Cordillera an undeformed basement descends westward beneath the belt of intense superficial folding and thrust faulting until lost in the metamorphic core complex to the west. Up to 200 km of superficial shortening is reported. Although the folds and thrusts are often interpreted as due to gravity sliding, the strong east-vergence and asymmetry argue against this hypothesis, and the 200 km of missing undeformed basement cannot be accounted for (Price & Mountjoy, 1970, 1971). These relationships suggest widespread compressive failure and telescoping in such belts are due to massive westward cratonic underthrusting. This amounts to an abortive, secondary, intra-orogen subduction zone. It perhaps resulted from back-arc diapiric thinning of lithosphere in a compressional stress field as a westward driving North America plate overrode east-dipping resistive subduction zones on the leading edge.

During Laramide orogeny (Fig. 2) in western United States the large thrust-bounded block uplifts and basement cored folds in the Rocky Mountains developed at about the same time as Laramide low-angle thrusting to the north in Canada and to the west in Wyoming. These classic structures were apparently produced by compressive failure of a stable cratonic shelf lacking stratigraphy or position conducive to thin-skinned detachment styles. They clearly developed during a period when North America plate motion was accelerated in a southwesterly direction (Fig. 3(b)). It seems inescapable that plate motion, particularly resisted plate motion, somehow transmits compressive stress through large widths of lithosphere and verifies what many structural geologists have argued for years, that tectogenic folds and thrusts are due to regional tangential compressive stress at deep crustal levels.

Had the North America plate, with oceanic lithosphere on its leading edge, moved westward into west-dipping subduction zones on the margin of Asia no tectogenesis would have resulted until a short-lived Himalaya-type collision terminated motion. Had North America plate, with oceanic lithosphere on its leading edge, moved westward into intra-oceanic 'Pacific' subduction zones no tectogenesis would have resulted until North America continental crust entered the subduction. The arc would then flip to east-dipping and drive everything westward ahead of it such as is apparently happening in New Guinea (Hamilton, this vol.). Had North America remained stationary (or moved slowly eastward) and consumed eastward moving Pacific sea-floor in east-dipping subduction zones the Cordillera would have been narrow, with off-shore arcs and marginal seas generated by back-arc spreading (Karig, 1971) resembling present day western Pacific arc-trench systems. A fourth alternative, that of west-dipping intra-oceanic arcs travelling eastward across the Pacific on the leading edge of 'Pacific' plate motion eventually colliding with North America to produce tectogenesis is here proposed impossible. It would seem that back-arc diapirs and thermal softening of the back-arc area would quickly cause the system to fail under active driving plate stress and flip the arc to east-dipping. In any event, no clear example of a leading-edge intra-oceanic arc facing in, and dipping opposite to, the direction of motion is obviously identifiable in present plate regimes.

Thus, widespread compressive failure, complex and wide plutonic patterns, and a prolonged history that characterize Cordilleran tectogenesis were dictated by variable rates and directions of an actively driving North America plate as it overrode a series of east-dipping subduction zones while rafting westward during the opening of the North Atlantic and Arctic Oceans. The Cordillera is the battered leading edge of these motions and the trailing edge magnetic anomalies in the North Atlantic and Arctic Oceans the passive tape recording of these motions. Timing and intensity of compressive tectonics in the Cordillera are inescapably linked to changes in these motions. Thus, compressive folding and thrusting, so characteristic of the world's mountain belts, are directly linked to the kinematics and dynamics of the unifying hypothesis of 'continental drift', sea-floor spreading, and plate tectonics.

References

Armstrong, F. C. and Oriel, S. S., 1965. Tectonic development of the Idaho–Wyoming thrust belt, *Am. Assoc. Petroleum Geol. Bull.*, **49**, 1847–66.

Armstrong, R. L., 1968. Sevier orogenic belt in Nevada and Utah, *Geol. Soc. Amer. Bull.*, **79**, 429–58.

Armstrong, R. L. and Besancon, J., 1970. A Triassic time scale dilemma: K-Ar dating of Upper Triassic mafic igneous rocks, eastern U.S.A. and Canada and post-Upper Triassic plutons, Western Idaho, U.S.A., *Eclogae geol. Helv.*, **63**, 15–28.

Atwater, Tanya, 1970. Implications of plate tectonics for the Cenozoic tectonic evolution of western North America, *Geol. Soc. Amer. Bull.*, **81**, 3513–36.

Baltz, E. H., 1965. Stratigraphy and history of Raton Basin and notes on San Luis Basin, Colorado–New Mexico, *Am. Assoc. Petroleum Geol. Bull.*, **49**, 2041–75.

Bell, J. S., Global tectonics in the southern Caribbean area (in press). Shagem, ed., *Geol. Soc. America Mem.* (Hess Vol.)

Berg, R. R., 1962. Mountain flank thrusting in Rocky Mountain foreland, Wyoming and Colorado, *Am. Assoc. Petroleum Geol. Bull.*, **46**, 2019–32.

Berg, R. R., 1963. Laramide sediments along Wind River thrust, Wyoming. *In:* Backbone of the Americas, edited by O. E. Childs and B. W. Beebe. *Amer. Assoc. Petroleum Geol., Mem.*, **2**, 220–30.

Burchfiel, B. C. and Davis, G. A., 1972. Structural framework and evolution of the southern part of the Cordilleran orogen, western United States, *Amer. J. Sci.* (in press).

Burk, C. A., 1965. Geology of the Alaska Peninsula–Island Arc and continental margin, *Geol. Soc. Amer. Mem.*, **99**, 250 pp.

Butterlin, J., 1956. La constitution geologique et la structure des Antilles. Centre Nat. de la Res. Scientifique, Paris, 453 pp.

Christiansen, R. L. and Lipman, P. W. Cenozoic volcanism and plate tectonic evolution of the western United States, pt. 2 (in press).

Churkin, M., Western boundary of the North American continental plate in Asia, *Geol. Soc. America Bull.* (in press).

Clark, Lorin, 1964. Stratigraphy and structure of part of the western Sierra Nevada metamorphic belt, California. U.S. Geol. Survey Prof. Paper 410, 70 pp.

Coney, P. J., 1970. The geotectonic cycle and the new global tectonics, *Geol. Soc. Amer. Bull.*, **81**, 739–48.

Coney, P. J., 1971a. Cordilleran tectonic transitions and motion of the North American plate, *Nature*, **233**, 462–5.

Coney, P. J., 1971b. Cordilleran tectonic transitions and North America plate motions, *Geol. Soc. Amer., Abs. with Prog.*, **3** (7), 529.

Coney, P. J., 1972. Cordilleran tectonics and North America plate motion, *Amer. J. Sci.* 272, 603–28.

Crowell, J. C., 1971. Tectonic problems of the Transverse Ranges, California, *Geol. Soc. Amer. Abs. with Prog.*, **3** (2), 106.

Davis, G. A., 1966. Metamorphic and granitic history of the Klamath Mountains. *In:* Geology of Northern California, edited by E. K. Bailey. *Calif. Dept. Nat. Res., Div. Mines Bull.*, **190**, 39–50.

de Cserna, Z., 1971. Development and structure of the Sierra Madre Oriental of Mexico *Geol. Soc. Amer., Abs. with Prog.*, **3** (6), 377–8.

Dengo, G., Bohnenberger, O. and Bonis, S., 1970. Tectonic and volcanism along the Pacific marginal zone of Central America, *Geol. Rundschau*, **59**, 1215–32.

Dewey, J. F. and Bird, J. M., 1970. Mountain belts and the new global tectonics, *J. Geophys. Res.* **75**, 2625–47.

Dickinson, W. R., 1962. Petrogenetic significance of geosynclinal andesitic volcanism along the Pacific margin of North America, *Geol. Soc. Amer. Bull.*, **73**, 1241–56.

Dickinson, W. R., 1970. Relations of andesitic volcanic chains, granitic batholith belts, and derivative graywacke-arkose facies to the tectonic framework of arc-trench systems, *Rev. Geophys. Space Phys.*, **8**, 813–60.

Dickinson, W. R., 1971a. Plate tectonic models of geosynclines, *Earth Planet. Sci. Lett.*, **10**, 165–74.

Dickinson, W. R., 1971b. Plate tectonics in geologic history, *Science*, **174**, 107–13.

Dott, R. H., 1961. Permo-Triassic diastrophism in the western Cordilleran region, *Am. Jour. Sci.*, **259**, 561–82.

Dott, R. H., 1969. Circum-Pacific Late Cenozoic structural rejuvenation: Implications for sea-floor spreading, *Science*, **166**, 874–6.

Douglas, R. F. W., Gabnelse, H., Wheeler, J. O., Stott, B. F. and Belyea, H. R., 1970. Geology of Western Canada. *In:* Geology and Economic Minerals of Canada, edited by R. J. W. Douglas. Geol. Surv. Canada Econ. Geol. Report no. 1, pp. 367–488.

Eardley, A. J., 1962. Structural Geology of North America (2nd ed.). Harper and Row, New York, 743 pp.

Elsasser, W. M., 1971. Sea-floor spreading as thermal convection, *J. Geophys. Res.*, **76**, 1101–12.

Ernst, W. G., 1970. Tectonic contact between the Franciscan melange and the Great Valley sequence—crustal expression of a Late Mesozoic Benioff zone, *J. Geophys. Res.*, **78**, 886–902.

Gansser, A., 1966. The Indian Ocean and the Himalayas—a geological interpretation, *Eclogae geol. Helv.*, **59** (2), 831–48.

Gilluly, J., 1969. Oceanic sediment volumes and continental drift, *Science*, **166**, 992–3.

Hamilton, W., 1969a. The volcanic central Andes—a modern model for the Cretaceous batholiths and tectonics of western North America. Oregon Dept. Geology and Mineral Industries, Bull. 65, pp. 175–84.

Hamilton, W., 1969b. Mesozoic California and the underflow of Pacific mantle, *Geol. Soc. Amer. Bull.*, **80**, 2409–30.

Hamilton, W., 1971. Continental drift in Arctic: Program with Abstracts. 2nd Int. Sym. on Arctic Geol., Am. Assoc. Petroleum Geologists, San Francisco, Calif., F.S 1–4, **71**, p. 24.

Hamilton, W., 1972. Tectonics of the Indonesian region. This volume.

Hess, H. H. and Maxwell, J. C., 1963. Caribbean research project, *Geol. Soc. Amer. Bull.*, **65**, 1–6.

Hutchison, W. W., 1970. Metamorphic framework and plutonic styles in the Prince Rupert region of the central Coast Mountains, B.C., *Can. J. Earth Sci.*, **7**, 376–405.

Irwin, W. D., 1966. Geology of the Klamath Mountains province. *In:* Geology of Northern California, edited by E. K. Bailey. *Calif. Dept. Nat. Resources, Div. Mines Bull.*, **190**, 19–37.

Karig, D. E., 1971. Origin and development of marginal basins in the western Pacific, *J. Geophys. Res.*, **76**, 2542–61.

Keefer, W. R., 1965. Stratigraphy and geologic history of the uppermost Cretaceous, Palaeocene and lower Eocene rocks in the Wind River Basin, Wyoming. U.S. Geol. Surv. Prof. Paper **495A**, 1–77.

Lanphere, M. A., Irwin, W. P. and Holtz, P. E., 1968. Isotopic age of the Nevadan orogeny and older plutonic and metamorphic events in the Klamath Mountains, Calif., *Geol. Soc. Amer. Bull.*, **79**, 1027–52.

Laughton, A. S., 1971. South Labrador Sea and the evolution of the North Atlantic, *Nature*, **232**, 612–17.

Le Pichon, X., Hyndman, R. D. and Pautot, G., 1971. Geophysical study of the opening of the Labrador Sea, *J. Geophys. Res.*, **76**, 4724–43.

Le Pichon, X. and Fox, P. J., 1971. Marginal offsets, fracture zones, and the early opening of the South Atlantic, *J. Geophys. Res.*, **76**, 6294–6308.

Lipman, P. W., Prostka, H. J. and Christiansen, R. L., 1971. Evolving subduction zones in the western United States, as interpreted from igneous rocks, *Science*, **174**, 821–5.

McKee, E. H. and Anderson, C. A., 1971. Age and chemistry of Tertiary volcanic rocks in north central Arizona and relation of the rocks to the Colorado Plateaus, *Geol. Soc. Amer. Bull.*, **82**, 2767–82.

McTaggart, K. C., 1970. Tectonic history of the northern Cascade Mountains. *In:* Structure of the Southern Canadian Cordillera. Canada Geol. Assoc. Spec. Paper 6, pp. 137–48.

Miller, G. M. and Misch, P., 1963. Early Eocene angular unconformity at western front of northern Cascades, Whatcom County, Washington, *Am. Assoc. Petroleum Geol. Bull.*, **47**, 163–74.

Misch, P., 1966. Tectonic evolution of the Northern Cascades of Washington State—A west-cordilleran case history. *In:* A Symposium on the Tectonic History and Mineral Deposits of the Western Cordillera, Vancouver, B.C., 1964. *Canadian Inst. Mining and Metallurgy Spec.*, **8**, 101–48.

Miyashiro, A., 1972. Metamorphism and related magmatism in plate tectonics, *Amer. J. Sci.* **272**, 629–56.

Molnar, P. and Sykes, L. R., 1969. Tectonics of the Caribbean and Middle America regions from focal mechanisms and seismicity, *Geol. Soc. Amer. Bull.*, **80**, 1639–84.

Monger, J. W. H. and Hutchison, W. W., 1971. Metamorphic map of the Canadian Cordillera, *Geol. Surv. Can.*, Paper 70-33, 61 pp.

Moore, J. C., 1971. Structural evolution of the Cretaceous pre-Aleutian Trench, Alaska *Geol. Soc. Amer.*, *Abs. with Prog*, **3** (7), 650–51.

Morgan, W. J., 1971. Convection plumes in the lower mantle, *Nature*, **230**, 42–43.

Morgan, W. J., 1972. Plate motions and deep mantle convection: Shagem, R., (ed.), *Geol. Soc. Amer. Mem.* (Hess mem. v.), in press.

Page, B. M., 1966. Geology of the Coast Ranges of California. *In:* Geology of Northern California, edited by E. K. Bailey. *Calif. Dept. Nat. Resources, Div. Mines Bull.*, **190,** 255–75.

Pitman, W. C. III and Talwani, M., 1972. Sea-floor spreading in the North Atlantic, *Geol. Soc. Amer. Bull.* (in press).

Plafker, G., 1960. Tectonics of the March 27, 1964, Alaska Earthquake. U.S. Geol. Survey Prof. Paper 543–I, 74 pp.

Price, R. A. and Mountjoy, E. W., 1970. Geologic Structure of the Canadian Rocky Mountains between Bow and Athabasca Rivers—A progress report. Geol. Assoc. Canada Spec. Paper 6, pp. 7–25.

Price, R. A. and Mountjoy, E. W., 1971. The Cordilleran foreland thrust and fold belt in the southern Canadian Rockies (abs.), *Geol. Soc. Amer.*, *Abs. with Prog.*, **3** (6), 404–5.

Reed, B. L. and Lanphere, M. A., 1969. Age and chemistry of Mesozoic and Tertiary plutonic rocks in south-central Alaska, *Geol. Soc. Amer. Bull.*, **80**, 23–43.

Robinson, G. D., Klappin, M. R. and Obradovich, J. D., 1968. Overlapping plutonism, volcanism, and tectonism in the Boulder Batholith region, western Montana. *In:* Studies in Volcanology, edited by R. R. Coats, R. L. Hay and C. A. Anderson. *Geol. Soc. Amer. Mem.*, **116**, 557–76.

Silberling, N. J. (in press). Geologic events during Permo-Triassic time along the Pacific margin of the United States.

Stanley, K. O., Jordan, W. M. and Dott, R. H., 1971. Early Jurassic Petrogeography, western United States, *Amer. Assoc. Petroleum Geol. Bull.*, **55**, 10–19.

Stoneley, R., 1967. The structural development of the Gulf of Alaska sedimentary provinces in southern Alaska, *Quart. J. of the Geol. Soc., London*, **123**, 25–57.

Suppe, J. and Armstrong, R. C., 1972. Potassium-argon dating of Franciscan metamorphic rocks, *Amer. J. Sci.*, **272**, 217–33.

Vine, F. J., 1972. Continental fragmentation and ocean floor evolution during the past 200 m.y. This volume.

Vogt, P. R., Ostenso, N. A. and Johnson G. L., 1970. Magnetic and bathymetric data bearing on sea-floor spreading north of Iceland, *J. Geophys. Res.*, **75**, 903–20.

Vogt, P. R., Johnson, G. L., Holcombe, T. L., Gilg, J. G. and Avery, O. E., 1971. Episodes of sea-floor spreading recorded by the North Atlantic basement, *Tectonophysics*, **12**, 211–34.

Vogt, P. R., Anderson, C. N. and Bracey, D. R., 1971. Mesozoic magnetic anomalies, sea-floor spreading, and geomagnetic reversals in the southwestern North Atlantic, *J. Geophys. Res.*, **76**, 4796–4823.

Williams, C. A. and McKenzie, D., 1971. The evolution of the northeast Atlantic, *Nature*, **232**, 168–73.

7

RIFTS AND OCEANS

7.1

H. FAURE

Laboratoire de Géologie Dynamique
Université de Paris VI
Paris, France

Vertical Movements and Horizontal Translation of the Lithosphere

---

In the African shield, positive vertical movements have been clearly shown by palaeogeographic, geomorphologic and palaeopedological studies. They affect present swells with a rate of up-lift of about one or two centimetres a century for some ten million years. Uplifts can also be proved during older periods when they affected not only present swells but also regions that are now basins. Negative vertical movements can be measured in recent and ancient large basins by the rates of accumulation of sediments, during a known interval of time. Results are of the same order of magnitude or smaller than those of positive movements. They have affected regions in the past that can now undergo positive deformation. Negative and positive deformations can be schematized as a kind of undulation of the crust through time (Amplitude = 10^3-10^4 m, wave length = $10^5-2\times10^6$ m, period = 10^7-10^8 yrs.)

Horizontal relative translation between lithosphere and asthenosphere can be a possible explanation of these vertical movements. The propagation of an internal wave

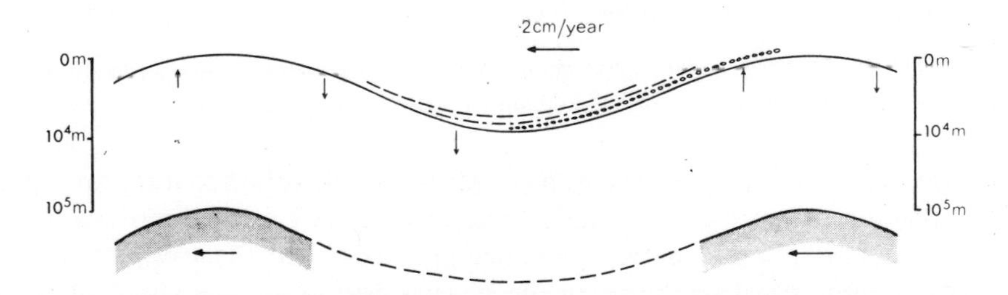

FIGURE 1 Diagram showing how the asymmetry of certain sedimentary basins, caused by vertical movements of the crust, can be interpreted as the result of a horizontal relative propagation or displacement between the lithosphere and the asthenosphere or upper-mantle.

The different discontinuous lines are successive sedimentary layers deposited in the bottom of the migrating basin (from right to left). The layers are progressively uplifted (on the right) by the rising swell. The disposition of sedimentary layers shows a transgression to the left and a regression on the right.

Black spots, on the lower part, indicate bumps on the asthenosphere, the relative displacement of which (horizontal lower black arrows) involves the horizontal migration of the basin and swells (horizontal top arrow) and the wave-like vertical deformation of the crust.

The horizontal scale is of about 2000 km. The vertical scale indicated is logarithmic. The diagram is established for a duration of about 10 to 100 m.y. of the epeirogenic wave.

All the black arrows indicate the *present* direction of the displacement.

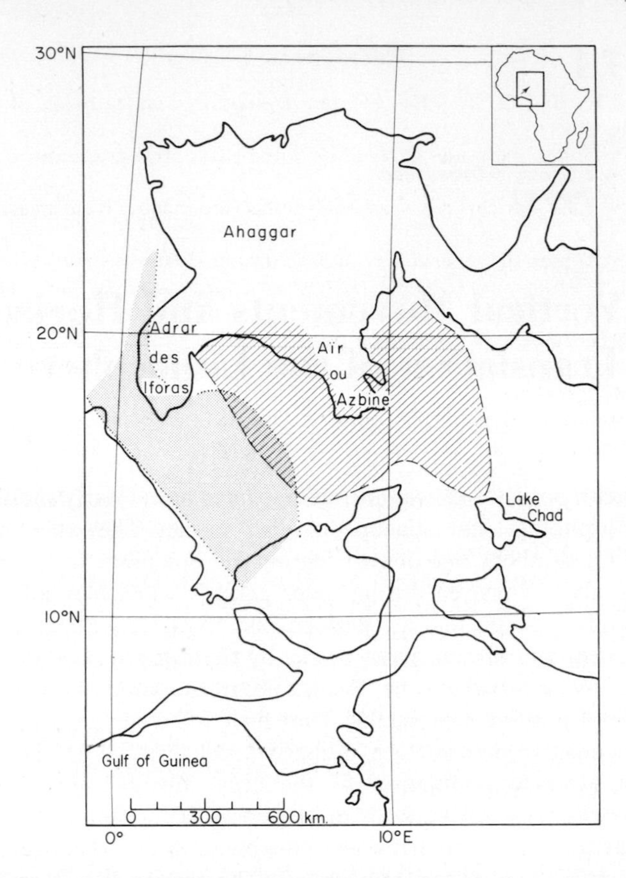

FIGURE 2 Migration of the marine basin between Upper Cenomanian and Lower Eocene in Niger. Between 100 m.y. and 60 m.y. the migration of the basin is progressive for about 700 km from the northeast to the southwest. The average horizontal displacement is 1·7 cm/year and could be interpreted as a translation (or rotation) of the African plate from southwest to northeast (arrow on the small map of Africa).

Diagonally shaded area = approximate limits of the Upper Cenomanian sea.
Dotted area = approximate limits of the Lower Eocene sea.

or the displacement of bumps in the asthenosphere as well as horizontal translation of the lithosphere over relief of the mantle could involve such vertical movements. The horizontal migration of the centre of sedimentary basins can effectively be measured using palaeogeographical methods. In the Niger–Chad basin, the speed of the displacement (1 or 2 cm/year) between Middle Cretaceous and Eocene is comparable with the rate of sea-floor spreading. The direction of migration, from northeast to southwest, is opposite to the translation of African plate during the same period.

The geology of large shields, therefore, show some possible effects that could be related to continental drift, sea-floor spreading and plate tectonics. These effects can sometimes be quantitatively evaluated and bring an independent method to study these mechanisms. The formation and localization of economical deposits is controlled by such palaeogeophysical evolutions and by their palaeogeographical consequences.

References

Brognon, G. P. and Verrier, G. R., 1966. Oil and geology in Cuanza basin of Angola. *Bull. A.A.P.G.*, **49** (11).

Busson, G., 1971. Principes, méthodes et résultats d'une étude stratigraphique du Mésozoïque saharien. Thèse, Paris.

Dixey, F., 1956. The East African rift system. Colon. Geol. Min. Resources, G. B., suppl. no. 1, 71 pp.

Fabre, J., 1969. Remarques sur la structure du Sahara occidental et central. (à propos de la Carte tectonique de l'Europe). *Bull. Soc. d'Hist. Nat. de l'Afr. du Nord*, **60,** fasc. 1 et 2, 43–73.

Faure, H., 1962. Reconnaissance géologique des formations sédimentaires post-paléozoïque du Niger oriental. Thèse, Paris, Public. no. 1, *Dir. Mines Géol. Niger et Mém. Bur. Rec. géol. min*, **47** (1966), 630 pp.

Faure, H., 1971. Relations dynamiques entre la croûte et le manteau d'après l'étude de l'évolution paléogéographique des bassins sédimentaires, *C.R. Acad. Sc. Paris*, **272,** 3239–42.

Faure, H., 1972. Paléodynamique du craton africain. *24ème Congrès géol. Internat.*, Montréal, sect. 3, 44–50.

Faure, H., Demoulin, D., Hebrard, L. and Nahon, D., 1970. Données sur la Néotectonique de l'extrême ouest de l'Afrique. Conf. on Afr. Geol. Ibadan, Déc. 11, p. 2 fig.

Furon, F., 1960. Géologie de l'Afrique. (Ed.), Payot, Paris.

Furon, R., 1965. Matériaux pour l'étude de la 'Houle crustale' et de la mégatectonique du socle africain, *Rev. Géogr. Phys. et Géol. dynam.* (1), **7**, 21–57.

Girot, M., 1968. Le Massif volcanique de l'Atakor (Hoggar, Sahara algérien). Etude pétrographique, structure et volcanologie. Thèse Sc. nat. Paris.

Glangeaud, L., 1970. La méthodologie géodynamique des ensembles bornés (NODS). Ses applications à l'évolution des grands ensembles mégamétriques terrestres. *Rev. Géogr. Phys. et Géol. dyn.* (2), **12**, fasc. 5, 465–92.

Hammond, A. L., 1970. Deep Sea Drilling: A Giant Step in Geological Research, *Science*, **170** (3957), 520–21.

Holmes, A., 1966. Principles of Physical geology, 2nd edition, fig. 763, p. 1288, Nelson.

Joulia, F., 1959. Précisions sur la discordance cambro-ordoviceinne d'In Azaouna (Niger), *C.R. Soc. géol. Fr.*, 177–8.

Klitsch, v. E., 1970. Die Strukturgeschichte der Zentralsahara, *Geol. Rundschau.*, **59**, 459–527.

Le Bas, M. J., 1971. Provinces ignées per-alcalines et alcalines, et gonflements tectoniques en Afrique. VI. Colloquium on Africa Geology, Leicester, April 1971.

Le Pichon, X., 1970. Structure et dynamique de la Lithosphère. Colloque de synthèse, Paris Hermann, p. 1–73.

Plauchut, S., 1959. Notice explicative sur la feuille Djado. B.R.G.M., Dakar.

Perrodon, A., 1969. Esquisse d'une géologie dynamique des bassins sédimentaires, *Sciences de la Terre*, **14**, 301–28.

Radier, H., 1953. Contribution à l'étude stratigraphique et structurale du détroit soudanais, *Bull. Soc. géol. Fr.*, **3**, 677–95.

Rapport de la, R. C. P. 180 (Afars) du CNRS pour l'année 1970–1971.

Vincent, P. M., 1960. Les volcans tertiaires et quaternaires du Tibesti occidental et central (Sahara du Tchad). Paris, Thèse Doctorat, 197 pp.

Vogt, J. et Black, R., 1963. Remarques sur la morphologie de l'Aïr. *Bull. B.R.G.M.*, **1**, 1–29.

7.2

K. BURKE
Erindale College
University of Toronto
Mississauga, Ontario
Canada

A. J. WHITEMAN
Department of Geology
University of Bergen
Bergen, Norway

Uplift, Rifting and the Break-up of Africa

Abstract

The history of 16 uplifts and 29 rifts of Mesozoic and Tertiary age in Africa reveals an evolutionary sequence from uplift to continental separation. Approximately 1 km high, 100 km wide, 200 km long uplifts are interpreted as isostatic responses to mass deficiencies produced by partial melting at the base of the lithosphere. Uplift has normally been followed by alkaline vulcanicity and crestal rift formation, but substantial crustal melting may stop the evolutionary process before the rifting stage. All but the youngest rifts bifurcate along their lengths forming rrr junctions (small letters indicate a distinction between these structures, produced by uplift, and RRR spreading junctions into which they may or may not evolve.)

The evolution of an rrr junction depends on how and whether motion across its rifts can be accommodated by the driving mechanism of the world-wide plate system. Of the 11 rrr junctions we recognize in Africa two are at too early a stage of development to spread. Two failed to be accommodated in the world system and their rifts became filled with sediment without spreading. One is developing through spreading on all its arms as an RRR junction. Five developed or are developing through spreading on two rifts, one rift having ceased to be active before the spreading stage. In this group the rift which fails normally trends close to east–west, permitting the poles of the new plate motion to be near the poles of the Earth's rotation axis. One junction has developed through spreading on one rift with transform motion on a second rift after pre-spreading cessation of activity on a third.

Seven of the rrr structures we consider were initiated between 180 and 130 m.y. ago and the other four in the last 25 m.y. During the intervening approximately 100 m.y. although existing structures continued to develop no new rrr junctions were established.

Introduction

Since Gondwanaland began to break up the processes of uplift, alkaline vulcanism, rifting, triple junction formation and spreading have dominated the development of Africa. The history of 56 African structures (16 swells, 29 rifts and 11 triple junctions) is outlined in terms of these processes. By comparing the Mesozoic structures of southeastern Africa and the Gulf of Guinea with the better known Neogene East African Rift System we demonstrate the existence of an evolutionary sequence from uplift to continental separation which has frequently failed before the stage of continental separation has been reached. We also distinguish divergent ways in which this sequence can develop after the stage of triple junction formation.

Within the last 200 m.y. there have been two periods during each of which the evolutionary sequence has started to develop at about the same time in different places. The events of the older phase (Jurassic–Cretaceous rifting) started between approximately 180 and 130 m.y. ago. Some of the systems put into operation by these events have failed and others are still active. Events of the younger (Neogene rifting) phase started within the last 25 m.y. and these systems are commonly still in early stages of development.

Uplift

Local uplift of 1 km or more commonly over areas of between 10^4 and 10^5 km^2 is the first event in the sequence. Although it has long been recognized as an important process in association with rifting and alkaline vulcanicity (e.g. Cloos, 1939; Shackleton, 1954; Dixey, 1965; Bailey, 1964; Black & Girod, 1970; Lebas, 1971) local uplift is difficult to identify unless it is associated with volcanic products or rifting. For this reason it may be widespread but unrecognized. Parts of the Cameroun zone in Adamawa (Fig. 1) which are over 2 km high but carry no volcanic material are believed to represent this first stage in the evolutionary sequence which may also be locally represented in the high ground along the north–south line of southern African epicentres distinguished by Fairhead & Girdler (1969).

A few African Neogene alkaline volcanic rocks do occur in low-lying areas (e.g. the Cape Verde peninsula; the Benue trough and the Afar) but these can almost all be related to known local peculiarities of crustal structure. For this reason although only indirect evidence is available of uplift as a component of the Jurassic–Cretaceous phase we regard local uplift as a general if not universal first manifestation of the sequence of events which leads to continental separation.

Uplift with Alkaline Volcanic Activity

The association in Africa of uplifts, rifts and alkaline igneous rocks has been reviewed by Lebas (1971). The volcanic rocks of uplifted areas (Fig. 1) are not essentially different from those of rift areas although they generally occur in less volume and in less compositional variety.

The petrology of the volcanic rocks of rift and uplifted areas in Africa has been extensively studied over the last 50 or more years. Review articles covering much of this work are included in *African Magmatism and Tectonics* (Gass, 1970) in which King discusses the East African Rift System; Black & Girod the Ahaggar, the Cameroun zone and the Jos plateau; Vincent the Tibesti; Woolley & Garson and Cox Malawi and areas farther south; and Gass the Afro-Arabian dome. Alkali basalts predominate in these areas, but many other alkaline and peralkaline rocks occur including some rocks of exceptional composition.

Geochemical, especially isotopic, studies indicate a mantle origin for the volcanics and associated intrusives of African rifts and uplifts with the exceptions of the Jurassic rocks of the Jos plateau (Bowden, 1970, and Bowden & van Breemen, 1972) and the Neogene rocks of Tibesti (Vincent, 1970). In both these areas there is evidence of substantial melting of the continental crust. There was no evolutionary development of the Jos structure beyond the stage of uplift with alkaline igneous (volcanic and sub-volcanic) activity although an entirely new uplift with alkaline vulcanism and a different orientation has been developing in the same area in Neogene times (Burke, in preparation). Despite a history that covers the whole of the Neogene there is as yet no evidence of rift development in Tibesti. For these reasons the evolutionary path of uplift to alkaline vulcanism with substantial partial melting of the crust is tentatively considered as ending with little or no further development.

Uplift and alkali vulcanism of Neogene age without substantial fusion of the crust is widely represented in an area of 10^7 km^2 between the Mediterranean and the Equator (Fig. 1 and Table 1). The Chad basin, filled with about 1 km of Neogene sediments, occupies a central position between the uplifted areas. Negative gravity anomalies (Fig. 2) are associated with the volcanic areas of the Jos plateau (Ajakaiye, 1970); the Cameroun zone (Burke, Dessauvagie & Whiteman, 1971, Fig. 5); Tibesti (Louis, 1970) and the Ahaggar (Tordi, 1971). Up to 1·5 km of basement uplift has been mapped beneath the basalts of the Ahaggar (Black & Girod, 1970).

The Cameroun zone (Fig. 1) is most likely to develop into a major rift system because it is a 1100 km long line of seven uplifts with intervening low areas. Annobon, Sao Tome and Principe rise from the ocean floor and Fernando Po lies on the con-

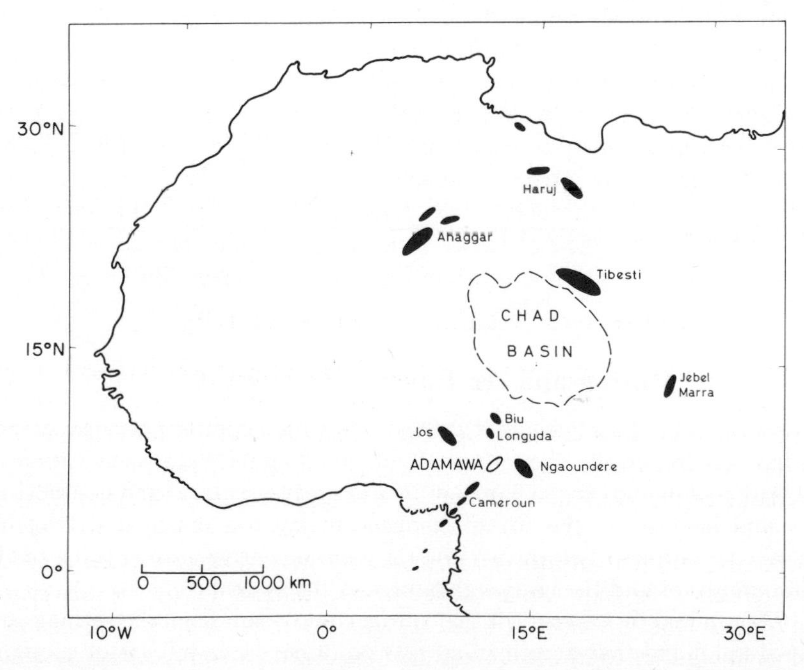

FIGURE 1 Neogene uplifts in North Africa. Black areas are uplifts with alkali basalt, the areas indicated are those of the uplifts not of basalt outcrop. The Adamawa uplift in the Cameroun zone is suggested to represent the earliest or pre-volcanic stage of uplift. Note that the uplifts, apart from those of the Cameroun zone, are tangentially disposed about the Chad basin.

TABLE 1 Physical characteristics of some African uplifts.
* indicates the influence of low density Jurassic granite

Uplift	Width (km)	Length (km)	Uplift of basement (km)	Peak gravity value (mgals)	Type of volcanic activity
Adamawa	50	150	1	−100	minor alkali basaltic
Jos (Neogene)	100	300?	1	− 90*	alkali basaltic
Atakor (Ahaggar)	60	300? } >400	1	−120	alkali basaltic
Ajjer (Ahaggar)	50	100 }	1		alkali basaltic
Bamenda (Cameroun)	120	250 } >500	1	−100	alkaline, mainly basalt
Cameroun Mt	120	250 }	1		alkaline, mainly basalt
Ngaoundere	50	150		−100	alkali basalt
Tibesti	100	300	1	− 70	alkali basalt to rhyolite
Jebel Marra	100	200	1		alkali basalt
Average Neogene Uplift	80	220	1	−100	alkali basaltic
Jos (Jurassic)	200	500?	1?		alkali, mainly acid subvolcanic

tinental slope. On the continent the Cameroun Mountain and Bamenda uplifts are separated by the Mamfe depression. The next uplift in the zone in Adamawa is taken to represent an early stage as it has an extensive 100 mgal negative Bouguer anomaly but relatively little volcanic material. 400 km farther north along the trend high ground on the Nigeria/Cameroun border near Mubi may represent an eighth uplift. Some accounts indicate that the Cameroun zone has been continuously active since the Cretaceous but there is no close link between the Cretaceous vulcanicity and the formation of the present zone of uplifts and alkaline volcanics which is an entirely Neogene and Recent feature (Hedberg, 1968).

Uplifted areas with volcanic rocks of the Jurassic–Cretaceous rifting phase cannot be directly recognized but their existence in Cretaceous times has been inferred on structural grounds for the Atlantic south of the Gulf of Guinea (Burke *et al.*, 1971; Lebas, 1971).

Rifting and rrr Junction Formation

The development of the East African Rift System has been complex because basement structures have controlled the orientation of individual faults; because vulcanicity has developed both within and on the flanks of rifts in great structural and compositional complexity and because, in the south, Neogene rifting has reactivated Cretaceous structures. King (1970) and Beloussov (1969) in comprehensive reviews bring out both the complex character and the unifying features of the system.

Cloos (1939) showed by experiment that rifting could result from the formation of a dome-shaped uplift and that experimental rifts on dome crests bifurcated along their lengths to produce the rrr (rift, rift, rift) pattern familiar in nature (Fig. 3). We use small letters to distinguish rrr junctions formed by uplift from the related RRR triple junctions formed by spreading of McKenzie & Morgan (1969) toward which they may

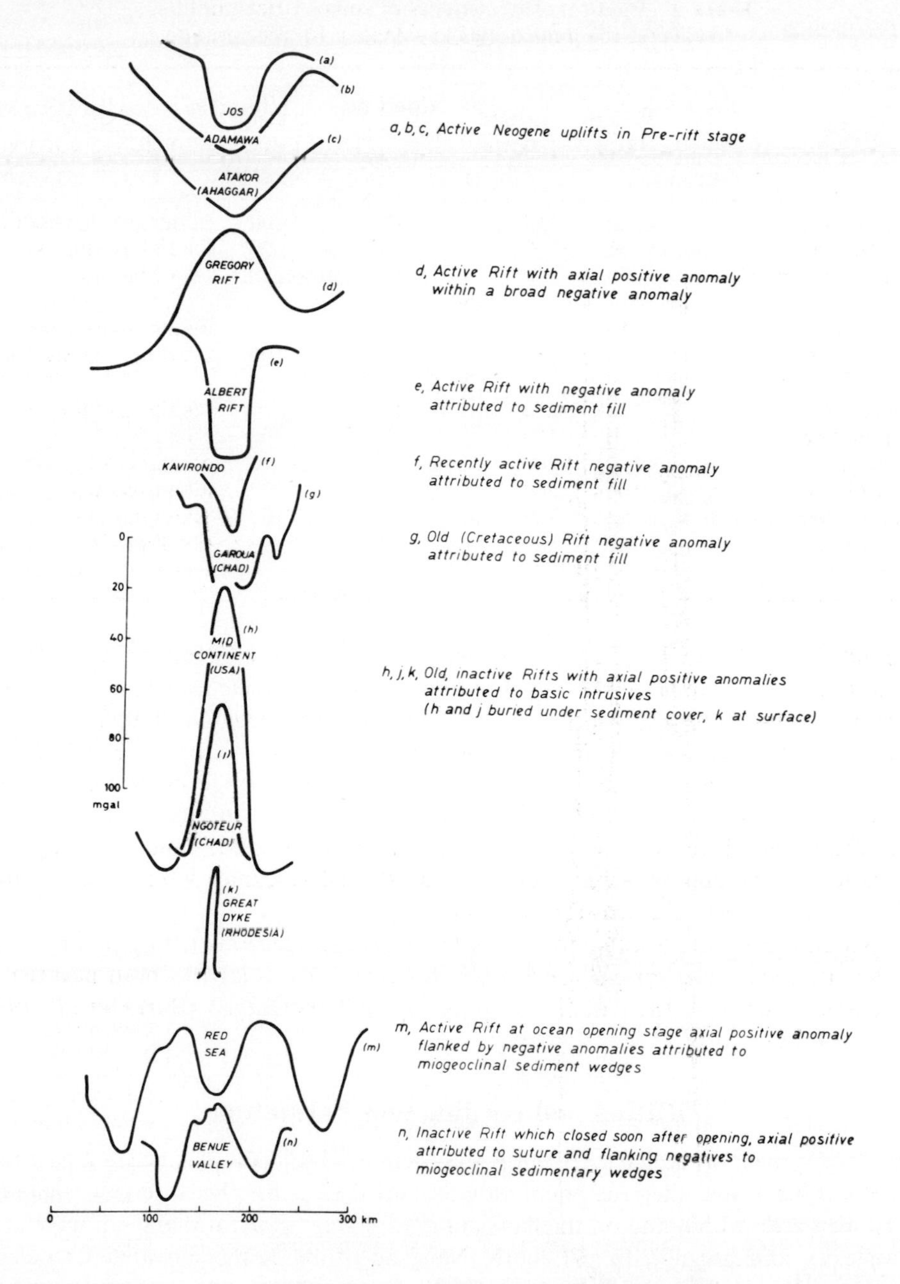

FIGURE 2 Selected gravity profiles across uplifts and rift structures. Active rifts and uplifts are associated with broad negative anomalies (a-f) which may be accentuated by sediment fill (e, f). An axial positive anomaly may also occur (d). Inactive rifts may display negative anomalies due to sediment fill (g) or narrow axial positives due to intrusives (h, j, k). Miogeoclinal sediment wedges cause flanking negative anomalies around the axial positive anomalies of an actively spreading young ocean (m) and a suture along which a young ocean has shut (n). a from Ajakaiye (1970), b and c from Tordi (1971), d and f from Khan & Mansfield (1971), e from Girdler (1964), g and j from Louis (1970), profiles 4, 5 and 17), h from Am. Geophys. Union and U.S. Geol. Survey (1964), j from Louis, k from Weiss (1940), m from Tramontini & Davies (1969) and n from Cratchley & Jones (1965).

TABLE 2 Properties of some African

rrr Junction	Rift Arms	Approx. Ages (in m.y.) Start	Stop	Uplift	Where and Whether Arms Extend
1 AFAR 12°N 42°E	*a* Red Sea	30	Not Yet	Completed shoulders	Reaches Dead Sea Transform
	b Gulf of Aden	20		Still Distinct	Reaches Owen F.Z.
	c Ethiopian	20		Completed Massive Shoulders	Ends S or Joins 2a
2 Nakuru 0°N 36°E	*a* Gregory (N)	20	Not	Probably very near completion	Ends N or Joins 1c
	b Gregory (S)	20	Yet		Reaches Junction 3
	c Kavirondo	20	?3		Ends Westward in Lake Victoria
3 Kilimanjaro 3°S 37·5°E	*a* Gregory (S)	10	Not	Probably Near Completion	Reaches Junction 2
	b L. Eyasi	10	Yet		Ends SW
	c Pangani	10			Ends SE
4 Rungwe 8°S 33°E	*a* Great Ruaha	10	Not	Probably Near Completion	Ends NE
	b Rukwa	10	Yet		Joins Western Rift
	c L. Malawi	10			Joins and Reactivates 10d
5 Mid-Zambesi Luangwa 16°S 29°E	*a* Upper Zambesi	160	130	Completed By Cretaceous	Ends SW
	b Mid Zambesi	160	130		Ends ESE
	c Luangwa	160	130		Ends NE Near Junction 4
6 Poli 9°N 14°E	*a* Yola	130	80	Completed In Cretaceous	Reaches Junction 7
	b Ft. Archambault	130	80		Ends E in Sudan
	c Ati	130	80		Ends N at Ounianga Kebir
7 Chum 9·5°N 12°E	*a* Yola	130	80	Completed In Cretaceous	Reaches Junction 6
	b Lower Benue	130	80		Reaches Junction 8
	c Upper Benue	130	80		Ends Under Chad Basin
8 Niger Delta 5°N 6°E	*a* Lower Benue	130	80	Completed In Cretaceous	Reaches Junction 7
	b Mid-Atlantic	130	Not Yet		Joins Gulf of Guinea Transforms
	c Mid-Atlantic	130			Extends into S. Atlantic
9 Takatu Uncertain Position	*a* Takutu	Cretaceous		Completed In Cretaceous	Ends S In Guyana
	b Mid-Atlantic	130	Not Yet		Extends Into N Atlantic
	c Mid-Atlantic	130			Joins Gulf of Guinea Transforms
10 Lower Zambesi 17°S 35°E	*a* Lr. Zambesi (Lupata)	130	100	Completed In Cretaceous	Ends NW
	b Mozambique Monocline	130	Not		? Joins Transform
	c Sabi Monocline	200	Yet		Reaches Junction 11
	d Shire–L. Malawi	130	100		Joins 4*c*
11 Lower Limpopo 21°S 36°E	*a* Limpopo	200	160	Completed In Cretaceous	Ends W
	b Sabi Monocline	200	Not		Reaches Junction 10
	c Lebombo Monocline	160	Yet		?

Rift Valleys and rrr Junctions

Rifting	Vulcanism	Spreading	Sediment Fill	References
Completed Completed Active in S	Alkaline Finished Axial Tholelltlc Mainly Alkaline Tholeiitic in N	Since 25 m.y. Slnce 20 m.y. Starting	Miogeoclinal Wedges >1 km Local <1 km	Falcon (1970) McKenzie *et al.*, (1970)
Active Active Completed	Active Active Completed	Possibly Starting None None	Local <1 km Local <1 km General <1 km	Baker & Wohlenberg (1971) King (1970)
Active ? Active ? Active ?	Active Close to Junction	None None None	Local <1 km Local <1 km None	
Active Active Active	Active Close to Junction None	None None None	None Local <1 km Local <1 km	
Completed Completed Completed	 Minor Alkaline Finished	None None None	Generally <1 km Karroo Fill	 Bailey (1961)
Completed Completed Completed	None None Cretaceous Basalt?	None None None	General <1 km General >1 km Little	Louis (1970)
Completed Completed Completed	None Cret. Alkaline Cret. Alkaline	None Opened 130 m.y. Shut 80 m.y. Little or None 130 – 80	General <1 km General >1 km General <1km	Burke & Dessauvagie (Unpubl.) Burke *et al.*, (1971)
Completed Completed Completed	Cret. Alkaline and calc-alkaline Axial Tholeiitic Axial Tholeiitic	Opened 130 m.y. Shut 80 m.y. Started 130 Started 130	General >1 km Miogeoclinal Wedges >1 km	Burke, Dessauvagie & Whiteman (1971)
Completed Completed Completed	None Axial Tholeiitic Axial Tholeiitic	None Started 130 Started 130	General <1 km Miogeoclinal Wedges >1 km	McConnell (1969)
Completed Completed Completed Reactivated	Alkaline Cret. ? Alkaline Jur.-Cret. Alkaline Cret.	None Spread Madagascar Away Till End Cret. None	General <1 km Miogeoclinal Wedges >1 km	Wooley & Garson (1970) Cox (1970)
Completed Completed Completed	Alkaline Jur. Alkaline Jur.-Cret. Tholeiitic Cret.	None Spread Madagascar Away Till End Cret.	 Miogeoclinal Wedges >1 km	Wooley & Garson (1970) Cox (1970)

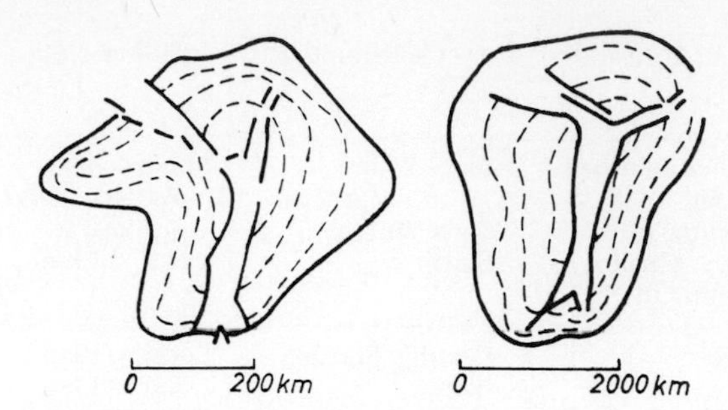

FIGURE 3 Sketches from Cloos (1939) to illustrate the bifurcation of the Rhine and Red Sea rifts along their lengths as analogues of the bifurcation of rifts produced in a clay layer by inflation of an underlying balloon.

or may not evolve. We present data (Table 1 and Fig. 4) on 11 African junctions in rift systems of both Neogene and Jurassic–Cretaceous ages which indicate that all rifts more than 5 m.y. old are associated with rrr junctions and suggest that the widespread occurrence of the distinctive rrr structure is strong evidence that the mechanism indicated by Cloos has operated in African rift development.

By considering seven rrr junctions formed in the Jurassic–Cretaceous phase and four formed in the Neogene we attempt an integrated view of the evolution of rift systems. 29 rifts meet in the 11 junctions of Table 2 (one junction has four arms and five rifts are common to two junctions each).

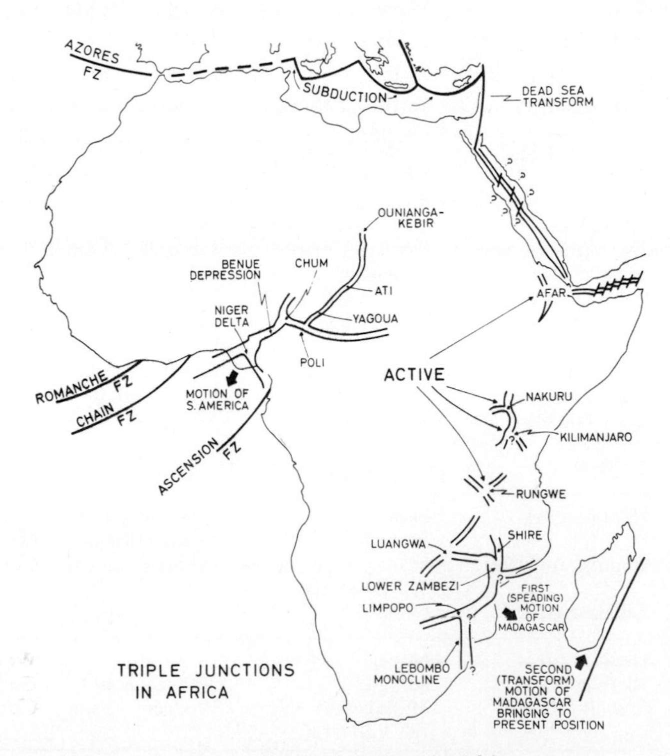

FIGURE 4 Ten rrr junctions in Africa.

East African Neogene Rifts

Afar

This structure has been designated an RRR junction by McKenzie *et al.* (1970) and there is a great deal about it in papers in the Royal Society's Discussion on the Red Sea (Falcon, 1970). Gass (1970) has interpreted the structural and magmatic development of the Afro-Arabian dome on the crest of which the Afar junction formed initially as an rrr structure as a response to a rising lithothermal system. He envisaged a sequence in which early alkali basalts associated with a dome gave way to intermediate and tholeiitic basalts as the dome evolved through a rift and rrr junction phase to the phase with new oceanic crust represented now in the Red Sea and Gulf of Aden. When spreading developed on all three rifts the rrr structure became an RRR triple junction. The Afar is important as the junction in Africa which illustrates that the three rift arms of rrr junctions do not all begin to develop or spread at the same time and that there is a transition both in structure and in magmatic composition from the continental rift to the ocean opening stage. The complex structures within the Afar, with the Danakil horst an isolated fragment of continental crust and the far from simple magmetic anomaly distribution, indicate that the pattern of structural development of the thin lithosphere at newly spreading junctions is likely to be very variable.

Nakuru

This rrr structure is formed by the junction of the east–west trending Kavirondo rift and the northern and southern arms of the Gregory rift (Fig. 5). Its position on the crest of an uplift deforming a pre-Miocene erosion surface (Baker & Wohlenberg, 1971, Fig. 2) and its gravity field (Fig. 2) with an axial positive anomaly and flanking negative anomalies (Khan & Mansfield, 1971) are distinctive. The Kavirondo rift arm of the Nakuru structure is full of sediment and volcanic activity ceased in it in Pliocene times (Baker & Wohlenberg, 1971). Volcanic and rift activity persist in the two Gregory rift arms and there is some evidence indicating that spreading may be starting on these structures. We interpret the Nakuru structures as an rrr junction in which one rift arm (the Kavirondo) has become inactive before reaching the spreading stage.

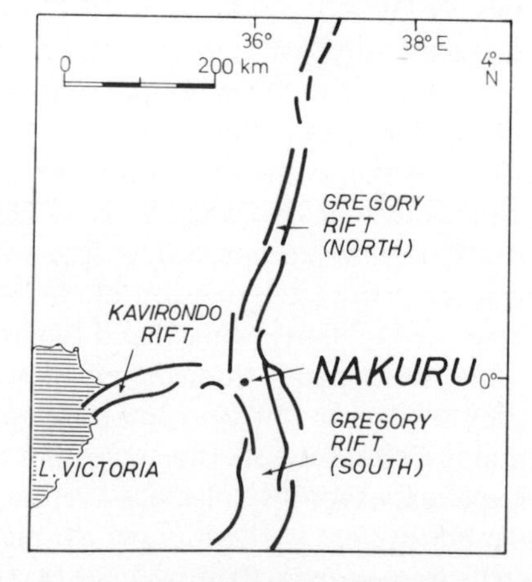

FIGURE 5 The Nakuru rrr junction at the meeting place of the two active arms of the Gregory rift and the Kavirondo rift which became inactive in the Pliocene (based on Baker & Wohlenberg, 1971).

Kilimanjaro

The evidence for an rrr structure at Kilimanjaro is far less strong than at the Afar and Nakuru and the suggestion is made because of topographic form, vulcanism and fault patterns. The fault pattern is formed by the junction of the projected Pangani structure with the Gregory rift southern extension. Continuing uplift in this area may lead to the development of a more obvious feature. Vulcanism is most intense at the junction itself and decreases rapidly along the southeastern and southwestern arms.

Rungwe

The Rungwe structure consists of the junction of the well developed Lake Rukwa and Lake Malawi rifts with the less obvious Great Ruaha rift which strikes northeastward into Tanzania (Fig. 4). The junction lies in a topographically high area and is marked by the Rungwe volcanics (Pliocene to Recent) whose activity is concentrated near the junction. The Rukwa rift extends northeast to join the Western rifts which were initiated within the last 5 m.y., much later than the Eastern rifts. We have distinguished no rrr junctions in the Western rift system. The active Lake Malawi rift reactivates a Cretaceous rift and provides a link between the Neogene East African rift system and the Jurassic–Cretaceous developments to the south.

Jurassic–Cretaceous Rifts of Southern East Africa

Zambesi-Luangwa

A 30° bend in the course of the Zambesi marks the junction of two rift arms and from this point the Luangwa rift strikes northeast (Fig. 6). Karroo sediment is found in all three arms which meet at this junction where igneous activity mainly took the form of carbonatite intrusion (Bailey, 1961). The junction ceased to be active and all the three arms were filled with sediment without spreading. Development appears to have been over by the end of the Jurassic.

Lower Zambesi

The Lower Zambesi structure (Woolley & Garson, 1970, Figs. 1 and 2) is unique in having four arms. The eastward directed arm cuts steeply across the trend of the Mozambique belt (Holmes, 1951) to the Mozambique channel. Because the basement falls steeply to a depth of 2000 m across this arm (ASGA/UNESCO International Tectonic Map of Africa, 1968) we suggest that it is a continental margin flexure like the Sabi and Lebombo monoclines (see below) and the East Greenland flexure (Wager, 1947) and refer to it as the Mozambique monocline. The south-trending arm of the Lower Zambesi structure is a northward extension of the Sabi monocline. Between these two monoclinal features lie the Shire–Malawi and Lupata (Lower Zambesi) rifts.

The Malawi rift is an area of abundant immediately subvolcanic alkaline igneous activity from which Woolley & Garson (1970) reported ten ages averaging about 125 m.y. The Lupata arm contains Cretaceous alkaline volcanics and Cretaceous sediment fill. Four ages from the Lupata Cretaceous volcanics average 115 m.y. The very large volumes of Cretaceous alkaline syenite in the Malawi rift may indicate an element of crustal melting, a process which we suggested on evidence from Jos and Tibesti impedes or brings to a halt the sequence of structural evolution. The Shire rift developed first

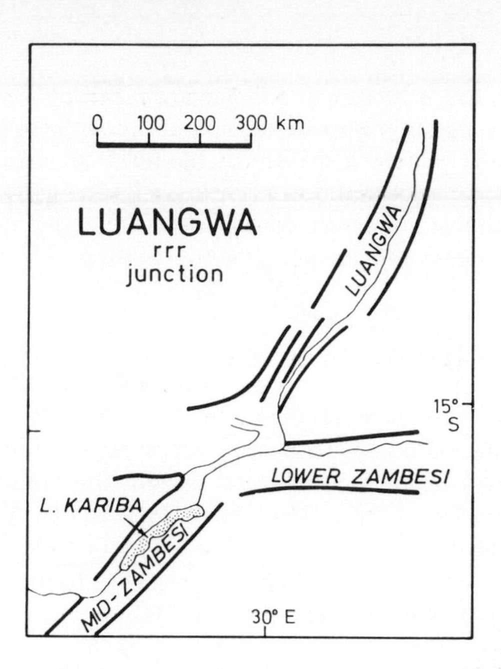

FIGURE 6 The Zambesi–Luangwa rrr junction linking three rifts which became inactive and did not spread in Jurassic times (based on Bailey, 1961).

and then failed (perhaps because of energy loss through crustal melting) and subsequent development of the Lupata rift began about 10 m.y. later.

Figure 7 illustrates the suggested relationships between the Lower Zambesi and Limpopo structures and Madagascar. Most Indian Ocean reconstructions start Madagascar from a place considerably to the north of its present situation. This is difficult to justify because late (post-Anomaly 31) transform motion in this part of the Indian Ocean is in a north-northeasterly direction so that the last motion of Madagascar has been northward (Wright & McCurry, 1970). No trace of a structural framework within which Madagascar can be moved first south and then north can be detected.

On our interpretation (Fig. 7) the Mozambique, Sabi and Lebombo monoclines mark the northwestern sides of three rift arms which spread to move the microcontinent toward the southeast. Because the Limpopo structure started to develop earlier than the Zambesi structure the southeasterly spreading also involved an element of anti-clockwise rotation. Later transform motion brought Madagascar to its present position.

In summary the Lower Zambesi junction first developed on the Shire rift and on rifts which were to become the Mozambique monocline and the northward extension of the Sabi monocline. The Shire rift ceased to function and its role was taken over by the Lower Zambesi–Lupata rift which went through the stages of uplift, rifting and vulcanism before itself ceasing to function. The other two arms continued to spread throughout the Cretaceous but during the Tertiary this movement had been replaced by transform motion. On this analysis the miogeoclinal wedges of Madagascar and Mozambique are predicted to lie on oceanic crust.

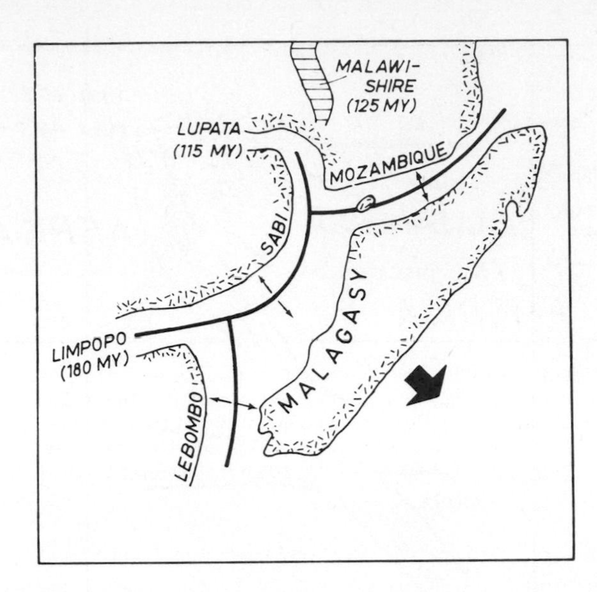

FIGURE 7 The Lower Zambesi and Limpopo rrr junctions. The Lupata and Limpopo rifts became inactive without spreading and spreading on three rifts led to the formation of the Lebombo, Sabi, Mozambique and Malagasy monoclines. Note that this reconstruction requires that parts of the sedimentary areas of Mozambique and Madagascar overlie oceanic crust.

Limpopo

The Limpopo structure consists of the Sabi and Lebombo monoclines (which with the Mozambique monocline are related to the drifting of Madagascar) and the Limpopo 'failed' rift which contains alkaline volcanic rocks in the Tuli Syncline and immediately subvolcanic rocks in the Nuanetsi and Marangudzi complexes (Cox, 1970; Woolley & Garson, 1970). The Limpopo igneous rocks yield Karroo ages. Like the Lower Zambesi rift the Limpopo structure ceased to function without spreading. Spreading motion related to the Sabi, Lebombo and Mozambique structures resulted in Madagascar drifting southeastwards.

Cretaceous Rifts of the Gulf of Guinea and the Benue Trough

On the western side of Africa the Gulf of Guinea opened in Cretaceous times about two rrr junctions, one situated north of the Takatu rift in Guyana and the other under the present site of the Niger delta (Burke *et al.*, 1971).

The Benue trough, extending northeastward from the Niger delta, formed part of this system and two more rrr junctions at Chum and Poli are distinguished along its length (Fig. 3). The four West African junctions have developed in rather different ways from those in East Africa and their histories complement the picture of junction development.

Niger Delta

Full analysis of the Niger delta structure awaits more complete magnetic and bathymetric mapping in the Gulf of Guinea. Three interpretations are current: (1)

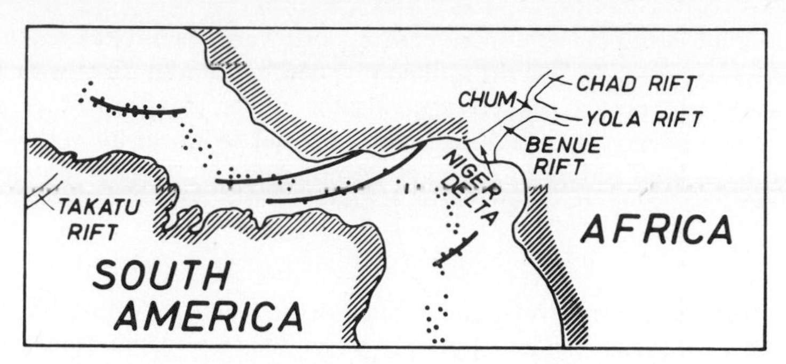

FIGURE 8 Rifts and giant fracture zones of the Gulf of Guinea rrr junctions under the Niger delta and north of the Takatu rift developed by spreading and transform motion. The Benue trough opened about 130 m.y. ago and closed again about 80 m.y. ago at about the time the south Atlantic rotation pole changed its position. From that time on the complex development of the area was reduced to almost east–west spreading on the mid-Atlantic ridge and transform motion on the giant fracture zones.

Burke *et al.* (1971) suggested that a rift developed and began to spread in the Benue trough early in the Cretaceous. Near the end of the Cretaceous the embryonic Benue Ocean closed because the spreading rate opposite the bulge of Africa was faster than that south of the Gulf of Guinea. Northwestern Africa overtook the southeastern part of the continent. This closure folded the Cretaceous sediments parallel to the length of the Benue trough. A south trending Early Cretaceous rift extending from the site of the Niger delta became the structure on which the South Atlantic opened and a west-trending structure became involved with the giant continental margin transform offsets which are now the Chain, Romanche and St. Paul's fracture zones (Fig. 8). (2) Grant (1971) brings a fracture zone to the northern flank of the Benue trough, slightly modifying the geometry of the previous interpretation and making the Niger delta junction an RRF rather than an RRR junction. This distinction he suggests to be important because McKenzie & Morgan (1969) showed that RRR junctions have vector motion stability which RRF junctions lack. (3) Francheteau & Le Pichon (1972) present a model with fracture zones on either side of the Benue trough but make no allowance for Cretaceous spreading in the trough.

Accurate mapping of magnetic anomaly patterns in the Gulf of Guinea may resolve the differences between these interpretations. The strike of the anomaly trend under the Niger delta is N 45° E (Burke *et al.*, 1971) and the results of Arens and others (1971, p. 68) indicate that this trend may extend for some hundreds of kilometres into the Gulf of Guinea.

Until more evidence is available we treat the Niger delta structure as an rrr feature initiated about 130 m.y. ago which became an RRR junction when all three rifts began to spread. The Benue rift closed again about 50 m.y. later and the south Atlantic arm has continued to spread, with minor transform offsets. The westward extending arm has developed through the giant transforms of the Chain, Romanche and St. Paul's fracture zones.

Takatu (*Guyana*)

The Takatu rift lies at the western end of the Gulf of Guinea where the mid-Atlantic ridge passes northward out of the zone of giant transforms. It is a southwesterly trending structure whose extension northward would take it to join the mid-Atlantic ridge (Fig. 8). The rift is filled to a depth of about 2 km with Jurassic or Cretaceous sandstones and underlying volcanic rocks (McConnell *et al.*, 1969). There is evidence of rrr shape, faulting, vulcanism and sediment fill in the Takatu rift which may have been linked to the mid-Atlantic ridge in an RRF junction through transform motion on the St. Paul's fracture zone.

Chum

600 km up the Benue trough from the Niger delta the rift structure bifurcates close to Chum. Eastward from the junction only about 300 m of gently folded Cretaceous sediments occupy the Yola rift (Burke & Dessauvagie, in preparation). The Chad arm which trends north from the junction contains a thicker section and more intense folds. Bouguer anomaly maps (Cratchley & Jones, 1965; Louis, 1970) show that the Benue structure extends along the north, trending rift to die out under the Chad Basin. The Chad and Benue arms spread and later closed in the Cretaceous but the Yola arm became inactive before spreading.

Poli

The Poli junction is ill-defined. The Yola rift extends eastward for 500 km from Chum as a series of shallow Cretaceous sediment-filled troughs to Doba (Fig. 2). Its continuation farther east is not known at outcrop but a 100 km wide belt of 50 mgal negative Bouguer anomalies (Louis, 1970) which extends for another 800 km through Fort Archambault to Birao on the Sudan border is believed to mark a rift continuation filled with Cretaceous sediment. Louis (1970) has also mapped a remarkable 1500 km long line of positive Bouguer anomalies from Poli to Ounianga-Kebir (Figs. 2 and 4) which we suggest marks a rift forming a third arm of the Poli structure. This 1500 km long feature lies almost entirely under the thin Neogene sediments of the Chad basin but because of its positive gravity anomalies we interpret it as a long series of basic intrusions marking the subsurface expression of a rift feature. A density contrast of 0·3 (basalt against granite for example) was used in model studies of the positive anomalies (Louis, 1970).

Sequences of Structural Development in Africa

Table 3 illustrates sequences of structural development which have followed the initial uplift stage in Africa during the last 200 m.y. Crustal melting appears to have inhibited development in Jos, Malawi and Tibesti. Jos stopped in Jurassic times at the uplift and vulcanism stage and Tibesti appears to be lingering a long time at the same stage. In Malawi development reached the rift stage and then stopped the Lupata rift taking over the role of the Shire rift.

Rift formation is considered as the stage following uplift because the topographic forms of uplifted surfaces are similar in the two types of structure and because of their similar volcanic rocks and gravity signatures. The experiment of Cloos (1939) in which he made rifts and rrr junctions by inflating a balloon under a sheet of wet clay is regarded as a fair analogy. We regard rrr junction formation as a normal consequence

TABLE 3 Sequences of African structural development.

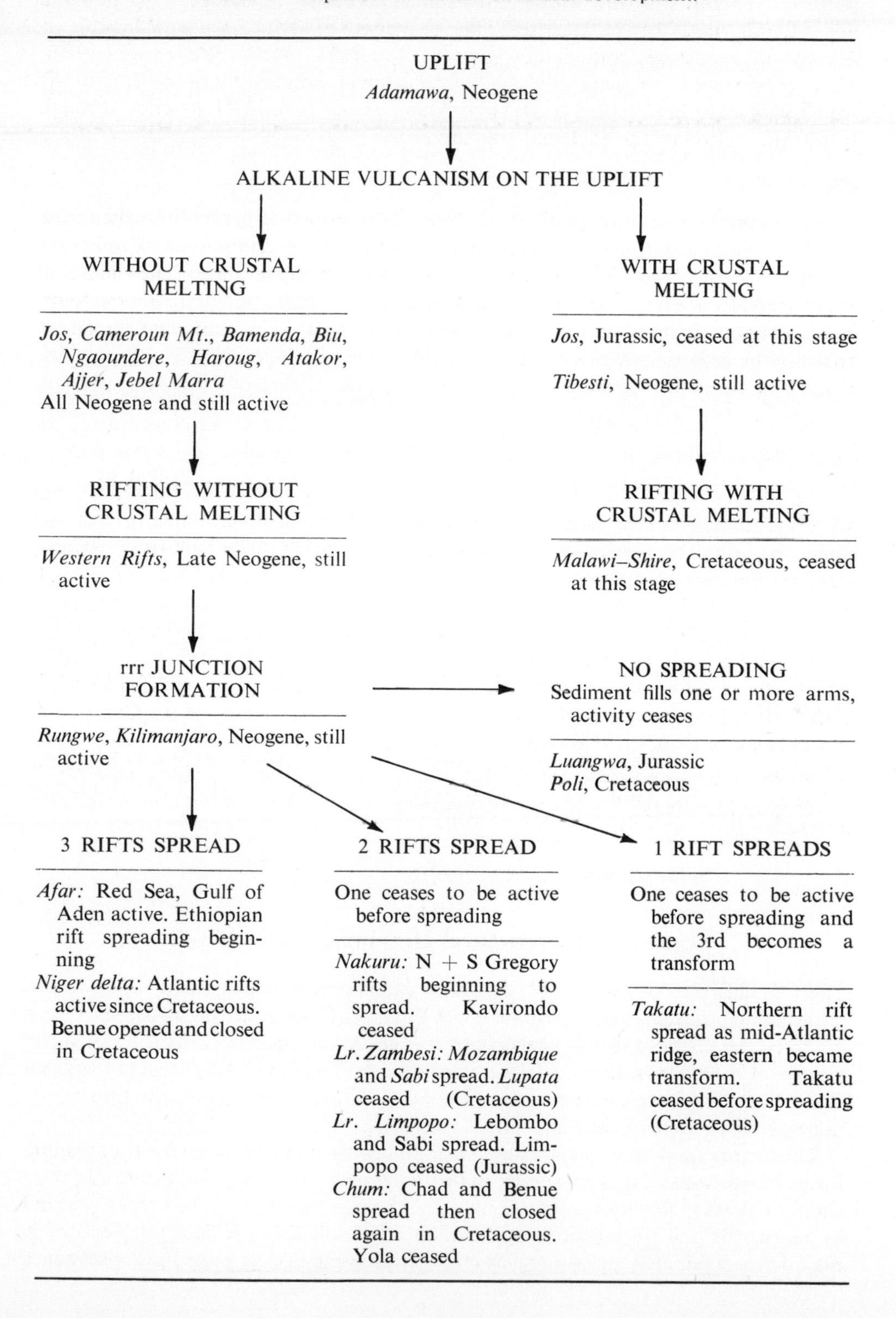

of rifting because all African rifts except the very young (< 5 m.y. old) structures of the Western rift are linked at one or both ends to a junction.

Four paths have followed the formation of rrr junctions:

(1) Activity ceases completely leaving a rift-faulted, sediment-filled rrr structure with volcanic or subvolcanic rocks. The Luangwa Jurassic structure developed to this extent. The poorly exposed Poli Cretaceous structure also appears not to have spread and two of its three arms are full of sediment.

(2) Spreading may develop on all three rifts and the structure passes from the rrr to the RRR state. This appears to be happening in the Afar. In view of the inherent stability of this relationship (McKenzie & Morgan, 1969) continental separation is likely to ensue. A Cretaceous development of this type under the Niger delta appears to have broken down when the Benue trough closed to accommodate fast spreading opposite the bulge of Africa.

(3) Spreading on two rifts seems to be starting at Nakuru and took place on the two structures which drifted Madagascar away from Africa. The Cretaceous feature at Chum also developed in this way. The rift which fails to develop when this path is taken appears more commonly to strike east–west than north–south (Table 4).

(4) The Takatu structure appears to have developed by spreading on one rift accompanied by transform motion on a second rift and cessation at the vulcanism-rifting-sediment fill stage on the third.

Although all post rrr formation paths except the first can lead to continental separation the third and fourth paths in which one rift ceases to function at the pre-spreading stage appear to have developed more often than the path in which all three rifts spread.

TABLE 4 Angles between rift trends and geographic north. Rifts which spread generally have meridional trends and those which do not have equatorial trends. The Shire rift perhaps failed to spread because of crustal melting. Junctions with no spreading not considered.

Spreading Rifts		Rifts which did not Spread	
Gregory N.	10°	Kavirondo	80°
Gregory S.	0°	Yola	90°
Benue	45°	Takatu	70°
Chad	40°	Shire	20°
Mid-Atlantic (N)	0°	Limpopo	70°
Mid-Atlantic (S)	0°	Lupata	45°
Mozambique	60°		
Lebombo	0°		
Sabi (S)	45°		
Sabi (N)	0°		
Ethiopian	30°		
Red Sea	30°		
Gulf of Aden	70°		
Average	25°		60°

Interpretation of the Development Sequence

The sequence illustrated in Table 3 can be related to the concepts of the new global tectonics. We suggest that the formation of an uplift is a response to a rising mantle plume (Morgan, 1971). The common association of uplifts with alkali basalts and negative gravity anomalies (Table 1) is interpreted as the result of partial melting of mantle material above the rising plume to form large volumes of alkali basalt magma at the base of the lithosphere. This magma is responsible for a mass deficiency to which the elongated domal uplift is an isostatic response (Fig. 9a). Because the erupted basalts are alkaline rather than tholeiitic the base of the lithosphere under the uplifts is taken to lie at 60–120 km (Harris, 1970, p. 428). Table 1 summarizes data on the dimensions of African uplifts. Widths cluster around 80 km and are probably comparable to depths to mass deficient material. Lengths are about three times widths, perhaps indicating asymmetry in the plumes which trigger uplift formation. Round structures 1000 km across which include both basins, e.g. the Chad basin (Fig. 1) and swells, e.g. the East African swell, embracing both the eastern and western rifts and the Lake Victoria area between are of problematic origin. Neogene uplifts are tangentially disposed about the Chad basin (Fig. 1) and the Eastern and Western Rifts are roughly tangentially disposed within the East African swell.

In Jurassic times in the area now occupied by the Jos plateau a volume of igneous rock in excess of 6×10^4 km^3 (assuming outcrop areas now to represent intrusions 10 km thick on average) was produced largely by crustal melting (Bowden, 1970). This is suggested to have impeded further development. Peak igneous activity in Jos was 160 m.y. ago (Grant, 1971) but the Jos topographic uplift had disappeared by Cenomanian times (Burke, in preparation). The Tibesti and Malawi areas are also suggested to have had restricted development because of crustal melting.

Rifts and rrr junctions are indicated in Table 2 as successive responses to lithospheric doming. It has been suggested that rift faults are too large in comparison with crustal thickness to form in this way (Freund, 1966) but the significant thickness is that of the lithosphere in the pre-rifting stage (perhaps about 100 km). Unfortunately crust and uppermost mantle in well developed rifts (Fig. 9b) like the Kenya rift (Griffiths *et al.*, 1971; Khan & Mansfield, 1971; Baker & Wohlenberg, 1971) are probably now in a very different state from their condition at the inception of rift faulting. There is need for seismic and magnetotelluric study of uplifts in the pre-rift stage. Gravity data are already available for several uplifts (e.g. Ajakaiye, 1970).

What happens after the formation of an rrr junction depends on whether and how the driving mechanism of the world-wide plate system can accommodate the new elements. Up to this stage tension and spreading are envisaged to have played no part in the development of uplifts, alkaline vulcanism and rifts which are responses to a thermal event in the mantle. If the structure cannot be accommodated then the existing plate motion will bring other parts of the lithosphere over the plume or the plume may subside. The Luangwa and Poli rift systems (Table 3) are interpreted as examples of structures resulting from this course of events. If the world system can accommodate the structure then plate boundaries form at the rifts and the structure develops by spreading. Table 3 indicates that spreading has taken place in Africa during the last 200 m.y. in different ways on different structures. Spreading on one, two and three rift arms have all happened. Spreading on three arms forms three plates and

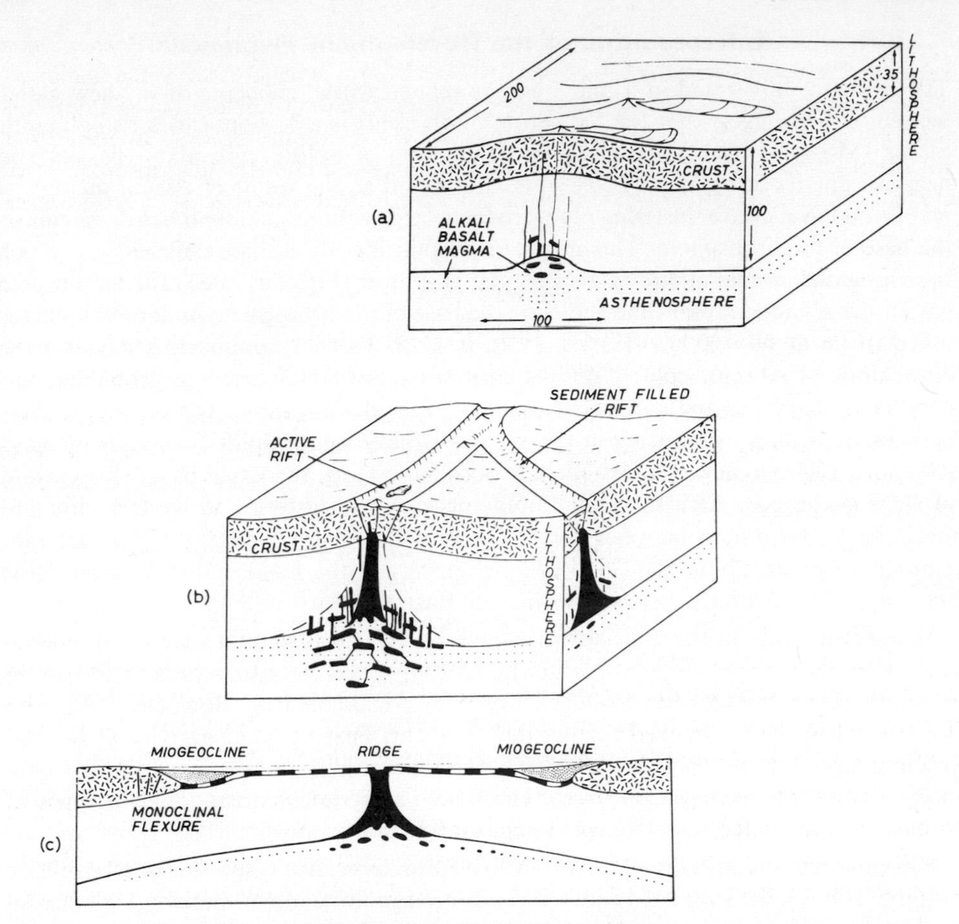

FIGURE 9 Uplift, rifting, spreading. (a) Sketch of an uplift with alkali basalt volcanoes resulting from a mass deficiency caused by partial melting at the base of the lithosphere (dimensions in km). (b) Rift development as a phase following uplift (based in part on the results of Griffiths *et al.*, 1971, and Khan & Mansfield, 1971). (c) Ridge and Miogeoclinal developments with marginal monoclinal flexures as developed across the Mozambique channel in the Cretaceous.

presumably follows the course described by McKenzie & Morgan (1969) but spreading on two rifts which forms only two plates requires one of the rrr junction arms to cease to function before reaching the spreading stage. Spreading on a single rift which again forms two plates requires transform motion on a second arm and cessation of activity on a third arm before the spreading stage. Single rift spreading could also be achieved in association with transform motion on two arms (FFR type of junction) but an example of this type has not yet been recognized in Africa.

The tendency of plate rotation poles to develop near the poles of the Earth's axis of rotation (Le Pichon, 1968) may be the reason why the rift arms which have failed to develop in Africa are those which strike close to east–west (Table 4). The same tendency is suggested to have influenced the development of the Gulf of Guinea giant transform faults between the Niger delta and Takatu rrr junctions. Spreading motion in relation to these two junctions was reduced to an approximately east–west opening of the Atlantic within 50 m.y. of their formation (Fig. 8).

A feature of the development of Africa has been the two peaks of rift formation in Jurassic–Cretaceous times (approximately 180–130 m.y. ago) and since the beginning of the Neogene (approximately 25 m.y. ago). The intervening 100 m.y. saw rift development but no rift inceptions. Coney has suggested (in conversation) that the episodes of rift inception may have resulted from slow absolute motion of the African plate which enabled thermal energy to be transferred from the asthenosphere into the lithosphere.

Conclusions

1. Uplifts, typically about 1 km high, 100 km wide and 200–300 km long have dominated the development of Africa since Gondwanaland began to break up.

2. These uplifts which are characterized by alkaline vulcanicity and negative gravity anomalies reaching about 100 mgal are interpreted as isostatic responses to mass deficiencies produced by partial melting of the mantle above rising plumes at the base of the lithosphere.

3. Rifts and rrr junctions normally develop on the crests of the uplifts but this process may be inhibited if substantial crustal melting takes place.

4. All rift systems more than 5 m.y. old have developed rrr junctions.

5. The rrr junctions, which originate as a result of uplift, have in some cases evolved through integration into the world-wide plate system. In cases in which the world-wide system has been unable to accommodate the rifts as plate boundaries, uplifts have collapsed and rifts have been filled with sediment.

6. African rrr junctions have evolved by spreading on one, two or three arms. A common path is for one rift to become inactive before spreading starts and for spreading to take place on two rifts which have tended to be aligned roughly north–south so that the poles of plate motion are close to the poles of the Earth's rotation axis.

7. Spreading on five of the 29 rifts considered has led to continental separation and spreading on three rifts has separated a micro-continent (Madagascar). Two rifts in the Benue trough opened and later shut and of the remaining 19, nine are active and may yet spread and ten have become inactive without spreading. Only the Afar junction has evolved from the rrr to the RRR condition. At the Takatu junction one arm became inactive before spreading, one arm continues to spread and the third arm has become a transform fault. The Niger delta junction may have developed in a similar way after the closing of the embryo ocean of the Benue rift.

8. Evolutionary sequences from uplift, through rifting and rrr junction formation to spreading, have been initiated in Africa between 180 and 130 m.y. ago and again within the last 25 m.y. During the 100 m.y. interval although existing structures continued to evolve no new structures developed.

References

Ajakaiye, D. E., 1970. Gravity measurements over the Nigerian Younger granite province, *Nature*, **225**, 50–52.

American Geophysical Union and U.S. Geological Survey, Bouguer gravity anomaly map of the United States (exclusive of Alaska and Hawaii) 2 sheets, scale 1: 2,500,000, 1964.

Arens, G., Delteil, J. R., Valery, P., Demotte, B., Montadert, L. and Patriat, P., 1971. The continental margin off the Ivory Coast and Ghana. *In:* The Geology of the East Atlantic Continental Margin, 4 Africa. IGS London, Report 70.16, 65–78.

ASGA/UNESCO, 1968. International Tectonic Map of Africa 1: 5M. Paris.

Baker, B. H. and Wohlenberg, J., 1971. Structure and evolution of the Kenya Rift Valley, *Nature*, **229**, 538–42.

Bailey, D. K., 1961. The Mid-Zambesi–Luangwa Rift and related carbonatite activity, *Geol. Mag.*, **98**, 227–38.

Bailey, D. K., 1964. Crustal warping—a possible tectonic control of alkaline magmatism, *J. Geophys. Res.*, **69**, 1103.

Beloussov, V. V., 1969. Continental Rifts. *In:* The Earth's Crust and Upper Mantle, Monograph 13, edited by Pembroke Hart, pp. 539–44. AGU, Washington D.C.

Black, R. and Girod, M., 1970. Late Palaeozoic to Recent igneous activity in West African and its relationship to basement structure. *In:* African Magmatism and Tectonics, edited by T. N. Clifford and I. G. Gass, pp. 185–210. Oliver and Boyd, Edinburgh.

Bowden, P., 1970. Origin of the younger granites of northern Nigeria, *Cont. Mineral and Petrol.* **25**, 153–62.

Bowden, P. and van Breemen, O. (1972). Isotopic and chemical studies on younger granites from Northern Nigeria, in *African Geology Ibadan*, 1970, pp. 105–20.

Burke, K., Dessauvagie, T. F. J. and Whiteman, A. J., 1971. Opening of the Gulf of Guinea and geological history of the Benue depression and Niger delta, *Nature Phys. Sci.*, **233**, 51–55.

Cloos, H., 1939. Hebung-Spaltung-Vulcanismus, *Geol. Rundsch.*, **30**, 405–527.

Cox, K. G., 1970. Tectonics and Vulcanism of the Karroo period. *In:* African Magmatism and Tectonics, edited by T. N. Clifford and I. G. Gass, pp. 211–36. Oliver and Boyd, Edinburgh.

Cratchley, C. R. and Jones, G. P., 1965. An interpretation of the geology and gravity anomalies of the Benue valley, Nigeria. Overseas geol. surveys, Geophys. paper no. 1. HMSO, London, 26 pp.

Dixey, F., 1965. The East African Rift System, *Colon. Geol. and Min. Res. Bull. No. 1.*

Fairhead, J. D. and Girdler, R. W., 1969. How far does the Rift System extend through Africa? *Nature*, **221**, 1018–20.

Falcon, N. L., Editor, 1970. A discussion on the structure and evolution of the Red Sea, *Phil. Trans. Roy. Soc. Lond.*, **267**A, 417 pp.

Francheteau, J. and Le Pichon, X., (1972) Marginal fracture zones as structural framework of continental margins in the south Atlantic Ocean, *Bull. American Assoc. Petrol Geol.*, **56**, 991.

Freund, R., 1966. Rift valleys in The World Rift System, *Can. Geol. Surv.*, pap. 66–14, 330–44.

Gass, I. G., 1970. Tectonics and magmatic evolution of the Afro-Arabian dome. *In:* African Magmatism and Tectonics, edited by T. N. Clifford and I. G. Gass, pp. 285–300. Oliver and Boyd, Edinburgh.

Girdler, R. W., 1964. Geophysical studies of rift valleys *Physics Chem. Earth*, **5**, 121–56.

Grant, N. K., 1971. A compilation of radiometric ages from Nigeria, *Journ. geol. Nigerian Mining, Geol. and Met. Soc.*, **6**, 37–54.

Grant, N. K., 1971. South Atlantic, Benue trough and Gulf of Guinea Cretaceous triple junction, *Bull. Geol. Soc. Amer.*, **82**, 2295–8.

Griffiths, D. H., King, R. F., Khan M. A. and Blundell, D. J., 1971. Seismic refraction line in the Gregory rift, *Nature Phys. Sci.*, **229**, 69–75.

Harris, P. G., 1970. Convection and magmatism with reference to the African continent. *In:* African Magmatism and Tectonics, edited by T. N. Clifford and I. G. Gass, pp. 419–37. Oliver and Boyd, Edinburgh.

Hedberg, J. D., 1968. A geological analysis of the Cameroun trend. Ph.D. Thesis, Princeton.

Holmes, A., 1951. The sequence of Precambrian orogenic belts in South and Central Africa, *18th Int. geol. Congr. London*, **14**, 254–9.

Khan, M. A. and Mansfield, J., 1971. Gravity measurements in the Gregory Rift, *Nature Phys. Sci.*, **229**, 72–75.

King, B. C., 1970. Vulcanicity and Rift Tectonics in East Africa. *In:* African Magmatism and Tectonics, edited by T. N. Clifford and I. G. Gass, pp. 263–82. Oliver and Boyd, Edinburgh.

Lebas, M. J., 1971. Peralkaline vulcanism, crustal swelling and rifting, *Nature Phys. Sci.*, **230,** 85.

Le Pichon, X., 1968. Sea-floor spreading and continental drift, *J. Geophys. Res.*, **73,** 3661–97.

Louis, P., 1970. Contribution geophysique à la connaissance geologique du bassin du lac Tchad. *Mem. ORSTOM*, **42,** 311 pp.

McConnell, R. B., 1969. Fundamental fault zones in the Guiana and West African shields. *Geol. Soc. Amer. Bull.*, **80,** pp. 1775–82.

McConnell, R. B., Masson Smith, D. and Berrange, J. P., 1969. Geological and geophysical evidence for a rift valley in the Guyana shield, *Geol. en Mijnbouw*, **48,** 189–99.

McKenzie, D. P. and Morgan, W. J., 1969. Evolution of triple junctions, *Nature*, **224,** 125–33.

McKenzie, D. P., Davies, D. and Molnar, P., 1970. Plate Tectonics of the Red Sea and East Africa, *Nature*, **226,** 243–8.

Morgan, W. J., 1971. Convection plumes in the Lower Mantle, *Nature*, **230,** 42–43.

Shackleton, R. M., 1954. The tectonic significance of alkaline igneous activity, *1st Inter. Univ. Geol. Cong. Univ. Leeds 21.*

Tordi, P., 1971. 1: 10M Bouguer Anomaly Map, Europe and Africa, *Inst. Int. Gravimetrie, Paris.*

Tramontini, C. and Davies, D., 1969. A seismic profile in the Red Sea, *Geophys. Journ.*, **17,** 225–41.

Vincent, P. M., 1970. The evolution of the Tibesti volcanic province. *In:* African Magmatism and Tectonics, edited by T. N. Clifford and I. G. Gass, pp. 301–320. Oliver and Boyd, Edinburgh.

Wager, L. R., 1947. Geological investigation in East Greenland Pt. IV, *Medd. Gronland*, **134.**

Weiss, O., 1940. Gravimetric and Earth magnetic measurements on the Great Dyke of Southern Rhodesia, *Trans. Geol. Soc. South Africa*, **43,** 143–53.

Woolley, A. R. and Garson, M. S., 1970. Petrochemical and tectonic relationship of the Malawi carbonatite–alkaline province and the Lupata–Lebombo volcanics. *In:* African Magmatism and Tectonics, edited by T. N. Clifford and I. G. Gass, pp. 237–62. Oliver and Boyd, Edinburgh.

Wright, J. B. and McCurry, P., 1970. The significance of sandstone inclusions in lavas of the Comores Archipalogo, *Earth Planet. Sci. Lett.* **8,** 267–8.

7.3

B. W. DARRACOTT*
J. D. FAIRHEAD**
School of Physics
University of Newcastle-upon-Tyne
U.K.

R. W. GIRDLER
S. A. HALL

The East African Rift System

Introduction

To a first order, the East African Rift System (Fig. 1) may be considered to be a consequence of movement along the boundaries of three plates, viz: the Arabia (A), Nubia (N) and Somalia (S) plates. The boundary between the Arabian and Somalia plates is the Gulf of Aden, the boundary between the Arabian and Nubian plates is the Red Sea and the boundary between the Nubian and Somalia plates is the rift in Ethiopia, Kenya and Tanzania. Possible minor plates (not discussed here) include the Danakil Horst of northern Ethiopia, and the region between the Eastern and Western Rifts (the Lake Victoria block). The theme of this contribution is that the three plate boundaries represent three different stages in the evolution of the break-up of a continent, from a rift to an ocean. The structures of the three plate boundaries are well documented, and only a brief summary of the main features is given here.

The Gulf of Aden

The form and evolution of the Gulf of Aden have been fully discussed by Laughton (1966) and Laughton *et al.* (1970). The oceanic nature of the crust beneath the Gulf has been confirmed by the seismic refraction work of Laughton & Tramontini (1970) and has been shown to exist to within at least 55 km of each shore at the western end of the Gulf, implying a minimum continental separation of 260 km. The Gulf of Aden is characterized by magnetic anomalies that can be interpreted in terms of sea-floor spreading. Figure 2 (from Laughton *et al.* (1970)) shows the increasing age of the sea floor away from the axial rift using anomaly numbers of Heirtzler *et al.* (1968). The anomalies can be traced as far as anomaly 5 (approx. 10 m.y.). The spreading rate varies from approximately 1·1 cm/yr/limb at the eastern end of the Gulf to about 0·9 cm/yr/limb at the western end, for the last 9 m.y. The Gulf of Aden may be regarded as truly oceanic.

The Red Sea

A review of the evolution and nature of the Red Sea has been given by Girdler (1969). Oceanic structure has been determined for at least the axial trough region (Drake &

*Now at Geol. Surv. S. Africa, Pretoria.
**Now at Dept. of Earth Sciences, University of Leeds, U.K.

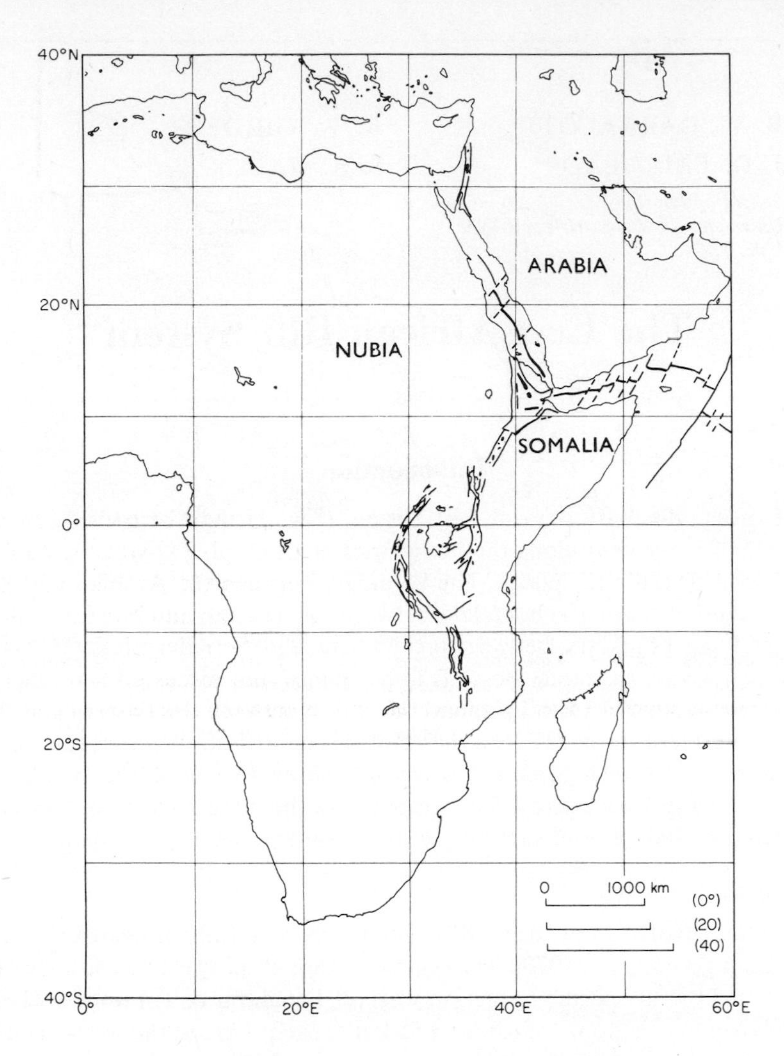

FIGURE 1 Outline of the East African Rift System.

Girdler, 1964; Girdler, 1969) and an intensive localized seismic refraction survey by Davies & Tramontini (1970) suggests that the region of oceanic crust may be wider than the axial trough. The position of the boundary between the continental and oceanic crust is still uncertain. Typical sea-floor spreading magnetic anomalies are found over the axial trough (Vine, 1966; Phillips, 1970), indicating spreading rates of approximately 1·0 cm/yr/limb for the last 3 to 4 m.y. Two magnetic anomaly profiles for two different altitudes across the Red Sea (Fig. 3) at about latitude 18° N, flown by PROJECT MAGNET, show sea-floor spreading anomalies over the axial trough. There is also an indication that they exist further away from the axial trough. If this is so, the implication is that the oceanic part of the Red Sea is considerably wider than just the axial trough region.

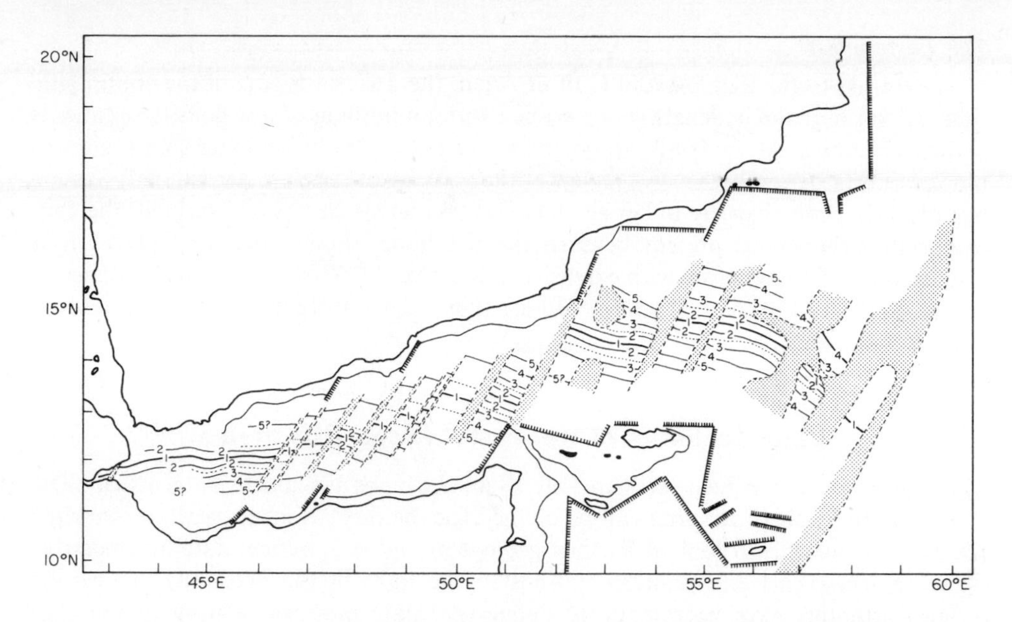

FIGURE 2 Isochron chart of the Gulf of Aden derived from the interpretation of magnetic anomalies, in terms of anomaly numbers. Solid lines represent the trends of the prominent anomalies out to anomaly 5 (about 10 m.y. ago); dashed lines show trend of anomaly $2\frac{1}{2}$ (about 3 m.y. ago). Zones of confusion are shown stippled. (After Laughton *et al.*, 1970; reproduced with permission of The Royal Society).

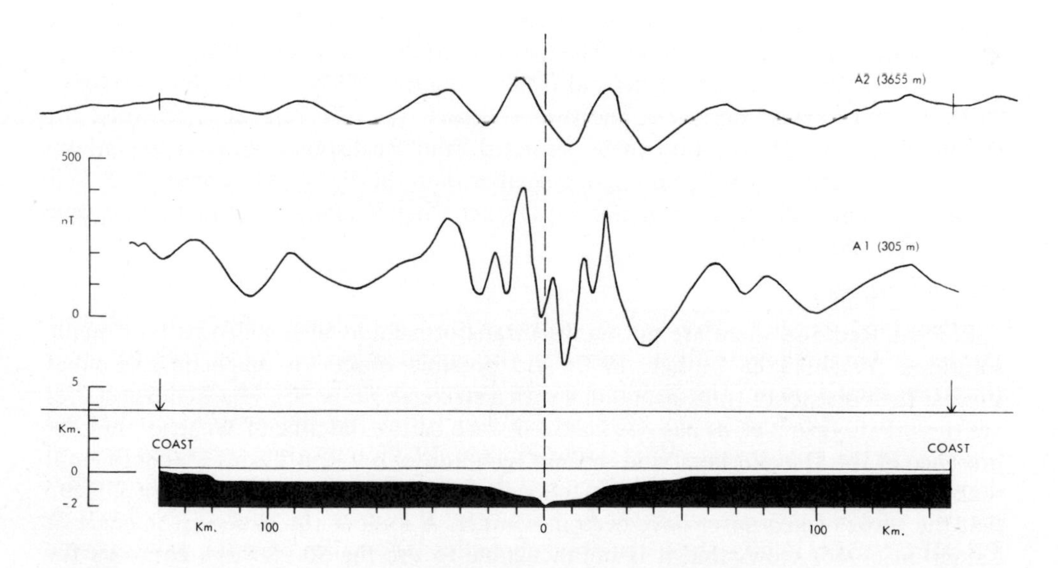

FIGURE 3 Two aeromagnetic profiles (flown as part of Project Magnet) across the Red Sea at latitude 18° N. The upper profile shows the magnetic anomalies observed at 3655 m; the lower those observed at 305 m.

The Eastern Rift

In contrast to the Red Sea and Gulf of Aden, the Eastern Rift is of the continental type. Along much of its length it is a graben with an infilling of low density sediments and/or volcanics. At the south, in northern Tanzania, the rift changes to a region of block faulting. A comprehensive review of the geology of the East African Rift System has recently been given by Baker *et al.* (1972). As far as can be ascertained, the Precambrian basement is present beneath the rift floor, though in many places it is fractured and fissured often with extensive volcanism. In short, there is no evidence of actual separation of the lithosphere though there seems to be evidence for extension and attenuation.

Plate Tectonics of the East African Rift System

The geometry of the East African Rift System can be discussed in terms of plate tectonics. If a pole of rotation can be defined for the displacement between two rigid plates, it is possible to look at various geological and geophysical data and possibly make reassessments and reinterpretations in the light of the predicted movement. Several attempts have been made to define the plate motions relating to the East African Rift System and these have been reviewed by Girdler & Darracott (1972). From this review, it is clear that there is a need for a detailed reappraisal of the pole position. Each plate boundary is considered, and the criteria which have to be satisfied are listed and an attempt made to estimate the best pole position and angles of rotation.

(i) *Gulf of Aden*

For the Gulf of Aden there are good bathymetric and magnetic charts (Laughton *et al.*, 1970). These indicate the presence of nine transform faults crossing the Gulf in a northeasterly direction. There is also one fault plane solution (Sykes, 1970) indicating shear along a plane striking N 30°. The fact that McKenzie *et al.* (1970) find that the pole obtained from the transforms and from fitting the 500 fathom contours to agree to within a degree is impressive, and their estimate, 26·5° N, 21·5° E, is accepted for this pole position. The rotation angle estimated from the distance between 500 fathom contours is found to be 6·3°, i.e. slightly smaller than the 7·6° of McKenzie *et al.* This seemingly small difference is of major importance when it comes to estimating the pole for the East African Rift.

(ii) *The Red Sea*

For the Red Sea, there are no mapped transform faults although two fault plane solutions (Fairhead & Girdler, 1970) and possible offsets of magnetic anomalies suggest the presence of transform faults with a strike about N 50°. The Red Sea shores are remarkably parallel, as has been noticed since before the time of Wegener, but the presence of the Danakil horst and seismic velocities of $5{\cdot}9 \pm 0{\cdot}2$ km s^{-1} (continental rocks) for eight refraction profiles (Drake & Girdler, 1964; Girdler, 1969) indicate that the shores could never have been in contact. A look at the bathymetric chart of Laughton (1966) shows that it is impracticable to use the 500 fathom contours for fitting as, towards the south, their position is more indicative of the extent of the sedimentary cover than the position of the continental edges. Two lines parallel to and 52 km seawards from the coasts were, therefore, fitted. This distance was obtained

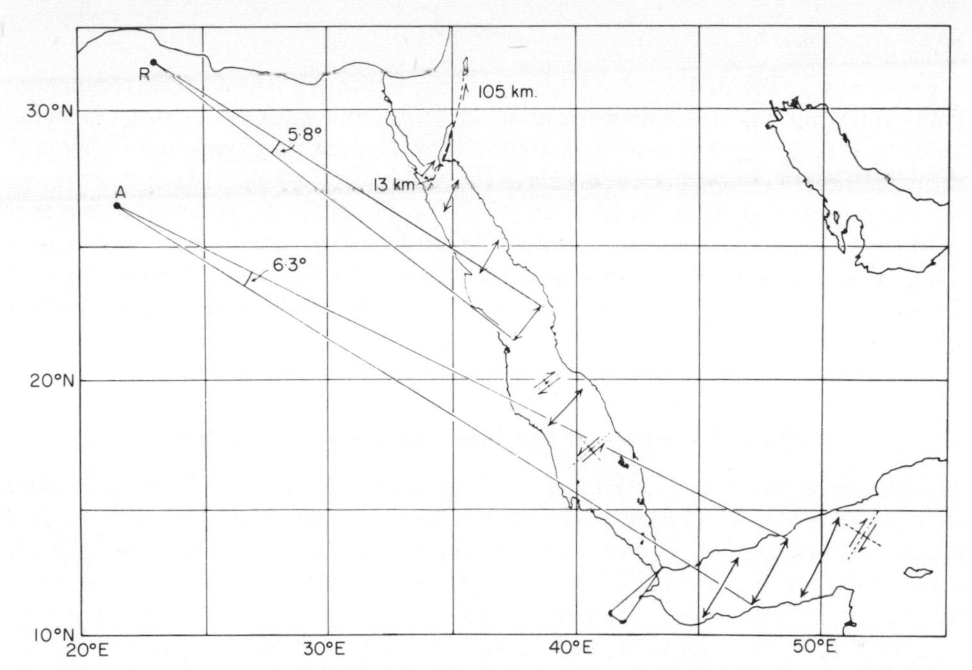

FIGURE 4 The Red Sea and Gulf of Aden with poles of rotation, directions of extension and estimates for the amounts of extension at various latitudes. The available fault plane solutions are also shown. (After Girdler & Darracott, 1972).

using the estimates for the horizontal displacement along the Dead Sea rift of 105 km found by Freund *et al.* (1970) and for the extension across the Gulf of Suez of 9 km found by Robson (1970), giving a total displacement of 118 km along a transform direction in continuity with the Gulf of Aqaba. This implies an opening of 82 km perpendicular to the Red Sea in the north where the corresponding distance between the coasts is 185 km. It is noted that 82 km is a little larger than the distance between the 500 fathom isobaths in this region and so it seems satisfactory to fit two lines 52 km seawards from the coasts. This gives a pole at 31·5° N, 23·0° E. The rotation angle is estimated to be 5·85°. This implies that at latitude 18° N, the oceanic part of the Red Sea is about 206 km wide where the distance between the coasts is about 310 km. The rotation poles for the Gulf of Aden and the Red Sea are shown in Fig. 4.

(iii) *The Eastern Rift*

For the Eastern Rift, there are no independant criteria for ascertaining the amount and direction of movement of the Nubia and Somalia plates, and the pole position has to be calculated from the other two poles of the three plate system using a finite rotational analysis (a programme written by J. Francheteau was used). This gives a pole at 19·6° S, 7·8° E (the '+' in Fig. 5) with a rotation angle of 0·7°. This pole is extremely sensitive to small errors in the rotation angles for the other two poles. For example, Fig. 5 shows the effect of various combinations of a ±5% error in the rotation angles for the Gulf of Aden and Red Sea, other parameters remaining constant. If the errors are in opposite directions then the pole (7, 8) can move by as much as 50° of latitude. A choice of a 5% error implies a 6 km error on the total displacement at the northern end

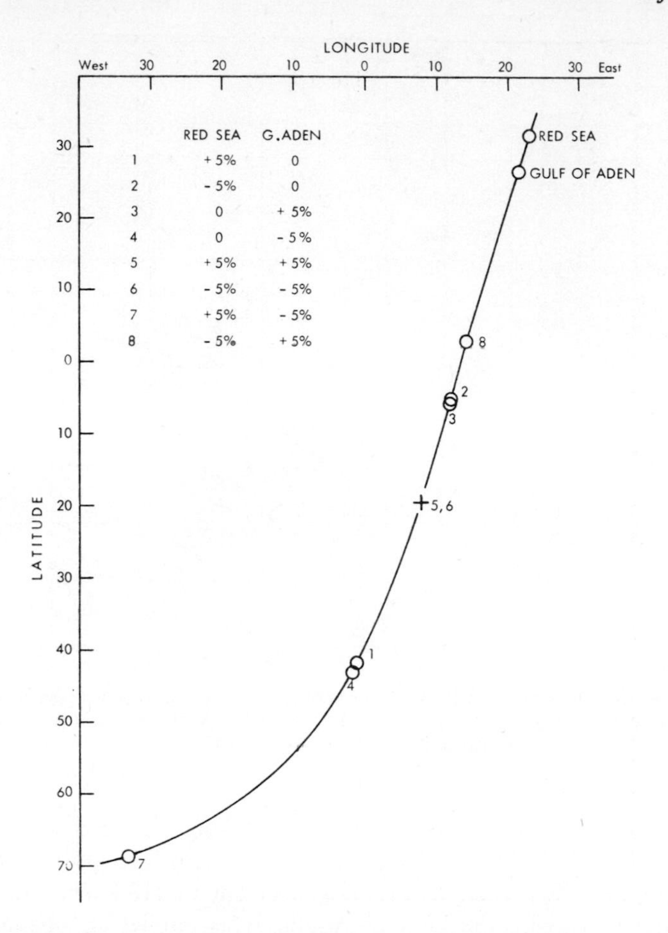

FIGURE 5 Pole positions for the three plate system Arabia–Nubia–Somalia. '+' is the pole position for Nubia–Somalia computed from the parameters given in the text; the other positions are for various errors in the rotational angles for the Red Sea and Gulf of Aden listed, the pole positions being fixed. (After Girdler & Darracott, 1972).

of the Red Sea, and the real error could well be in excess of this. If the pole position errors are also considered, the curve in Fig. 5 would become a band of possible positions. The difficulty of estimating a pole position in this way is obvious and it is dangerous to make predictions concerning the amount of extension across the Eastern Rift using this method.

Seismic Studies

Seismic studies, especially fault plane solutions, can give further information on the relative motions of the plates. In particular, it is possible to constrain the possible position of the pole for the Nubia–Somalia plate boundary by considering the stress field in East Africa (Fairhead & Girdler, 1970, 1971). Figure 6 shows 12 fault plane solutions for earthquakes from the period 1963–1970 and their most likely interpretations. The mechanisms indicate either strike-slip or normal faulting; there is

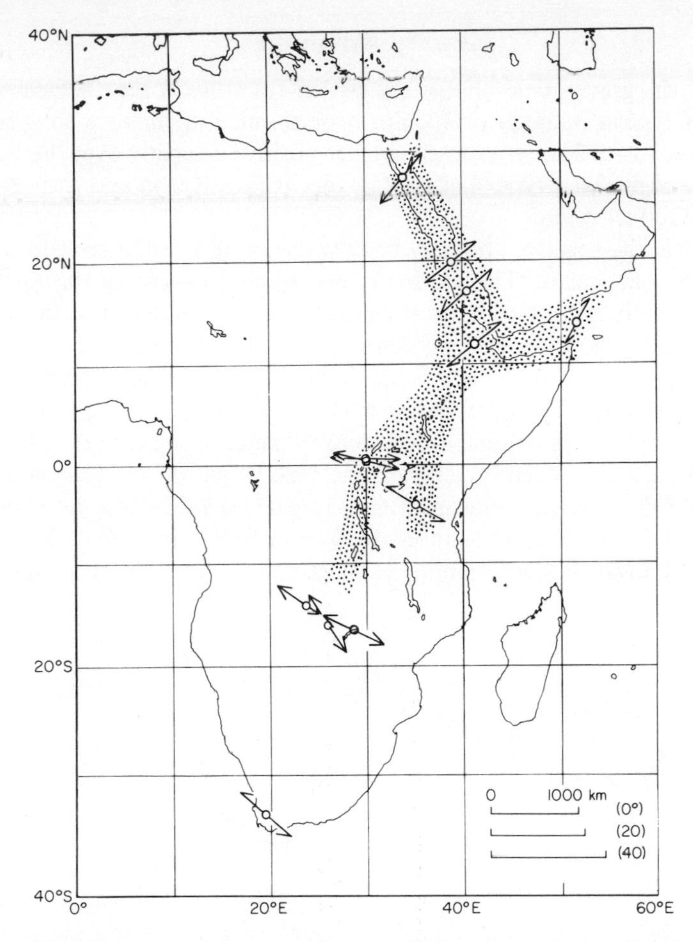

FIGURE 6 Map showing possible stress fields with which various parts of the Rift System may be associated and tentative mapping of the extent of the region of slow P propagation and thinning of the lithosphere (stippled). (After Fairhead & Girdler (1971); reproduced by courtesy of the Royal Astronomical Society).

no evidence of compression. For the Gulf of Aden and Red Sea, the stress fields are approximately northeasterly, i.e. consistent with the separation of the Arabia–Somalia and Nubia–Arabia plates about rotation poles in North Africa. For Nubia–Somalia, the tensional stress field seems to have a direction of about N 120° (Fig. 6). This suggests a pole for the Nubia–Somalia plates somewhere southwest of Africa, which would lie near positions 1, 4 in Fig. 5.

From a study of station travel time corrections (Fairhead & Girdler, 1971), it has been noticed that for seismic recording stations near the rift, P waves are slowed, sometimes by as much as two seconds. The stippled region in Fig. 6 is an attempt to map out the region of slowing down. The region correlates with the region where S_n waves are attenuated and interpreted by Gumper & Pomeroy (1970) to represent a gap in the upper mantle part of the lithosphere. The implication is that there is low velocity material beneath a considerable region of the rift zone (Fig. 6).

Gravity Anomalies

The nature of the gravity field characterizing the rift zone in East Africa is now well established. A typical Bouguer profile north of about 5° S shows a long wavelength negative anomaly with a superimposed smaller positive anomaly over the Eastern Rift axis. This type of profile is typical of the Gregory Rift in Kenya and Ethiopia, and the Red Sea and Gulf of Aden.

In East Africa, the negative anomaly has a width of up to 1000 km and a maximum amplitude of −150 mgals. The gradients are small, suggesting the light material causing the anomaly is at considerable depth and has small density contrast with the surrounding rocks. The anomaly has been interpreted by Girdler *et al.* (1969), Girdler & Sowerbutts (1970) and Baker & Wohlenberg (1971) as being due to lower density asthenosphere replacing the upper mantle part of the lithosphere (s.g. contrast −0·12). Such a model with a large volume of low density material replacing the lower part of the lithospheric plate can readily explain the travel time delays associated with the rifting. The axial positive anomaly has been mapped in considerable detail in Kenya by Searle (1969, 1970) who found the anomaly to be 40 to 80 km wide with an amplitude of +30 to +60 mgal. His interpretation suggests the presence of a mantle derived

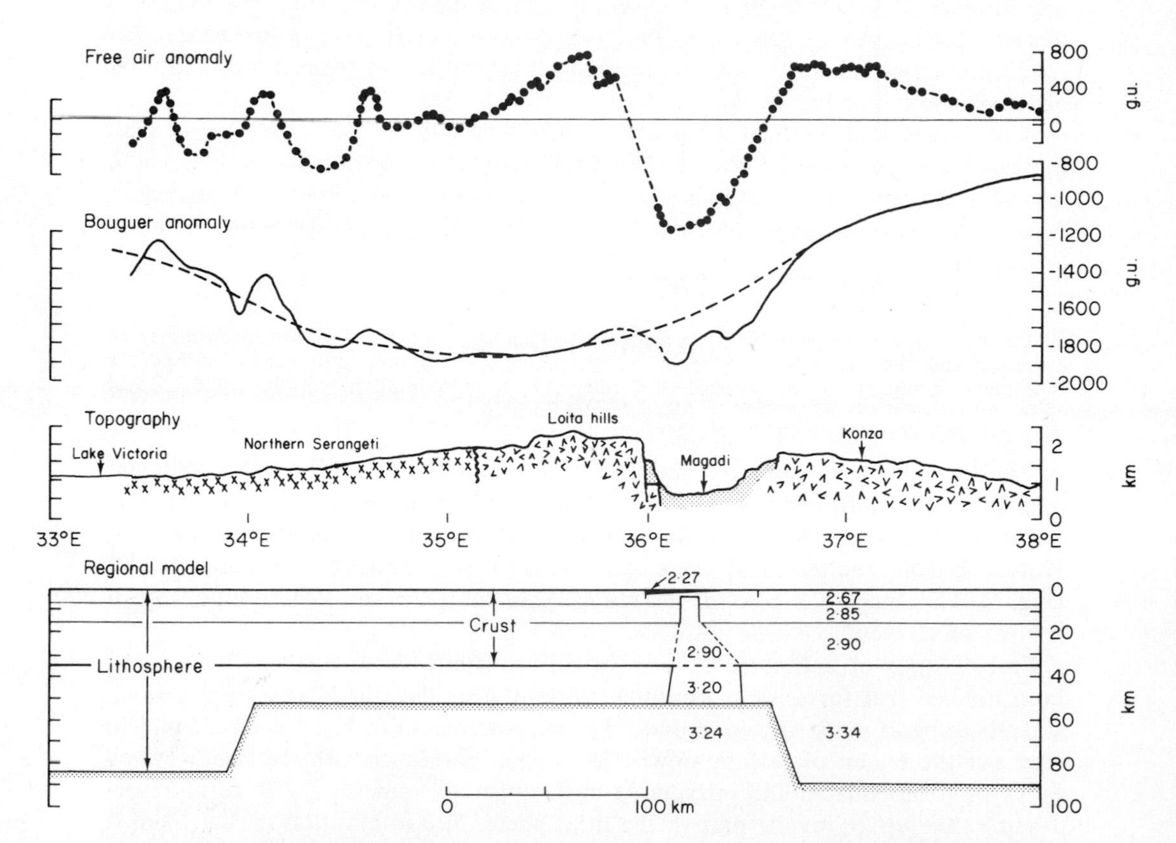

FIGURE 7 Gravity anomalies, geology and model of the lithosphere and asthenosphere satisfying the gravity for a section at the latitude of Lake Magadi (1·8°S). The dashed line in the Bouguer profile is the assumed regional anomaly. (After Darracott *et al.*, 1972).

intrusive zone about 20 km wide reaching in places to within 2 km of the rift floor. Later survey work (Darracott *et al.*, 1972) showed the axial positive anomaly to die out at about 2°S, near Lake Magadi. Figure 7 shows a west–east Bouguer anomaly profile at this latitude, where the intrusive zone is 10 km wide and reaches to within 4 km of the rift floor. There is thus extreme thinning of the lithosphere beneath the rift, accounting for some extension and the extensive volcanism.

The width and amplitude of the positive anomaly is dependent on the extent to which the axial intrusive zone has developed. Over the Gulf of Aden and the Red Sea, the positive anomaly is much larger (e.g. reaching +150 mgals over parts of the Red Sea) and here the intrusive zones are much larger, reaching to the surface (Girdler, 1958) with complete separation of the continental lithosphere.

Conclusions

The Red Sea, Gulf of Aden and the Eastern Rift maybe regarded as three stages in the evolution of an ocean. In the Gulf of Aden and Red Sea, there is complete separation of the continental lithosphere, but in the Eastern Rift the plates do not appear to have separated, though there may be extension and thinning of the lithosphere beneath the rift zone. All three branches of the East African Rift System are characterized by broad negative gravity anomalies with superimposed axial positive anomalies. The width and amplitude of the positive anomaly is dependant on the extent to which the axial intrusive zone has developed.

The geometry of the East African Rift System can be discussed in terms of plate tectonics. Poles of rotation can be defined for the Arabia–Somalia and Nubia–Arabia plate boundaries, but it is more difficult to assign a pole of rotation to the Eastern Rift, where there is no actual separation of the plates, and a relatively small amount of extension.

Acknowledgements

We wish to thank the Natural Environment Research Council, Shell International Petroleum Company and the Royal Society for financial support during the course of these studies. We are grateful to Dr. Henry Stockard (US Naval Oceanographic Office) for making available the Project Magnet data.

References

Baker, B. H. and Wohlenberg, J., 1971. Structure and evolution of the Kenya Rift Valley, *Nature*, **229**, 538–42.

Baker, B. H., Mohr, P. A. and Williams, L. A. J., 1972. Geology of the Eastern Rift System of Africa, *Bull. Geol. Soc. Amer. Spec. Pub.* No. 136.

Darracott, B. W., Fairhead, J. D. and Girdler, R. W., 1972. Gravity and magnetic surveys in northern Tanzania and southern Kenya, *Tectonophys.*, **15**, 131–41.

Davies, D. and Tramontini, C., 1970. The deep structure of the Red Sea, *Phil. Trans. Roy. Soc.* **A267**, 181–9.

Drake, C. L. and Girdler, R. W., 1964. A geophysical study of the Red Sea, *Geophys. J. Roy. Astron. Soc.*, **8**, 473–95.

Fairhead, J. D. and Girdler, R. W., 1970. The seismicity of the Red Sea, Gulf of Aden and Afar triangle. *Phil. Trans. Roy. Soc.* **A267**, 49–74.

Fairhead, J. D. and Girdler, R. W., 1971. The seismicity of Africa, *Geophys. J. Roy. Astron. Soc.*, **24**, 271–301.

Freund, R., Garfunkel, Z., Zak, I., Goldberg, M., Weissbrod, T. and Derin, B., 1970. The shear along the Dead Sea Rift, *Phil. Trans. Roy. Soc.* **A267**, 107–85.

Girdler, R. W., 1958. The relationship of the Red Sea to the East African Rift System. *Quart. J. Geol. Soc. Lond.*, **114**, 79–105.

Girdler, R. W., 1969. The Red Sea—a geophysical background. *In:* Hot Brines and Recent Heavy Metal Deposits in the Red Sea, edited by Degens and Ross, p. 38. Springer-Verlag, N.Y.

Girdler, R. W. and Darracott, B. W., 1972. African poles of rotation. *Comments on the Earth Sciences: Geophysics*, **2** (5), 7–15.

Girdler, R. W., Fairhead, J. D., Searle, R. C. and Sowerbutts, W. T. C., 1969. Evolution of rifting in Africa, *Nature*, **224**, 1178–82.

Girdler, R. W. and Sowerbutts, W. T. C., 1970. Some recent geophysical studies of the East African Rift System, *J. Geomagn. and Geo. elec.*, **22**, 153–63.

Gumper, F. and Pomeroy, P. W., 1970. Seismic waves and earth structure on the African continent, *Bull. Seism. Soc. Amer.*, **60**, 651–68.

Heirtzler, J. R., Dickson, G. O., Herron, E. N., Pitman, W. C. and Le Pichon, X., 1968. Marine magnetic anomalies, geomagnetic field reversals and motions of the ocean floor and continents, *J. Geophys. Res.*, **73**, 2119–36.

Laughton, A. S., 1966. The Gulf of Aden, *Phil. Trans. Roy. Soc.*, **A259**, 150–71.

Laughton, A. S. and Tramontini, C., 1969. The structure of the Gulf of Aden, *Tectonophysics*, **8**, 359–75.

Laughton, A. S., Whitmarsh, R. B. and Jones, M. T., 1970. The evolution of the Gulf of Aden, *Phil. Trans. Roy. Soc.*, **A.267**, 227–66.

McKenzie, D. P., Davies, D. and Molnar, P., 1970. Plate tectonics of the Red Sea and East Africa, *Nature*, **226**, 243–8.

Phillips, J. D., 1970. Magnetic anomalies in the Red Sea. *Phil. Trans. Roy. Soc.*, **A.267**, 205–17.

Robson, D. A., 1970. The Suez Rift, *Nature*, **228**, 1237.

Searle, R. C., 1969. Barometric hypsometry and a geophysical study of part of the Gregory Rift. Ph.D. Thesis, Univ. of Newcastle, U.K.

Searle, R. C., 1970. Evidence from gravity anomalies for thinning of the lithosphere beneath the Rift Valley in Kenya, *Geophys. J. Roy. Astron. Soc.*, **21**, 13–31.

Sykes, L. R., 1970. Focal mechanism solutions for earthquakes along the world rift system. *Bull. Seism. Soc. Amer.*, **60**, 1749–52.

Vine, F. J., 1966. Spreading of the ocean floor: new evidence, *Science*, **154**, 1405–15.

7.4

P. A. MOHR
Astrophysical Lab.
Smithsonian Institute, Cambridge
Mass., U.S.A.

Crustal Deformation Rate and the Evolution of the Ethiopian Rift

Introduction

The African rift system is probably the result of exceedingly slow dilatation of continental lithosphere (Baker *et al.*, 1972). In central Kenya, geological and geophysical data suggest that the Gregory rift results from crustal extension of about 10 km since the Middle Miocene, an average extension rate of 0·5 mm/yr, but in northern Kenya and Tanzania the rate has been of the order of only 0·1 mm/yr (Baker & Wohlenberg, 1971).

Ethiopia is further than Kenya from the presumed plate tectonic rotation pole for the opening of the African rift system (McKenzie *et al.*, 1970; Mohr, 1970), and the extension rates should be correspondingly faster. However, plate geometry does not seem to apply as rigorously to continental as to oceanic regions, owing to the greater importance of non-elastic deformation in sialic crust (Atwater, 1970). In the African rift system, dilatation appears to have been greatest where vertical, domal uplift has been greatest, and lessens towards the swell margins.

The present study concerns the northern sector of the main Ethiopian rift, at the very centre of the Ethiopian swell, and where the rift begins to widen out into the Afar depression (Fig. 1).

The Fault Pattern

The main Ethiopian rift shows abnormally wide development near the latitude of Addis Ababa: 110 km at latitude 8½° N compared with 90 km at 9° N, 80 km at 8° N and 65 km south of 7° N. This excessive width is related to the Addis Ababa structural embayment (Mohr, 1967b). Although perhaps atypical of the main Ethiopian rift, because this sector is better studied and also because it is now traversed by a geodimeter network (Mohr, 1971b) its faulting is selected for examination in some detail here.

All faults manifested by a preserved topographic displacement of 5 m or more are indicated on Fig. 2. The distribution of faults between Addis Ababa and Siri allows the rift floor to be discussed in terms of structural sub-units.

Between Addis Ababa (0 km) and Akaki (14 km direct to the southeast), preserved faulting is virtually absent. Pliocene(?)trachytic ash-flow tuffs, with patches of overlying olivine basalt, dip gently riftwards. The hydrothermally active Filwoha fault in Addis Ababa itself has little topographic expression along its northeast strike. Any rift dilatation in this sector must pre-date the ash-flow tuffs.

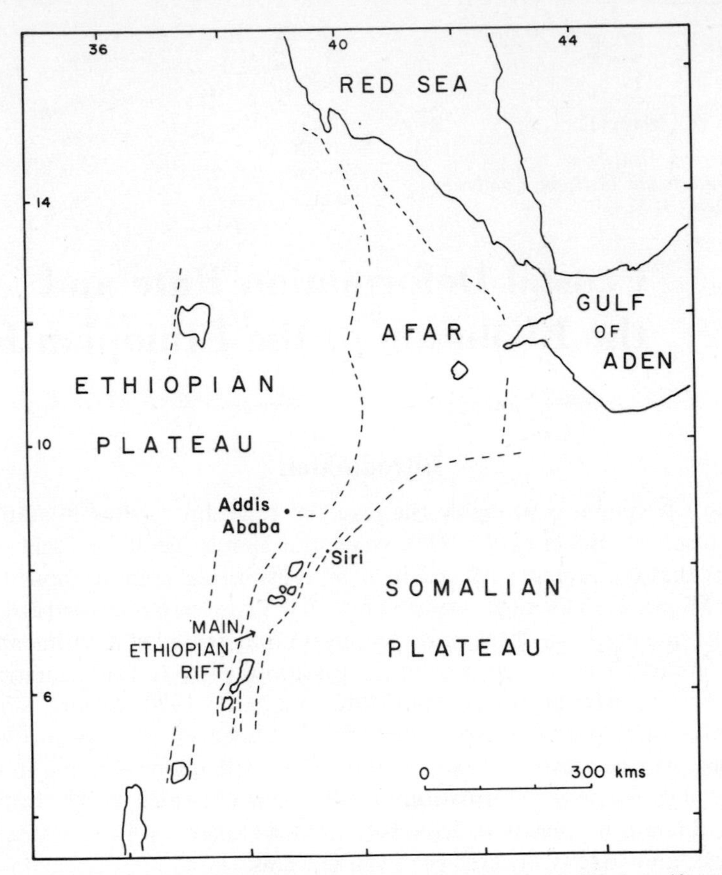

FIGURE 1 Location map of Addis Ababa–Siri traverse in the Ethiopian rift system.

From Akaki (14 km) to Dukam (18 km) some denuded fault-scarps strike north-east–north-northeast and throw an order of 20 m each down to the rift. It is important to realize that any use of denudation as an age criterion for faulting must take into account the fact that rainfall is greater and vegetation more abundant near the rift margins than in the 'cleanly' faulted centre of the rift floor. Nevertheless, the Akaki–Dukam faults may be relatively old (Mid-Quaternary?) judging from the sedimentary thickness which has accumulated behind the upwarped, upthrown sides of the faults. Denuded basaltic cinder cones are situated on or near these faults.

Between Dukam (18 km) and Mojjo (60 km), fluvial and lacustrine rift sediments are cut by rare faults. However, the common occurrence of well-preserved basalt cinder cones, in places aligned along a rift trend, suggests the presence of buried faults. In particular, the much denuded Socora silicic centre (caldera?) is associated with a north-northeast line of maars and small basalt cones which Mohr (1961) considers to have formed about 7000 B.P. The ejecta from these maars effectively obscure most of the related faulting, though farther to the north-northeast some antithetic rift faults, upthrown east, are developed. This maar lineament near Bishoftu is a major rift feature, and may be a northerly continuation, via Zuquala volcano and its satellite cones, of the Koshi fault of the Galla lakes basin (Mohr & Gouin, 1967).

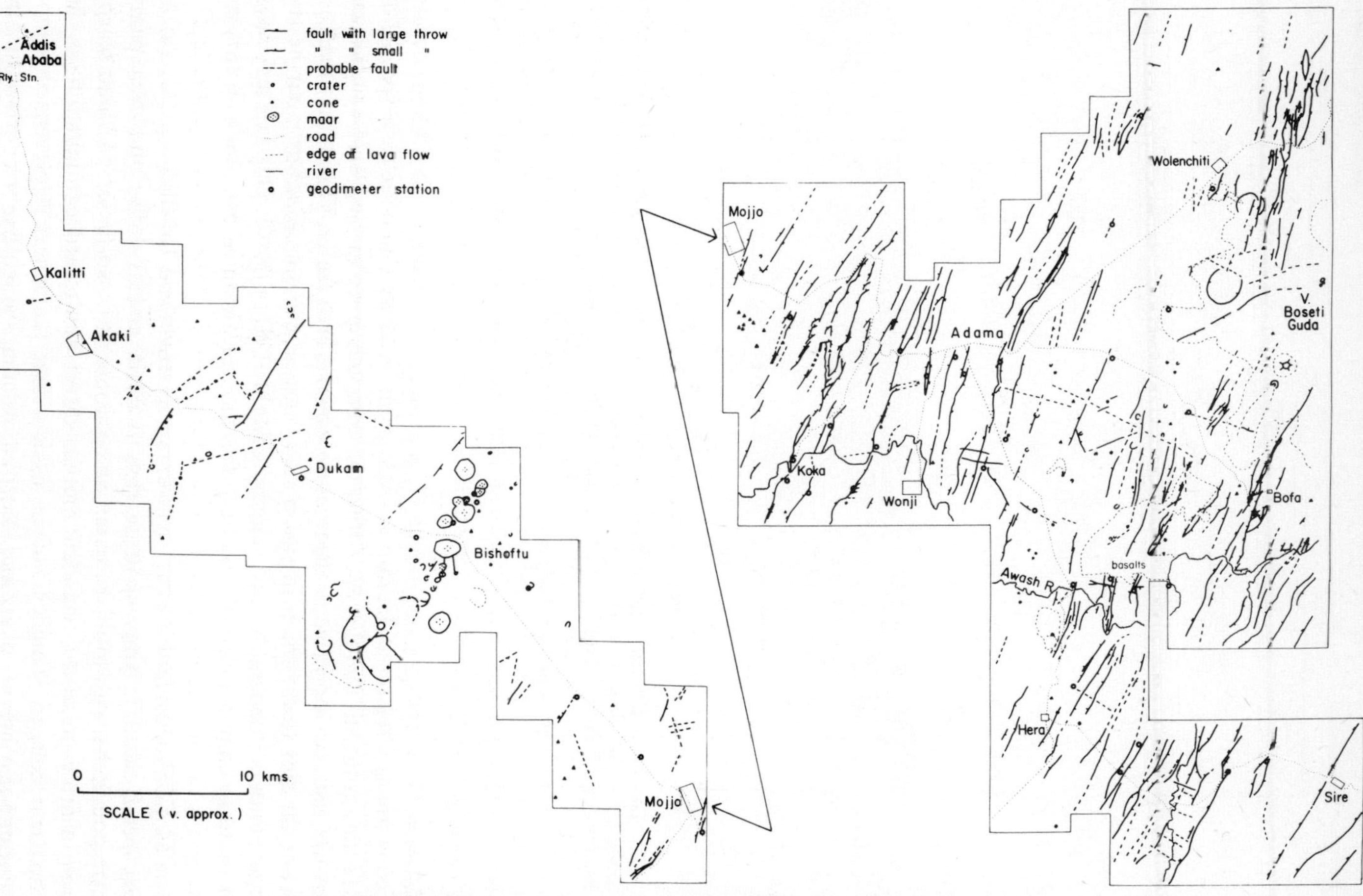

FIGURE 2 Fault map of the Addis Ababa–Siri sector of the main Ethiopian rift.

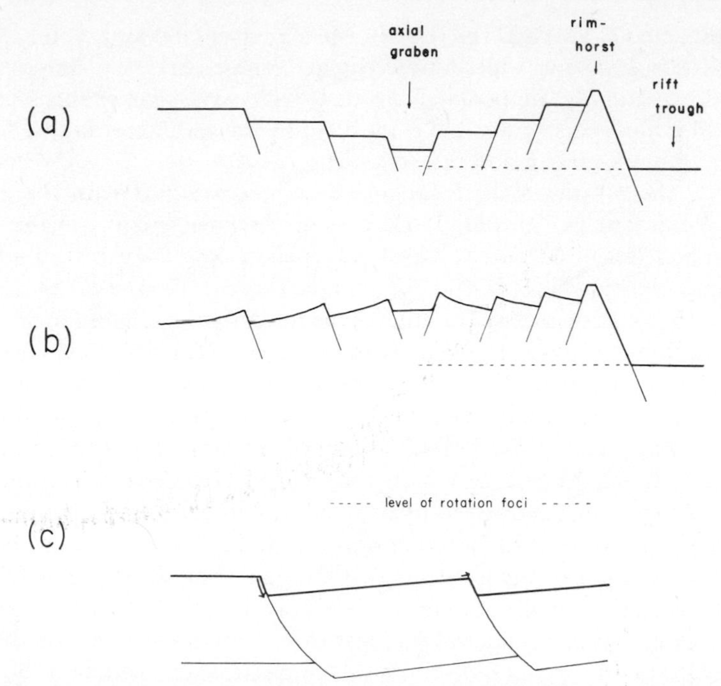

FIGURE 3 The role of different types of faulting-tilting-warping from crustal extension in determining rift topography: (*a*) planar normal faulting, (*b*) planar normal faulting and upwarping of faulted blocks, (*c*) curved normal faulting. Vertical scale in these cross-sections is grossly exaggerated, except that fault-hades are kept at 20°. Note that warping does not significantly affect the computation of crustal extension from fault hade and displacement. Neither does curvature of the fault-planes (with surface hades of 20°) significantly alter the extension data, though it does impose certain mechanical constraints. Curved fault-planes are particularly suited to rachet-type faulting of rigid upper crust.

Between Mojjo (60 km) and Adama (77 km) an uplifted zone of the rift floor occurs which is strongly fractured. Upthrows are both east and west, and the close spacing of the faults results in horst-graben formation, precluding upwarping of the upthrown blocks except near Mojjo where the fracture intervals are greater. The uplift is marked by regional dips to the trachytic ignimbrites, away either side of a north-northeast–south-southwest topographic axis (Gouin & Mohr, 1964, Fig. 2i). This axis may have originated in conjunction with the Late Tertiary volcanism of Mt. Bokkam (Mohr, 1967a).

The Mojjo–Adama faulting cuts sediments of Garbien and Tabellien age (M. Taieb, 1969, pers. comm.) lying upon severely welded crystal tuffs, and was pene-contemporaneous with localized basalt and rhyolite obsidian flows. The fault scarps remain sharply preserved. This faulting has direct southward continuation, as the Wonji fault belt, to Wonji, Gadamsa caldera and the intense Holocene volcano-tectonism northeast of Lake Zway. North of Adama, the faulting strengthens and its eastern margin (the western margin of the 4 km-wide Adama graben) is formed by a 100–150 m east-facing scarp heightened by horst formation on the very rim of the upthrown block (Fig. 3; see also Mohr, 1971a, p. 80). Thirteen km north-northeast of Adama this scarp reaches its greatest elevation at an east–west axis fringed by antithetic faults.

Between Adama (77 km) and the Kaletta/Mennez river (100 km), young pyroclastic and ash-flow tuffs, and some superimposed basalt cones and rhyolite domes, are cut by intense Holocene faulting. Immediately north of the Awash river in this sector a fresh basaltic shield buries older faults, but is itself cut by the very latest faults. The type of faulting is both horst-graben and ratchet.

Northwards the Adama–Kaletta faulting becomes identified with the en echelon transposed Wonji fault belt (Mohr, 1960), passing via the dormant volcanic centres of Boseti Guda and Boseti Bericha to Gariboldi caldera. Southwards it dies out at the Badda–Chilalo transposition of the rift margin (Mohr, 1966, 1967b). East of the Kaletta river, the massive, stepped faulting of the rift margin is reached at Siri (105 km). This margin faulting throws Palaeogene Trap Series basalts at least 750 m in four major steps, with block tilting away from the rift being splendidly developed (Mohr, 1966).

Despite the similarity in fault density for the Mojjo–Adama and Adama–Kaletta sectors, there are notable differences in fracture behaviour. Fault-throws tend to be smaller in the Adama–Kaletta sector, but associated Holocene volcanism has been much more intense, and basalt-filled fissures indicate direct crustal dilatation.

In summary, the Addis Ababa–Siri traverse of the rift is of unusual length, and involves a dual crossing of the transposed Wonji fault belt. This is possible because these lateral transpositions are not abrupt (Gibson, 1969; Mohr, 1960, 1967b). Progressively younger volcanic and tectonic features, from margin to axis of the main Ethiopian rift, is not the general case. This is emphasized at latitude 8° N where the most recent brittle deformation of the rift is concentrated at the margins, leaving the wide interior of the rift largely unfractured.

Fault Displacements

Displacements of faulted topography have been measured along the Addis Ababa–Siri traverse. The values can show rapid lateral variations along the fault-strike, such that it is misleading to quote precise figures. Indeed, the short lengths of the faults, usually less than 10 km, indicate a percentage of remaining strain or else non-elastic deformation. The faults are considered here in two broad categories, according to whether their throw is nearer 20 m or 50 m average along their length. An average hade of 20° is taken, though the bigger rift margin faults hade slightly more than this, and some Wonji belt faults hade less or occasionally very much more. Illies (1967) took observed values of 25–30° in computing the faulted extension of the Rhine graben.

(*a*) *Total horizontal displacement*

Relevant data for the traverse are summarized in Tables 1 and 2. The youngest faulting, between Mojjo and the Kaletta river, has produced an extension of 500 m across a distance of 40 km, which is 1 part in 80. If, as the sediment ages suggest, this extension has taken place since a maximum of 100,000 B.P., then a minimum extension rate of 5 mm/yr is obtained.

For the whole width of the rift, the observed surficial extension due to Quaternary faulting is 900 m across 110 km, or 1 part in 120, at a minimum rate for the past 3 m.y. of 0·3 mm/yr. This is precisely the order of magnitude predicted from plate analysis of the African rift system (Baker *et al.*, 1972). In considering these values, two points must be borne in mind: (i) that earlier Quaternary faulting, especially of direct dilatational

TABLE 1 Quaternary fault densities across the northern part of the main Ethiopian rift.

Traverse	Distance (km)	Elevations (W–E) (metres)	Number of faults Downthrown W 20 m	50 m	Downthrown E 20 m	50 m
Addis Ababa–Mojjo	60	2440–1775	8	0	7	1
Mojjo–Adama	17	1775–1910–1690	14	1	5	5*
Adama–Kaletta	25	1690–1590	11	3	18	1
Kaletta–Siri	10	1590–2440	0	5†	0	1

* The large fault west of Adama is thrown 100–150 m

† Four of these rift margin faults throw more than 100 m each

TABLE 2 Cumulative magnitudes of fault displacements for various sectors across the northern part of the main Ethiopian rift.

Traverse	Vertical displacement (m) Down W	Down E	Horizontal displacement (m)
Addis Ababa–Mojjo	160	190	115
Mojjo–Adama	330	425	250
Adama–Kaletta	370	410	260
Kaletta–Siri	c. 800	50	280
Total (with margin)	1660	1025	905
(without margin)	860	975	625

type, is likely to have been obscured by later sediments: the localized cracks which periodically appear in the rift floor are filled in by slumping and flood-water deposition within a few months (Gouin & Mohr, 1967); and (ii) that the preserved faulting is preferentially concentrated into belts, such that extension due to brittle fracture is not a process uniformly affecting the entire width of the rift floor.

The extension rates quoted here are given as minima because they are based on only one mechanism of crustal extension: viz. normal faulting. Other possible mechanisms include horizontal separation of the type preserved in hard ash-flow tuff near Fant-ali volcano (Gibson, 1967), and crustal necking due to aseismic slip (creep) with accompanying upper crustal subsidence. In both these cases, rift floor sedimentation is likely to obscure the surficial evidence of extension, and geodetic measurements are required to detect contemporary movement.

The values for Quaternary extension rates of the Ethiopian rift calculated here are appreciably higher than for the Kenya rift (Baker & Wohlenberg, 1971), which in turn has higher values than the Rhine graben (Illies, 1970) and Baikal rift (Artemjev & Artyushkov, 1971).

(b) Total vertical displacement

Plate tectonic considerations, in which horizontal crustal displacements are the essence of the matter, have tended to detract from the importance and significance of vertical displacements (Sclater *et al.*, 1971). The impressive rift valleys of Africa are in most places bordered by normal faults, which have produced a vertical displacement to match the estimated Quaternary horizontal extension of these rifts.

In a symmetrically faulted rift valley, the vertical throws of the two sets of opposing faults essentially cancel out. The main Ethiopian rift at Addis Ababa is asymmetrically faulted (Mohr, 1962, 1967b), with a Quaternary total of 1660 m westerly downthrow and 1025 m easterly downthrow. The fact that the bordering plateaux lie at approximately the same elevation (2450–2500 m), east and west of the rift, requires either: (i) the presence of buried east-downthrown faults near Addis Ababa, and/or (ii) the accomplishment of vertical displacement by tilting in addition to faulting. The latter case is more probable. Indeed, care must be taken in relating rift topography to the amount of vertical slip on crustal fractures (Fig. 3).

In the axial zone of the rift, between Mojjo and the Kaletta river, there has been a slight excess of west-upthrown over east-upthrown displacement, though not sufficient to match the elevation difference (Table 1). This indicates a gentle tilting component down to the east.

Absolute vertical displacement rate for the Ethiopian rift cannot be estimated because (i) of the unknown effect of regional uplift of the Ethiopian swell, and (ii) even using a datum on the plateau, displacement on the rift floor is complicated by warping and tilting. However, values of the order of 1 mm/yr seem reasonable, and are comparable with levelling data from the Icelandic rift zone (Tryggvason, 1970), the Levantine rift (Karcz & Kafri, 1971), and the Rhine graben (Illies, 1970), though they are much less than the $1\frac{1}{2}$ cm/yr uplift rate for some Afar graben (Tazieff *et al.*, 1972).

Evolution of the Ethiopian Rift

Only the Quaternary history of the rift can be discussed here, subject to the serious handicap of the present lack of reliable age-determinations on the rocks concerned: crude estimates have perforce to be made from stratigraphy and degree of denudation.

A first important question in rift evolution is whether structures are younger at the rift axis than at the rift margins. If affirmed, this in turn leads to the question of whether the crustal rocks themselves are younger at the axis than the margins: such younger crust would be essentially volcanic except where graben subsidence along the rift axis has caused sedimentary infilling (Baker & Wohlenberg, 1971).

The present study permits a reassertion that the bulk of Holocene faulting in the rift is confined to the sub-axial Wonji fault belt (Mohr, 1960, 1962). However, this prime fact must not be used to overlook the occurrence of equally young tectonism nearer or at the rift margins (e.g. at Bishoftu, Butajira, Awasa), perhaps in response to a crustal boundary stress system (Bott & Dean, 1972). Gibson & Tazieff (1970, p. 332) consider that the floor faulting becomes 'progressively older as one approaches the rift margins' in the Addis Ababa region, but the present study cannot unequivocally confirm this. Nor indeed is it clearly the case for any other part of the rift. Whatever the nature and origin of the Ethiopian rift, it appears to have developed within a zone of crustal deformation whose width has remained relatively constant since its inception.

Relatively denuded, presumably older faults are admittedly not common in the axial zone of the rift. The young volcano-tectonism and sedimentary cover militate against their preservation, but instances are known for example from 8 km west of Wolenchiti and 5 km west of Bofa (Fig. 2). Presumably, if during the Quaternary new crust has been injected into the base of a sutured rift block, its amount has matched the 1 km of surface extension taken up by normal faulting across the Wonji fault belt and the rift margin zones. Yet the Wonji fault belt, like the axial fault zone in the Kenya rift, is marked by positive gravity anomalies whose causative sub-surface mass is considered to have a width similar to the 5–15 km width of the surface fault belts (Searle, 1970; Searle & Gouin, 1972).

Either the axial intrusive mass is largely older than the surface fault belts above it, or else massive stoping has occurred such that assimilation or anatexis and eruption of crustal magmas is required (R. C. Searle, personal communication). Mohr (1971c), on the basis of volcanic petrochemistry, has argued against the latter possibility involving sialic crust, but nevertheless the anatexis of previously differentiated 'new' crust remains feasible. Thus the rift crustal block is envisaged as having, both in its volcanism and tectonism, a history of self-ingestion through which much of the evidence for its earlier history is destroyed. This helps explain the disparity noted by Baker & Wohlenberg (1971) between their estimated 10 km width of the axial intrusion in the Kenya rift block and the 3 km extension evidenced in the observed surface geology.

Searle & Gouin (1972) have made a very detailed gravity survey of part of the Ethiopian rift, and conclude that virtually the entire width of the rift floor is underlain by a shallow, high density intrusion. *If* the width of this intrusion represents the amount of lithospheric dilatation, an average extension rate of 3 mm/yr since the early Miocene is obtained. This is much more than the Quaternary average rate obtained in this paper, but is of the same order as the Holocene rate for which the surface geological evidence is best preserved. A 70 km crustal extension across the rift near Addis Ababa agrees well with plate theory (McKenzie *et al.*, 1970), but is far greater than some geologists have found acceptable (Mohr, 1967; Baker & Wohlenberg, 1971; Baker *et al.*, 1972). Seismic refraction work now in progress (Berckhemer, personal communication), may help to distinguish between rift intrusion due to dilatation and through stoping.

A second question concerning rift evolution is whether the average deformation rates calculated here apply on a scale of centuries or even decades.

The present very low level of seismicity in the main Ethiopian rift (Gouin, 1960, 1970; Molnar *et al.*, 1970) is difficult to reconcile with the observed Holocene volcano-tectonism. Gouin (1970) notes however that the pattern of Ethiopian seismicity has changed quite drastically even since the beginning of this century, and on this time scale cannot be expected to mirror the regional rift tectonics. Nevertheless, low seismicity in the rift could be related to dissipation of stress by creep in a hot lower crust, and stoping rather than forcible injection by the rift intrusion.

Brittle fracture of the rift upper crust is likely a stop-go phenomenon responding to presumed progressive stress acting at its base. In view of the slow accumulation of strain indicated by the deformation rates calculated here, the stop-go periodicity will probably be a long-term one, except where the strength of the rift crust is being reduced by thinning (Makris *et al.*, 1972). Certainly the 800 m of vertical displacement of the rift margin in the Early Quaternary indicates a very considerable prior accumulation of strain, though the energy involved depends on the fault-plane areas which may have been relatively small.

The Wonji fault belt represents short period deformation of thin crust, and the variable throw-direction of its faults could in part be an isostatic response to hydro-dynamic stress. Because of this, and also the steeper hades compared with the rift margin faults, some of the horizontal strain at the centre of the rift floor has perforce been released by open tensional fissuring.

If the driving mechanism for global plate motions is push from the ridge-rifts (Baker *et al.*, 1972), then an acceleration of rifting with time (prior to the complete oceanization of the rift floor and margins) might be expected in the African rift system. As the lithosphere thins due to faulting and igneous intrusion in the rift zone, so less stress is required to rupture it, and the rise of geotherms will facilitate aseismic slip up to progressively shallower depths. Strain release will concentrate in the necked zone. On the other hand, a driving mechanism acting outside the swell-rift zone will presumably act at rates independent of the degree of rifting and crustal dilatation, though the ratio of ductile through cataclastic deformation to brittle fracture will still increase with time.

In view of the likely long-term periodicity of rift fracturing, there is presently insufficient evidence to determine whether or not the rate of crustal deformation has changed during the evolution of the Ethiopian rift. Neglecting margin and near-margin faulting, however, the intensity of Holocene volcano-tectonism along the rift axis is suggestive of a quickening deformation rate for this zone at least.

Geodetic Monitoring

The uncertainties of rift structural evolution can partly be resolved by precise geodetic measurements of crustal deformation. It is important to seek both strain which is being released by aseismic slip (fault-creep) and creep, and accumulating strain to be released in future seismic slip.

The S.A.O. geodimeter network at latitude 8½° N in the Ethiopian rift was set up in 1969, and is capable of detecting crustal movements of 1–2 cm or more, depending on the length of the line concerned. This network has been remeasured and extended both in 1970 and 1971, and other networks have been established on the Wonji fault belt at latitudes 7½° and 6° N. The networks are designed for detection of horizontal tension and longitudinal shear. Vertical movement is not yet being monitored, excepting the 1969 triangulation network established by the National Technical University of Athens.

The results of the geodimeter programme to date are to be published in Bulletin Geophys. Obs. Addis Ababa, no. 14 (in press).

Acknowledgement

The author is grateful to Dr. R. C. Searle for a review of this paper which led to considerable improvements in its presentation.

References

Artemjev, M. E. and Artyushkov, E. V., 1971. Structure and isostasy of the Baikal rift and the mechanism of rifting, *J. Geophys. Res.*, **76**, 1197–1211.

Atwater, T., 1970. Implications of plate tectonics for the Cenozoic tectonic evolution of western North America, *Bull. Geol. Soc. Amer.*, **81**, 3513–36.

Baker, B. H., Mohr, P. A. and Williams, L. A. J., 1972. Geology of the Eastern Rift System of Africa, *Spec. Paper Geol. Soc. Amer.*, **136**, 67 pp.

Baker, B. H. and Wohlenberg, J., 1971. Structure and evolution of the Kenya rift valley, *Nature*, **229**, 538–42.

Bott, M. H. P. and Dean, D. S., 1972. Stress systems at young continental margins, *Nature Phys. Sci.*, **235**, 23–25.

Gibson, I. L., 1967. Preliminary account of the volcanic geology of Fantale, Shoa, *Bull. Geophys. Obs. Addis Ababa*, **10**, 59–68.

Gibson, I. L., 1969. The structure and volcanic geology of an axial portion of the main Ethiopian rift, *Tectonophysics*, **8**, 561–5.

Gibson, I. L. and Tazieff, H., 1970. The structure of Afar and the northern part of the Ethiopian rift, *Phil. Trans. Roy. Soc. London*, **A267**, 331–8.

Gouin, P., 1960. Seismological notes: a historical survey of the seismicity of Ethiopia and the first seismological report of the University College Observatory (March–June 1959), *Bull. Geophys. Obs. Addis Ababa*, **3**, 23–35.

Gouin, P., 1970. Seismic and gravity data from Afar and surrounding areas, *Phil. Trans. Roy. Soc. London*, **A267**, 339–58.

Gouin, P. and Mohr, P. A., 1964. Gravity traverses in Ethiopia (interim report), *Bull. Geophys. Obs. Addis Ababa*, **7**, 185–239.

Gouin, P. and Mohr, P. A., 1967. Recent effects possibly due to tensional separation in the Ethiopian rift system, *Bull. Geophys. Obs. Addis Ababa*, **10**, 69–78.

Illies, H., 1967. Development and tectonic pattern of the Rhinegraben, *Rhinegraben Prog. Rep.*, **6**, 7–9.

Illies, H., 1970. Graben tectonics as related to crust-mantle interaction. *In:* Graben Problems, edited by J. H. Illies and St. Mueller, pp. 4–27. Schweizerbart, Stuttgart.

Karcz, I. and Kafri, U., 1971. Geodetic evidence of possible recent crustal movements in the Negev, southern Israel, *J. Geophys. Res.*, **76**, 8056–65.

Makris, J., Menzel, H. and Zimmermann, J., 1972. A preliminary interpretation of the gravity field of Afar, northeast Ethiopia, *Tectonophysics*, **15**, 31–9.

McKenzie, D. P., Davies, D. and Molnar, P., 1970. Plate tectonics of the Red Sea and East Africa, *Nature*, **226**, 243–8.

Mohr, P. A., 1960. Report on a geological excursion through southern Ethiopia, *Bull. Geophys. Obs. Addis Ababa*, **3**, 9–20.

Mohr, P. A., 1961. The geology, structure and origin of the Bishoftu explosion craters, *Bull. Geophys. Obs. Addis Ababa*, **4**, 65–101.

Mohr, P. A., 1962. The Ethiopian rift system, *Bull. Geophys. Obs. Addis Ababa*, **5**, 33–62.

Mohr, P. A., 1966. Geological report on the Lake Langano and adjacent plateau regions, *Bull. Geophys. Obs. Addis Ababa*, **9**, 59–75.

Mohr, P. A., 1967a. Further notes on the explosion craters of Bishoftu (Debra Zeit), *Bull. Geophys. Obs. Addis Ababa*, **10**, 99–112.

Mohr, P. A., 1967b. The Ethiopian rift system, *Bull. Geophys. Obs. Addis Ababa*, **11**, 1–65

Mohr, P. A., 1970. Plate tectonics of the Red Sea and East Africa, *Nature*, **228**, 547–8.

Mohr, P. A., 1971a. Tectonics of the Dobi graben region, central Afar, *Bull. Geophys. Obs. Addis Ababa*, **13**, 73–89.

Mohr, P. A., 1971b. Smithsonian geodimeter survey, *Bull. Geophys. Obs. Addis Ababa*, **13**, 121.

Mohr, P. A., 1971c. Ethiopian rift and plateaus: some volcanic petrochemical differences, *J. Geophys. Res.*, **76**, 1967–84.

Mohr, P. A. and Gouin, P., 1967. Gravity traverses in Ethiopia (third interim report), *Bull. Geophys. Obs., Addis Ababa*, **10**, 15–52.

Molnar, P., Fitch, T. J. and Asfaw, L. M., 1970. A micro-earthquake survey in the Ethiopian rift, *Earthquake Notes*, **41**, 37–44.

Sclater, J. G., Anderson, R. N. and Bell, M. L., 1971. Elevation of ridges and evolution of the central Eastern Pacific, *J. Geophys. Res.*, **76**, 7888–7915.

Searle, R. C., 1970. Evidence from gravity anomalies for thinning of the lithosphere beneath the rift valley in Kenya, *Geophys. J.*, **21**, 13–31.

Searle, R. C. and Gouin, P., 1972. A gravity survey of the central part of the Ethiopian rift valley, *Tectonophysics*, **15**, 41–52.

Taieb, M., 1969. Différents aspects du Quaternaire de la vallée de l'Aouache (Ethiopie), *C.R. Acad. Sci. Paris*, **269**, 289–92.

Tazieff, H., Varet, J., Barberi, F. and Giglia, G., 1972. Tectonic significance of the Afar (or Danakil) depression, *Nature*, **235**, 144–7.

Tryggvason, E., 1970. Surface deformation and fault displacement associated with an earthquake swarm in Iceland. *J. Geophys. Res.*, **75**, 4407–22.

7.5

R. BLACK

Department of Geology
University of Haile Selassie
Addis Ababa, Ethiopia

Structures of the Afar Floor (Ethiopia)

Field mapping near the southeastern and western margins of the Afar Depression has led to the following conclusions:

The Aisha 'horst' is not a rigid horst block but an integral part of an attenuated sialic floor extending beneath the post-Miocene flood basalts of southern Afar. This attenuation is expressed at the surface by fields of tilted blocks with bedding dip angles which may exceed 40°. Likewise the Danakil Alps are believed to be attenuated crust which has stabilized at an early stage.

In the northern part of the Aisha region the following tectonomagmatic evolution has occurred:

(1) East–west crustal extension by block faulting and tilting of pre-Miocene rocks on north and north-northwest-trending faults preceded and followed by widespread acid volcanism.

(2) Extrusion of flood basalts, the upper part of the sequence having been affected only by west-northwest block faulting which curves into the Gulf of Tadjura–Gulf of Aden structures.

The area is also cut by northeast-trending wrench faults running parallel to the transform faults of the Gulf of Aden. Recent basaltic volcanism occurs along the southern escarpment related to the rejuvenation of east–west faults.

Triple plate separation has given rise to extremely complex interference fault patterns. Furthermore the fault patterns observed in recent lavas may be totally different to those developed in the older underlying rocks.

Eastward block tilting and marginal graben formation along the western Afar margin needs reinterpretation in terms of extension tectonics, and a model for sialic crustal extension and attenuation is proposed. Evidence from the Gulf of Aden and Red Sea points to extensive sialic crustal separation. In the Afar a similar amount of plate separation appears to be accommodated by a complex interplay of attenuation mechanisms with only minor oceanic crust creation.

7.6

I. G. GASS
Department of Earth Sciences
The Open University
Wilton Hall, Wilton
Bletchley, Bucks., UK

The Red Sea Depression: Causes and Consequences

If sea-floor spreading and its corollary plate tectonics are accepted then continents that are now separated by oceanic floor were once contiguous. But why and how did transcontinental rupture take place and how did this lead to the formation of new oceanic crust? These questions are most likely to be answered in an area where oceanic crust has been created in the geologically recent past. It is proposed therefore to examine the Red Sea and the continental plates adjacent to it for, irrespective of which geophysical authority is accepted (Girdler, 1958; Vine, 1966; Davies & Tramontini, 1970 and Darracott *et al.*, this volume), this 'proto'-ocean formed in the last 25 m.y. However, in using the Red Sea region as a model, it is necessary to enquire why extensive Pliocene–Recent alkali volcanics mantle much of the western part of this supposedly aseismic (McKenzie, 1970) plate and if these magmatic products are related, or unrelated to the spreading processes in the Red Sea.

It is proposed that the tensile strength of the lithosphere is such that transcontinental fracturing is unlikely unless the lithosphere is weakened first—but what is this weakening process? Thermodynamic considerations suggest that below continental lithosphere, the lithosphere/asthenosphere boundary is one of thermal instability and that the plate itself, with its high content of radioactive isotopes in the sialic crust, acts as a thermal blanket and further accentuates the thermal instability between the two layers. In such a situation the lithosphere/asthenosphere boundary would be regionally elevated should the underlying mantle, for some reason, become hotter than its surroundings. Several workers have proposed various models, ranging from lower mantle thermal plumes (Morgan, 1971), excess volatiles in the asthenosphere (Wyllie, 1971), penetrative convection as consequence of uniform heating from below (Elder, 1970) to the consideration of the dynamics of viscous layers with contrasting densities (Ramberg, 1972) to explain such thermal perturbations.

The primary aim of this paper is to enquire into the consequences of these thermal perturbations and not into the means by which they were produced. The thermal gradient within the perturbation would be steeper than in the surrounding colder areas. This would (i) allow partial melting of the mantle peridotite to occur higher in the mantle and be more extensive than elsewhere in the asthenosphere (Gass, 1970) and (ii) necessitate the downward movement of the main phase boundaries in the mantle Magnitsky & Kalashnikova, 1970; Wyllie, 1971, pp. 233–52) which in turn would result in an increase in volume as low temperature/high pressure minerals convert to their less dense polymorphs. The conversion of mantle peridotite of density 3·5 g/cm^{-3} to density 3·3 g/cm^{-3} over a layer of some 50 km would result in an uplift in the order of 3·0 km (Magnitsky & Kalashnikova, 1970). This increase in volume would be most easily relieved by the uplift of the Earth's surface and fracturing would occur as the brittle crust, and mantle, overlying the thermal perturbation was domed. Such fractures would then afford easy egress to the surface for the magma already generated within the thermal upwarp.

So, the surface manifestations indicating the presence of an underlying thermal eminence in the asthenosphere/lithosphere interface should be the elevation of the Earth's surface associated with crustal fracturing and the products of volcanic activity; logically, the uplift and fracturing should precede the eruptive processes. In Africa, much of whose surface was peneplained by erosion during the Mesozoic, there are several areas of Late Mesozoic and Early Tertiary domal uplift surmounted by subsequent volcanic rocks (Black & Girod, 1970; Vincent, 1970; Burke & Whiteman, this volume; Faure, this volume).

Although the majority of these domes are isolated and therefore the associated fractures are of limited extent, there are others which overlap and produce seeming near transcontinental fractures. The best documented case is that of the East African rift systems where domal uplift centred on Ethiopia, Kenya–North Tanzania and Malawi, with a smaller dome athwart the Red Sea at latitude 22° N, has produced a fracture system extending for some 3500 km, but, as oceanic crust has only been produced in the Red Sea and the Gulf of Aden, it is this region that is now considered in more detail.

The region now trisected by the Red Sea, Gulf of Aden and the Ethiopian Rift was one of magmatic and tectonic quiescence from the uppermost Precambrian until the Early Tertiary. In the Early Tertiary, two things happened. First, vast quantities of alkalic basalts were extruded upon a peneplain surface of crystalline basement and Jurassic calcareous sediments. Second, this peneplain was upwarped to produce the large surface feature now termed the Afro-Arabian Dome and also a smaller dome further to the north. With the passage of time, volcanism became concentrated in the Gulf of Aden, the Red Sea and the Ethiopian Rift, the primary zones of weakness produced by the crustal uplift. Within these major fracture zones, the volcanic products became less undersaturated and a basalt type intermediate between true tholeiite and true alkalic basalt was erupted in association with abundant peralkalic silicic differentiates. Finally, in the floor of the Red Sea and the Gulf of Aden, and possibly also in the north central Afar depression, once continental separation had taken place, tholeiitic basalts were erupted.

A petrogenetic model has been erected (Gass, 1970, pp. 375–9) to explain this temporal and spatial variation in volcanic products. Briefly, it was suggested that as the saturated/undersaturated nature of basaltic melts is seemingly related to the pressures and thereby the depth within the mantle at which they form (undersaturated basalts being generated at greater depths than oversaturated tholeiites), the sequence of events could be related to equilibration of magmas at various levels (pressures) within the Earth. So, as partial melting advanced in response to the elevation of the thermal gradient within the lithosphere/asthenosphere perturbation, the first body of magma would be generated at a depth of some 60 km. This magma would be alkalic in com-

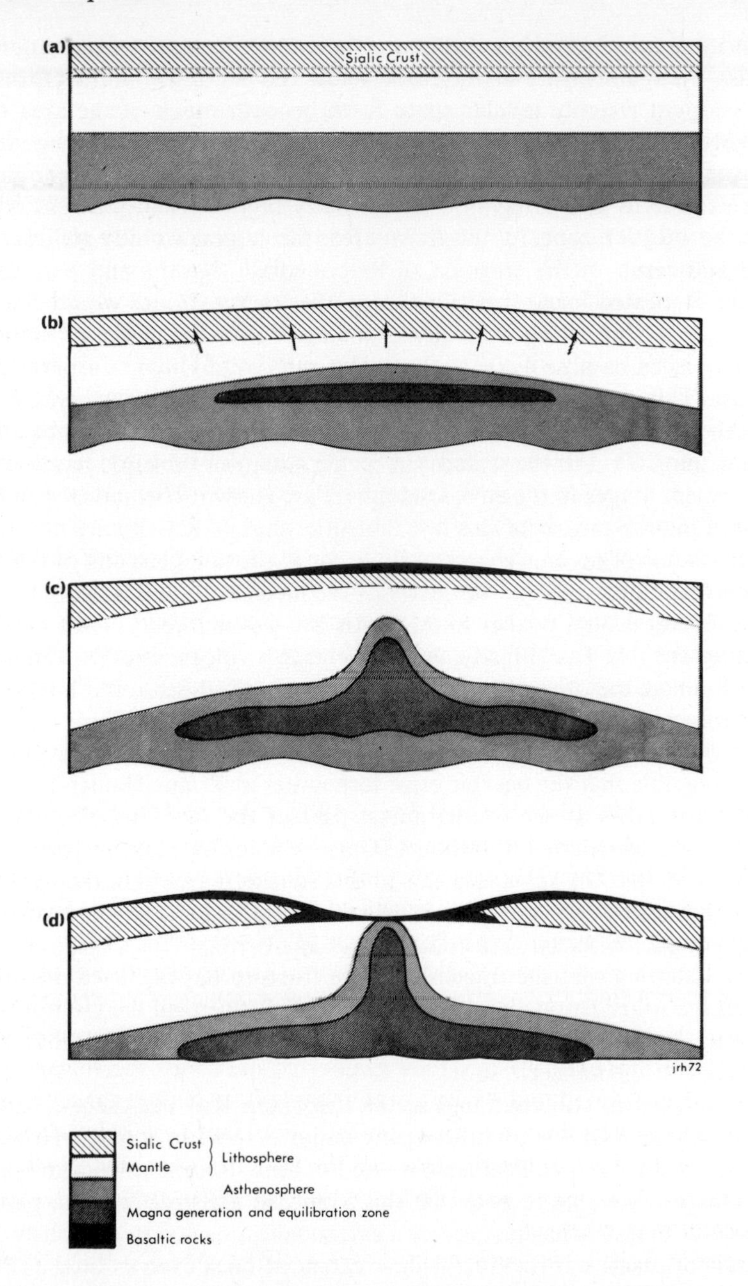

FIGURE 1 Diagrammatic representation of magmatic and structural processes related to a thermal perturbation in the lithosphere/asthenosphere boundary. (a) No thermal perturbation. (b) Thermal perturbation allows a tabular magma genetic zone to develop and causes upwarp in the overlying lithosphere. (c) Primary zones of weakness produced in the lithosphere allow concentration of magma injection which, in turn, distorts the isotherms upwards and allows magma generation and equilibration higher in the mantle. (d) Repeated basaltic injections cause the separation of sialic crust and further distort the isotherms so that magma can equilibrate within 10 km of the surface.

position and, because more heat is required to extend the zone of melting vertically than to raise the temperature of rock still below the melting point (McBirney, 1963, p. 6351), occupy a roughly tabular space form beneath much of the area of uplift. Magma injection and surface volcanism would be most intense along the major fracture zones and this would distort the isothermal surfaces from their original near-horizontal attitude to give a region of particularly high thermal gradient above and adjacent to the injection zone. In such linear areas the magma would equilibrate high in the mantle and result in the eruption of 'intermediate' basalts and peralkali silicic differentiates. Repeated linear injections along the fracture zones would result in the moving apart of the bordering sialic blocks and eventually the host rock would be the previously emplaced basaltic dykes so that sialic crust would now be separated by new basaltic crust. This repeated dyke injection would steepen, still further, the thermal gradient in the linear belt beneath the fracture zones and magma would be allowed to equilibrate within 10 km of the surface; in such a situation tholeiitic basaltic magmas would be formed. Stages in the envisaged model are shown schematically in Fig. 1.

The surface manifestations of this last tholeiitic stage of activity are not abundant, but volcanic islands of oceanic character do occur at the southern end of the Red Sea. Of these, Jebel At Tair rises from the centre of the median rift and is strongly tholeiitic, whereas the Zubair group, further to the south, are less markedly tholeiitic and more undersaturated than At Tair. Finally, the southernmost volcanoes of the Zukur-Hanish group, which unlike the other islands, do not express a Red Sea trend but are erupted along a northeasterly lineament, are strongly alkalic.

Although the position at the southern end of the Red Sea is by no means clear, it does appear possible that the oceanic crust terminates at Zukur-Hanish Line and that continental crust exists at the southernmost part of the Red Sea where the ocean depths are always less than 100 fathoms (Gass, Mallick & Cox, in press). It seems distinctly possible that the spreading axis in this southern area is in the Ert-Ale range of the Afar depression which both petrochemical and structural data suggests is a terrestrial analogue of oceanic crust (Barberi *et al.*, 1970).

However, assuming that the translithosphere fracture has occurred down the Red Sea, through the Afar Depression and into the Gulf of Aden via the Gulf of Tadjoura, it is interesting to enquire into what has happened to the plates on the other side of the Red Sea since continental separation.

In Africa, volcanism still continues in the Ethiopian Rift but there is little on the western flank of the Red Sea to indicate any major magmatic activity. However, this is not the case in the Arabian Peninsula where Pliocene–Recent volcanism is extensive. It is particularly interesting to note that this supposed aseismic Arabian plate, which McKenzie (1970) reports has no recorded earthquakes, has, very obviously, been far from aseismic or inactive if the time scale is extended back even as little as 10 m.y. A glance at a geological map of western Arabia (see Fig. 2) indicates that there are large areas covered by the volcanic products primarily of Pliocene, Pleistocene and Recent age; the most recent recorded volcanism being near the Holy City of Medina in 1250 A.D.

This is no minor volcanic episode, for the volcanic products cover an area stretching from Syria in the north to south central Saudi Arabia in the south and extend over an area exceeding 100,000 km^2 (Fig. 2). It is difficult to estimate the volume of these volcanics as they vary from thin veneers of only a few metres to sequences which

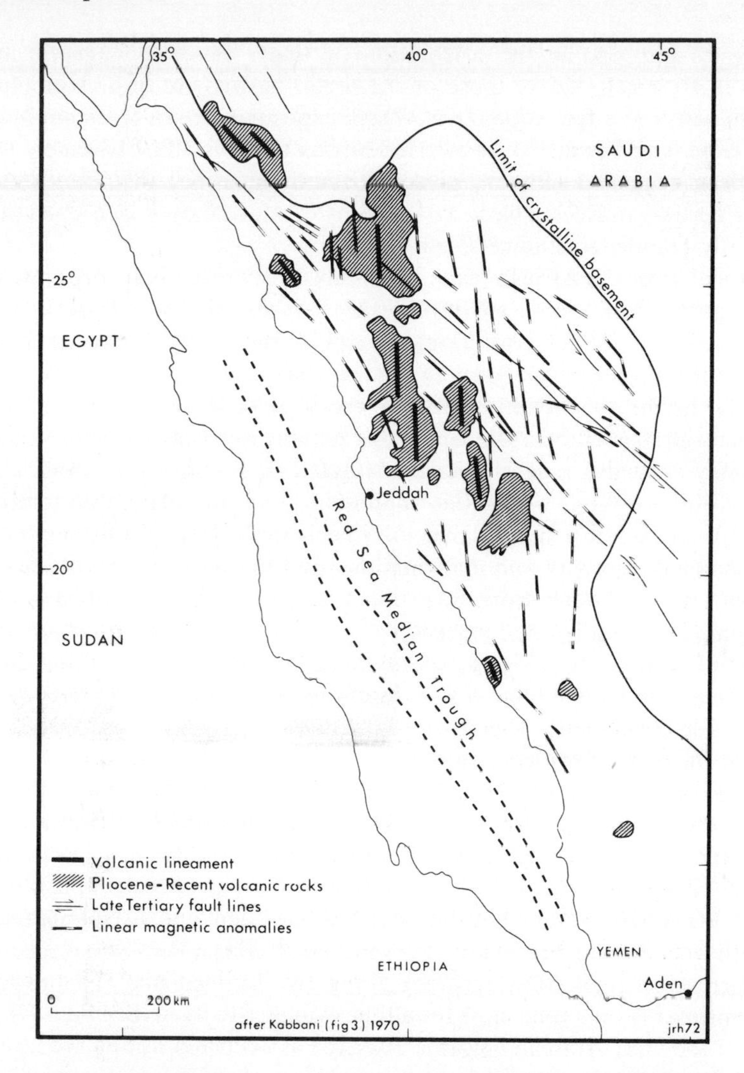

FIGURE 2 Sketch map of western Arabia showing generalized position of Pliocene–Recent volcanic rocks, volcanic lineaments, magnetic anomalies and Late Tertiary faults.

borehole data reveal to be in excess of 500 m. However, if we accept that the average thickness throughout this area is in the order of 350 metres, then the total volume of eruptive rock is 35,000 km^3. This volume, it should be stressed, is equivalent to that of the Thulian province and greater than that of the classic Otago province of South Island, New Zealand. It is therefore an episode that is quantitatively significant and must be taken into consideration when discussing the tectonic and magmatic evolution of the area.

These Recent volcanics are, so far as the meagre evidence we have to hand indicates (Baker *et al.*, in preparation), strongly alkalic in their petrochemistry and occur along markedly north–south lineaments except in Syria and Jordan where a northwesterly trend is dominant. These lineaments are not only expressed in the trend lines of the volcanic cones but also in the orientation of linear positive magnetic anomalies (Kabbani, 1970) which, unlike those of the ocean basins, are of high amplitude and restricted in width to a few kilometres. Where exposures occur, these anomalies have been correlated with Tertiary dyke swarms cutting the crystalline basement and it has therefore been suggested (Brown, personal communication) that they indicate the presence of Tertiary to Recent basic dykes, many of which have not reached the surface of this mainly granitic crystalline shield.

In the global sense the African plate, which until very recently incorporated Arabia, has for the past 100 m.y. been moving northeastwards relative to Eurasia in response to the production of oceanic crust along the axes of the oceanic ridges of the southern oceans. This movement of the Arabian segment has been accentuated during the last few (25) m.y. by the production of oceanic crust under the Red Sea and the Gulf of Aden. As a result, the Arabian segment is now moving northeastwards at an accelerated rate, probably around a pole of rotation situated in northeastern Libya (Darracott *et al.*, this volume). Why, within this supposedly aseismic plate, well removed from destructive or constructive margins, has there been, in the geologically very recent past, extensive magmatic activity and deformation resulting in a dominantly north–south fracture pattern? Although work into the structure of the crystalline rocks of the Arabian peninsula has advanced markedly during the last decade, primarily due to the efforts of the United States Geological Survey and the Bureau Researches Geologique et Miniere in Saudi Arabia, there is still insufficient evidence to propose a finite model to explain these phenomena. Therefore, two tentative models are proposed below and discussed in the light of evidence to hand.

(1) That north–south fractures were produced as the Earth's crust was elevated in response to the upward perturbation of the lithosphere/asthenosphere boundary and that magmatic liquids ascended from depth within the mantle to erupt at the surface as undersaturated alkalic basalts. This initially attractive proposal implies that major crustal fracture occurred along both the Red Sea lineament and also along the Arabian north–south line. Subsequent unknown events selected the Red Sea as a constructive margin whereas no separation occurred along the Arabian line. This model is unacceptable primarily on a time basis for all the evidence to hand (Brown, 1970) suggests that most of the intra-Arabian volcanic activity has occurred during the last 7 m.y. It is therefore probably younger than the separation of Africa and Arabia along the Red Sea.

(2) That the fractures were formed by stresses produced when the Arabian plate collided with Eurasia. In this model it is suggested that the northern margin of the Arabian plate impacted against the continental plate of Eurasia during the Upper Miocene whereas further to the south and east, the collision did not occur until later. In support of this proposal it can be shown that the main climax of the Alpine orogeny in Cyprus (Gass & Masson-Smith, 1963) and southern Turkey occurred prior to the Pontian (Upper Miocene) whereas seismic evidence suggests that subduction is active beneath the Zagros Mountains (Falcon, 1968). Even further to the southeast, although the Persian Gulf is probably underlain by sialic crust of the Arabian plate, the floor to the Gulf of Oman appears to be oceanic in character and the two continental plates have not yet met. It is therefore tentatively suggested, and admittedly on slender and inconclusive evidence, that the northern apex of the Arabian plate impacted into Eurasia some 10 m.y. ago and, as the southern part was still able to move in a north-

easterly direction, a northeasterly regional shear couple was generated across the peninsula. In response to this lateral stress the brittle lithospheric carapace ruptured producing a series of north–south dilational fissures which afforded lines of easy egress for undersaturated alkalic magmas generated deep within the mantle.

It must be emphasized that although the second proposal is currently preferred, it can be no more than a model that needs to be investigated by further detailed study of the Pliocene–Recent volcanic rocks and structures of the Arabian peninsula.

References

Baker, P. E., Brosset, R., Gass, I. G. and Neary, C. R. (in preparation). The recent acid-basic volcanic complex of Jebel al Abyad, Saudi Arabia.

Barberi, F., Borsi, S., Ferrara, G., Marinelli, G. and Varet, J., 1970. Relations between tectonics and magmatology in the northern Danakil Depression, *Phil. Trans. Roy. Soc. Lond.*, **A 267**, 293.

Black, R. and Girod, M., 1970. Late Palaeozoic to Recent igneous activity in West Africa and its relationship to basement structure. *In* African Magmatism and Tectonics, edited by T. N. Clifford and I. G. Gass. Oliver and Boyd, Edinburgh.

Brown, G. F., 1970. Eastern margin of the Red Sea and the coastal structures in Saudi Arabia, *Phil. Trans. Roy. Soc. Lond.*, **A 267**, 75.

Davies, D. and Tramontini, C., 1970. The deep structure of the Red Sea, *Phil. Trans. Roy. Soc. Lond.*, **A 267**, 181.

Elder, J. W., 1970. Quantitative laboratory studies of dynamic models of igneous intrusions. *In* Mechanism of Igneous Intrusions, edited by G. Newall and N. Rast, *Geol. J.*, sp. issue 2.

Falcon, N. L., 1968. The geology of the northeast margin of the Arabian basement shield, *Advancement of Science*, **24**, 1.

Gass, I. G., 1970. The evolution of volcanism in the junction area of the Red Sea, Gulf of Aden and Ethiopian Rifts, *Phil. Trans. Roy. Soc. Lond.*, **A 267**, 369.

Gass, I. G. and Masson-Smith, D., 1963. The geology and gravity anomalies of the Troodos Massif, Cyprus, *Phil. Trans. Roy. Soc. Lond.*, **A 225**, 417.

Gass, I. G., Mallick, D. I. J. and Cox, K. G. (in press). Volcanic islands of the Red Sea. *Q. Journ. Geol. Soc. Lond.*

Girdler, R. W., 1958. The relation of the Red Sea to the East African Rift System, *Q. Journ. Geol. Soc. Lond.*, **114**, 79.

Kabbani, F., 1970. Geophysical and structural aspects of the central Red Sea rift valley, *Phil. Trans. Roy. Soc. Lond.*, **A.267**, 89.

Magnitsky, V. A. and Kalashnikova, I. V., 1970. Problems of phase transitions in the upper mantle and its connection with the Earth's crustal structure, *J. Geophys. Res.*, **75**, 877.

McBirney, A. R., 1963. Conductivity variations and terrestrial heat-flow distribution, *J. Geophys. Res.*, **68**, 6323.

McKenzie, D. P., 1970. The development of the Red Sea and Gulf of Aden in relation to plate tectonics, *Phil. Trans. Roy. Soc. Lond.*, **A 267**, 393.

Morgan, W. J., 1971. Convection plumes in lower mantle, *Nature*, **230**, 42.

Ramberg, H., 1972. Mantle diapirism and its tectonic and magmagenetic consequences, *Phys. Earth Planet. Interiors*, **5**, 45.

Vincent, P. M., 1970. The evolution of the Tibesti volcanic province, Eastern Sahara. *In* African Magmatism and Tectonics, edited by T. N. Clifford, and I. G. Gass, Oliver and Boyd, Edinburgh.

Vine, F. J., 1966. Spreading of the ocean floor: new evidence, *Science, N.Y.*, **154**, 1405.

Wyllie, P. J., 1971. Role of water in magma generation and initiation of diapiric uprise in mantle, *J. Geophys. Res.*, **76**, 1328.

Wyllie, P. J., 1971. The Dynamic Earth. Wiley, New York.

Comment

R. FREUND
Department of Geology
The Hebrew University
Jerusalem, Israel

Professor Gass's explanation of the origin of the north–south alignment of volcanoes in Arabia seems to apply as well to the volcanic alignment in Transjordan and Syria. The 40° divergence of the latter from the former corresponds well to the change in the movement vector of the Arabian plate between the two localities. Yet in both cases there is a small (c. 10°) deviation of the actual alignment from that expected at 45° to the movement vectors.

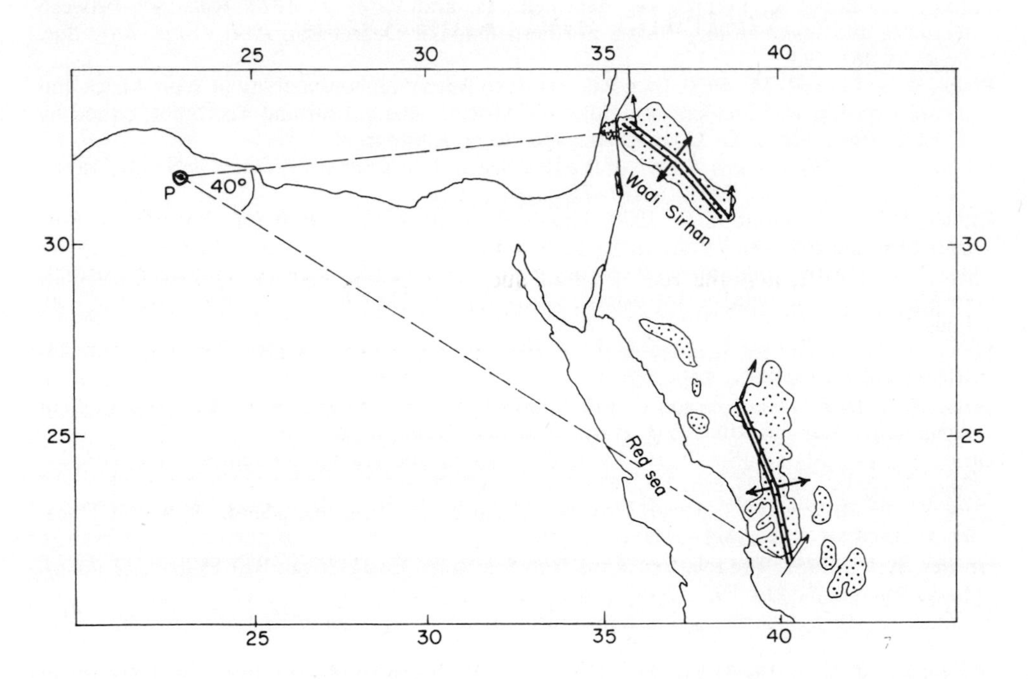

The volcanic activity of the Transjordanian–Syrian alignment and its offset continuation in Israel is of the same Late Tertiary to Recent age. However, the depression of Wadi Sirhan along the southern side of this alignment dates already from Late Cretaceous, and its presumed offset continuation along Mount Carmel is a location of faulting and volcanicity in the Lower and Upper Cretaceous.

7.7

N. D. WATKINS and J. P. KENNETT

Graduate School of Oceanography
University of Rhode Island
Kingston, R.I., USA

Response of Deep-Sea Sediments to changes in Physical Oceanography resulting from Separation of Australia and Antarctica

Introduction

Attention is drawn to the changing boundaries in the hydrosphere which result during continental drift, and the recording of such changes as particle size variation and disconformities in deep-sea sediments. Specific consideration is given to the Tertiary sedimentary regime between Australia and Antarctica. It is considered probable that in the Middle Cenozoic, vast regional disconformities were common south of Australia, and equivalent-age coarse sediments may occur south of the Indian–Antarctic ridge.

The proof of continental drift has yielded three main implications for the Earth Sciences: the separation and collision of continental and oceanic blocks; the creation and consumption of oceanic crust; and changes in the hydrosphere. The first two of these topics have recently received a great deal of attention, but the effects of changes in the hydrosphere have yet to be debated in any detail. It is clear that even for the Upper Tertiary alone, an understanding of variation in the size, shape, and circulation pattern of the various oceans will have a profound role in clarifying causes of the major palaeoclimatic and palaeo-oceanographic changes which are only now becoming defined. Deep-sea sediments provide the essential record of these varying parameters.

One of the well-known results of sedimentation over a spreading mid-oceanic ridge is an increase in thickness and maximum age of sediments with distance from the spreading centre. Other effects include the migration of sea floor away from zones of high biological productivity, and depth changes which cause the role of carbonate dissolution to be regionally variable. Frakes & Kemp (1972) have proposed models of sedimentary facies changes which may result from sedimentation on ridges with different spreading rates. Here attention is drawn to another effect, involving the consideration of marine sediments as fossil bottom current meters, and those changes in bottom water velocity and direction which must result from some forms of continental separation and morphology development. Jones *et al.* (1970, Fig. 18) have considered the effects that Early Tertiary and Cretaceous circulation in the North Atlantic may have had on sediment distribution, but the effect on sediment particle

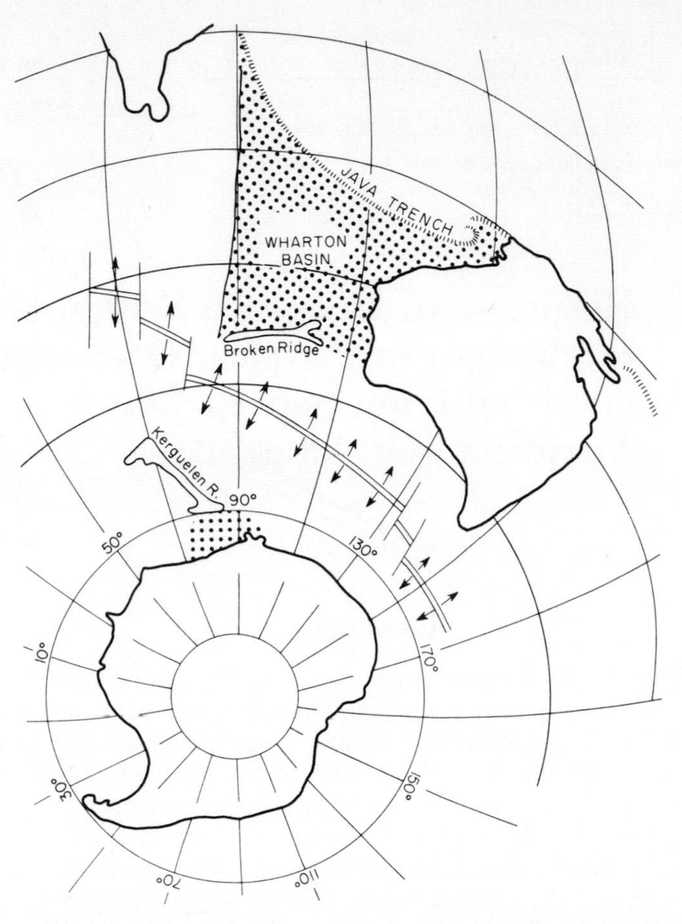

FIGURE 1 Sketch diagram showing position of Australia and Antarctica and the spreading Indian–Pacific ridge segment. The suspected microcontinents of Kerguelen Ridge and Broken Ridge are shown. The diagram is chosen to show the regime of the circum-Antarctic current (clockwise around the Antarctic continents). From Dietz & Holden, 1971.

size was not considered. This may be best presented by examining the history of the opening of the Australian–Antarctic segment of the Southern Ocean (Fig. 1).

Water Velocity : Particle Size Relationships

Figure 2 shows the known experimental relationships between water velocity and sediment particle size, as summarized by Heezen & Hollister (1964). This diagram shows, for example, that a fine to medium sand is eroded by a water velocity of about 10 to 20 cm sec^{-1}, and is in transport until the velocity drops to 2 cm sec^{-1}. Another way of interpreting the diagram is to realize the inferred hiatus requirements: for example, if currents of 10 cm sec^{-1} are common, only sediments coarser than medium sand will be deposited. While doubt must exist about the exact applicability of these data to in-situ deep-sea sediments, particularly for the finest particles, they are unlikely to be substantially misleading. Continuing experimental refinements (example by Southard

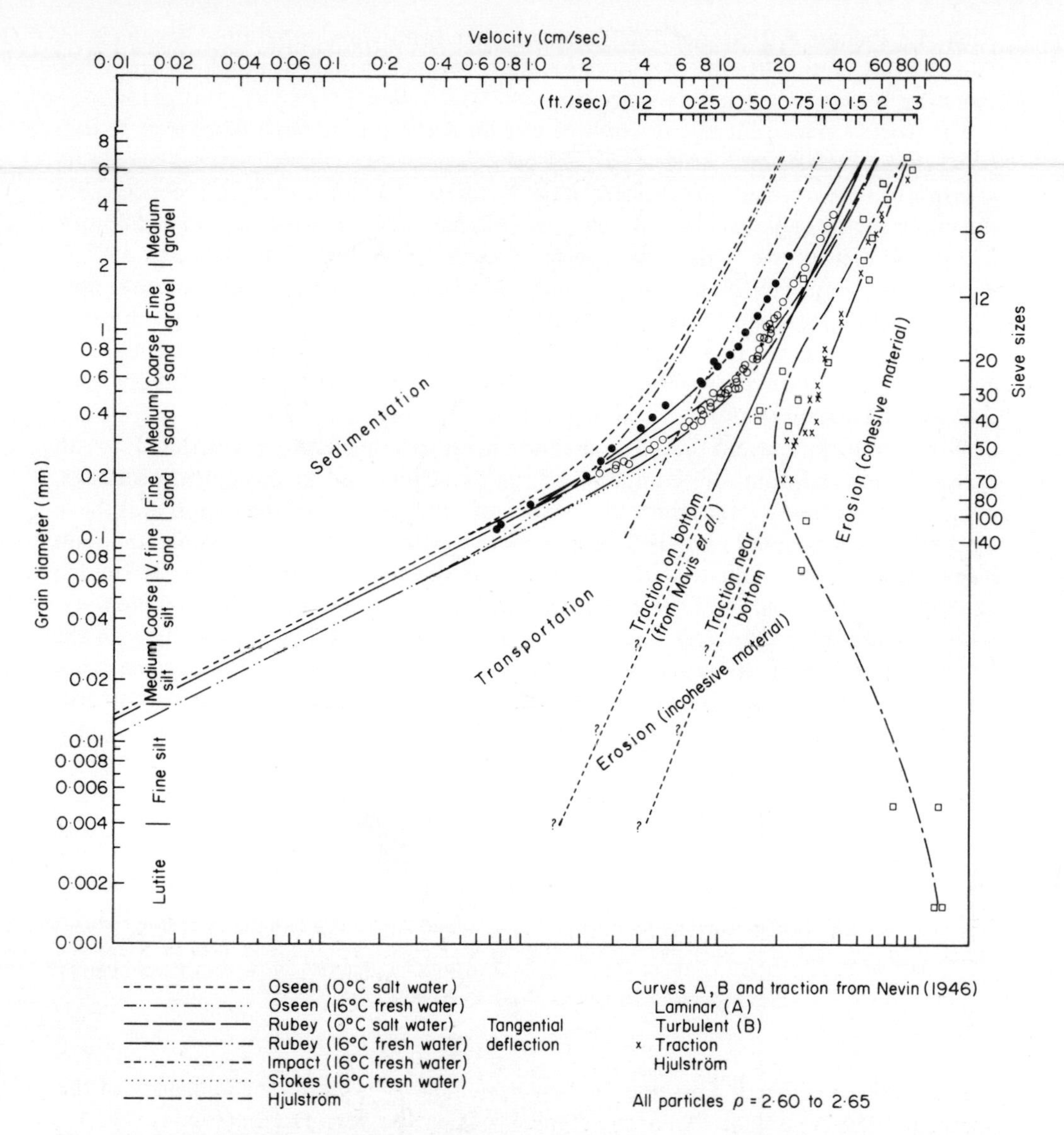

FIGURE 2 Experimentally derived relationships between particle size and current velocities required for erosion, transportation, and deposition of stream bed materials. From Heezen & Hollister (1964). For sources of original experimental results, see Heezen & Hollister (1964).

et al., 1971) will clarify the effect of salinity, cohesiveness, and other variables in particle size : water velocity relationships.

The Present Australian–Antarctic Sea Floor

As Fig. 1 implies, Australia is at present actively drifting away from Antarctica. The most recent analysis of the associated sea-floor spreading is by Weissel & Hayes (1971). This study, as well as earlier analyses of magnetic anomaly patterns (Heirtzler *et al.*,

1968; Le Pichon & Heirtzler 1968), suggest that continental separation began about 50 to 55 m.y. ago. Virtually the entire sea floor between Antarctica and Australia has therefore been produced since that time.

The world's major current system (the circum-Antarctic current) flows west to east between Australasia and Antarctica. The behaviour of this current, particularly in its transport to the northern hemisphere in the Pacific and Atlantic oceans, is conceivably the major factor in global climatic changes. Callahan (1971) has measured a net flow of 233×10^6 metres^3 sec^{-1} through a section south of Australia at latitude 132° E. This is very similar to the net flow through the Drake Passage. In detail, current flow through the Australian–Antarctic gap is, of course, complex. Very high bottom velocities are observed on the north side of the Indian Antarctic ridge (Callahan, 1971, Fig. 4). Present current directions of bottom waters in the region (Fig. 3) have been defined by Gordon (1972).

While bottom current velocity measurements are necessarily restricted to an extremely limited duration, compared to the times involved in geological processes, evidence exists to show that the high bottom current velocities of the region must have existed for a substantial period. By use of bottom photographs, it is possible to obtain a semi-quantitative evaluation of bottom current velocities, as summarized by Heezen & Hollister, 1964, and Hollister & Heezen, 1967. As shown in Fig. 3, high bottom water velocities are inferrable over much of the South Tasman basin, particularly on the north flank of the ridge. Watkins & Kennett (1972) have described the dominance of coarse residual particles in the surface sediments of the same region: a fast bottom current has been at least a spasmodic feature of the region for perhaps the last 3·3 m.y.

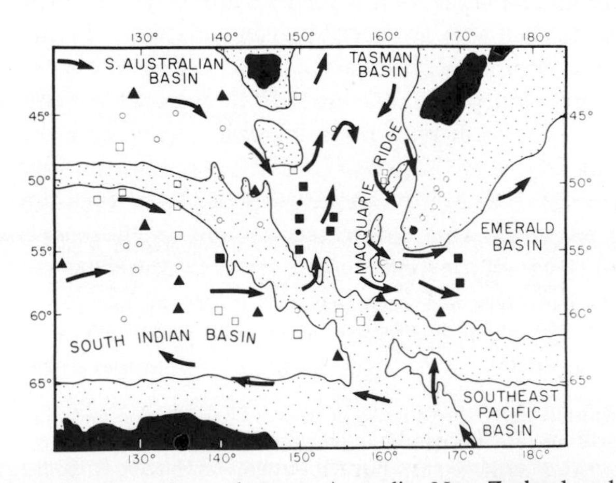

FIGURE 3 Map of the Southern Ocean between Australia, New Zealand and Antarctic showing circulation patterns of bottom waters and bottom photograph (Jacobs *et al.*, 1970) characteristics. Arrows represent bottom current directions from Gordon (1972). Submarine ridge system and shallow plateaus stippled. Bathymetric contour is 2000 m depth. Bottom photograph symbols are shown at camera station sites and indicate the following features (in order of increasing inferred bottom current velocities):

□ = abundant bioturbation
○ = bioturbation common or absent
● = ripples and scour marks present
▲ = manganese nodules common
■ = manganese nodules abundant

From Watkins & Kennett (1971).

The Early Teritiary Australian–Antarctic 'Sluice'

Consideration is now given to the particle size which may have dominated in those sediments (if any) between Australia and Antarctica, during past stages of continental separation. As discussed above, this separation has occupied only 50 to 60 m.y. A form of circum-Antarctic circulation has undoubtedly been a feature of Southern Hemisphere circulation since the Cretaceous, even prior to separation of the two continents (see Fig. 3B of Dietz & Holden, 1971, for the continental reconstruction). It is certain that at that time, most of the current would have flowed northwards, along the west coast of Australia. After only a short period of continental separation and associated production of sea floor, however, a long west to east channel must have come into existence. An increasing fraction of the Early to Middle Tertiary circum-Antarctic current would be funnelled through the gap, under the very dominant influence of the Earth's rotation and wind systems.

Estimation of the resulting changes in net transport through this widening Australian–Antarctic 'sluice' requires knowledge of a large number of variables, none of which can be determined precisely. These variables include the average vertical velocity gradient; the fractions of the circum-Antarctic current flowing north of Western Australia and through the 'sluice'; the average water depth; and several other conditions. Given that a circum-Antarctic current existed, with a net flow equal to that of the present day, then limits can be placed on the possible net flow through the region throughout the Cenozoic. The most simple model involves maintaining present transport, with all water being funnelled through the 'sluice', which has a constant average depth of 4000 m, while continental separation is steady from 55 m.y. ago until the present. The resulting velocity increase with age (or decreasing continental separation) can then be used to estimate the limiting sedimentary particle size to be expected for the region if the ratio between net transport and average bottom water velocity is known. This is, again, a variable which is not known for the past, but if the present net transport is considered as being related to the present high average bottom velocity (of the order of 10 cm sec^{-1}), then use of this ratio can yield a predictable dominant sediment particle size (or bottom velocity). This parameter, as well as those mentioned above, can obviously be readily varied, to yield a whole series of predicted particle sizes. Such a 'family' of curves can have a bearing on the interpretation of depositional history of Early to Middle Cenozoic sediments, when they become available, as will soon be the case for the region (Hayes & Edgar, 1972).

Fig. 4 shows *one* of the family of curves which can be constructed. It employs the simple assumptions described above. The whole family of curves for the Cenozoic can vary from a first order relationship parallel to the time axis (representing no change in particle size with age during opening of the 'sluice'), to one which is third or fourth order and overall of steeper gradient than that shown in Fig. 4, during the Upper Tertiary. These more complex curves are not presented here, because of their large number, variability and complexity. Such an exercise will become more rewarding when the relevant sedimentary data become available. Figure 4 is therefore presented aa *concept* rather than as being with any firm quantitative value. In other words, it is believed that *given a source of variable sedimentary particle size* an overall increase in average particle size can be expected with increasing age in the region between Australia and Antarctica, and this increase will become more noticeable prior to

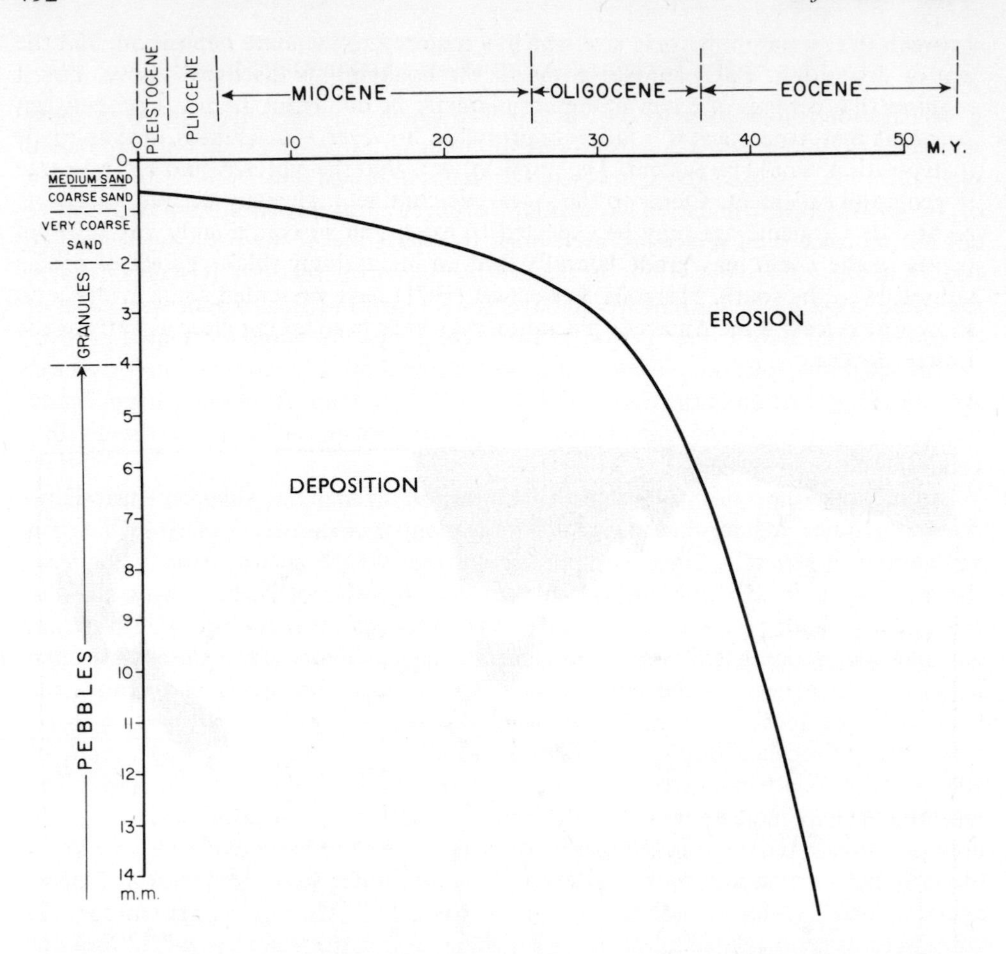

FIGURE 4 'Hiatus' diagram showing one of a series of curves relating average minimum particle size to sediment age in the Australian–Antarctic 'sluice'. See text for assumptions involved in derivation. NOTES: (*a*) This is one of a series of curves, which can range from one parallel to the time axis (i.e. no change in sediment particle size with increasing age) to complex non-linear steeper curves involving diverse assumptions (see text). (*b*) Since coarse sediments are rare in the deep seas the diagram is also designed to represent the times when a depositional hiatus might be expected, given the assumptions discussed in the text: for example, 30 m.y. ago particle sizes would need to be coarser than 3 mm to remain deposited.

Middle Cenozoic times. While fine sediments would be expected to be preserved in local velocity minima (pockets), higher velocities would preclude their consistent dominance, assuming sources of coarse sediment were available.

The 'Hiatus' or Disconformity Diagram

Since coarse fractions are relatively rare in deep-sea sediments, it is stressed that (as with Fig. 2) the curve in Fig. 4 should not be interpreted literally as showing a particle size versus age (or velocity) relationship. Rather, it should be viewed as a relation between the minimum particle size which is required to facilitate deposition, and the age of deposition. For example, given all the assumptions discussed above, Fig. 4 requires that pebbles of 6 mm minimum diameter be dominant 35 m.y. ago, between Australia and Antarctica. It is far more probable, however, that a hiatus, or simple halt in deposition, would be evident. The implication is that the hiatus would be very large in geographical extent. Closer to the Antarctic continent, glacially-derived sediments of Middle Cenozoic age may be expected to exist, and so conceivably any regional hiatus to the north may grade laterally into an increasingly thick deposit of coarse sediments to the south. Margolis & Kennett (1971) have presented some evidence to show that at least some Antarctic glaciation may have been taking place as early as the Lower Eocene.

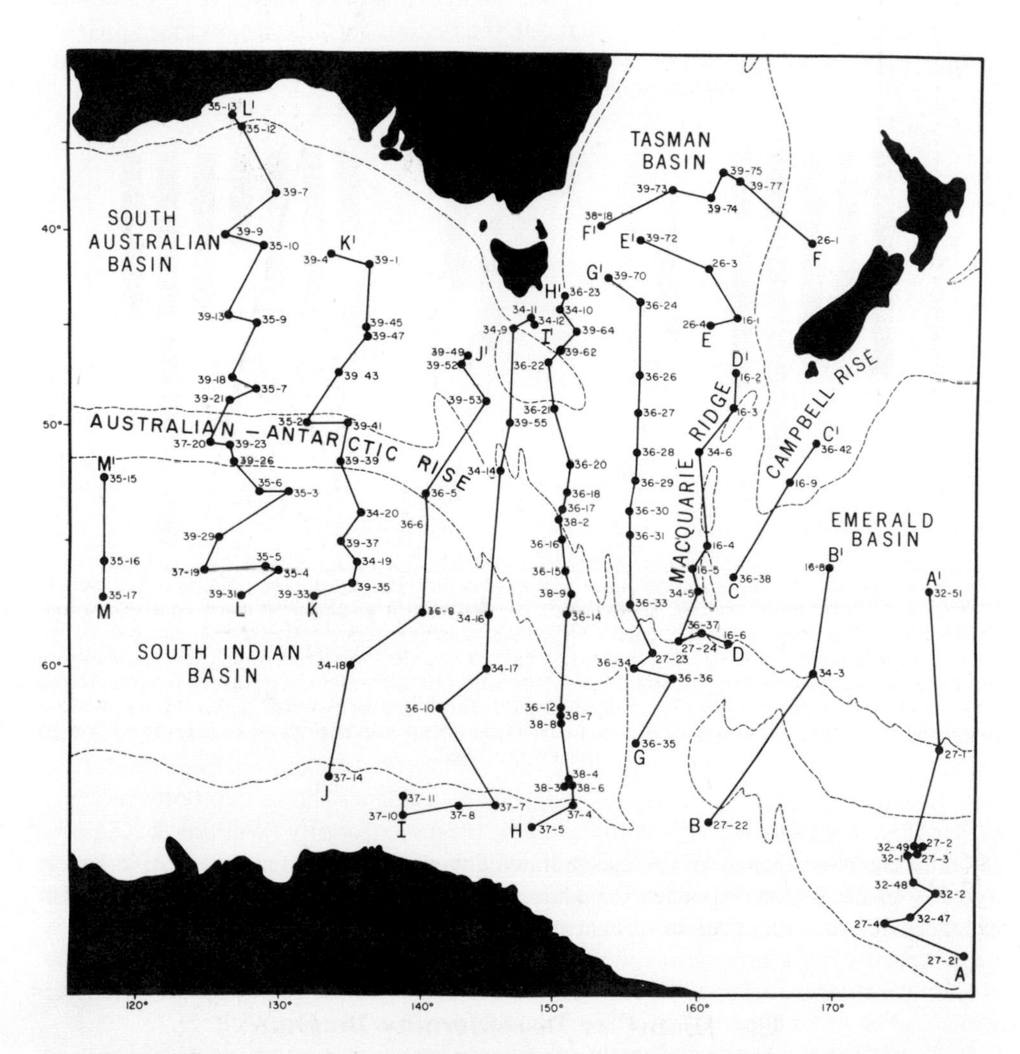

FIGURE 5 Map showing location of *Eltanin* piston cores. Number next to each core site is cruise number followed by core number. Bathymetric contour is 2000 m depth. Selected traverses are labelled A-A′ to M-M′. From Watkins & Kennett (1972).

Additional Factors

Regular geometry and a forced simplicity of assumptions has been used to derive the concept shown in Fig. 4. A few of the natural complexities (in addition to those already discussed) which can be expected to distort this simple argument are given below:

Variation in Antarctic Bottom Water Production

Figure 5 shows the locations of 126 sedimentary cores recently dated by Watkins & Kennett (1972), using palaeomagnetic and micropalaeontological methods. Figure 6 shows the palaeomagnetic data as a function of depth in the core for traverse H′–H, between the Tasmanian and Antarctic continental shelves (Fig. 5). The classical geomagnetic polarity history (Cox, 1969) is not found in any of the cores. When combined with micropalaeontological dating methods, the reasons for this fact become apparent (Fig. 7): substantial segments of the sedimentary sequences are simply absent. A second-

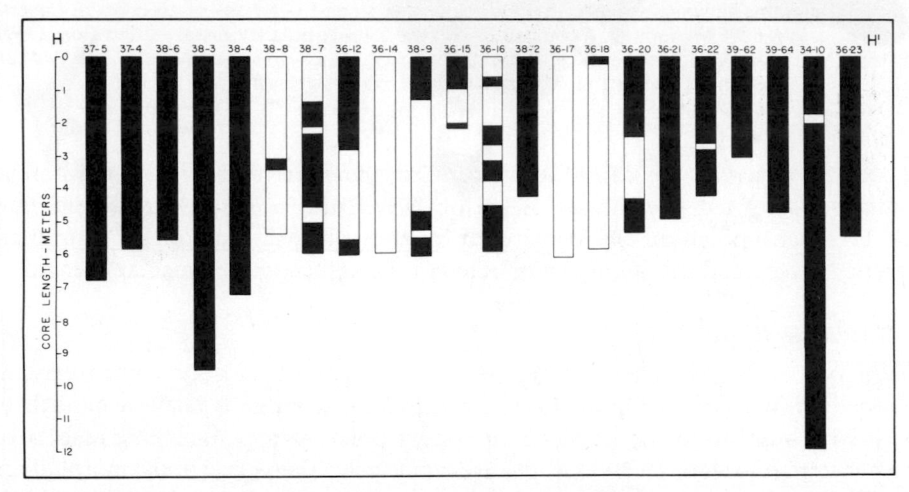

FIGURE 6 Palaeomagnetic polarities for each core in traverse H′-H as a function of depth below the top of each core. In each log, black = normal polarity, clear = reversed polarity. See Fig. 5 for location of each core. From Watkins & Kennett (1972).

order trend surface of the age of the surface of the sediments (Fig. 8) expresses the regional extent of the hiatus or disconformity. A vast volume of sediment has been eroded by bottom current activity in the region. All available evidence shows that bottom scour, rather than any other activity, is the cause of the disconformity (Watkins & Kennett, 1971, 1972). Why, then, does there exist much fine-grained in-situ sediment of an older age still in-situ? How was such deposition possible in view of the evidence for the existence of a fast, scouring, bottom current? It is clear that a finite increase in bottom water velocity must have occurred since the Gilbert epoch (Fig. 7), to facilitate the erosional activity. Increase of cohesion of sediments with compaction and age would result in an attenuation of erosion with depth in the core (Fig. 2).

It is suspected that the indirectly observed increase in bottom water velocity is associated with an increase in Antarctic bottom water productivity, which in turn reflects increased Antarctic glaciation including ice shelf growth (Watkins & Kennett,

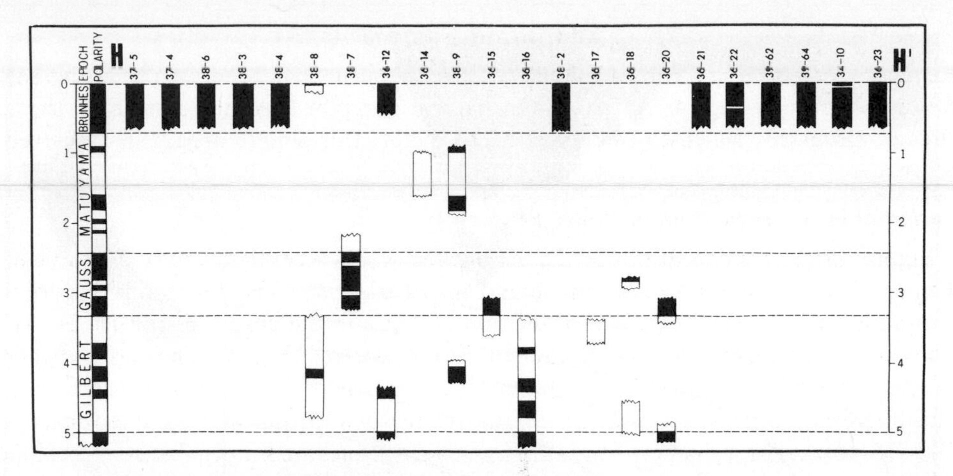

FIGURE 7 Diagrams showing assigned time ranges and measured polarities of all cores in traverse H′-H (Figs. 5 and 6). The results are compared with the known polarity history (after Cox, 1969), which is at the left of each traverse. Black = normal polarity; clear = reversed polarity. See Watkins & Kennett (1972) for details of age assignments.

1971, 1972). Whatever the cause, it is clear that the effect of changes in Antarctic glaciation and related fluctuations of bottom water and circum-polar current flow would be superimposed on any velocity variation due to widening of the Australian–Antarctic 'sluice', and this would be reflected in the associated sedimentary record.

The Macquarie Ridge

The origin of the Macquarie Ridge (Fig. 5) is a subject of some controversy, as mentioned by Watkins & Gunn (1971). At present, the ridge is shallow enough to cause a large southward deflection of the circum-polar current, including associated bottom waters (Gordon, 1972). Any changes that have taken place in the morphology of the ridge could result in the path of the circum-Antarctic current being drastically modified. Griffiths & Varne (1972) consider that the Macquarie Ridge has been largely uplifted from oceanic depths during the Late Cenozoic, being marked prior to this as only a shear zone, extending south of the Alpine Fault in New Zealand. No evidence exists for any positive element in association with the shear zone, but if such a feature did exist, it would have deflected circum-polar flow through the Tasman Sea region or to the south as in the present situation. Very high velocities would have existed in the Tasman Sea. Present bottom velocities (Laird & Ryan, 1969) are not noticeably large, and recent sedimentation rates are normal (Watkins & Kennett, 1972). The remoteness of this region from the coarse-sized sedimentary sources is such that a regional hiatus rather than a coarse sediment would be created. It is tempting to surmise that any fluctuation in elevation of the Macquarie Ridge due to tectonic activity may be reflected in the sedimentary history of the Tasman Basin.

The Drake Passage

Speculation of the effect of shallowing of the Drake Passage on circulation of the southern ocean has recently been made by Gill (1971), on theoretical grounds. It is

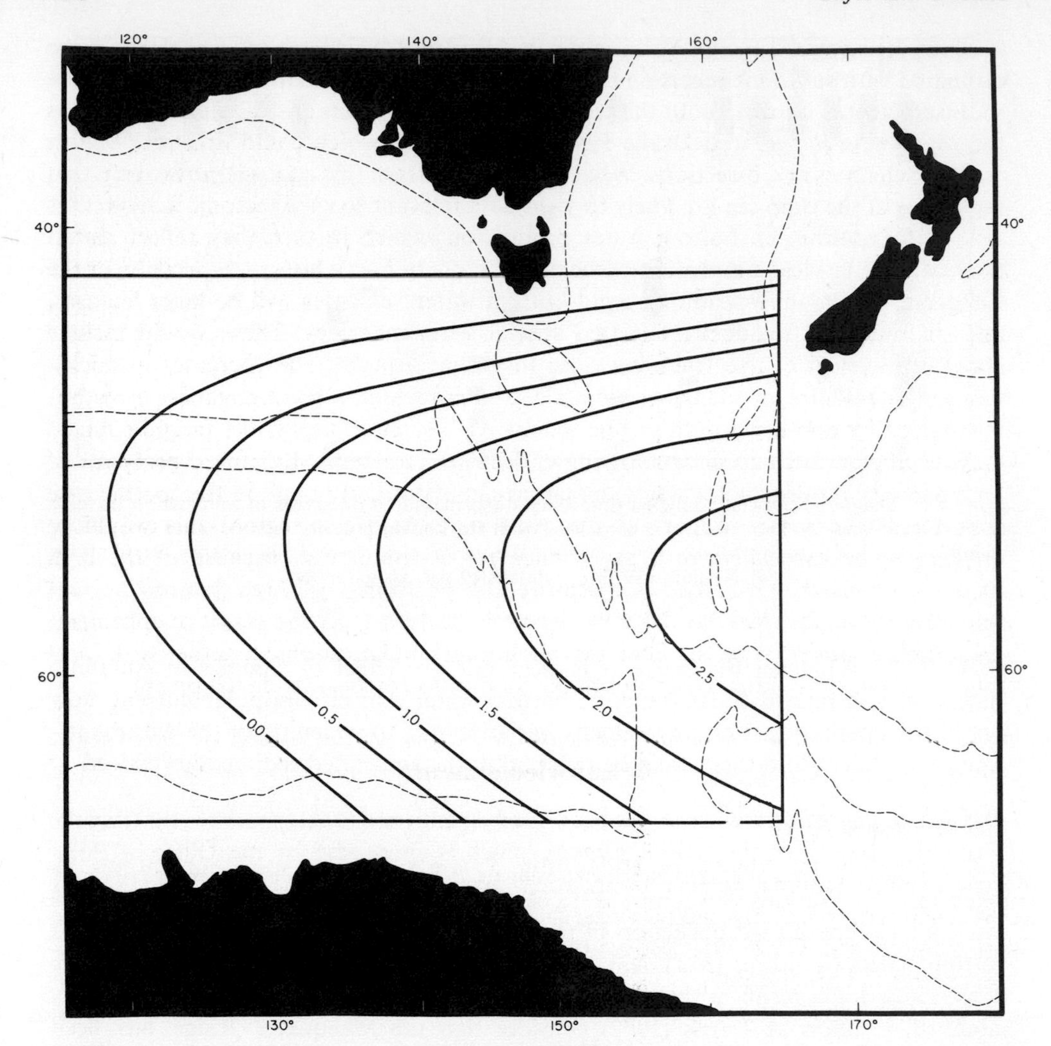

FIGURE 8 Map of second order trend surface on the age (in millions of years) of the sediment at the top of each sedimentary core, or the age at the top of the regional erosional disconformity if a core happens to have a short upper segment of young sediment (for example, core 38–8 in Fig. 7). See Watkins & Kennett (1972) for details of computation and limitations. Note that the trend surface limits are dictated by the data coverage (Figure 5): the erosional feature has now been detected to the west of the area shown.

virtually certain that the velocity of the circum-Antarctic current would be effected by such a change. In turn, sediments in the Southern Indian and Pacific Oceans and the Australian–Antarctic 'sluice' should record the event.

Conclusions

There can be no doubt that continental drift has modified the oceanic current systems, through the inherent boundary modifications. Such modifications must include changes in bottom water velocities, and therefore, changes in dominant sediment particle sizes.

It is not yet known if the many variables involved and their inter-relationships can be estimated with sufficient precision to lead to meaningful predictions. Even if this were so, insufficient is known about the tectonic history of various critical elements such as the Macquarie Ridge, and Drake Passage, changes in which could drastically effect current velocities and directions. What does emerge from this analysis, however, is that sediments of the deep sea are likely to yield data relevant to such tectonic activities, as well as fluctuations in bottom water production (which in turn may reflect glacial fluctuations), to yield a more comprehensive picture of Earth history, especially for the Cenozoic. The manifestation of rapid bottom water velocities will be large hiatuses, and (in restricted regions) associated special sediment types. These would include concentrations of coarse fractions close to glacial sources; and increases in micro-manganese nodules, or manganese nodules and pavement, since manganese growth is maximized by non-deposition of fine sediments. Therefore a possibly tangible return may result from such an understanding of deep-sea sediments. Enhanced prospects of high porosity result from coarse sediment accumulation, although in the specific case considered (the southern part of the Australian–Antarctic region) this would be unlikely to be associated with gas condensate accumulation, because of the high latitudes involved. This does not remove the possibility of high bottom current velocities in much lower latitudes in the geological past, as the result of ephemeral favourable continental or sea floor morphological configurations, together with local sources of coarse sediments, or optimum conditions for pisolite development.

Acknowledgements

The palaeomagnetic data which has been briefly mentioned in this paper results from the continuing cruises of the USNS *Eltanin*, which is sponsored by the Office of Polar Programmes, National Science Foundation. Grant numbers GA1678, GA13132 and GV25400 are gratefully acknowledged for continuation of the palaeomagnetic programme, and GV28305, for continuation of the micropalaeontological palaeo-environmental programme.

References

Callahan, J. R., 1971. Velocity structure and flux of the Antarctic circum-polar current south of Australia, *J. Geophys. Res.*, **76**, 5859–64.

Cox, A., 1969. Geomagnetic reversals, *Science*, **163**, 237–45.

Dietz, R. A. and Holden, J. C., 1971. Pre-Mesozoic oceanic crust in the Eastern Indian Ocean (Wharton Basin?) *Nature*, **229**, 309–12.

Frakes, L. A. and Kemp, E. M., 1972. Generation of sedimentary facies on a spreading ocean ridge, *Nature*, **236**, 114–7.

Gill, A. E., 1971. Ocean models, *Phil. Trans. Royal Soc. London*, **A 270**, 391–413.

Gordon, A. L., 1972. Spreading of Antarctic bottom waters, II. *In* Studies in Physical Oceanography—A Tribute to G. Wust on his 80th Birthday, edited by A. L. Gordon. Gordon and Breach Publ., New York (in press).

Griffiths, J. R. and Varne, R., 1972. Evolution of the Tasman Sea, Macquarie Ridge and Alpine Fault, *Nature*, **235**, 83–86.

Hayes, D. E. and Edgar, N. J., 1972. Extensive drilling program planned for *Glomar Challenger* in Antarctic waters, *Antarctic Journ. United States*, **7**, 1–4.

Heezen, B. C. and Hollister, C., 1964. Deep-sea current evidence from abyssal sediments, *Marine Geology*, **1**, 141–74.

Heirtzler, J. R., Dickson, G. O., Herron, E. M., Pitman, W. C. and Le Pichon, X., 1968. Marine magnetic anomalies, geomagnetic field reversals, and motions of the ocean floor and continents, *J. Geophys. Res.*, **73**, 2119–36.

Hollister, C. and Heezen, B. C., 1967. The floor of the Bellingshausen Sea. *In* Deep Sea Photography, edited by J. B. Hersey, pp. 177–89. Johns Hopkins Press, Baltimore.

Jacobs, S. S., Bruchausen, P. M. and Bauer, E. B., 1970. *Eltanin* reports: hydrographic status, bottom photographs, current measurements, cruises 32–36, 1968. *Published by Lamont Doherty Geological Observatory (Columbia University) for the U.S. Antarctic Research Program*, 460 pp.

Jones, E. J. W., Ewing, J. I. and Eittrem, S. L., 1970. Influences of Norwegian Sea overflow water on sedimentation in the North Atlantic and Labrador Sea, *J. Geophys. Res.*, **75**, 1655–80.

Laird, N. P. and Ryan, T. V., 1969. Bottom current measurements in the Tasman Sea, *J. Geophys. Res.*, **74**, 5433–8.

Le Pichon, X. and Heirtzler, J. R., 1968. Magnetic anomalies in the Indian Ocean and sea-floor spreading, *J. Geophys. Res.*, **73**, 2101–17.

Margolis, S. V. and Kennett, J. P., 1971. Cenozoic palaeoglacial history of Antarctica recorded in sub-Antarctic deep-sea cores, *Amer. J. Sci.*, **271**, 1–36.

Reid, J. L. and Nowlin, W. D., 1971. Transport of water through the Drake passage, *Deep Sea Res.*, **18**, 51–64.

Southard, J. B., Young, R. A. and Hollister, C. D., 1971. Experimental erosion of calcareous ooze, *J. Geophys. Res.*, **76**, 5903–9.

Watkins, N. D. and Gunn, B. M., 1971. Petrology, geochemistry, and magnetic properties of some rocks dredged from the Macquarie Ridge, *New Zealand J. Geol. Geophys.*, **14**, 153–68.

Watkins, N. D. and Kennett, J. P., 1971. Antarctic bottom water: major change in velocity during the Late Cenozoic between Australia and Antarctica, *Science*, **173**, 813–8.

Watkins, N. D. and Kennett, J. P., 1972. Regional sedimentary disconformities and Upper Cenozoic changes in bottom water velocities between Australia and Antarctica, *American Geophys. Un. Monograph (Antarctic Res. Series)* 19, edited by D. E. Hayes, 273–94.

Weissel, J. K. and Hayes, D. E., 1971. Asymmetric sea-floor spreading south of Australia, *Nature*, **231**, 518–22.

7.8

A. E. M. NAIRN
Case Western Reserve University
Cleveland, Ohio, USA

The Argument for a Southern Uratlantic*

Introduction

When we consider the geological evolution of the South Atlantic in Phanerozoic times, it becomes clear that the critical area in Africa is the coastal region lying between the Walvis ridge and the Guinea rise. Along the West African coast, west of the Guinea rise, there is evidence, both onshore and offshore, of marine Devonian in Ghana (Machens, 1972). South of the Walvis ridge in addition to the Lower Palaeozoic marine history of Cape Province (Rust, 1972) there is also evidence of Upper Palaeozoic marine horizons (Caster, 1952), with Permian marine horizons reaching into southwest Africa (Martin, 1972). In South America what emerges from the study of the Parnaiba and Amazon basins is a pattern of epicontinental marine conditions during much of the Palaeozoic including marine Permian with evaporites in the Parnaiba basin (Bigarella, 1972). In the Argentine Devonian faunas are found similar to those in Cape Province (Caster, 1952). Between the two ridges, however, insofar as is known, no marine horizons older than Lower Cretaceous have been penetrated in drilling off the African coast, and the same appears to be true in the corresponding regions of Brazil. In the onshore regions older horizons are demonstrably absent.

In West Africa a Late Jurassic and a ?Cenomanian–Turonian marine horizon are recorded in the Congo basin (Cahen, 1954) but the ostracod evidence (Grekoff, 1960; Grossdidier, 1967) and the pattern of sedimentation do not suggest continuity with the coastal regions.

It is for these reasons that particular attention is concentrated on the marginal basins of Brazil and equatorial Africa in discussing the separation of Africa and South America (Fig. 1). The marginal basins, which are in general grabens or semi-grabens (Asmus & Ponte, 1972), show remarkable similarities in the sedimentary sequences contained within them. These similarities, which in the simplest terms lie in a succession beginning with continental sedimentation, passing through a phase of evaporite formation until finally normal marine conditions become established, have been recognized by many writers (e.g. Fernandes, 1966) and will be discussed elsewhere in Volume I (Douglas *et al.*, 1973). Similarities in the non-marine faunas are also apparent (Krommelbein, 1965, 1966; Grosdidier, 1967), palaeontological similarities are less marked in marine faunas of Later Cretaceous times.

* Contribution No. 83, Case Western Reserve University, Cleveland, Ohio.

FIGURE 1 The marginal basins of the South Atlantic. These basins contain principally sediments of Cretaceous age or younger. The Turonian marine link between the Gulf of Guinea and the Tethys (through southern Tunisia) is not shown, nor is the location of the Congo marine horizon since their limits are poorly known (Figure reproduced by permission of Plenum Press, N.Y.)

In most cases the trend of the grabens is oblique to the coast so that the deeper parts of the basins are off-shore. Not all are equally well-known, however the Cuanza basin seems to be typical and detailed sections and descriptions have been published by Brognon & Verrier (1966).

An Aborted Fracture?

In the following section, the attempt will be made to point to some possible consequences of the tectonic model adopted to explain the evolution of the South Atlantic. The model used is the now current model, in which a combined continental land mass, fitted in the manner suggested by Bullard, Everett & Smith (1965) and Dietz & Holden (1968), was fractured with the present continents, separating about 140–135 m.y. ago according to Le Pichon & Hayes (1971) and Dickson *et al.* (1972). In Fig. 2 an attempt has been made to indicate the principal tectonic cycles and the location of the grabens. From the figure it is apparent that, as Kennedy (1965) and others have indicated, the grabens follow structural trends established during the Late Precambrian Pan-African

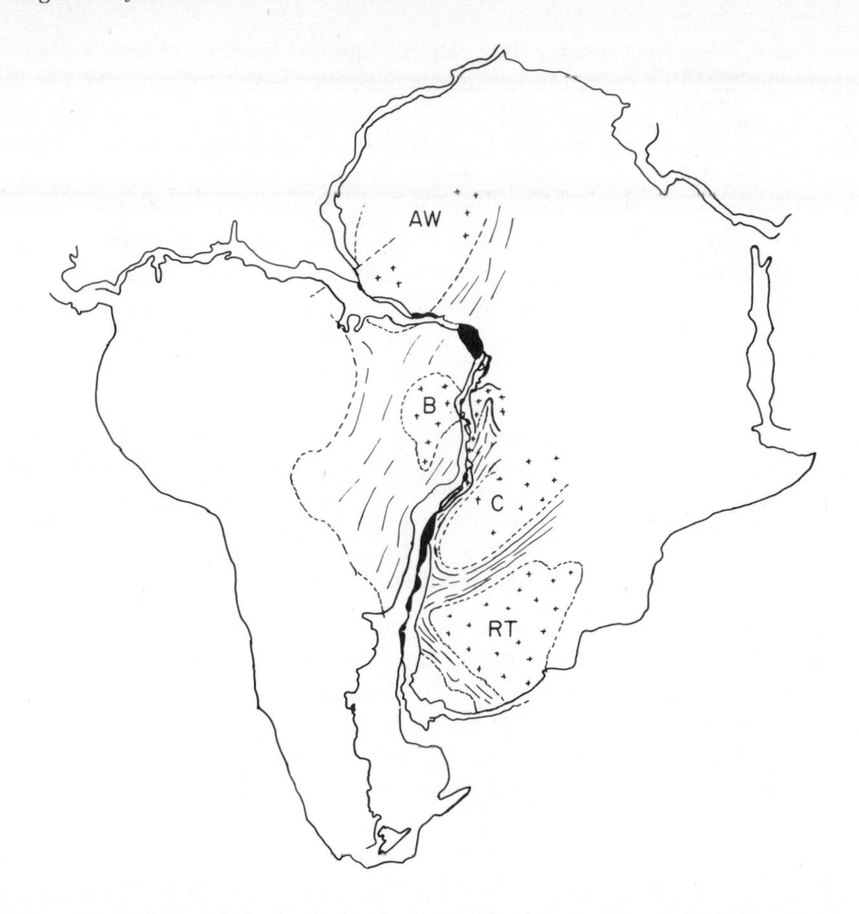

FIGURE 2 The distribution of Precambrian rocks in the South Atlantic region. The approximate structural trend lines of the Late Precambrian (Pan-African thermo-tectonic event, Braziliano cycles) are shown. The older cratonic nucleii, A. W. West Africa, B Brazil, C Congo, RT Rhodesia Transvaal are indicated with symbols representing outcrops with radiometric ages in excess of $2{\cdot}5 \times 10^9$ years. Areas in black represent areas of misfit on the Bullard *et al* reconstruction. (Modified from Nairn & Stehli by permission of Plenum Press).

thermo-tectonic event and its South American equivalent, the Braziliano cycle (Almeida *et al.*, 1972). Hence there is the natural assumption that the location of these grabens was controlled by the trend established during this earlier phase of tectonic activity.

On the reconstruction the grabens on both sides of the Atlantic show a remarkable match (see also Beurlen, 1961). Two graben deserve particular attention, the Reconcavo–Tucano graben of Brazil and the Goa graben of Nigeria, both of which follow the general trend respectively south and north of the sudden change in direction of the appropriate coastlines. As they line up with the coastal grabens to the south it is tempting to suggest that in the earlier phase of development, the general north–south pattern tended to follow a line of weakness along the eastern margin of the West African craton, and that it was aborted by the fracturing represented by the development of the transform faults now referred to the Chain, Romanche and St. Paul's fracture zones. These fractures are thus younger than the coastal grabens to the south.

Onshore the trend of these fracture zones appears to be continued by the Benue graben (Fail *et al.*, 1970). If this is so then their development dates from at least Middle Albian times at which time the Benue graben began receiving a flood of sediments (Machens, 1972). It is at this time or somewhat earlier that the marginal ridges (Le Pichon & Hayes, 1971) began acting part at least, as barriers controlling the influx of seawater into the marginal basins further to the south where evaporites were forming (Douglas, Moullade & Nairn, this volume). There are no obvious reasons for this change of trend. The west-northwest strike and location of the fractures suggest that it may be a control imposed by trends associated with the Propria geosyncline (Humphrey & Allard, 1968; Allard & Hurst, 1969) perhaps aided by tensional effects as suggested by Funnell & Smith (1968). But for this breakthrough West Africa might well have become an 'Atlantis.' At any rate by Late Albian times separation was complete with normal marine condition along both African and Brazilian coasts.

An Uratlantic

Irrespective of whether or not a general north–south fracture pattern was aborted, there is a very clear parallel with the situation in the North Atlantic where the lines of fracture also closely parallel older structural trends, even if, in the South Atlantic, the structural trends were older. We may thus refer to the suggestion of Wilson (1966) of an earlier opening of the North Atlantic, an idea given substance by the detailed geological descriptions and interpretation of Bird & Dewey (1970), and conclude that a case can equally well be made for a Southern Uratlantic. That is, an oceanic area existed in the south, which seaway was closed up during the tectonic events of the Late Precambrian. Subsequently, as numerous writers have suggested, during Panerozoic time South America and Africa behaved as a unit until the Late Mesozoic fragmentation.

These latter-named authors (Creer, 1964, 1970; Briden, 1967; McElhinny & Luck, 1970; McElhinny & Briden, 1971) not only show the uniform behaviour, but by means of the history of the movement give an indication of ridge activity in Early Phanerozoic time different from that subsequently active. The suggested periods of rapid motion, during Late Ordovician and Carboniferous times, are periods characterized by major orogenic phases elsewhere in the world.

If we follow this idea and consider the spatial distribution of the tectonic events of the Late Precambrian then the Late Precambrian appears to stand out as one of the most significant times in earth history in every way comparable with the Late Mesozoic. Tectonic events of Late Precambrian age appear to encompass older cratonic nucleii in Africa, the West African craton, and the Congo and Transvaal–Rhodesian blocks. To this number can be added the older nucleus of eastern Brazil (Almeida *et al.*, 1972). We can thus argue that this Late Precambrian event welded into a single unit at least four older cratonic nucleii. This idea was expressed by Sutton some years ago in his chelogonic cycle concept (Sutton, 1963), and by Vail (1968). It is a concept which can be critically tested by palaeomagnetism, by the examination of rocks from either the Congo or West Africa to test against Precambrian results already available from the Transvaal–Rhodesian block.

References

Almeida, F. F. M. de., Amaral, G., Cordani, U. G. and Kawashita, K., 1973. The Precambrian evolution of the South American cratonic margin south of the Amazon river. *In:* The Ocean Basins and Margins, 1. The South Atlantic. Edited by A. E. M. Nairn and F. G. Stehli, Plenum Press N.Y.

Allard, G. O. and Hurst, V. J., 1969. Brazil–Gabon geologic link supports continental drift, *Science*, **163**, 528–32.

Asmus, H. E. and Ponte, F. C., 1972. The Brazilian marginal basins. *In:* The Ocean Basins and Margins, 1. The South Atlantic. Edited by A. E. M. Nairn and F. G. Stehli. Plenum Press, N.Y.

Beurlen, K., 1961. Die paläogeographische Entwickling des sudatlantischen Ozeans, *Nova Acta Leopoldina*, **24**, 3–56.

Bigarella, J. J., 1972. The geology of the Amazon and Parniaba basins, *In:* The Ocean Basins and margins, 1. The South Atlantic. Edited by A. E. M. Nairn and F. G. Stehli. Plenum Press, N.Y.

Bird, J. M. and Dewey, J. F., 1970. Lithosphere plate-continental margin tectonics and the evolution of the Appalachian orogen, *Bull. Geol. Soc. America*, **81**, 1031–60.

Briden, J. G., 1967. Recurrent continental drift of Gondwanaland, *Nature*, **215**, 1334–9.

Brognon, G. and Verrier, G., 1966. Teconique et sedimentation dans le bassin du Cuanza (Angola). *In:* Sedimentary Basins of the African coasts, edited by D. Reyre, (New Delhi, 1964 symp.) 207–52.

Bullard, E., Everett, J. E. and Smith, A. G., 1965. The fit of continents around the Atlantic, *Phil. Trans. Roy. Soc.* **A.258**, 41–51.

Cahen, L., 1954. Geologie du Congo belge. Vaillant-Carmanne, Liege, 577 pp.

Caster, K., 1952. Stratigraphic and palaeontologic data revelent to the problem of Afro–American ligation during the Palaeozoic and Mesozoic, *Bull. Am. Mus. Nat. Hist.*, **99**, 105.

Creer, K. M., 1964. A reconstruction of the continents for the Upper Palaeozoic from palaeomagnetic data, *Nature*, **203**, 1115–20.

Creer, K. M., 1970. A review of palaeomagnetism, *Earth Sci. Rev.*, **6**, 369–466.

Dickson, G. O., Ladd, J. W. and Pitman, W. C., 1973. The Age of the South Atlantic. *In:* The Ocean Basins and Margins, 1. The South Atlantic. Edited by A. E. M. Nairn and F. G. Stehli. Plenum Press, N.Y.

Dietz, R. S. and Holden, J. C., 1970. Reconstruction of Pangaea: Break-up and dispersion of continents, Permain to present, *J. Geophys. Res.*, **75**, 4939–56.

Douglas, R. G., Moullade, M. and Nairn, A. E. M., 1973. This conference, Volume 1.

Fail, J. P., Montadert, L., Delteil, J. R., Valery, P., Patriat, P. and Schlich, R., 1970. Prolongation des zones de fractures de l'ocean Atlantique dans le Golfe de Guinée, *Earth Planet. Sci. Letters*, **7**, 413–19.

Fernandes, A., 1966. Analogia des bacias saliferas de Sergipe, gabão congo e Angola, *Bol. tec. Petrobras*, **9**, 349–65.

Funnell, B. M. and Smith, A. G., 1968. Opening of the Atlantic Ocean, *Nature*, **219**, 1328–33.

Grekoff, N., 1960. Ostracodes du Basin du Congo II, Cretacé *Ann. mus. roy. Congo belge. Sc. geol.*, **35**, Tervuren.

Grosdidier, E., 1967. Quelques ostracodes nouveaux de la serie ante-saliferè ('Wealdienne') des bassins côrtiers du Gabon et du Congo, *Rev. Micropal.*, **10**, 107–48.

Humphrey, F. L. and Allard, G. O., 1968. The Propria geosyncline, a newly recognized Precambrian tectonic province in the Brazilian Shield, XXIII. *Intern. Geol. Congr.* (*Prague*), **4**, 123–39.

Kennedy, W. Q., 1965. The influence of basement structure on the evolution of the coastal (Mesozoic and Tertiary) basins of Africa. *In:* Salt Basins around Africa. Inst. Petrol. London, pp. 7–15.

Krommelbein, K., 1965. Ostracoden aus der nicht-marinen Unter-Kreide ('Westafrikanischer Wealden') des Congo-Kustenbeckens, *Meyniana*, **15**, 59–74.

Krommelbein, K., 1966. On 'Gondwana Wealden' Ostracoda from NE Brazil and West Africa. Proc. 2nd W. African micropal. Coll. (Ibadan 1965), 113–18.

Le Pichon, X. and Hayes, D. E., 1971. Marginal offsets, fracture zones, and the early opening of the South Atlantic, *J. Geophys. Res.*, **76**, 6283–93.

McElhinny, M. W. and Briden, J. C., 1971. Continental drift during the Palaeozoic, *Earth Planet. Sci. Letters*, **10**, 407–16.

McElhinny, M. W. and Luck, G. R., 1970. Palaeomagnetism and Gondwanaland, *Science*, **168**, 830–2.

Machens, E., 1973. The geological history of the marginal basins along the north shore of the Gulf of Guinea. *In*: The Ocean Basins and Margins, 1. The South Atlantic. Edited by A. E. M. Nairn and F. G. Stehli. Plenum Press, N.Y.

Martin, H., 1973. Geology of Southwest Africa. *In*: The Ocean Basins and Margins, 1. The South Atlantic. Edited by A. E. M. Nairn and F. G. Stehli. Plenum Press, N.Y.

Rust, I., 1973. The evolution of the Palaeozoic Cape Basin, southern margin of Africa. *In:* The Oceans Basins and Margins. Edited by A. E. M. Nairn and F. G. Stehli. Plenum Press, N.Y.

Sutton, J., 1963. Long term cycles in the evolution of continents, *Nature*, **198**, 731–5.

Vail, J. R., 1968. Significance of the tectonic pattern of southern Africa, *Tectonophysics*, **6**, 403–11.

Wilson, J. T., 1966. Did the Atlantic Ocean close and then re-open? *Nature*, **211**, 676–81.

7.9

R. A. REYMENT
Paleontologiska Institut
University of Uppsala, Uppsala, Sweden

Cretaceous History of the South Atlantic Ocean

Introduction

In 1964 I undertook a preliminary survey of the Upper Cretaceous molluscs of the South Atlantic region with the end in view of marshalling palaeogeographical evidence against drift interpretations of the origin of the Atlantic Ocean (Reyment, 1965). Far from achieving this aim, I was, at the close of the project, well on the way to becoming a convert to 'mobilism'. Further studies on other groups of invertebrates showed that the question of the distribution of the Cretaceous invertebrates merited special study. This study was initiated in 1969, in collaboration with E. A. Tait of the University of Aberdeen, with an extensive field programme in Gabon, South Africa, Brazil and the Caribbean. Both of us had previously worked for many years in the Cretaceous of West Africa, particularly Nigeria and Cameroun. The results of this study appeared recently (Reyment & Tait, 1972b). A detailed field programme has just been concluded in the Cretaceous of the state of Sergipe, Brazil, by two postgraduate students of the University of Uppsala, P. and S. Bengtson. Their work forms a continuation of our project in that area.

The present paper gives a summary of the results of the afore-mentioned field programme, a review of the results achieved by other palaeontologists and stratigraphers who have worked in the area under consideration, plans for future research, and an outline of problems awaiting solution.

Cretaceous Deposits of the South Atlantic Margins

Cretaceous deposits are known from most of the countries bordering the South Atlantic, although not all of these have been studied in detail. It is also quite possible that significant discoveries remain to be made. The Cretaceous successions of the southern Atlantic margins are all narrow, rather elongated occurrences, with the exception of Nigeria. There are pronounced faunal agreements of a nature such as to suggest a special zoological province up to about the middle of the Cretaceous period.

Angola to Zaire

The Cretaceous sequences of the marginal deposits of Angola, northwards to the Congo River, show many marked similarities.

(1) They rest usually on crystalline basement, perhaps locally on continental Palaeozoic.

(2) They begin with a continental lacustrine series. Although the palaeontology of

these beds is still imperfectly known, it seems clear that they range from Late Jurassic to Neocomian* in age.

(3) There are evaporites of Aptian age. The northerly salt beds are quite homogeneous, whereas the evaporitic sequence of Angola (Cuanza) displays a more complex lithology (Hourcq, 1966), beginning with homogeneous salt beds which are followed by black shales, carbonates, anhydrite, a further salt sequence and finally anhydrite, and impure carbonates.

(4) Fossiliferous Upper Albian.

(5) Poorly characterized Cenomanian which, in Angola, occurs in a carbonate facies. Turonian, also poorly marked, follows.

(6) Upper Turonian and younger beds.

Gabon

The Gabonese sequence is one of the keys to the solution of the Cretaceous history of the southern Atlantic. Although, in general, the same scheme as recorded above applies to it, much more detailed information is available for it.

The Neocomian deposits are of freshwater, probably lacustrine origin. They are similar in development to the non-marine Neocomian of Zaire. The ostracods have been studied in some detail mainly by Krömmelbein (1966) and there are also reports on other aspects of the animal palaeontology (de Klasz, 1965) and palynology (Freake, 1966). The Neocomian beds lie in the middle and lower parts of the Cocobeach Formation of Gabon. The upper part of this formation is placed tentatively in the Lower Aptian by French workers.

The Upper Aptian is marked by the occurrence of homogeneous salt deposits, overlain by marls in which *Neodeshayesites*? has been found. The Lower Albian contains species of *Douvilleiceras*, Middle Albian has rare *Oxytropidoceras*, and Upper Albian *Elobiceras* and *Mortoniceras*. Cenomanian has been established by means of foraminifers and the lower part of the Lower Turonian by the presence of *Bauchioceras nigeriense* (Woods) and *Wrightoceras wallsi* Reyment.

Ivory Coast and Ghana

The Ivory Coast and Ghana contain a narrow coastal strip of Cretaceous and Tertiary sediments. Fragments of *Elobiceras* and *Dipoloceras* have been obtained from a borehole in the Ivory Coast and the host sediments of Upper Albian age lie on non-marine shales and conglomerate. The Cenomanian and Turonian of the area contain no ammonites and it is difficult to arrive at a definite stratigraphical conclusion for these beds. There is some information on the ostracods of Early Cretaceous freshwater sediments; these differ from those of Sergipe-Bahia-Gabon (Krömmelbein, 1968).

Rio de Oro Basin

This markedly elongate basin is known best from its northern end in the Tarfaya area (Collignon, 1966, Choubert & Faure-Muret, 1962; Hottinger, 1966). The importance of its Cretaceous ammonite associations for the understanding of the palaeobiogeographical development of the Atlantic Ocean has been reviewed at length by Reyment

* Although it is now generally accepted, by specialists of Lower Cretaceous stratigraphy, that the term 'Neocomian' can no longer be retained for serious work in Europe, it is practically useful in areas such as the South Atlantic, where one cannot usually separate Berriasian, Valanginian, Hauterivian and Barremian.

& Tait (1972b). The stratigraphical information available on the southern extent of this basin is slight. Salt deposits, assumed to be of Triassic age, have been reported from the basal part of the sequence (Furon, 1968, p. 139). There is a thick sequence of what has been said to be marine Jurassic sediments, overlain by an apparently Cretaceous non-marine succession, followed by anhydrite deposits and poorly fossiliferous limestones. The Tarfayan section of the basin is well known and of exceptional importance for studies on the history of the Atlantic Ocean. Triassic sediments may overlie the crystalline basement and there is Jurassic between these beds and the Cretaceous. The Lower Cretaceous consists of sandstones and marls which are overlain by Upper Cretaceous marls, shales and limestones. The Upper Albian beds contain typical North American and European species of *Dipoloceras*, *Hysteroceras* and *Mortoniceras*. The oxytropidoceratids are represented by species of the subgenera of *Oxytropidoceras*, *Venezoliceras* and *Tarfayites*. The Cenomanian contains calycoceratids of northern aspect as well as the North American *Tarranteoceras*. This northern aspect of the fauna continues into the Lowermost Turonian and there are species of *Metoicoceras*, *Mammites*, *Neoptychites* and *Selwynoceras*, but none of the vascoceratids so typical of the Iberian Peninsula, Saharan Africa and West Africa. Collignon (1966, 1967) made special mention of the missing vascoceratids element in the lowermost Turonian of Tarfaya.

Higher up in the Turonian, the position is entirely reversed and a great number of species in common with Nigeria and Cameroun have been recorded, for example, species of the genera *Watinoceras*, *Benueites*, *Glebosoceras*, etc. It is worthwhile remembering at this stage that the genus *Watinoceras* is more typically a northern entity and there is some phylogenetical evidence to suggest that it is the forerunner of *Benueites*.

Collignon (1967) was able to correlate the Moroccan Albian with the standard European successions, but was impressed by the slight agreement with the Upper Albian of the southern reaches of the ocean. He pointed out the lack of the characteristic Upper Cenomanian genus *Neolobites* in Morocco, species of which are well known in the western Tethys and trans-Saharan regions.

Senegalese Basin

Information on the pre-Campanian stratigraphy of this basin derives from off-shore drilling (Furon, 1968). A Jurassic age has been assigned to algae and foraminifers, and ostracods and foraminifers have been determined as belonging to the Neocomian and Aptian; all species recorded hitherto are typically marine. The evidence for Lower Albian is less secure. Calcareous beds, interfingered with gypsum layers, have, on their content of poorly preserved fossil plants, been referred to this time. Upper Albian and Cenomanian have been tentatively noted, as also poorly defined Turonian.

Nigerian Basin

I have purposely let the Nigerian basin wait until last. It is the only basin along the West African seaboard that lacks deposits older than upper Middle Albian, as far as is known at the present time. Although terrestrial deposits referred to the 'continentale intercalaire' are known from Northern Nigeria, there is no evidence for sediments of the lacustrine 'Cocobeach type' in the coastal sedimentary basin. Although it has been suggested now and then that salt beds could underlie the Albian deposits of Eastern Nigeria, no direct evidence in support of this hypothesis has yet been forthcoming.

Careful examination of the biostratigraphical information available discloses that up to the lowermost Lower Turonian, the affinities of the molluscan associations lay with those of the Atlantic to the south of the Niger Delta and, for a short period of the lowermost Turonian, with the Tethys via the trans-Saharan region (Reyment & Tait, 1972a). After this time, in the Lower Turonian, the ammonite associations, as well as other molluscan groups, become widely spread throughout the Atlantic Ocean and the distributional pattern is in general the same as is found today (Reyment & Tait, 1972b).

Trinidad

Upper Lower Turonian ammonites occur in the Plaisance member of the San Fernando Formation of southwestern Trinidad. There is close agreement in this association with those of Nigeria, Cameroun and Colombia (Reyment & Tait, 1972b).

Potiguar Basin

The Potiguar Basin of the state of Rio Grande do Norte in northeastern Brazil was initiated by earth movements during the Lower Cretaceous. Although it contains a fairly rich association of molluscs, there is no agreement whatsoever with the associations of the same age in the Sergipe-Alagoas basin, on the other side of the 'horn' of Brazil. This fact was discussed at some length by Beurlen (1961). The Pernambuco coastal basin, structurally a continuation of the Potiguar basin, also contains Cenomanian and Lower Turonian fossils, none of which is known from the nearby Sergipe-Alagoas sequence.

Sergipe-Alagoas Basin

The faunistic relationships of this basin, and of the Brazilian basins further south, lie, during the Lower Cretaceous and Upper Cretaceous until Lower Turonian, with West Africa and not with the areas of sedimentation in the Brazilian coastal areas to the north (Beurlen, 1961). The geology of the state of Sergipe in northeastern Brazil has been studied extensively by the geologists of Petrobrás and by Karl Beurlen and more recently, in detail, by a team from the Universities of Aberdeen and Uppsala. The field project, begun in November 1969, was concluded in May, 1972. The Late Jurassic-Neocomian sedimentary sequence begins with a thick series of continental lacustrine sediments, located in a 'half graben'. These beds contain ostracods. There are hypersaline beds separating the non-marine and typically marine successions and these form a thick series of possibly commercially exploitable salt deposits to which an Aptian age has been given. The Albian overlies marine Upper Aptian with a few fossils, and appears to be reasonably complete. Lower Albian is represented by abundant *Douvilleiceras* and Upper Albian by typical *Elobiceras*, the species of which genus show close agreement with the associations of Angola.

The Recôncavo Basin

It was first in the Neocomian sediments of this basin that the important associations of ostracods were found by Krömmelbein and it was he who recognized their significance for correlational studies in the southern Atlantic region (for example, see

Krömmelbein, 1966). This large basin has been described by Fonseca (1966) as a 'half graben'. A synoptic discussion of the geology of the Recôncavo basin was given in Reyment & Tait (1972b). The basin is lined with continental sediments of Palaeozoic age. Sedimentation became general throughout the entire basin with the deposition of a thick series of continental sediments of Late Jurassic to Neocomian age. During the Neocomian to lowermost Aptian, connexions existed at times between the Recôncavo and Sergipe-Alagoas basins and the Almada basin to the south (Fonseca, 1966, p. 69). The first marine incursion of the sea took place during the Aptian, but no salt deposits appear to have been formed.

Trans-Atlantic Correlations

Any reasonable synthesis of the geological development of the South Atlantic region during the Cretaceous Period must take account of the following.

(1) The occurrences of thick series of freshwater deposits, located in 'half grabens' in some of the basins, which are open to the present-day ocean, and the agreement in a significant part of the sedimentary sequences.

(2) The close agreement in the non-marine ostracod faunas and microfloras of northeastern Brazil and those of Gabon and Zaire, and the distinctness of the microfauna from other associations.

(3) The occurrence of thick series of salt deposits in sedimentary basins, open to the ocean; the present-day configuration of the South Atlantic Ocean is difficult to reconcile with the accumulation of evaporites in such impressive amounts. It is significant that the salt deposits were only formed during a relatively short period of time in the Upper Aptian, neither before nor after.

(4) The palaeobiogeography of the ammonites of the Upper Albian, Cenomanian and lowermost Turonian of key areas in West Africa, from Angola to Southern Morocco, along the eastern margin of the South Atlantic, and Brazil, Trinidad, along the western margin of this ocean, indicates that the ammonite associations of the northern and southern regions of the southern Atlantic Ocean of the Cretaceous Period were almost entirely different. It requires to be explained why a profound difference is to be seen in the dispersal of the ammonite species from the latter part of the Lower Turonian and onwards. The distributions of the pelecypod and ostracod species adhere to the same pattern as shown by the ammonites, although in a less sharply outlined fashion.

(5) The evolutionary divergence of many living groups of animals, for example, the chironomid midges (Insecta). Entomologists, who have long been at a loss to explain the global pattern of distribution of certain insect groups, have reached the conclusion that the South American and African species found today must have derived from common ancestors (Brundin, 1967).

(6) The occurrence of non-tropical shallow-water Cretaceous (Austral) foraminifers as far north as 6°02·4′ N (Scheibnerova, 1972).

(7) The occurrence of the hyperalkaline suite of intrusions, known as the 'Younger Granites', in West Africa (Nigeria and Cameroun in particular) and north-eastern Brazil (Almeida & Black, 1966).

The Neocomian freshwater deposits

Let us first look at the general structure of the basins of Gabon and northeastern Brazil. They are largely in the form of 'half grabens' with minor horsts within the boundary faults. The very conformation of these basins is such as to provoke serious misgivings about a sedimentary environment analogous to that pertaining at the present time in the areas under consideration. As Martin (1968, p. 43) has pointed out, it is difficult to see how 5000 metres of non-marine sediments, containing solely non-marine fossils, could have been deposited in the subsiding Recôncavo-Tucano graben, without marine ingressions, if there had been a Neocomian Atlantic Ocean.

In a recent review of research results on the Neocomian of northeastern Brazil and Gabon, Krömmelbein (1971) has summed up what is known at present on the ostracods. A very high percentage of the species from the non-marine Neocomian of northeastern Brazil are found in the corresponding beds in Gabon and Zaire. Moreover, in addition to the agreement in associations, whole assemblages of the ostracods occur in the same stratigraphical order. The significance of this observation by Krömmelbein may be further brought out by means of the readily calculated Koch Index of systematic biology, a variant of the well known coefficient of Jaccard. The results for the assemblages of three correlated formations from each of northeastern Brazil and Gabon, based on the information in Krömmelbein (1966), are shown in Table 1. Here, we see the interesting fact that not only are the values of the Koch Index of Biotal Dispersity exceptionally high, but they tend to diminish upwards in the section.

TABLE 1. Koch Index of Biotal Dispersity for Krömmelbein's (1966) ostracod data.

Formations	Index	
São Sebastião and Benguié	40	(youngest)
Upper Ilhas and Bifoune	32	
Lower Ilhas and Fourou Plage	60	
Upper Middle Candeias and corresponding beds in Gabon	61	(oldest)

Krömmelbein (1971) also reminds us that the sedimentary strata of Bahia-Sergipe and Gabon are essentially the same. This was first pointed out by Krömmelbein and Wenger (1966) and later by Allard & Hurst (1969). These latter authors were concerned with an analysis of ancient tectonic structures in the two areas, and their conformability, but also pointed out the close agreement in lithological properties and sequence of sedimentary events in the Aliança Formation of northeastern Brazil and its equivalent in Gabon, the M'Vone red beds. Recent studies have shown that these beds in both regions, previously thought to be devoid of animal fossils, now contain the same associations of ostracods and other crustaceans.

The question of the marine Neocomian must now be briefly considered. Hallam (1967) pointed out that the Neocomian trigoniids of South America and southern Africa are remarkably similar to each other and that they are distinct from those in other parts of the world. He also reminded us that this fact was the mainspring of Uhlig's (1911) reasoning for including the 'Uitenhage Beds' of South Africa in his South Andine faunal province and the basis for his deduction that there must have been shelf communication between the two continents. I commend Hallam's thoughtful paper to the attention of all students of drift problems. Whitehouse (1947) noted that there are similarities between the trigoniids of Queensland and the Uitenhage Formation of South Africa.

Reyment & Tait (1972b) reported on a collection of marine Valanginian molluscs from the Uitenhage Formation and pointed out that the closest affinities of this well known fauna lie with the fossils of the Mulichno and Agrio Formations of western Argentina and Chile.

The Aptian salt deposits

Marine Upper Aptian has been documented for the west coast of Africa from Angola to Gabon and, in Brazil, for the State of Sergipe. In all of these places, the Aptian is characterized by the occurrence of extensive evaporite deposits and in Gabon and Brazil (Sergipe) these are of almost identical development. It would be difficult, although not impossible, to explain the formation of these salt bodies by postulating the existence of barriers of some kind or other, behind which the necessary environment for salt accumulation developed. If one accepts that during Aptian time, a long narrow sea existed in the southern Atlantic region, akin to the Red Sea of today, the implication of these barriers becomes minor.

Ammonite palaeobiogeography

Up until the Turonian, the distribution of the ammonites is virtually impossible to explain if one assumes that the South Atlantic had the same configuration as today. Attempts to do so have led to rather strange distributional charts (e.g. Freund & Raab, 1969). The majority of ammonite shells floated after the death of the animal, and for shells of the type represented by the elobiceratids, vascoceratids, pseudotissotiids, the floating position was higher in the water than the living *Nautilus*, a well known constituent of the nekroplankton (Mutvei & Reyment, 1972). Any reasonable palaeogeographical interpretation of the South Atlantic must therefore take account of this fact. It must, firstly, tell us why the nekroplanktonically drifting ammonite shells, as well as the living animals, could not pass Brazil-Nigeria until the latter part of the Lower Turonian and why the ammonites kept managing to do this successfully from then on until they died out, and after them, numerous Paleogene nautiloids.

To be sure, the neat and satisfying picture presented by the ammonite palaeogeography is not without flaws, the main one of these involving the distribution of the lowermost Turonian species in the western Tethys and the interpretation of the oceanological conditions that prevailed in northwestern Morocco at that time. More work requires to be done on this aspect of the problem, an enterprise that would appear to be faced with difficulties owing to the paucity of good outcrops in the crucial area. Nevertheless, L. Hottinger, a geologist and palaeontologist who has worked many years in Morocco, was led to remark (Hottinger, 1966, p. 97) 'l'ancien schéma paléogéographique du double bras de mer réunissant la Méditerranée et l'Atlantique le long du géosynclinal rifo-bétique, tel qu'il est reproduit dans tous les traités, ne correspond plus à l'état actuel des connaissances'.

Evolutionary divergence of living animals

As a case history of information of this kind I have selected Brundin's (1967) magnificent treatise on the chironomid midges. In this massive work, Brundin was, amongst other things, concerned with explaining the interrelationships between the midge faunas of Africa and South America. He came to the conclusion that the relationships between the midges in various parts of their present-day distribution are such as to indicate a common centre of origin in the Upper Jurassic and that this postulated centre of origin can only be reconstructed if one invoke the idea of a Gondwana landmass (Brundin, 1967, p. 452). He also concluded that only the oldest components among the sister group systems of the chironomid midges that display transantarctic relationships are represented in Africa. I refer the interested reader to Brundin's treatise, in which not only the midges are discussed in the light of continental displacement, but also many other groups of animals, as well as plants.

Austral foraminiferal elements in the South Atlantic

Scheibnerova (1972) noted the abundant occurrence of certain Cretaceous foraminifers in a bottom core from a latitude of 6°02·4′N in the southern Atlantic Ocean and explained this by postulating the existence of a cold current, from the southwest and south, or simple influx of water from the south, and a northward shift of Africa. She believes that the mode of distribution of the foraminifers indicates that it was first in post-Turonian time that the pattern of circulation of currents in the South Atlantic began to resemble that of the present day. This agrees well with the ammonite dating of the union of the southern and northern arms of the Atlantic (Reyment, 1969). A further important fact that should not be overlooked in this connexion is that the oldest sea-bottom sediments found in the South Atlantic to date by the JOIDES project are Upper Cretaceous (Berggren & Phillips, 1970).

The Younger Granites

The well known Younger-Granite hyperalkaline petrological province of West Africa, best known from northern Nigeria, has been identified in the Cabo area of Pernambuco, northeastern Brazil (Almeida & Black, 1966). The probability of two separate petrological provinces with such a complicated petrology and mineralogy occurring on both sides of the Atlantic was judged by Almeida & Black to be very low indeed. The radioactive age for the Cabo granite fits in with the Lower Turonian date for the separation of the continents. Almeida & Black could demonstrate that, using the Bullard fit of South America to Africa, the faulting pattern of Pernambuco (Brazil) continues exactly into Cameroun by the Ngaoundéré fault.

Concluding Remarks

None of the pieces of 'evidence' presented in this paper on its own can constitute a proof of the former oneness of the South American and African continents, but when viewed collectively, and together with the evidence of oceanologists, geophysicists and structural geologists, they can be seen to fit into a general intellectual framework of considerable plausibility.

To be sure, one may think up alternative interpretations for the bricks in the structure, considered one at a time and not taken as a group. For example, oceanic currents are known to be able to transport surface plankton and nekroplankton from southwestern Africa to northeastern South America during certain times of the year. This could be invoked to explain the dispersal of the ammonite shells across the

Atlantic, and perhaps the transport of some other groups of animals, but it is only useful up to a point, as there is still the matter of the restriction of the southern Atlantic genera to their realm and the northern Atlantic ones to theirs until late Lower Turonian and the question of how the 'trans-Atlantic barrier' suffered dissolution at this time.

I have also touched upon an alternative interpretation of the salt accumulations, and perhaps, analogously, the lacustrine beds can be 'explained' in the same manner, although with less credibility. Nevertheless, the fact remains that in all of these counter-arguments, the dissenter is in the weaker position of having to grope among less likely alternatives.

Wyllie (1971), in his book, has made the point that geologists are now in the position of finding themselves in possession of a 'ruling theory'. The situation has changed radically from what it was a couple of decades ago and the opponents of drift find themselves confined to attacks on isolated weak links in the chain of evidence, rather than being able to mount full frontal assaults. I maintain, however, that Meyerhoff and collaborators (cf. Wyllie, 1971, p. 376) have a very important rôle to play at the present time, that of the devil's advocate. The very fact that the weight of evidence is greatly in favour of the 'mobilists' may tend to make some of them careless in their claims, dogmatic in their assertions and overambitious in interpreting their results. An active opposition is certainly going to do more, in the long run, to help straighten out the twisty problems still remaining than to hinder this. There is no doubt, however, that a point will be reached sooner or later in the discussions and counter-arguments when it will have to be accepted that the much stretched arm of coincidence can be stretched no further.

Questions in search of an answer

Why is the heteromorphic *Labeceras-Myloniceras* fauna of the Upper Albian, composed of species with nekroplanktonic shells, only found in Moçambique, central Australia and New Guinea?

Why do the Valanginian trigoniid associations of Argentina, South Africa and central Australia have so much in common?

References

Allard, G. O. and Hurst, V. J., 1969. Brazil-Gabon geologic link supports continental drift, *Science*, **163** (3867), 528–32.

Almeida, F. F. M. de and Black, R., 1966. Comparaison structurale entre le N.E. du Brésil et l'ouest africain. Proc. 1st IUGS Gondwana Symposium (Montevideo, 1966).

Berggren, W. A. and Phillips, J. D., 1970. Influence of continental drift on the distribution of Tertiary benthonic Foraminifera in the Caribbean and Mediterranean regions, *Symp. Geo. Libya* (1969), pp. 1–91.

Beurlen, K., 1961. Die paläogeographische Entwicklung des südatlantischen Ozeans, *Nova Acta Leopoldina*, **24**, 1–36.

Brundin, L., 1967. Transantarctic relationships and their significance as evidenced by chironomid midges, *Kgl. Vetenskapsakad. Handl*, **4**, 11, 1–472.

Choubert, G. and Faure-Muret, A., 1960–1962. Evolution du domaine atlasique marocain depuis les temps paléozoiques, *Mém. Soc. géol. Fr.* (Livre-mémoire P. Fallot), **1**, 447–527.

Collignon, M., 1966. Les céphalopodes crétacés du bassin côtier de Tarfaya, *Notes Mém. Serv. géol. Maroc*, **175**, 7–148.

Collignon, M., 1967. Les ammonites crétacées du bassin côtier de Tarfaya, Sud-Marocain, *C.R. Acad. Sci. Paris*, **264**, 1390–92.
Freake, J. R., 1966. Palynology: a summary of results obtained during the Second West African Micropaleontological Colloquium, *Proc. 2nd. W. African Micropal. Coll.* (Ibadan 1965), p. 269.
Furon, R., 1968. *Géologie de l'Afrique*. Payot, Paris.
Hallam, A., 1967. The bearing of certain palaeozoogeographic data on continental drift, *Palaeocl., Palaeogeogr., Palaeoec.*, **3**, 201–41.
Hottinger, L., 1966. Résumé de la stratigraphie micropaléontologique du Mésozoique et du Tertiaire marocain, *Proc. 2nd W. African Micropal. Coll.* (Ibadan 1965), 92–104.
Hourcq, V., 1966. Les grands traits de la géologie des bassins côtiers du group équatorial. IUGS Symposium on West African Cretaceous Sedimentary Basins of the African coasts (ed. D. Reyre) Pt. 1: Atlantic Coast, 171–78.
Klasz, I. de, 1965. Biostratigraphie du Bassin Gabonais, *Bur. Rech. Géol. Min.*, **32**, 277–303.
Krömmelbein, K., 1966. On 'Gondwana-Wealden' Ostracoda from NE Brazil and West Africa, *Proc. 2nd W. African Micropal. Coll.* (Ibadan, 1965), 113–18.
Krömmelbein, K., 1968. The first non-marine Lower Cretaceous ostracods from Ghana, West Africa, *Palaeontology*, **11**, 259–63.
Krömmelbein, K., 1971. Non-marine Cretaceous ostracodes and their importance for the hypothesis of 'Gondwanaland', *Proc. 2nd IUGS Gondwana Symposium* (South Africa, 1970), 617–19.
Krömmelbein, K. and Wenger, R., 1966. Sur quelques analogies remarquables dans les microfaunes crétacées du Gabon et du Brésil oriental (Bahia et Sergipe). IUGS Symposium on 'Sedimentary Basins of the African coasts' (ed. D. Reyre) Pt. 1: Atlantic Coast, 193–96.
Martin, H., 1968. A critical review of the evidence for a former direct connection of South America with Africa. *In:* Biogeography and Ecology in South America. (Ed.), Fittkay *et al.* W. Junk, The Hague, 25–53.
Mutvei, H. and Reyment, R. A., 1973. Buoyancy control and siphuncle function in ammonites, *Palaeontology* (in press).
Reyment, R. A., 1965. Upper Cretaceous fossil molluscs in South America and West Africa, *Nature*, **207**, p. 1384.
Reyment, R. A., 1969. Ammonite biostratigraphy, continental drift and oscillatory transgressions, *Nature*, **5215**, 137–40.
Reyment, R. A. and Tait, E. A., 1972a. Biostratigraphic dating of the evolution of the South Atlantic rift. *Sect. 7—Paleontology*, *24th Int. Geol. Congr.*, (Montréal, 1972).
Reyment, R. A. and Tait, E. A., 1972b. Biostratigraphical dating of the early history of the South Atlantic Ocean, *Phil. Trans. roy. Soc.* (*London*), **264**, 55–95.
Scheibnerova, V., 1972. Non-tropical Cretaceous Foraminifera in Atlantic deep sea cores and their implications for continental drift and palaeoceanography of the southern Atlantic Ocean, *J. foram. Res.* (in press).
Uhlig, V., 1911. Die marinen Reiche des Jura und der Unterkreide, *Mitt. geol. Ges. Wien*, **3**, 329–448.
Whitehouse, F. W., 1947. A marine Early Cretaceous fauna from Stanwell (Rockhampton District), *Proc. roy. Soc. Qld.*, **57**, 7–20.
Wyllie, P. J., 1971. *The Dynamic Earth*, Wiley & Sons, New York, London.

7.10

P. J. BUREK

Institut für Geologie-Paläontologie
Universität Tübingen
Sigwartstr, 10, Germany

Structural Deduction of the Initial Age of the Atlantic Rift Systems

Introduction and structural relations between sea-floor spreading and continental features in Arabia

The Arabian Shield is surrounded by rift troughs (Gulf of Aden and Red Sea) on its southern and western sides and by Alpino-type orogenic mountain belts (Tauros and Zagros) on its northern and eastern sides. One may suggest that the tectonic phases recorded in the mountain belts may be used to date the spreading events within the rift troughs. Major volcanic eruptions are also associated with the rifting in this area: the Palaeocene Trap-Series, occurring only in the neighbourhood of the Gulf of Aden and the younger Aden volcanics which occur in the vicinity of the Red Sea along the western side of Arabia. The association of sea-floor spreading, orogeny and volcanism is obvious.

Less conspicuous epeirogenic movements, which control the deposition and erosion and thereby the conditions of sedimentation on continental plates have been suggested to be a result of sea-floor spreading too (Burek, 1969a). In the Arabian Shield region two sets of epeirogenic undulations are found: one parallel to the Sheba rift and a somewhat younger one trending parallel to the Red Sea. Both have several features in common: They are considerably disarranged near the continental shelves, reverse faulting, overthrusting and folding as well as normal faulting associated with isostatic upward movements have been reported (Beydoun, 1966). A similar style of continental margin tectonics was observed within the Danakil fragment, situated between the 'oceanized' Afar depression and the 'oceanic' Red Sea trough (Burek, 1973). Another characteristic feature of both Arabian undulation systems is the intense warping close to the rift troughs and the fact that the wavelength for the warping increases systematically with distance from the rift. This implies that these undulations are folding reactions of consolidated (cratonic) plates, caused by directed stresses normal to and originating within, the rift troughs. (The plates are pushed from the rift sides, Burek, 1970b.) The approximate geological dates for the orogenic phases of the 'Adenean' undulation are: Tithonian—Neocomian—Albian—Cenomanian/Turonian—Coniacian/Santonian—Lower Maastrichtian—Eocene—Oligocene. The epeirogenic phases correlate with the orogenic movements in the Tauros mountains. Further, even the relative intensity of epeirogenic and orogenic movements are correlative: the main phases were during the Neocomian and Palaeogene. The Trap volcanism around the Gulf of Aden also occurred during the latter phases.

The approximate dates for the tectonic phases of the 'Eritrean' elements (epeirogenic, as well as orogenic) are as follows: Upper Maestrichtian—Danian—Upper Paleocene—Eocene—Oligocene/Miocene—Lower Miocene—Miocene/Pliocene. The intensity of the tectonic events within the 'Eritrean' undulations correlates with the intensity of those in the Zagros mountains; the main phases are: Maestrichtian and Miocene/Pliocene. However, in the Oman area an intensity inversion has been reported (Wilson, 1969), which cannot be explained at present. The Aden Volcanics Belt extending over large areas of western Arabia is associated with the Neogene movements. Essentially the movements within the 'Adenean' elements, predate the 'Eritrean' movements, i.e. sea-floor spreading in the Gulf of Aden is older than in the Red Sea. It is difficult to visualize a perpendicular superposition of two undulation systems over one another. Thus it is not surprising that the 'Eritrean' undulations are only found in North Arabia, where the older 'Adenean' undulations fade away.

The post-Jurassic structural history of Arabia seems to provide a good example of the interactions of crustal spreading, epeirogenic movements, orogenesis and magmatism. Sea-floor spreading is a world wide phenomenon. One should therefore expect the same relations in other parts of the world (cf. Burek, 1970b). Figure 1 presents an attempt to plot the relations described above in a diagram. The 'epeirogenic' curve of transgressions and regressions is derived mainly from Schuchert's 'Palaeogeographic Atlas of North America' (Schuchert, 1955) and from Bubnoff's work on the geologic

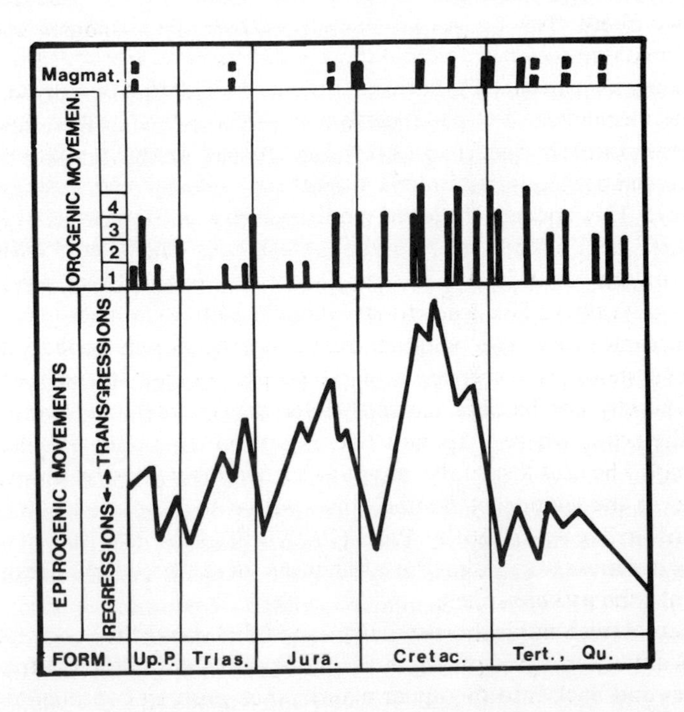

FIGURE 1 Schematic diagram, showing interaction of epeirogenic and orogenic movements and magmatic activities since Permian times.

history of Europe (Bubnoff, 1935a, 1936). The orogenic events are the phases as defined by Stille. Information on the period of magmatic activity has been collected from several sources. The diagram, derived from North American and European data, is similar to other data compilations published previously (cf. Brinkmann, 1964). Thus, it seems, the interaction of tectonic phenomena is basically the same all over the world. This suggests that the time sequence of tectonic events may be used to date initial stages of other rift features if their trends are new, differ structurally from preceeding features and are obviously parallel to the new rift-structures.

Before we attempt to deduce the initial age of the different Atlantic rift systems using an interpretation of the adjacent structures, a short review of the possible mechanisms causing plate movements and the associated structures will be given.

Continental Push or Mantle Diapirism as dominant Driving Mechanism for plate movements and the associated structural Process

At present three major mechanisms are considered as possible causes for plate movements: *Convection* in the upper mantle or asthenosphere (cf. Elsasser, 1971), *gravitational gliding* (Hales, 1969), allowing uncoupling of the rigid and dense lithosphere along the upper parts of the asthenosphere (Gutenberg's low velocity zone) which is tectonically weak due to reduced viscosity. In places the asthenosphere is less dense than the lithosphere (Press, 1970). Thirdly *continental push* or *mantle diapirism* is a possible mechanism (Burek, 1969, 1970a; Lliboutry, 1969).

The first two alternatives do not satisfactorily explain all structural features caused by compressional stresses observed in Arabia and elsewhere. For this and several other reasons including arguments of symmetry (Burek, 1973), they are not likely to represent the dominant mechanism for plate movements. The writer prefers the concept of continental push (mantle diapirism). Like gravitational gliding it requires uncoupling of lithosphere and asthenosphere along Gutenberg's low velocity zone (Magnitsky & Zharkov, 1969). This concept does not require gravity as the main energy source, but suggests that the heat of the upper mantle is the source of potential energy. Thermal conductivity of the crust is too low to reduce the temperature gradient mantle-atmosphere significantly. This leads to an exchange of thermal energy by a convective overturn of oceanic crust. The potential mantle energy, concentrated below the mid-oceanic ridge systems, is converted into mechanical energy in the form of 'mantle diapirism', whereby hot basaltic material differentiates from the mantle and is emplaced into the crust, where it spreads laterally, producing cold oceanic plates (sea-floor spreading). The heat loss of the mantle is further augmented by absorption of cold oceanic plates in the vicinity of Benioff zones within the upper mantle (subduction). The density inversion described by Press (1970) within the low velocity layer under tectonically active areas may reflect the 'absorption' of reheated light oceanic materials, reemplaced into the asthenosphere.

The continental push model is essentially a modified convective concept, but differs from mantle convection by assuming direct transport of oceanic crust from the ridges to the trenches and back into the upper mantle, thus pushing continental plates away from the rift sides. Mantle convection implies that the rising parts of the convection cells bend underneath the lithosphere and, being frictionally coupled to the lithosphere, pull the crust away from the ridges. The continental push concept has the advantage that it satisfies the observed compressional tectonic features in cratonic and orogenic areas and creates no difficulties with regard to the observed symmetry of structural elements. Arabia, for example, would at present require two simultaneously active cells of mantle convection or elongated asthenospheric cells overlapping each other in a perpendicular fashion. Such a convective pattern is mechanically unlikely. Injection of oceanic material into perpendicularly arranged zones of crustal weakness ('crustal spreading') and compensation of the crustal extension by folding in alpinotype structures ('crustal shortening') seems tectonically reasonable.

Theoretically, epeirogenic uplift of cratonic areas could be produced by increasing the temperature gradient in a boundary of layers with different density (e.g. Mohorovicic discontinuity) but identical composition, for example, gabbro and eclogite (McKenzie, 1969).

On the other hand, the sialic low velocity layer (8–15 km and 20–30 km) deduced from seismic and magnetotelluric evidence (Landisman *et al.*, 1971 and v. Zijl *et al.*, 1970) indicate that the sialic part of the lithosphere is stratified, at least in the vicinity of areas of tectonic activity. Intergranular water (vapour under increased temperature and pressure) reduces electrical resistivity (Lebedev & Khitarov, 1964) and increases seismic attenuation (Nur & Simmons, 1969). Intergranular vapour also substantially facilitates translational gliding of polycrystals (Griggs & Blacic, 1965). Thus the lithosphere cannot be regarded as a compact, rigid, consolidated block of sial and sima but rather as a stratified body with layers enriched in interstitial vapour. These layers are zones of tectonic adjustment (Burek, 1970a; Landisman *et al.*, 1971) which, in analogy to sediments, facilitate gliding along bedding planes during folding; consolidated, cratonic areas react only by gentle epeirogenic warping. Interstitial vapour also facilitates partial melting and recrystallization by reducing melting points (Fyfe *et al.*, 1958). For this reason at least the lower sialic inversion-channel could, under tectonic activation, produce intrusions of silica-rich magma or contaminated basic magmas. We may conclude that a layered sialic crust with horizons enriched in interstitial water vapour facilitates cratonic warping and magmatism under directed compressional stress. Therefore cratonic uplift may not only be produced by an increase in the temperature gradient below an uplifted area, which could cause a downward migration of the Mohorovicic discontinuity, but also, it is suspected that changes in isostatic equilibrium, caused by compressional warping (uplift), are adjusted by (downward) migrations of a layer boundary (Moho) between materials of different density but similar chemical composition.

The three concepts described here must be considered as simplified models. However, they bear a reasonable probability and can explain structural features observed in Arabia and elsewhere, especially gentle epeirogenic movements that control palaeogeography on consolidated cratonic plates.

Undulation-Systems around the Atlantic

The tectonic diagram (Fig. 1) of the relation and interaction of the main structural elements was derived from palaeogeographic and structural evolution data of the 'Laurasian' continent and indicates, that the structural interactions (germanotype, alpinotype and magmatism) observed in Arabia, are a global phenomenon. This would

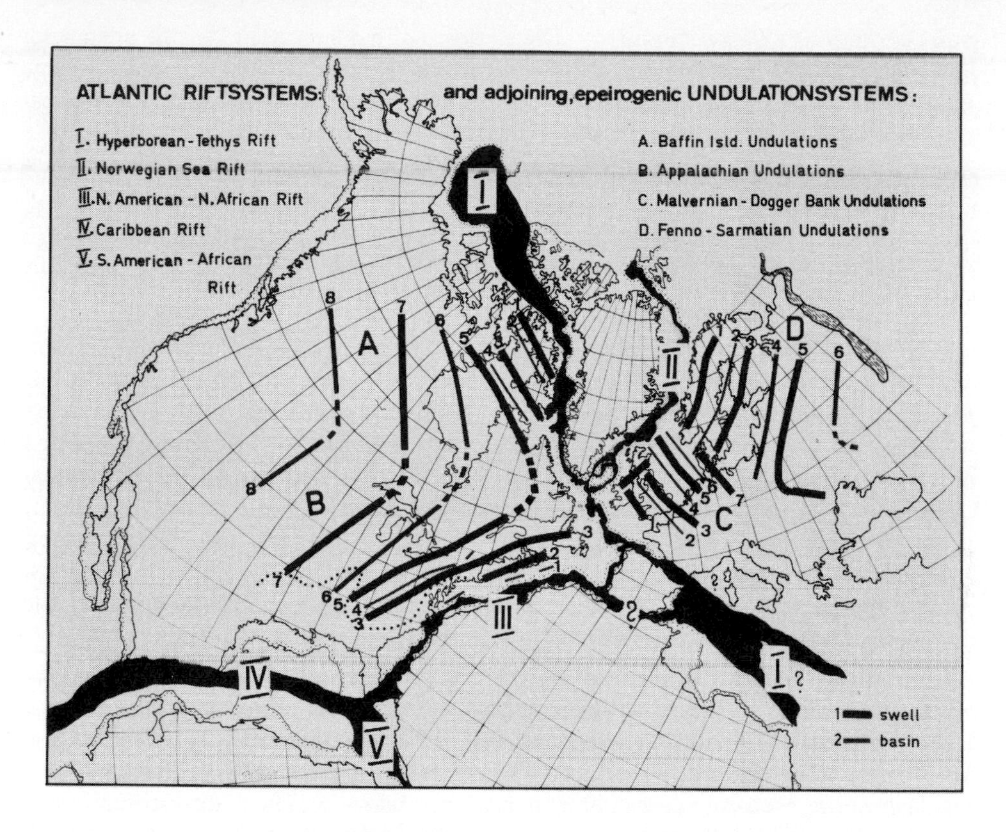

FIGURE 2 Atlantic rift systems and adjoining epeirogenic undulation systems.

imply that rift-parallel undulations and alpinotype orogenic structures can be located in other parts of the world too, for example around the Atlantic.

Figure 2 shows the computer fit of the continents around the Atlantic, slightly modified from Bullard *et al.* (1965). Three changes have been made: The Queen Elizabeth and Ellesmere Islands are readjusted on to Greenland along the Wegener fault (Dawes, Vol. II, p. 925). Further, an opening between Spain and Europe in the Pyrenean area is introduced. These two changes permit a connection of the ancient oceanic Arctic basin or Hyperborean Sea (Ostenso, Vol. I, p. 165), through Lancaster Sound (Burek, 1970; Henderson, pers. communication), Baffin Bay, Labrador Sea, Biscay and thus via the Pyrenees to the ancient Tethys oceanic basin. A possible connection between the Atlantic and the Pacific rift systems (Vogt *et al.*, 1971) within the Caribbean Sea and Gulf of Mexico is indicated. This modified continental fit indicates at least five possible rift structures in the Atlantic: I. Labrador–Biscay (Hyperborean–Tethys) Rift, II. Norwegian Sea Rift, III. North African–North American Rift, IV. Caribbean–Gulf of Mexico Rift, V. African–South American Rift. At the moment, the dominant structure is the Mid-Atlantic Ridge. This, however, does not exclude that the rift systems above mentioned were independently active at different times. In this case one should expect several undulation systems parallel to the rift troughs, similar to the features observed in the Arabian craton. Rift-parallel undulations can indeed be recognized in the following areas (Burek, 1969b): A. the Canadian Shield, B. the North American Craton. C. in Northwest Europe and on D. the Scandinavian Shield (Fig. 2).

A. The undulation system on the Canadian Shield parallels the Labrador–Biscay (Hyperborean Sea–Tethys) lineament or rift system. The different swells and basins are: 1. The Baffin Island; 2. The Foxe Basin; 3. The Melville–Meta Incognita Peninsulas separated by 4. The Foxe Channel depression from 5. The Southampton Islands–Ungavia Mts.–Labrador Uplift; 6. The Hudson Bay; 7. The Slave–Superior Province Swell and 8. The Alberta–Williston Depression (Fig. 2).

B. In the United States part of North America the epeirogenic undulations are parallel to the Mid-Atlantic Rift between Northwest Africa and North America: 1. There are two off-shore basins, with continental Newark Series type bottom sediments separated by 2. The Atlantic Coastal Platform from 3. The post-Triassic uplifted Piedmont–Appalachian mountain belt, 4. The Allegheny Basin, 5. The Nashville–Cincinatti Arch, 6. The Mississippi Embayment–Illinois Basin–Michigan Basin, 7. The Midcontinent Gravity High, in the vicinity of the Missouri–Wisconsin and Nemaha-Sioux uplifts and 8. The Denver–Williston Basin (Fig. 2). These undulations, parallel to the Mid-Atlantic Ridge, can be connected with the undulations of the Canadian Shield, which strike parallel to the Baffin–Labrador trough.

C. In Northwestern Europe, covering mainly the area of the British Isles and the North Sea, another set of undulations approximately parallel to the oceanic troughs of Labrador–Biscay (Hyperborean Sea-Tethys) can be recognized; its strike is predominantly NNW to SSE. These features are: 1. The Irish Uplift, 2. The Irish Sea, with a possible Mesozoic connection to the Paris Basin, 3. The Late Carboniferous Pennines–Brabant Swell, 4. The West–North Sea Depression separated by 5. The Mid–Northsea Gravity High/Dogger Bank area from 6. The East–North Sea Depression, 7. The Jütland Arch and 8. The Kattegatt Basin (Fig. 2). It should be noted that the Uralides follow the same trend.

D. On the Scandinavian Shield the undulations trend parallel to the continental slope of the Norwegian Sea, i.e. parallel to the initial rift trough (Rockall trough, East Norwegian Sea). They are: 1. The Norwegian Highlands, possibly extending to North-Scotland and North-Ireland, 2. The Gulf of Bothnia–Central Swedish Driftland–Skagerrak Depression, 3. The Finnish Granite Upland–Swedish Driftland–Fin Swell, 4. The Baltic Syncline, South of the present Baltic Gulf and Gulf of Finland, 5. The Scythian Wall, turning towards the Podolian Mass in the Brest area (Bubnoff, 1935b). 6. The Moscow Basin, possibly turning toward the Don Basin (Bubnoff, 1935b). All these undulations can be recognized on geological or physiographic maps. Their general trend is also reflected in 1° by 1° Bouguer anomaly maps (Woollard, 1972).

The undulations listed for the continents around the Atlantic have several features in common. They are fractured and often injected by dykes close to the continental slope. Warping is pronounced and has a relatively short wavelength near the rift trough. The wavelength increases with distance from its corresponding rift system. It is assumed, that these undulations are gentle reactions of the shield to directed stresses associated with initial and subsequent rifting. It is interesting to note that the northernmost undulations were warped downwards during Quaternary times and are presently uplifted. Besides these vertical movements, caused by vertical stresses (ice-loading and unloading), the undulations also seem to reflect a horizontal stress component, caused by large scale tectonic forces (sea-floor spreading), i.e. they represent a minor shortening of consolidated areas. The first appearence in geologic history of these new, rift-parallel, palaeogeographic trends should indicate the age of the initial stages of rift formation. These first stages of 'crustal spreading' (e.g. in the East African Rift System —or the Red Sea—and the Gulf of Aden-stage) are often too inconspicuous to be detected by palaeomagnetic or oceanographic methods, but they are strongly reflected by their associated structures, as can be seen in the example of Arabia.

The Palaeographic pattern of Europe during Late Palaeozoic times and its significance for initial rifting

During the Early Palaeozoic an extensive geosyncline or oceanic basin existed between Greenland and Fennoscandia, allowing an exchange of Arctic and European faunas. This Caledonian geosyncline was folded during the Late Ordovician and Silurian and closed completely at the end of Silurian (Brinkmann, 1960). Later, faunal invasions into the Devonian Central European geosynclines and Carboniferous Variscan basins came from the Tethys or through the Uralian geosyncline. There was definitely no subaqueous Norwegian Sea-Rift during the Devonian and most of the Carboniferous (Brinkmann, 1960). This may reflect the Early Palaeozoic closing of a Proto-Atlantic (Wilson, 1966) and a subsequent period of minimal activity.

The Variscan fold belts of Central Europe and Spain formed during the uppermost Devonian and the Carboniferous do not show any parallel trends to the Atlantic rift system. This suggests that their cause is associated with plate-movements in a different direction, which preceeds and is independant of the post-Carboniferous large scale stress configurations. After the last orogenic phase of the Variscan (Westphalian), which still affected the Southwest of the British Isles, completely new structures and trends appeared. They run almost North–South or NNW–SSE, and diagonally cut the older Caledonian and Variscian folds and thrusts.

(a) The 'Malvernian–Dogger Bank' Undulation System and the Formation of the Uralides

During post-Westphalian and Pre-Permian times, the Malvernian or Dogger Bank undulation system began to develop in Great Britain. Its trend is parallel to the Labrador–Biscay (Hyperborean–Tethys Sea) Rift. In Great Britain these structures are the Malvern- and Abberly Hills, the Pennines, E–W compressed structures in Northumbria and further North. Associated with these movements is the intrusion of the Great Whin Sill (280 m.y.). The Scottish Midland Valley Sill is also very likely to be a result of Malvernian activities (Bennison, 1969). In Ireland and Scotland, regional uplift, tholeiite intrusions and geofractures trending North–South occur at the boundary Westphalian–Stephanian (Russell, Vol. I, p. 577). The Pennines, forming the backbone of Great Britain, can be extended to the South and connected with the London–Brabant Massive (Schönenberg, 1971). Further to the East, in the North Sea, the Mid–Northsea Gravity High in the Dogger Bank area was also formed during Late Carboniferous times (Bartenstein, 1968). Again the undulation within the North Sea trend NNW–SSE, are diagonally superimposed on older Carboniferous swells and separate the North Sea into a western and eastern basin. The Dogger Bank

Swell is marked by a pronounced free air anomaly high (Hinz, 1968). In Denmark the Fünn Swell still trends NW–SE (Schönenberg, 1971; Bartenstein, 1968) and, at least during the Lower Jurassic, the present Jütland Arch began to develop (Hinz, 1968).

Thus at the end of the Carboniferous (Stephanian) the Western part of the Malvernian Dogger Bank undulation system had begun to form. It developed further, especially during Lower Permian times (Bartenstein, 1968) and dominated the palaeogeography of the British Isles and the North Sea throughout the Permian (Fig. 5) and most of the Mesozoic (Hinz, 1968).

In Central Europe the Permian (Rotliegend) basins were still predominantly influenced by Variscan trends. However, during the Saalian orogenic phase (Lower/Upper Rotiegendes), a large part of Europe as characterized by volcanic activity (tholeiites, rhyolites, melaphyrs, quarz-porphyries and spilites). During this phase the Uralian Geosyncline was almost closed (Figs. 3 and 5) and the essential structures of the Uralides were formed (Brinkmann, 1960; Hamilton, 1970).

It is important to note two points: 1. The Ural-geosyncline, or oceanic basin, separates the Meso–European plate and the Siberean plate (Fig. 4); 2. The Uralides parallel the Hyperborean–Tethys (Labrador–Biscay) rift (Fig. 2). The Late Carboniferous and Early Permian development of the Pennines–Dogger Bank undulation system in NW-Europe, the Lower Permian volcanism over wide parts of Central Europe (Saalian magmatic phase) and the closing of the Ural trough along with the formation of the Uralides (Saalian orogenic phase, Fig. 3, 5) are indicative of two

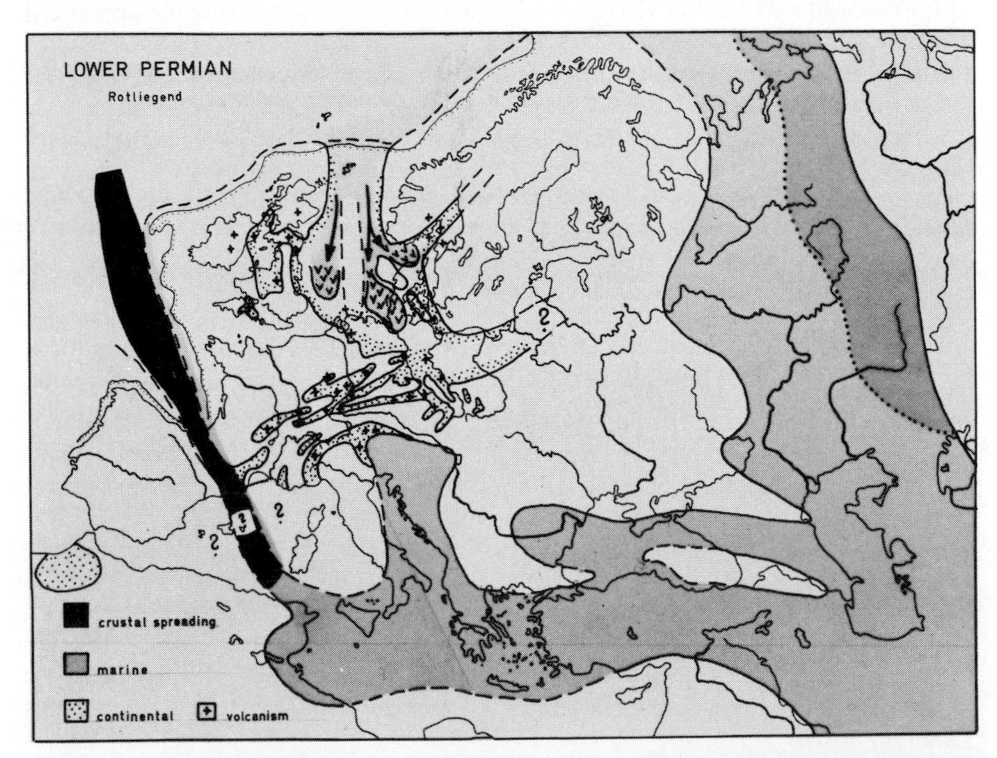

FIGURE 3 Stephanian and Lower Permian (Rotliegend) palaeogeography of Europe compiled after Bartenstein, 1968, Brinkmann, 1960, Schoenenberg, 1971, and ETH-demonstration maps.

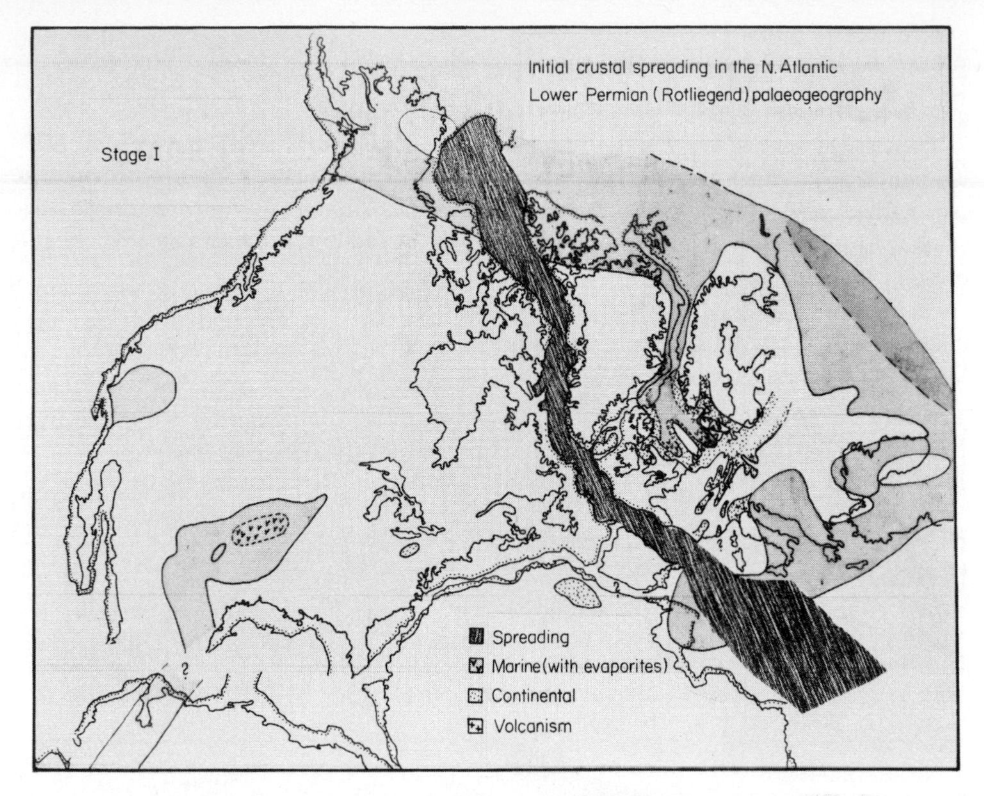

FIGURE 4 Initial crustal spreading (I. stage) in the N-Atlantic (Hyperborean Sea–Tethys Rift) and Lower Permian palaeogeography of the adjoining continental plates.

events: (i) the Hyperborean–Tethys rift was actively spreading during latest Carboniferous and Lower Permian time, causing epirogenic warping, magmatism in Europe and folding of the Uralides; (ii) The Meso–European plate advanced towards, and finally collided with, the Siberean plate.

Palaeomagnetic evidence supports the collision of the European and Siberean plate (Khramov & Sholpo, 1967; Hamilton, 1970). McElhinny (Vol. I, p. 77) has summarized the Russian data from the Siberean plate and compared them with palaeomagnetic data from Europe and concludes that the two apparent polar wandering paths of Europe and Siberia are far apart during the Lower Palaeozoic, approach one another during Silurian, but the Silurian to Permian path for Europe remains consistently eastwards of its Siberean counterpart (McElhinny, Vol. I, p. 77, Fig. 1 and Table I and II) until they ultimately converge in the Triassic. The observations are in accord with the geology of the Uralides (Hamilton, 1970) and suggest that Europe and the Siberean platform were far apart during Early Palaeozoic, approached one another during the Middle Palaeozoic, and collided near the end of the Upper Palaeozoic such that they were welded together by Triassic times. Henderson (Vol. I, p. 599, and pers. communication) assumes spreading activity in the Baffin Bay–Labrador Biscay rift before the Late Carboniferous. North–South trending depressions and rises are superimposed perpendicularly on Variscan fold structures (Eifel

North South Zone, Hessian Depression), as reported by Cloos (1948). These may reflect the interference of two spreading systems active at the same time. However, the geological evidence clearly indicates pronounced spreading along the Hyperborean–Tethys Rift during the Stephanian and Lower Permian, although this does not exclude earlier activity (in the Devonian or earlier).

(*b*) *The 'Fennosarmatian' Undulation System*

In the Upper Carboniferous, the Arctic Sea intruded the North European plate and divided the Old Red continent into a Greenlandian and a Fennoscandian block (Brinkmann, 1960). However, this transgression was not very extensive and only indicates minor structural activities (Graben stage). The Stephanian transgression may also be caused by a NNW-trending depression in the wider area of the present Nansen Fracture Zone-transform faults which, according to the computer fit, corresponds to an initial rift (Fig. 2). This Upper Carboniferous transgression was followed by a regression and subsequent Lower Permian transgression (Brinkmann, 1960).

More obvious indications for Lower Permian initial rifting in the Norwegian Sea are two palaeogeographic events in NW-Europe: 1. Lower Permian marine evaporites are found in the Western and Eastern part of the North Sea Basin (Figs. 3, 4). They indicate a first connection between the Arctic Ocean and the Norwegian Sea, with a SSE-trending side transgression into both North Sea Basins (part of the Dogger Bank

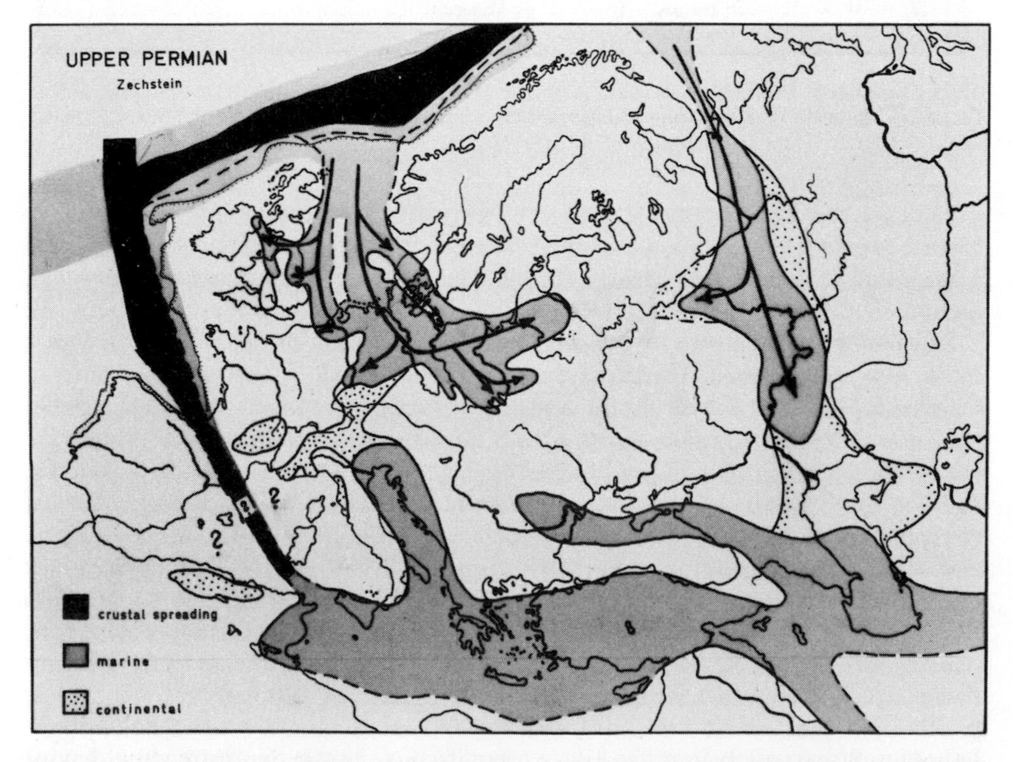

FIGURE 5 Upper Permian (Zechstein) palaeogeography of Europe compiled after Bartenstein, 1968, Brinkmann, 1960, Schoenenberg, 1971, and ETH-demonstration maps.

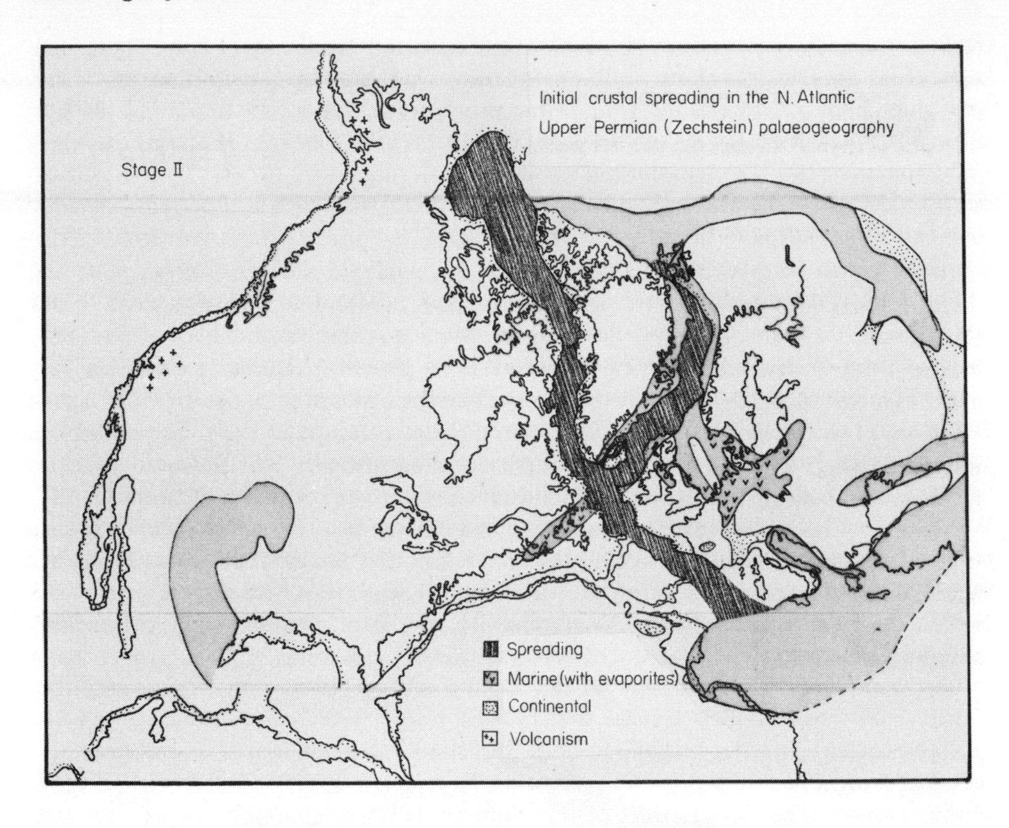

FIGURE 6 Initial crustal spreading (II. stage) in the N-Atlantic (Norwegian Sea) and Upper Permian palaeogeography of the adjoining continental plates.

Undulation System, Figs. 2, 3 and 4). 2. The Oslo Graben, with its extensive rhomb porphyries, alkalisyenitic intrusion and ringdykes, was formed during Lower Permian times. The Oslo Graben roughly parallels the Norwegian continental slope (Fig. 3).

During Upper Permian times the evidence for spreading in the Norwegian Sea is much more pronounced. During this time a major transgression extended outward from the Arctic Basin. The marine invasion extended into Russia as the 'Sea of the Kazanian Stage' (foredeep-basin West of the Uralides) and into Germany as the 'Zechstein Sea' (Fig. 5) and even reached Nova Scotia as part of the Norwegian Sea Basin (Fig. 6) (Brinkmann, 1960); Zechstein evaporites are also reported from East Greenland (Perch-Nielson, pers. communication). The extension into Nova Scotia is important, since it indicates that the entire length of the Norwegian Sea Rift was affected by the marine transgression proceeding from the Arctic Sea. The connection through both North Sea basins (Dogger Bank trend) into Central Europe is even more significant since, in Central Europe, the Zechstein Basin reflects a new trend superimposed on the Variscan and Lower Permian features. In Germany the North Sea Basin curves from a SSE direction into the SW-trending 'Zechstein Basin', which extends into the area around Heidelberg (pre-Rhenian trend). North of the Island of Rügen it curves into the NE-trending 'Baltic Syncline' (Fig. 5), extending almost into the Riga area. The connections of these two embayments parallel the 'Fennosarmatian' trend (Fig. 5). The Nordhorn Embayment (near the German–Dutch boundary), the Yorkshire Embayment and the Tyne-connection towards the NNW-trending Irish Sea trough also reflect the same directions. By Upper Permian times the 'Fennosarmatian' undulation system was completely developed as indicated by the coastal facies distribution of the Zechstein sediments. The Scythian wall does not permit a South-eastern connection with the Moscow Basin (Fig. 2), which therefore was invaded via the Uralian foredeep basin (Fig. 5).

The Zechstein evaporites were deposited in four cycles, each corresponding to an opening and closing of the sea-connection via the North Sea (Brinkmann, 1960). This cyclic deposition may perhaps be associated with phases of rifting: Closing of the North Sea connection would correspond to a crustal spreading phase in the Norwegian Sea, which for isostatic reasons caused a rise of the continental shelf. Similarly, the opening of the Northsea connection may perhaps be associated with spreading phases in the Biscay–Labrador Sea, which would reactivate and downwarp the Eastern and Western North Sea basins, thus allowing sea invasions. The subsequent Triassic palaeogeography in Central Europe, the North Sea and the British Isles reflects the inherited Permian epeirogenic swells and troughs. The palaeogeographic pattern of NW-Europe is dominated by the 'Malvernian–Dogger Bank' and the 'Fennosarmatian' undulation systems.

To summarize: The evaporites in the vicinity of the Norwegian Sea trough, the tectonic activity in the Oslo Graben area (Vokes, Vol. I, p. 573) and the close relations of the Zechstein palaeogeography to both European sets of structural undulations are taken as evidence for initial movements of the Norwegian Sea (Rockall trough) during Permian times. The Eastern part of the Norwegian Sea is characterized by few and low-amplitude magnetic anomalies (Avery *et al.*, 1968); in fact it seems to represent a magnetically smooth zone. The Permian is almost homogeneously reversely magnetized, corresponding to Irving's 'Kiaman Interval'. The Norwegian Sea 'magnetic quiet zone' partly may reflect a period of crustal spreading during Permian and thus would correlate with the Permian Interval of predominantly reversed magnetization.

(*c*) *Plate movements and initial spreading in the North Atlantic rift troughs*

An opening of the Hyperborean-Tethys lineament would result in an Eastward movement of Meso–Europe towards the Siberean plate. Meso–Europe and the Siberean plate would eventually collide (Saalian orogenic phase), forming the Uralides. The existence of this movement is supported by palaeomagnetic evidence (Hamilton, 1970; McElhinny, Vol. I, p. 77). An opening of the Norwegian Sea would move Greenland Northward with respect to Canada and Europe (Fig. 6); Deutsch & Murthy (Vol. II, p. 987) support a Northward movement and initial separation of North America relative to Europe at least by Triassic times.

Palaeogeographic pattern of North America during Early Mesozoic times and its significance for initial rifting

The 'Appalachian' undulations

The Appalachians were folded and metamorphozed (Piedmont) during Palaeozoic times. The last minor movements occurred during the lowermost Permian. Subsequently there was a long period of tectonic inactivity and the formation of a peneplain during the Permo–Triassic. The present morphological pattern (the undulation system shown in Fig. 2) began to develop during the Late Triassic (Rhaetic). At this time the Newark Graben Belt was formed and filled with clastic, continental redbeds (Fig. 7). These events were associated with extensive volcanism (tholeiites) of Rhaetic to Lower Jurassic age (de Boer, 1967). The basal sediments found by drilling in one of the shelf troughs are redbeds and probably correlate with the Newark Series. The Mississippi Embayment was also formed during lowermost Jurassic or Late Triassic times. The initiation or rejuvenation of the North American undulation system seems to be indicative for rifting in the North African–North American rift system (Vine, pers. communication). Both sides of the Atlantic are characterized by extensive magnetically smooth zones. The Newark Series are normally magnetized. This led the writer to assume that not only the Late Triassic, but also parts of the Lower Jurassic are predominantly normally magnetized (Burek, 1970b) (this does not exclude several short lived reversals). A summary of worldwide palaeomagnetic data (McElhinny & Burek, 1971) showed that this indeed is the case. Therefore the predominantly normal 'Graham Interval' was introduced within the Mesozoic reversal time scale. This Late Triassic to Lower Jurassic 'Graham Interval' would correlate with the magnetic quiet zones of the Middle Atlantic (Pitman & Talwani, 1972; Vogt, pers. communication). In three coastal basins of Northwest Africa Rhaetic evaporites (Fig. 7) have been found (Reyre, 1964). These appear to correlate

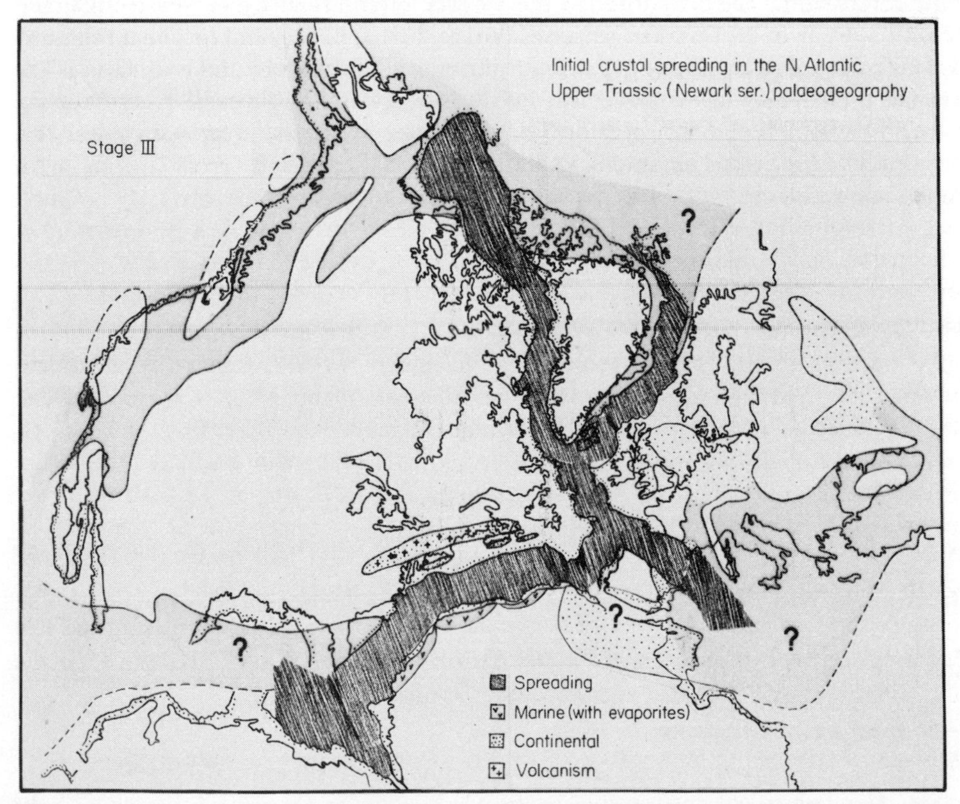

FIGURE 7 Crustal spreading (III. stage) in the North- and Middle Atlantic (North American–N-African Rift) and Late Triassic palaeogeography of the adjoining continental plates.

with diapiric structures rising from the bottom sediments of the abyssal plains in the Canary Island area (Rona, 1969). This may be an indication for Late Triassic crustal spreading from the Northwest African plate.

To summarize: The origin of the North American undulation system (associated with tholeiitic volcanism in the Newark Graben Belt) in the Late Triassic, the correlation of the predominantly normal 'Graham Interval' with the magnetic quiet zones on both sides of the Atlantic and the occurrence of Rhaetic evaporites in the African shelf sediments (probably having the same age as the abyssal evaporites), are evidence for Late Triassic initial spreading in the North African–North American rift system. The 'Keathley Sequence' of magnetic oceanic anomalies (Vogt *et al.*, 1971), correlates with the Upper Jurassic to Lower Cretaceous 'Mixed Polarity Interval' (McElhinny & Burek, 1971). Vogt suggests it is connected through the Caribbean–Gulf of Mexico with the Pacific (Vogt *et al.*, 1971), while Reyment (Vol. II, p. 799) suggests that the Afro–South American rift system became active still later, during Cretaceous times.

Summary

A combination of palaeogeographic, structural and palaeomagnetic data and evidence from oceanic magnetic anomalies suggests that the North Atlantic rift systems began to open from the North to South. The Hyperborean–Tethys lineament appears to indicate spreading activities during Latest Carboniferous and Lower Permian times. The Norwegian Sea rift probably opened during the Late Permian; the North African–North American rift is still younger and opened during the Late Triassic. The Caribbean rift may have joined the Pacific rift systems during the Late Jurassic. Finally, the Afro–South American rift definitely was opened at Turonian times. It is emphasized that palaeogeographical and other structural evidence are indicative of large scale stress configurations caused by sea-floor spreading.

Acknowledgements

This work was supported by grants of Deutsche Forschungsgemeinschaft (Bu-183, 1, 2, 5, 9a,b) and the National Science Foundation/Washington (GA-1311 and GA-1601). I would like to thank Dr H. U. Nissen, Mr K. Kelts and Dr D. H. Tarling for improving the English in the manuscript and for valuable discussions.

References

Avery, O. E., Burton, G. D. and Heirtzler, T. R., 1968. An Aeromagnetic Survey of the Norwegian Sea, *J. Geophys. Res.*, **74**, 4583–4600.

Burek, P. J., 1969a. Structural effects of sea-floor spreading in the Gulf of Aden and the Red Sea on the Arabian Shield. *In:* Hot Brines and Recent Heavy Metal Deposits in the Red Sea. Edited by E. T. Degens and D. A. Ross, pp. 59–70. Springer-Verlag, New York.

Burek, P. J., 1969b. The palaeogeographic pattern of Europe and North America around the Palaeo-Mesozoic boundary and its significance for continental drift (abs.), *Geol. Soc. America Ann. Meeting Prog.*, **7**, 24–25.

Burek, P. J., 1970a. Tectonic effects of sea-floor spreading on the Arabian Shield. *Geol. Rdsch.*, **59**, (2), 382–90.

Burek, P. J., 1970b. Magnetic Reversals: Their Application to Stratigraphic Problems. *The Am. Assoc. of Petrol. Geol. Bull.*, **54**, (7), 1120–39.

Burek, P. J., 1973 Plattentektonische Problem im Bereich Arabiens und der Afar-Senke. Grotekt, Forsch. (in press).

Beydoun, Z. R., 1966. Geology of the Arabian Peninsula, Eastern Aden Protectorate and Part of Dhufar. U.S. Geol. Survey, Prof. Paper 560-H.

v. Bubnoff, S., 1936. Geologie von Europa. Gebr. Borntraeger Verl., Berlin.

v. Bubnoff, S., 1935. Neue Angaben über den scythischen (Polessje) Wall. Geol. Rdsch., **26**, 258–66.

Brinkmann, R., 1960. Geologic evolution of Europe. Enke Verl. Stuttgart, 1–158.

Brinkmann, R., 1964. Abriss der Geologie. Enke Verl. Stuttgart, 1–360.

de Boer, J., 1967. Palaeomagnetic-tectonic study of Mesozoic dike swarms in the Appalachians. *J. Geophys. Res.*, **72**, 2237–80.

Bennison, G. M. and Wright, A. E., 1969. The Geological History of the British Isles. Edward Arnold (Publishers) Ltd., London, p. 385.

Bartenstein, H., 1968. Present status of the Palaeozoic palaeogeography of northern Germany and adjacent parts of north-west Europe, in Donovan, D. T., ed., Geology of shelf seas. Oliver and Boyd, Edinburgh–London, 31–54.

Bullard, E. C., Everett, J. E. and Smith, A. G., 1965. The fit of the continents around the Atlantic. *In:* Symposium on Continental Drift: *Royal Soc. London Philos. Trans.*, ser. A,**258**, (1088), 41–51.

Cloos, H., 1948. Grundschollen und Erdnähte, Entwurf eines konservativen Erdbildes, *Geol. Rdsch.*, **35**, 133–54.

Elsasser, W. M., 1971. Two-Layer Model of Upper-Mantle Circulation, *J. Geophys. Res.*, **76**, (20), 4744–53.

Fyfe, F., Turner, F. J. & Verhoogen, J., 1958. Metamorphic reactions and metamorphic facies, *Geol. Soc. Am.*, Memoir 73.

Griggs, D. T. and Blacic, J. D., 1965. Quartz: Anomalous Weakness of Synthetic Crystals, *Science*, **147** (3655), 292–95.

Hales, A. L., 1969. Gravitational Sliding and Continental Drift, *Earth Planet. Sci. Lett.*, **6**, (1) 31–34.

Hamilton, W., 1970. The Uralides and the Motion of the Russian and Siberian Platforms. *Geol. Soc. Am.*, **81**, 2553–76.

Hinz, K., 1968. A contribution to the geology of the Palaeozoic of the North Sea according to geophysical and adjacent investigations by the Geol. Surv. of Germany. *In:* Geology of Shelf Seas, edited by D. T. Donovan, pp. 55–68. Oliver and Boyd, Edinburgh/London.

Khramov, A. N. and Sholpo, L. Ye., 1967. Palaeomagnetism, *Leningrad Izdat. Nedra*, **251**

Landisman, M., Mueller, S. and Mitchell, B. J., 1971. Review of evidence for Velocity Inversions in the Continental Crust. *In:* The Structure and Physical Properties of the Earth's Crust, edited by J. G. Heacock, pp. 11–35, *Am. Geophys. Un., Washington, Geophys. Mon.*, **14**.

Lebedev, E. B. and Khitarov, N. I., 1964. Dependence of the beginning of melting of granite and the electrical conductivity of its melt on high water vapour pressure, *Geo-chem. Int.*, **1**, 193–97.

Lliboutry, L., 1969. Sea-floor spreading, continental drift and lithosphere sinking with an asthenosphere at melting point, *J. Geophys. Res.*, **74**, 6525–40.

Magnitsky, V. A. and Zharkov, V. N., 1969. Low-Velocity Layers in the Upper Mantle. *In:* The Earth's Crust and Upper Mantle, edited by P. J. Hart, pp. 664–75, *Am. Geophys. Un., Geophys. Mon.*, **13**.

McKenzie, D. P., 1969. The Mohorovicic Discontinuity. *In:* The Earth's Crust and Upper Mantle, edited by P. J. Hart, pp. 660–64, *Am. Geophys. Un., Geophys. Monogr.*, **13**.

McElhinny, M. W. and Burek, P. J., 1971. Mesozoic Palaeomagnetic Stratigraphy. *Nature*, **232**, 98.

Nur, A. and Simmons, G., 1969. The effect of saturation on velocity in low porosity rocks. *Earth Planet. Sci. Lett.*, **7**, 183–93.

Press, F., 1970. The earth's interior as inferred from a family of models. *In:* The Nature of the Solid Earth, edited by E. C. Robertson, pp. 147–71. McGraw-Hill, New York,

Pitman, W. C. and Talwani, M., 1972. Sea Floor Spreading in the North Atlantic. *Bull. Geol. Soc. Am.*, **83,** 619–46.

Rona, P. A., 1969. Possible Salt domes in the deep Atlantic of North-West Africa. *Nature*, **224** (5215), 141–43.

Reyre, D., 1964. Sedimentary basins of the African coast, pt. 1: Atlantic coast. *Assoc. African Geol. Surveys, Paris*, **304**.

Roy, J. L., 1972. A pattern of rupture of the Eastern North American–Western European Palaeoblock. *Earth Planet. Sci. Lett.*, **14**, 103–14.

Schoenenberg, R., 1971. Einführung in die Geologie Europas. *Verl. Rombach, Freiburg*, 1–300.

Schuchert, C., 1955. Atlas of Palaeogeographic Maps of North America. John Wiley & Sons, Inc., New York.

Vogt, P. R., Anderson, C. N. and Bracey, D. R., 1971. Mesozoic Magnetic Anomalies, Sea-Floor Spreading and Geomagnetic Reversals in the Southwestern North Atlantic. *J. Geophys. Res.*, **76** (20), 4796–4823.

Woollard, G. P., 1972. Regional variations in gravity. *In:* The Nature of the Solid Earth, edited by E. C. Robertson, pp. 463–505. McGraw-Hill, New York.

Wilson, H. H., 1969. Late Cretaceous Eugeosynclinal Sedimentation, Gravity Tectonics, and Ophiolite Emplacement in Oman Mountains, Southeast Arabia. *The Am. Assoc. of Petrol. Geol. Bull.*, **53** (3), 626–71.

Wilson, J. T., 1966. Did the Atlantic close and then reopen? *Nature*, **211**, 676–81.

Van Zijl, J. S. V., Hugo, P. L. V. and de Belloco, J. H., 1970. Ultra Deep Schlumberger Sounding and Crustal Conductivity Structure in South America. *Geophys. Prospecting*. **18** (4), 615–34.

7.11

F. J. VINE
School of Environmental Sciences
University of East Anglia
Norwich, NOR 88C, England

Continental Fragmentation and Ocean Floor Evolution during the past 200 m.y.

It is now commonly believed that the deep ocean basins of the present day were formed by a process of sea-floor spreading about mid-ocean ridge crests during the past 200 m.y. (Vine & Hess, 1971). Thus, beneath the sedimentary cover, the 'stratigraphy' of the ocean floor is laid down horizontally and is revealed by the sequence of linear oceanic magnetic anomalies which result from a combination of lateral accretion and reversals of the Earth's magnetic field. Relative ages and rates of spreading may be assigned to all oceanic areas in which the anomaly sequence has been recognized; furthermore if a rate (or rates) of spreading as a function of time is assumed for any one area, the 'absolute' age of the oceanic crust in all areas may be predicted. Heirtzler *et al.* (1968) made such predictions by assuming a constant rate of spreading in the South Atlantic for the past 75 m.y. Subsequently the results of the JOIDES Deep Sea Drilling Project have essentially confirmed these predictions (e.g. Maxwell *et al.*, 1970; JOIDES, 1969–1972); more than 200 sites have now been drilled, several specifically chosen in order to sample some of the oldest oceanic crust. The age of the oldest sediment recovered to date, at a truly oceanic site, is estimated to be 162 m.y., and in no case does the age seriously conflict with predictions made on the basis of the sea-floor spreading hypothesis.

Figure 1 summarizes the results obtained by JOIDES drilling at sites coincident with previously recognized and dated oceanic magnetic anomalies. It will be seen that relatively few sites to date satisfy this criterion, but that the oldest material recovered at these sites is either younger or slightly older than the predicted age. The age assigned to the basement on the basis of drilling is the palaeontological age of the sediment immediately overlying the basement; this is because of the difficulty of obtaining a reliable radiometric age from deep-sea extrusives (e.g. Dymond, 1970). Clearly it is not possible to assess the validity of the magnetic anomaly time-scale in detail because of the difficulty of correlating faunal zones with an 'absolute' time-scale (e.g. Berggren, 1969). Thus one is not justified, as yet, in suggesting changes to the magnetic anomaly time-scale of Heirtzler *et al.* (1968), but in view of the importance attached to this in terms of dating the ocean floor, deducing spreading rates, and correlating variations and discontinuities in these with various geological phenomena, it is regrettable that so few JOIDES sites, either drilled or planned, contribute to the calibration of this time-scale.

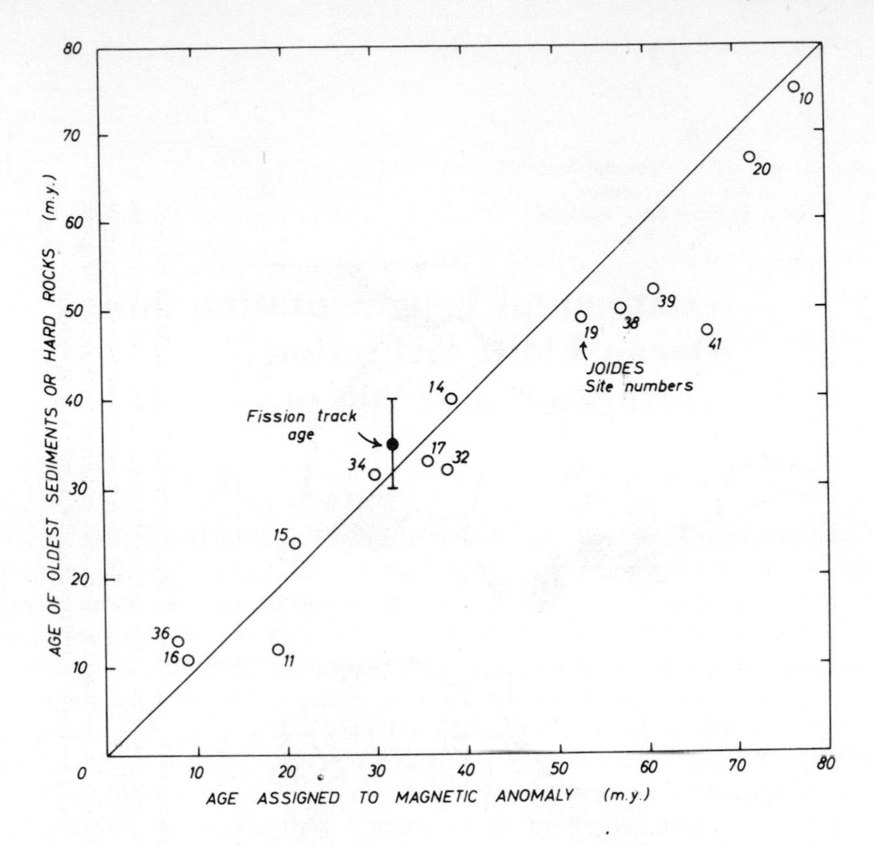

FIGURE 1 Graph showing the extent to which the magnetic anomaly time-scale of Heirtzler *et al* (1968) has been calibrated by dating of the volcanic basement beneath dateable anomalies. Fission track age from Luyendyk & Fisher (1969), JOIDES results from the Initial Reports of Legs 2, 3 and 5 (1970).

Once accepted, a calibration of the oceanic magnetic lineations, such as that suggested by Heirtzler *et al.*, enables one to map 10 m.y. 'growth' lines about ridge crests in all areas where the anomaly sequence out to 75 m.y. ago has been recognized (see Fig. 2). By adding to this areas of presumed Cenozoic oceanic crust in the Equatorial Atlantic and Pacific, where the ridge crests run north–south and the magnetic anomalies are consequently poorly developed, and the areas of Tertiary 'back-arc' spreading, documented in the western Pacific by Karig (1971) and in the Scotia Sea by Barker (1970), it is possible to summarize the extent of the present deep-sea floor which has been created during Cenozoic time (see Fig. 2). It will be seen that rather more than 50% of the area of the present ocean basins or one third of the surface area of the Earth has been created during Cenozoic time, less than 1·5% of geological time. Unless one is prepared to accept large-scale expansion of the Earth the corollary of this is that a comparable area of oceanic crust has been destroyed beneath the active trench systems, island arcs and mountain ranges during this period, with all its implications for the igneous and tectonic activity associated with these subduction zones.

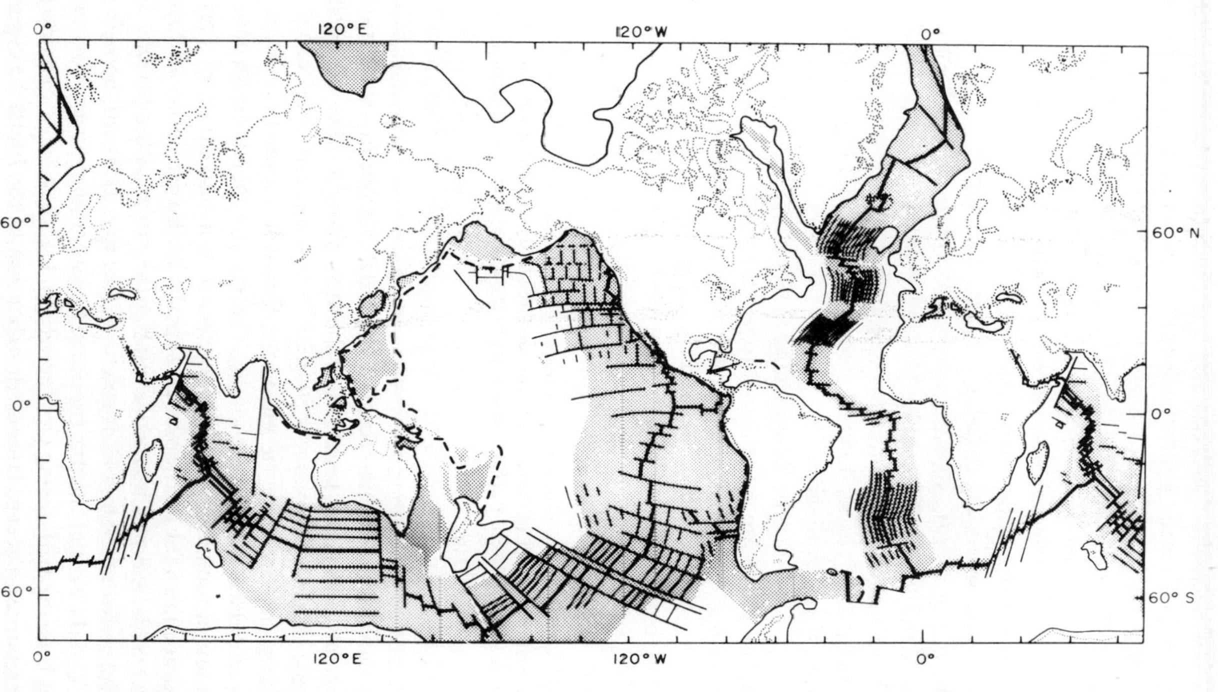

FIGURE 2 Provisional attempt to delineate areas of continental and oceanic crust. Within the ocean basins, trenches are indicated by thick dashed lines, ridge crests by thick solid lines, and fractures (transverse to the ridge crests) and correlateable linear magnetic anomalies (parallel to the ridge crests) by thin solid lines. In addition to the references cited in the text, papers by Atwater & Menard (1970) and Herron (1971) have been utilized in summarizing the areas in which correlateable magnetic anomalies have been recognized. Oceanic crust thought to have been formed within the past 65 m.y. is shaded.

The details regarding the age of the ocean floor revealed by the oceanic magnetic anomalies enable one to assign dates for the initiation of sea-floor spreading, following continental rifting, in various areas; notably the extreme North Atlantic, and south of Australia and New Zealand. However in many areas of the Atlantic and Indian Oceans, i.e. those in which spreading was initiated prior to 80 m.y. ago, it is clearly not possible to do this on the basis of magnetic lineations and one must resort to other criteria based on the geological record on the appropriate continental margins. For the three major areas involved, Vine & Hess (1971) suggested an Early Jurassic age for the initiation of spreading between Africa and North America, and Africa and Antarctica, and a Mid-Cretaceous age for that between Africa and South America. These dates were based on radiometric ages obtained on igneous extrusives and intrusives on the appropriate continental margins which were considered to be associated with the initial rifting and to be precursors of spreading proper. The suggested age for the initiation of spreading in the North Atlantic, south of the Azores, has been corroborated by the results of the JOIDES Deep Sea Drilling Project (JOIDES, 1970), and the age suggested for the South Atlantic agrees with other geological evidence from the continental margins (Allard & Hurst, 1969; Reyment, 1969); that suggested for the southwest Indian Ocean is still equivocal.

In Fig. 3 these dates for the initiation of spreading in the North and South Atlantic and the southwest Indian Ocean are summarized together with those documented by the dated anomaly sequence, i.e. south of Australia and New Zealand, and between Greenland and Europe. The age of 80 m.y. ago indicated for the initiation of spreading in the Labrador and Arabian Seas is uncertain since the magnetic lineations are not traceable up to the continental slopes. In the light of igneous activity on the adjacent continental margins it seems unlikely that spreading occurred any earlier in the Arabian Sea area, but this could be so in the Labrador Sea area (see Pitman & Talwani, 1972). The opening of the Gulf of Aden, Red Sea and Gulf of California in the Late Tertiary has been omitted from Fig. 3, although clearly these events are very instructive as regards the delay between the onset of igneous activity, rifting and evaporite deposition, and the initiation of spreading. It would appear that the duration of this delay is very variable and may be anything from a few million to a few tens of m.y. Moreover spreading may not be recorded initially by magnetic lineations because of metamorphic effects beneath thick sequences of evaporites or sediments developed on the first formed oceanic crust.

Figure 3 also attempts to summarize the spreading rates which have been operative during the evolution of the various oceanic areas. The rates for the past 80 m.y. are based on the recognition of dated magnetic anomalies, with the exception of that for the central Pacific; earlier rates for the north and south Atlantic are based on various other criteria as discussed below. Clearly the rates given only strictly apply to a particular latitude with respect to the pole of opening for a given spreading boundary between lithospheric plates (Morgan, 1968). Ideally perhaps one should specify angular or equatorial rates of opening with respect to the instantaneous poles of opening and calculate the area of new oceanic crust formed during a period for which the opening rate was essentially constant. However for the present purpose of defining world-wide discontinuities and systematic variations in spreading rate it seemed adequate to summarize the more familiar rates of spreading per ridge flank. This can be further justified by the fact that most areas of ocean-floor spreading appear to be approxi-

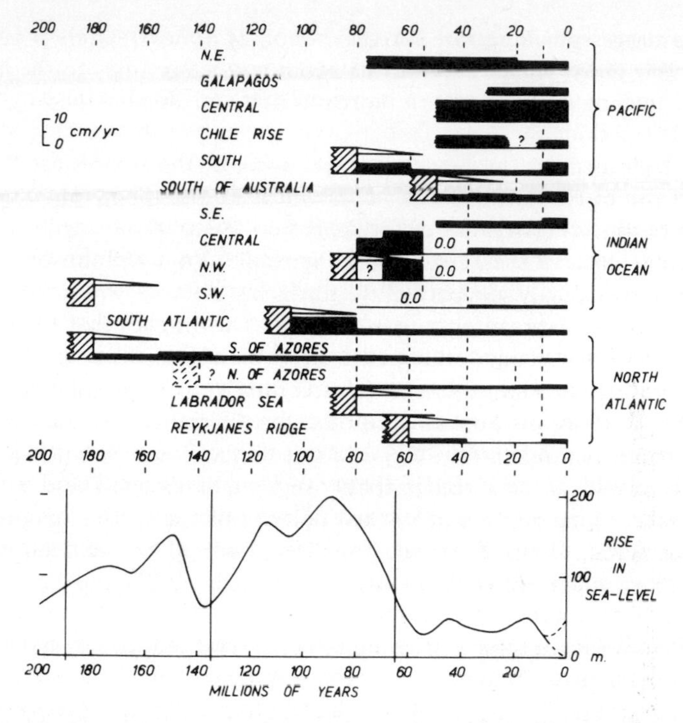

FIGURE 3 Summary of spreading rates and times of initiation of spreading in various oceanic areas together with possible world-wide variations in mean sea level during the past 200 m.y. The mean sea level curve is derived from Hallam (1963, 1969) and Grasty (1967). The effect on sea level of melting all the ice of the Late Tertiary and Quaternary glaciations is indicated schematically.

mately equatorial with respect to their pole of opening and there is therefore no great variation in spreading rate along their length; the main exceptions to this are noted below.

The rates of spreading indicated for the northeast and south Pacific are taken from Heirtzler *et al.* (1968) and apply to 41° N and 50° S respectively. The rate of spreading decreases markedly to the southwest along the Pacific–Antarctic ridge, towards the pole of opening, as noted by Morgan (1968). The rate for the Galapagos Rise is that deduced by Herron & Heirtzler (1967) for the area immediately north of the Galapagos Islands. Again there would appear to be considerable variation in the spreading rate along this ridge because of the proximity of the pole of opening to the west (Hey *et al.*, 1972). The rate shown for the central Pacific, west of the East Pacific Rise, is suggested on the basis of JOIDES drilling results in this area (JOIDES, 1971). The rates for the Chile Rise are taken from Morgan *et al.* (1969) and Herron & Hayes (1969).

The history of spreading south of Australia has been summarized by Weissel & Hayes (1971) and in the Indian Ocean by McKenzie & Sclater (1971). The recent spreading rate in the southwest Indian Ocean predicted by McKenzie & Sclater has been confirmed by the work of Bergh (1971). The rate given for the south Atlantic for the past 80 m.y. is that deduced by Dickson *et al.* (1968) for 30° S. If the Mid-Cretaceous, or specifically post-Albian, age for the initiation of the spreading in the South Atlantic is accepted, the implication is that approximately half of the present area of the South Atlantic basin was formed within the first 25 m.y. of its spreading history. This in turn implies an average spreading rate for this period of approximately 6 cm per year at 30 °S, with very much higher rates to the south and lower rates to the north because of a pole of opening in the southern North Atlantic, as documented by Le Pichon & Hayes (1971).

The rates indicated for the North Atlantic south of the Azores are those given by Pitman & Talwani (1972) for 35° N. These authors also point out that the implied average rate of spreading in this area for the period 180 to 80 m.y. ago is approximately 2 cm/yr. On the basis of the JOIDES drilling results Vogt & Johnson (1971) assume that the Keathley anomaly sequence of the northwest Atlantic was formed between 155 and 135 m.y. ago at a spreading rate of 3 cm/yr. This implies that the spreading rate between 180 and 155 m.y. ago was approximately 2 cm/yr and the rate between 135 and 80 m.y. ago rather less than 2 cm/yr. This comparatively low rate of spreading for Early and Mid-Cretaceous time may account for the rather poor development of the Cretaceous magnetic 'quiet zones', in contrast to the South Atlantic and the North Pacific where spreading rates were probably very much higher. These 'quiet zones' are thought to reflect long periods of normal polarity of the earth's magnetic field (e.g. McElhinny & Burek, 1971). The spreading rates indicated for the Atlantic north of the Azores are those suggested by Williams & McKenzie (1971) and Pitman & Talwani (1972).

From the above discussion and compilation several reasonably hard facts emerge with respect to the past 80 m.y.:

1. The time of the initiation of sea-floor spreading between continental areas which are being drifted apart,

2. The rate of spreading at a particular time in a given geographic area, and

3. Times at which there appear to have been world-wide changes in spreading rates and presumably plate geometry. Such events occur in many areas at 80, 60–55, 38 and 10 m.y. ago, and in certain areas at 70, 50 and 20 m.y. ago.

Much work to date has been concerned with the geometry of spreading and plate movements, and with continental reconstructions. It seems important however to try to ascertain the net rate at which sea-floor spreading has taken place in the past and to identify any systematic changes in spreading rates. For if the net rate of accretion of oceanic crust at ridge crests is equal to the rate of consumption in subduction zones, as is commonly assumed, then clearly any variation in rate, or complete stoppage of this process as suggested by Ewing & Ewing (1967) and Le Pichon (1968), will have profound effects in terms of tectonism and, in all probability, magma generation at destructive plate boundaries.

A further corollary of variations or stoppages in ridge activity is their effect on mean sea-level. The elevation of oceanic crust associated with a mid-ocean ridge, relative to the flanking deep ocean basins, is a function of its age (Sclater *et al.*, 1971), the younger its age the greater the elevation. The 'decay' of this topographic excess with time is quasi-exponential with a 'half-life' of approximately 28 m.y. Thus all ridges displace water over the continents and without the present day ridges, i.e. if there had been no ridge activity throughout Cenozoic time, the oceans would be 500 m shallower and, allowing for isostatic readjustment, sea level would stand 350 m lower. Similarly if present day rates of accretion at ridge crests had been maintained throughout the Tertiary and Quaternary, and none of the resulting new crust destroyed in trench systems, sea level would stand 200 m higher than it does today. In fact one could perhaps argue that the net accretion rate *was* the same throughout the past 65 m.y. but that much of the new crust formed during this time has been lost, in the marginal trench systems of the north and east Pacific and beneath the Indonesian Arc (JOIDES, 1972). However there is also an indication from Fig. 3 that spreading rates were systematically lower throughout much of the Tertiary. Clearly this line of reasoning is amenable, to some extent, to more quantitative analysis although there will always be some ambiguity as to the precise area and age of the oceanic crust which has been lost.

The suggestion that ridge activity might be an important control on sea level was perhaps first suggested by Hallam (1963) and has been revived more recently by Russell (1968), Armstrong (1969) and Valentine & Moores (1970). Basically it invokes changes in the net rate of accretion at ridge crests, transgressions following increases and regressions decreases in this rate, but the situation is complicated by the age and hence depth of the oceanic areas being resorbed in trench systems. For the process to be most effective old and hence deep oceanic crust must be resorbed at the expense of the new shallow ridge crests. To consider a further example, if the initial phase of spreading in the South Atlantic occurred at a rate of 6 cm/yr/ridge flank for 25 m.y. as suggested above, and this represented additional crustal accretion at the expense of deep-sea floor, then sea level would have risen by 85 m during this period. It is clearly tempting to suggest that the peak of the Cretaceous transgression was due to this event (see Fig. 3).

Of all the processes capable of producing significant eustatic changes in sea level in the Phanerozoic the elevation and subsidence of mid-ocean ridges would appear to be the most potent, as emphasized previously by Armstrong (1969). Clearly the amount of water locked up in ice caps and glaciers must be taken into consideration but this is not relevant to much of geologic time. Of the other mechanisms suggested to account for eustatic changes in sea-level, several only relate to the progressive regression of the epicontinental seas, i.e. lowering in sea-level, throughout the Phanerozoic. These include increasing the freeboard of the continents by progressive thickening or underplating (Hallam, 1971), increasing the area of the ocean basins by earth expansion (Egyed, 1956), and invoking a progressive decline in ridge activity due to decreased heat production within the Earth (Armstrong, 1969).

Fluctuations in sea level with time have been linked with basically two possible causes: orogenic cycles (Grasty, 1967) and epeirogenic movement of land and sea areas (Hallam, 1963). The former postulates that the crustal shortening associated with mountain building and the resulting increase in area of the ocean basins produces a world-wide lowering of sea level. Presumably this is followed by increased rates of erosion and oceanic sedimentation which result in a gradual transgression. It is estimated that both processes are only capable of producing changes in sea level of a few tens of metres at the most in contrast to the inferred range of a few hundred metres during Phanerozoic time. Epeirogenic movements in land areas will have an important effect on sea level locally but epeirogenic movements of the sea floor will produce world-wide changes in sea level.

It is argued here that the major and world-wide eustatic changes in sea level in the past are the result of the net elevation and subsidence of the deep-sea floor as a function of ridge activity. Moreover because of the inherent ambiguity in determining the detailed palaeobathymetry and hence volume of the ocean basins in the past it is suggested that a mean sea-level curve determined from the geological record of continental areas might be the best indicator of palaeobathymetry and hence ridge activity available to us. If this is true a mean sea level curve such as that shown in the lower part of Fig. 3 should also correlate with the net rate of destruction of oceanic crust at any particular time and hence with the intensity of tectonic, igneous, metamorphic and certain sedimentary processes. In view of the possible significance of the mean sea level curve suggested here it seems desirable to define such a curve as accurately as possible for, say, Phanerozoic time on the basis of world-wide data. Its correlation with the above parameters should then be investigated, in the first instance for, say, Tertiary time, in order to test its validity.

It is suggested, therefore, that since one cannot predict the implied changes in sea level from sea-floor spreading data alone, because they are incomplete, one might turn the argument around and suggest that the fact that there *are* significant world-wide changes in sea level indicates that there have been systematic changes in spreading rates with time. Moreover since ridge activity is directly related to the rate of under-thrusting in the trench systems, and transgressions and regressions may also influence organic diversity and even account for certain faunal extinctions (Hallam, 1972), the definition of a reliable mean sea level curve as a function of time may be of fundamental interest and importance to various aspects of geology.

References

Allard, G. O. and Hurst, V. J., 1969. Brazil–Gabon geologic link supports continental drift, *Science*, **163**, 528–32.

Armstrong, R. L., 1969. Control of sea level relative to the continents, *Nature*, **221**, 1042–3.

Atwater, T. and Menard, H. W., 1970. Magnetic lineations in the northeast Pacific, *Earth and Planet. Sci. Letters*, **7**, 445–50.

Barker, P. F., 1970. Plate tectonics of the Scotia Sea region, *Nature*, **228**, 1293–6.

Berggren, W. A., 1969. Cenozoic chronostratigraphy, planktonic foraminiferal zonation and the radiometric time scale, *Nature*, **224**, 1072–5.

Bergh, H. W., 1971. Sea-floor spreading in the southwest Indian Ocean, *J. geophys. Res.*, **76**, 6276–82.

Dickson, G. O., Pitman, W. C. and Heirtzler, J. R., 1968. Magnetic anomalies in the South Atlantic and ocean floor spreading, *J. geophys. Res.*, **73**, 2087–2100.

Dymond, J., 1970. Excess argon in submarine basalt pillows, *Bull. geol. Soc. Amer.*, **81**, 1229–32.

Egyed, L., 1956. The change of the Earth's dimensions determined from palaeogeographical data, *Geofisica Pura e Applicata*, **33**, 42–48.

Ewing, J. and Ewing, M., 1967. Sediment distribution on the mid-ocean ridges with respect to spreading on the sea floor, *Science*, **156**, 1590–2.

Grasty, R. L., 1967. Orogeny, a cause of world-wide regression of the seas, *Nature*, **216**, 779–80.

Hallam, A., 1963. Major epeirogenic and eustatic changes since the Cretaceous, and their possible relationship to crustal structure, *Amer. J. Sci.*, **261**, 397–423.

Hallam, A., 1969. Tectonism and eustasy in the Jurassic, *Earth Sci. Rev.*, **5**, 45–68.

Hallam, A., 1971. Re-evaluation of the palaeogeographic argument for an expanding Earth, *Nature*, **232**, 180–82.

Hallam, A., 1972. Diversity, provinciality and extinction of Mesozoic invertebrates in relation to plate movements. *In:* Implications of Continental Drift to the Earth Sciences, Vol. 1, pp. 287–94, edited by D. H. Tarling and S. K. Runcorn. Academic Press, London and New York.

Heirtzler, J. R., Dickson, G. O., Herron, E. M., Pitman, W. C. and Le Pichon, X., 1968. Marine magnetic anomalies, geomagnetic field reversals and motions of the ocean floor and continents, *J. geophys. Res.*, **73**, 2119–36.

Herron, E. M., 1971. Crustal plates and sea-floor spreading in the southeastern Pacific, *Antarctic Res. Ser., Amer. Geophys. Un.*, **15**, 229–37.

Herron, E. M. and Hayes, D. E., 1969. A geophysical study of the Chile Ridge, *Earth Planet. Sci. Letters*, **6**, 77–83.

Herron, E. M. and Heirtzler, J. R., 1967. Sea floor spreading near the Galapagos, *Science*, **158**, 775–80.

Hey, R. N., Deffeyes, K. S., Johnson, G. L. and Lowrie, A., 1972. The Galapagos triple junction and plate motions in the East Pacific, *Nature*, **237**, 20–22.

JOIDES, 1969–1972. Initial Reports of the Deep Sea Drilling Project, **1–9**, U.S. Government Printing Office, Washington D.C.

JOIDES, 1970, Deep Sea Drilling Project: Leg 11, *Geotimes*, **15** (7), 14–16.

JOIDES, 1971, Deep Sea Drilling Project: Leg 16, *Geotimes*, **16** (6), 12–14.

JOIDES, 1972, Deep Sea Drilling Project: Leg 22, *Geotimes*, **17** (6), 15–17.

Karig, D. E., 1971. Origin and development of marginal basins in the Western Pacific, *J. geophys. Res.*, **76**, 2542–61.

Le Pichon, X., 1968. Sea-floor spreading and continental drift, *J. geophys. Res.*, **73**, 3661–97.

Le Pichon, X. and Hayes, D. E., 1971. Marginal offsets, fracture zones, and the early opening of the South Atlantic, *J. geophys. Res.*, **76**, 6283–93.

Luyendyk, B. P. and Fisher, D. E., 1969. Fission track age of magnetic anomaly 10: A new point on the sea-floor spreading curve, *Science*, **164**, 1516–17.

McElhinny, M. W. and Burek, P. J., 1971. Mesozoic palaeomagnetic stratigraphy, *Nature*, **232**, 98–102.

McKenzie, D. and Sclater, J. G., 1971. The evolution of the Indian Ocean since the Late Cretaceous, *Geophys. J.*, **24**, 437–528.

Maxwell, A. E., Von Herzen, R. P., Hsu, K. J., Andrews, J. E., Saito, T., Percival, S. F., Milow, E. D. and Boyce, R. E., 1970. Deep Sea Drilling in the South Atlantic, *Science*, **168**, 1047–59.

Morgan, W. J., 1968. Rises, trenches, great faults and crustal blocks. *J. geophys. Res.*, **73**, 1959–82.

Morgan, W. J., Vogt, P. R. and Falls, D. R., 1969. Magnetic anomalies and sea-floor spreading on the Chile Rise, *Nature*, **222**, 137–42.

Pitman, W. C. and Talwani, M., 1972. Sea-floor spreading in the North Atlantic. *Bull. Geol. Soc. Amer.*, **83**, 619–46.

Reyment, R. A., 1969. Ammonite biostratigraphy, continental drift and oscillatory transgressions, *Nature*, **224**, 137–40.

Russell, K. L., 1968. Oceanic ridges and eustatic changes in sea level, *Nature*, **218**, 861–2.

Sclater, J. G., Anderson, R. N. and Bell, M. L., 1971. Elevation of ridges and evolution of the central Eastern Pacific, *J. geophys. Res.*, **76**, 7888–7915.

Valentine, J. W. and Moores, E. M., 1970. Plate-tectonic regulation of faunal diversity and sea-level: a model, *Nature*, **228**, 657–9.

Vine, F. J. and Hess, H. H., 1971. Sea-floor spreading. *In:* The Sea, Vol. IV, Pt. 2, pp. 587–622, edited by A. E. Maxwell. Wiley-Interscience, N.Y.

Vogt, P. R. and Johnson, G. L., 1971. Cretaceous sea-floor spreading in the western North–Atlantic, *Nature*, **234**, 22–25.

Weissel, J. K. and Hayes, D. E., 1971. Asymmetric sea-floor spreading south of Australia, *Nature*, **231**, 518–22.

Williams, C. A. and McKenzie, D., 1971. The evolution of the northeast Atlantic, *Nature*, **232**, 168–73.

8

PALAEOGEOGRAPHIC IMPLICATIONS

8.1

G. A. L. JOHNSON
University of Durham
England

Closing of the Carboniferous Sea in Western Europe

Introduction

Carboniferous plate tectonics in Europe has been introduced in papers by Hamilton (1970), Nicolas (1972) and Johnson (in press) who describe the Uralide and Hercynian orogens. Application of the global tectonics theory to the Carboniferous is providing a sound explanation for previous stratigraphical inconsistencies, space problems and faunal provinces. Movement of continental plates causing closure of seaways and the formation of orogenic belts is shown to be the main feature of Carboniferous times in Europe. The Eurasian Sea closed by the rafting together of the Siberian plate and the European plate to form the Uralide orogenic belt according to Hamilton. According to Johnson, the southern European plate and the northern European plate converged, closing the Mid-European sea and forming the Hercynian fold-belt. The sequence of phases of continental plate movement which lead finally to the elimination of the Mid-European sea and the building of the Hercynides is the subject of this paper.

The Mid-European Sea

Detailed analysis of Ordovician trilobite provinces by Wittington & Hughes (1972) lead to the introduction of a Mid-European sea between northern and southern Europe; the east–west line of separation runs to the north of Brittany (Fig. 1). These authors trace the pattern of changes in palaeogeography through the Ordovician and find that the Mid-European sea is required during the successive stages from the Arenig to the Ashgill. During the Silurian a sea lay to the south of Britain according to the recent interpretation of Ziegler (1970). Following the evidence provided by the Ordovician trilobite provinces it is an easy presumption that this was the Mid-European sea, particularly as this sea persisted through the Silurian, through the Devonian and into Carboniferous times. A reconstruction of the Old Red Sandstone continent by House (1968) shows the sea in Devonian times lying to the south of Britain and stretching eastwards across Europe. The fit of Africa and southern Europe used by this author is based on Bullard *et al.* (1965) and is now known to be inapplicable to the Devonian, but the northern shoreline and shelf of a Mid-European sea is clearly indicated. More recently Johnson & Dasch (1972) have recognized a narrow Appalachian seaway in the Devonian of eastern USA which could be the continuation of the Mid-European sea.

At the beginning of the Carboniferous the level of the Mid-European sea appears to have risen so that it transgressed widely over the northern Old Red Sandstone coastal

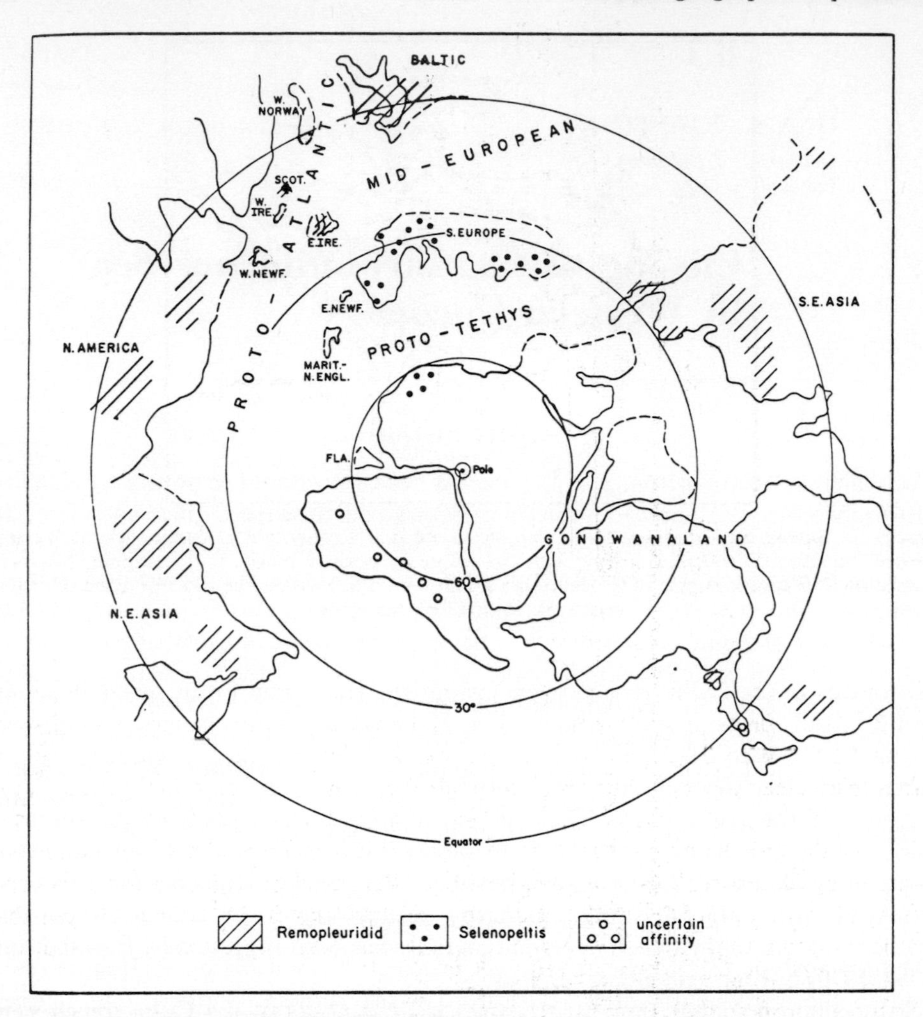

FIGURE 1 Stereographic projection of the southern hemisphere showing Ordovician, Caradoc Series, palaeogeography and trilobite faunal provinces. E. Ire., southeastern Ireland, England and Wales; E. Newf., east Newfoundland; Fla., Florida State, USA; Marit.-N. Eng., maritime provinces of Canada and New England states of USA; Scot., Scotland; W. Ire., northwestern Ireland; W. Newf., western Newfoundland; W. Norway, western Norway. Reproduced from Whittington & Hughes (1972, fig. 9).

plain and into the intermontane basins of the interior. Initially the uplands stood out as islands forming an archipelago over Britain and marine sedimentary basins developed in what were previously interior basins (George, 1969). The upland cores were progressively inundated by the Carboniferous sea so that by Late Carboniferous times only the larger islands, such as the Southern Uplands of Scotland and St. George's Land in Wales and the Midlands, remained as islands.

The sedimentary record of the Carboniferous Mid-European sea is best seen in the paralic deposits of Britain and northern Europe. From a shoreline running through southern Scotland and Denmark changing facies across an unstable shelf can be traced

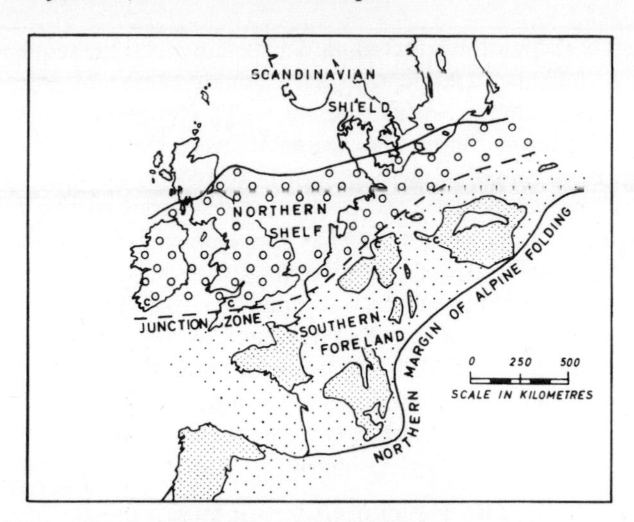

FIGURE 2 Sketch map showing Hercynian structural and stratigraphical framework of Western Europe. Southern Foreland in stipple with contemporary positive blocks in fine stipple. Northern paralic shelf in coarse stipple. The Hercynian Front suture lies within the junction zone. C marks regions of thick Culm deposition.

southwards to the Culm trench system against the Hercynian Front suture in south-west England, Belgium and Germany (Fig. 2). Dinantian facies changes in south-west England have been described by Ramsbottom (1970) who traces a sequence southwards from clear-water marine shelf, through Walsortian reef marginal facies, to the sequence of the deeper water Culm trench. No evidence remains of the southern margin of the trench and all that lay to the south of it is missing. The Hercynian Front suture brings contorted Devonian and possibly older rocks with igneous intrusives and extrusives with ophiolites abutting sharply against the Culm trench. A possible southern margin to the trench in Namurian times has been suggested by Ramsbottom (1969, 1970) but the evidence is tenuous.

Sedimentation studies show that coarse clastic deposits of the Culm trench were derived from the south and west during the Dinantian; turbidites entered from the same direction during the Namurian and Westphalian with deltaic deposits entering from the north (Prentice, 1962). A source of sediment from the north and from the south and west is indicated.

In western Europe sediments of the northern paralic shelf are seen in the Franco-Belgian basin where the Carboniferous sequences are directly comparable with Britain. The marine Dinantian contains Walsortian reefs and is much like the deposits found in southern Ireland and south Wales; the deltaic Namurian and Westphalian sequences are also similar to those of south Britain. Further east in Germany Dinantian marine limestones are restricted to the north, elsewhere the Dinantian follows the Devonian in a Culm facies without unconformity. The Namurian and Westphalian are deltaic facies with productive coal seams and the Stephanian is present in the northern and eastern part of the paralic shelf.

The shelf at the southern side of the Mid-European sea appears to be preserved in Brittany where the base of the Dinantian is transgressive over Devonian and pre-Devonian beds. The Dinantian is composed of a thick sequence of shallow water marine sediments with limestones including Walsortian reefs; the sequence thins southwards towards a shoreline. The deposits are strongly folded by Mid-Carboniferous (Sudetic) earth movements at which time the older Hercynides were formed. Continental facies of Upper Carboniferous sediments were laid down in basins between these young mountain ranges in the Southern Foreland region (Fig. 2).

Stratigraphical contrast between the Upper Carboniferous paralic and limnic facies across the Hercynian Front is striking and has been studied by a succession of European workers, particularly with respect to the varying coal deposits; a map showing the distribution of paralic and limnic facies was published by Kukuk (1938). In the Dinantian a northern and southern shelf covered with shallow water marine sediments can be identified at the margins of the Mid-European sea. But, after the Mid-Carboniferous earth movements which folded and uplifted the Southern Foreland, contrasting Upper Carboniferous limnic and paralic deposits were laid at the margins of the sea.

The Hercynian Front Suture

Using structural and palaeomagnetic data, Rutten (1969) writes that the elements of the mosiac now forming Hercynian Europe were at quite different positions relative to each other at the time of folding, but his data was not sufficiently advanced to supply a reliable palaeogeography. Now, the presence of the Hercynian Front suture, separating northern and southern Europe, provides a key to the understanding of the regional palaeogeography.

Along much of the junction zone between the northern and southern European plates the Hercynian Front suture is masked by overlying younger deposits, particularly Quaternary glacial drift, and complicated by major thrusts and severe folding. In Britain, Belgium and France estimates of the main south to north movement of the thrust zone varies from 14 to at least 30 km (Waterschoot van der Gracht, 1938; Bott *et al.*, 1958). Thrusting of this magnitude is capable of disguising the junction zone and possibly covering the line of suture with a cloak of derived material. Possibly because of this the Hercynian Front suture has only recently been recognized.

The distinction of two very different paralic and limnic sedimentary environments in the Westphalian and Stephanian of Europe is clearly displayed in the structural history of the region. The northern paralic shelf was generally quiescent tectonically until the onset of the Late Carboniferous (Asturic) earth movements whereas the Southern Foreland was strongly affected by the Late Devonian (Bretonic) and Mid-Carboniferous (Sudetic) orogenic phases. It is difficult to reconcile the strikingly different tectonic histories of the two regions which are now closely adjacent without invoking an intervening sea and plate margin to cause separation prior to the Late Carboniferous collision.

Radiometric dating of slates and phyllites from southwest England provides evidence of the age of folding on both sides of the Hercynian Front suture (Dodson & Rex, 1971). On the south, metamorphic rocks of Cornwall give dates between 365 and 345 m.y. corresponding to a Late Devonian or Lower Carboniferous period of folding. North of this, the Devonian slate belt gives dates ranging from 340 to 320 m.y., suggesting a Namurian phase of tectonism. Further to the north and east the southern margin of the Culm trench, near to the north side of the suture, gives dates from 310 to 270 m.y. and represent minimum ages for the Late Carboniferous (Asturic) folding. These dates are supported to some extent by detailed structural evidence which indicates that the main orogenic phase is much earlier in the south than it is in the north in this region (Roberts & Sanderson, 1971).

An important feature of sutures formed by collision of plate margins is the presence of ophiolite complexes. The ophiolite suite is a gross sequence from variably serpentinized ultrabasics through gabbros to pillow lavas capped by marine sediment, they are believed to be fragments of the oceanic crust and mantle tectonically emplaced in orogenic belts (Dewey & Bird, 1971). The pillow lava–black shale–chert association is particularly well developed in the Culm trench of southwest England. It is first found in the region in the Middle and Upper Devonian where widespread and thick pillow lava sequences are developed, cut by basic sills and dykes. Thick spilites and tuffs are widespread in the early Culm of north Cornwall, the south border of the Culm trench, and are associated with radiolarian cherts and black shales; manganese ores have been worked in the lava. An intrusive ophiolite suite forms the Lizard complex of south Cornwall in which serpentines, gabbros and dolerites are intruded into metamorphosed country rock believed to be originally marine muds and slates. Large intrusive masses of serpentine and gabbro have risen through the schists and granulites and, at a later stage, there was renewed injection of basic dykes. The Lizard complex has been radiometrically dated at about 350 m.y., corresponding to the Late Devonian–Early Carboniferous phase of earth movements (Dodson & Rex, 1971). It is noteworthy that long before the global tectonics theory comparisons had been made between the Lizard Complex and the very similar Ordovician Ballantrae Volcanic Series of south Scotland which is now recognized as a Caledonian ophiolite (Pringle, 1948, p. 12; Bird *et al.*, 1971).

A concensus of stratigraphical, structural and radiometric evidence suggests that the Hercynian Front suture lies at the southern margin of the Culm trench in southwest England. In this region the position of a Carboniferous Benioff or subduction zone might be anticipated. It is noteworthy that a belt of high magnetic anomaly has been recorded along the southern margin of the trench (Bott *et al.*, 1958) which they interpret as the presence of magnetic rocks extending to considerable depths. These observations are of considerable interest, but the interpretation is made uncertain owing to the near proximity of the Dartmoor granite and the presence of magnetic metamorphic rocks at the surface (Bott, personal communication, 1972).

Closing Pattern of the Mid-European Sea

In Early Carboniferous times the margins of the Mid-European sea spread northwards over the coastal plain of the Old Red Sandstone continent and southwards over eroded Devonian and pre-Devonian rocks on the Southern Foreland. This widespread transgression northwards over the paralic shalf and southwards over foreland suggests a eustatic rise in sea-level though the picture is complicated by known tectonic earth movements. The first phase of the closing of the Mid-European sea seems to have culminated at this time with the final separation from the African landmass of the southern European plate. This plate was then rafted northwards and westwards on the flank of the opening Tethys sea (Fig. 3a, b). The shape of the southern European plate during the Dinantian is uncertain owing to crustal shortening at the northern and southern sides during Hercynian and Alpine folding episodes. In Fig. 3 the plate is

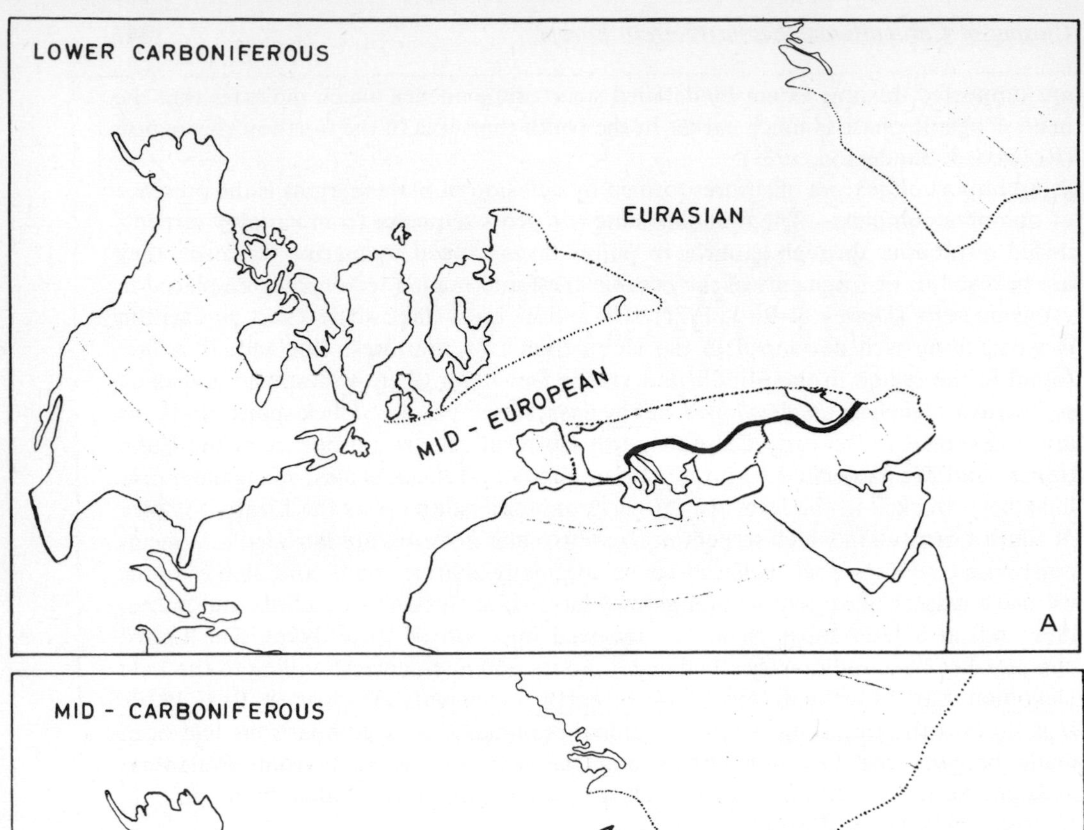

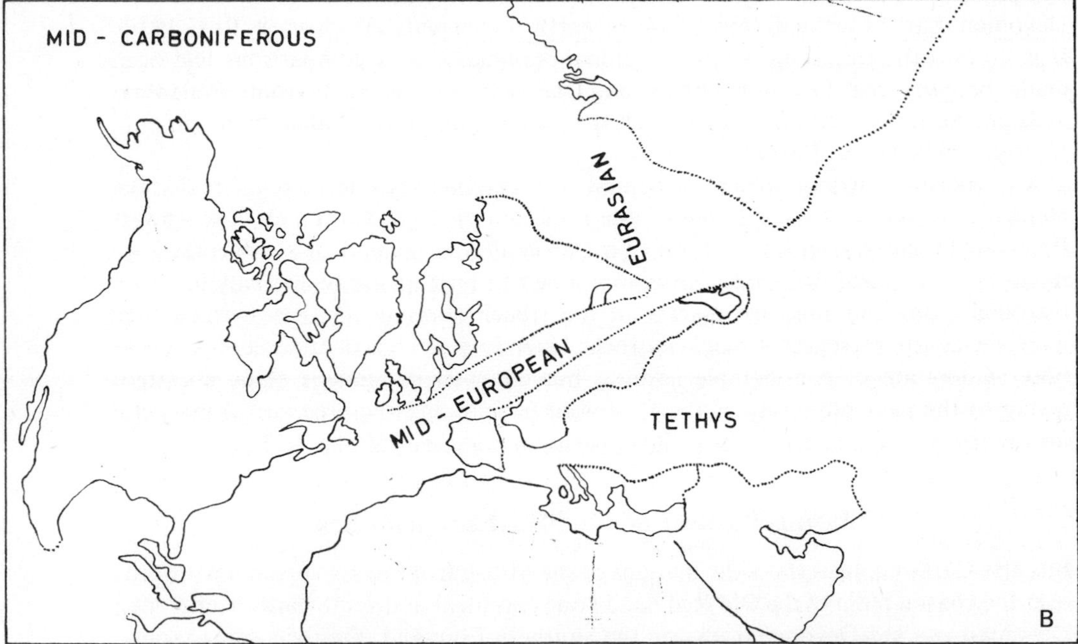

FIGURE 3 Palaeogeographical sketch maps showing the pattern of closure of the Mid-European sea. Data from many sources, particularly 'Putative Phanerozoic World Maps' privately circulated by J. C. Briden (Leeds), A. G. Smith and G. E. Drewry (Cambridge) December 1971*. A. Lower Carboniferous palaeogeography showing the disposition of continental masses about the Mid-European sea. Fracture along which the southern European plate separated from the African landmass indicated by a thick line. B. Mid-Carboniferous palaeogeography showing the southern European plate between the closing Mid-European sea and the opening Tethys. At this time the Sudetic cordilleran or leading edge margin orogeny took place on the northern and western flanks of the moving plate.

* (Smith *et al.*, 1973.)

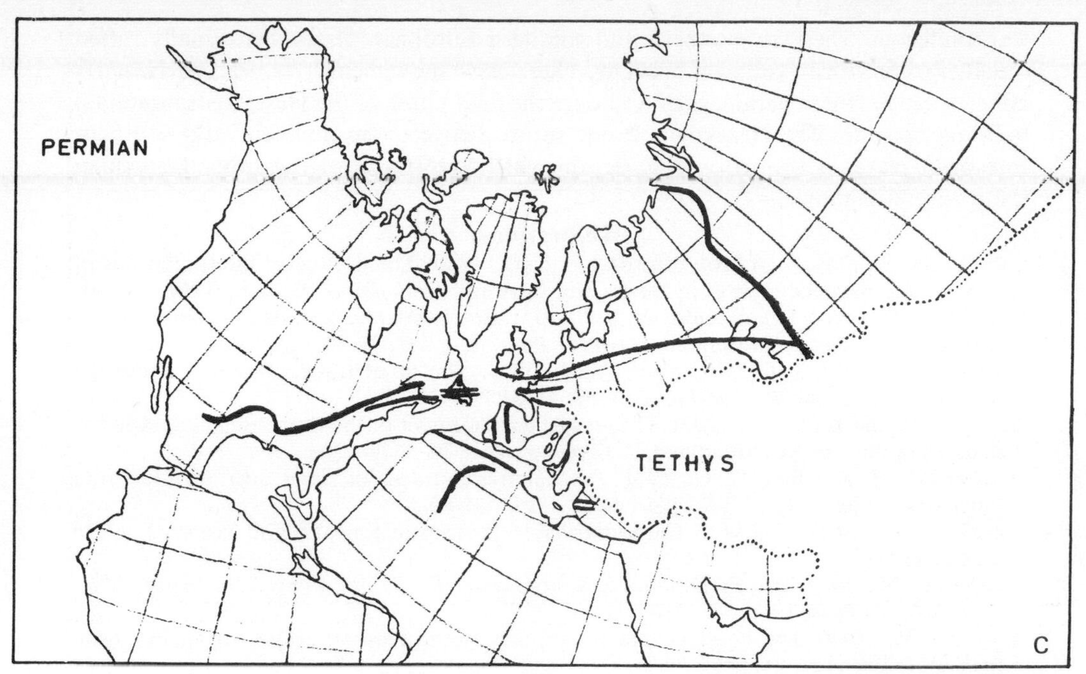

C. Permian palaeogeography showing the trend of the Hercynian ranges and the Hercynian Front suture at the junction of the northern and southern European plates after the final phase of collision orogeny.

shown bounded on the north and south by the Hercynian and Alpine sutures. Instability of the southern European plate, possibly during the period of separation from the African landmass, is indicated by the Late Devonian to Early Carboniferous Bretonnic earth movements. It is possible that at this period a major mid-ocean ridge system was forming beneath the Tethys and that displacement of oceanic water caused the Early Carboniferous eustatic rise in sea-level.

The second phase of closing of the Mid-European sea is dated by the Sudetic (Mid-Carboniferous) earth movements which formed the main European Hercynide chains. The earth movements were accompanied by submarine and subaerial intermediate and acid volcanism of explosive type and have been shown to be similar in structure and igneous provenance to the contemporary Andean cordillera of South America (Nicolas, 1972). It follows that on the evidence available the Sudetic earth movements belong to a cordilleran or leading edge margin orogeny associated with the northern and western margin of the south European plate as it was rafted towards the north (Fig. 3b). Carboniferous fauna provinces in western Europe indicate that the seaway connection with the Eurasian sea was lost during Mid-Carboniferous times. The goniatite distribution has been described by Ramsbottom (1971) who shows that the Dinantian faunas were almost cosmopolitan, but in the Namurian the seaway passage between Europe and the Eurasian sea was closed, possibly in Poland. As a result of this the goniatite fauna of western Europe declined in variety. The Middle and Upper Carboniferous fusulinids are similarly restricted and occur only in the Eurasian sea and Tethys, they do not enter the western European sea.

The last phase of closure of the Mid-European sea came late in the Upper

Carboniferous when the northern and southern European plates were finally rafted together to produce a collision orogeny. These earth movements (Asturic) particularly affected the northern paralic shelf and were the final phase of the Hercynian mountain building episode. The Hercynian Front suture between the northern and southern European plates is shown on Fig. 3c, on which the trends of the other Hercynian chains are also shown.

References

Bott, M. H. P., Day, A. A. and Masson-Smith, D., 1958. The geological interpretations of gravity and magnetic surveys in Devon and Cornwall, *Phil. Trans. R. Soc.*, **A251,** 161–91.

Bird, J. M., Dewey, J. F. and Kidd, W. S. F., 1971. Proto-Atlantic oceanic crust and mantle: Appalachian/Caledonian ophiolites, *Nature Phys. Sci.*, **231,** 28–31.

Bullard, E., Everett, J. E. and Smith, A. G., 1965. The fit of the continents around the Atlantic, *Phil. Trans. R. Soc. Lond.*, **A258,** 41–51.

Dewey, J. F. and Bird, J. M., 1971. Origin and emplacement of the ophiolite suite: Appalachian ophiolites in Newfoundland, *J. Geophys. Res.*, **76,** 3179–3206.

Dodson, M. H. and Rex, D. C., 1971. Potassium-argon ages of slates and phyllites from southwest England, *Q. Jl. geol. Soc. Lond.*, **126,** 465–99.

Francis, E. H., 1970. Review of Carboniferous volcanism in England and Wales, *J. Earth Sci.*, **8,** 41–56.

George, T. N., 1969. British Dinantian Stratigraphy. *C. R. 6th Congr. Int. Strat. Geol. Carbonif., Sheffield*, 1967, 193–218.

Hamilton, W., 1970. The Uralides and the motion of the Russian and Siberian Platform, *Bull. geol. Soc. Am.*, **81,** 2553–76.

House, M. R., 1968. Continental drift and the Devonian system. Inaugural lecture series, University of Hull, Yorks., 24 pp.

Johnson, G. A. L. (in press). Crustal margins and plate tectonics during the Carboniferous, *C. R. 7th Int. Kongr. für Stratigr. und Geol. des Karbons, Krefeld* (1971).

Johnson, J. G. and Dasch, E. J., 1972. Origin of the Appalachian Faunal Province of the Devonian, *Nature*, **236,** 125–6.

Kukuk, P., 1938. Geologie des Neiderrheinisch-Westfälischen Steinkohlengebietes. Springer, Berlin, 706 pp.

Nicolas, A., 1972. Was the Hercynian orogenic belt of Europe of the Andean type? *Nature*, **236,** 221–3.

Prentice, J. E., 1962. The sedimentation history of the Carboniferous in Devon. *In:* Some aspects of the Variscan fold belt, edited by K. Coe. 163 pp., Manchester.

Pringle, J., 1948. The South of Scotland. British Regional Geology, 2nd Ed., H.M.S.O., Edinburgh, 87 pp.

Ramsbottom, W. H. C., 1969. The Namurian of Britain. *C. R. 6th Congr. Int. Strat. Geol. Carbonif., Sheffield*, 1967, 219–32.

Ramsbottom, W. H. C., 1970. Carboniferous faunas and palaeogeography of the southwest of England region, *Proc. Ussher Soc.*, **2,** 144–57.

Ramsbottom, W. H. C., 1971. Palaeogeography and goniatite distribution in the Namurian and Early Westphalian, *C. R. 6th Congr. Int. Strat. Geol. Carbonif., Sheffield*, 1967, 1395–1400.

Roberts, J. L. and Sanderson, D. J., 1971. Polyphase developments of slaty cleavage and the confrontation of facing directions in the Devonian rocks of north Cornwall, *Nature Phys. Sci.*, **230,** 87–89.

Rutten, M. G., 1969. The Geology of Western Europe. Elsevier, Amsterdam, 520 pp.

Smith, A. G., Briden, J. C. and Drewry, G. E., 1973. Phanerozoic World Maps. Spec. Pap. in *Palaeontol.*, **12,** 1–42.

Waterschoot van der Gracht, W. A. J. M., 1938. A structural outline of the Variscan front and its foreland from south-central England to eastern Westphalia and Hessen, *C. R. 2nd Congress Avanc. Etud. Stratigr. Carb.*, **3,** 1485–1565.

Whittington, H. B. and Hughes, C. P., 1972. Ordovician geography and faunal provinces deduced from trilobite distribution, *Phil. Trans. R. Soc. Lond.*, **B263,** 235–78.

Ziegler, A. M., 1970. Geosynclinal development of the British Isles during the Silurian period, *J. Geol.*, **78,** 445–79.

8.2

LESTER KING
University of Natal
Durban, South Africa

An Improved Reconstruction of Gondwanaland

Introduction

A. L. du Toit died a quarter of a century ago, on 26 February 1948. He left this first reconstruction of Gondwanaland, dating from 1937 (Fig. 1). In 1970 a computer, arbitrarily assessing continental outlines at the 500 m isobath, produced a reassembly (Fig. 2), ironically spelling out the same continental positions, but jamming the whole tightly together.

Let us first examine the reasons for the differences of 'fit' between the reassemblies of du Toit and of the computer. du Toit's synthesis was founded upon long and detailed field studies of geological phenomena in the relevant territories. In particular, when he made geological comparisons between South America and Africa as early as 1927,

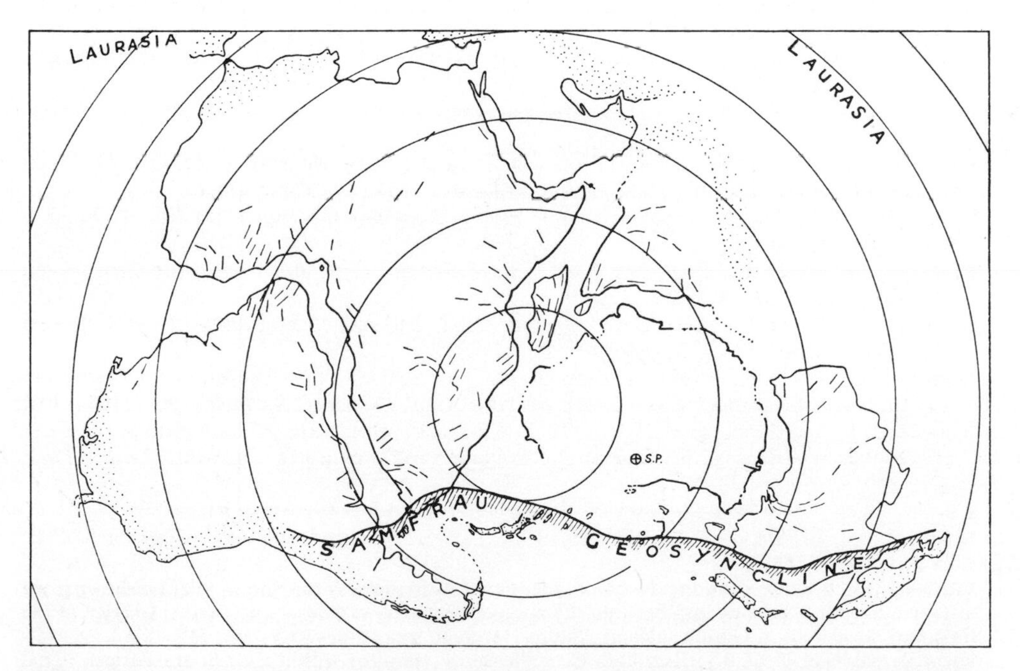

FIGURE 1 A. L. du Toit's reconstruction of Gondwanaland published in 1937, now requiring modification.

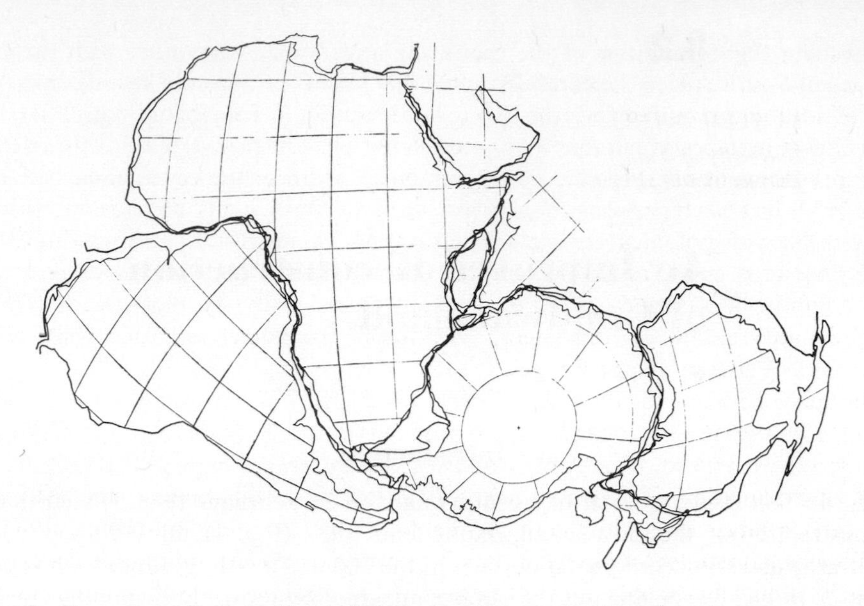

FIGURE 2 Computerized reassembly of Gondwanaland, using the 500 fathom isobath for 'best fit', obtained by Smith & Hallam (1970). The computer overlooked the geological data on the refit which require that continents be not so closely apposed (Fig. 3).

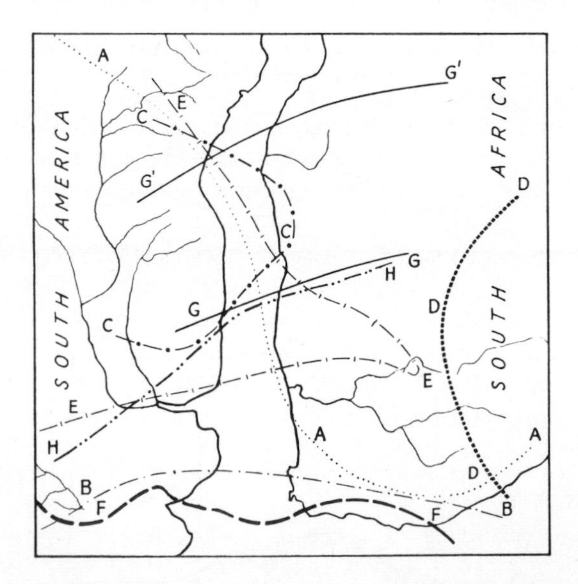

FIGURE 3 A. L. du Toit's comparison of geological features across the South Atlantic showing the distance that must have existed between the apposed segments of Africa and South America at the time they were both in Gondwanaland. 'Siluro-Devonian sandstones to W and SW of AA; glacials unconformable to N of BB; Brazilian 'Coal Measures' (Bonito) within CC; South African 'Coal Measures' (Ecca) to E of DD; Ecca (blue-green) between DD and EE; Ecca (red) to NW of EE; Gondwanide Foldings FF; Gondwanide Unwarpings GG, G′G′; Triassic unconformable to N of HH'.

establishing the correlation of the Sierra de la Ventana, Argentina, with the Cape Ranges of South Africa, he carefully noted that lateral variation of sedimentary types and of structure required that the two regions should, in Palaeozoic times, not have been almost juxtaposed but that a distance of 500–600 km was required between them. This measurement notably exceeds the combined width of the continental shelves.

In 1937, in *Our Wandering Continents*, du Toit verified his conclusion with ten different lines of geological correlation between Africa and South America (Fig. 3). As these lines criss cross, only one position was obtained where all these correlative factors could be satisfied—and this position agreed with the measure of 500–800 km previously obtained from lateral variation in the Palaeozoic rock series of the Sierra de la Ventana—Cape Ranges.

The space between the continental edges on the du Toit reassembly was found by King (1958, 1967) to have special significance in the pattern of many of the major sub-oceanic ridges. Indeed, it not only explained the pattern of these global features but afforded explanation of why continental type rocks are sometimes found associated with certain ridges, and why the ridges stand high, in isostatic equilibrium, above the neighbouring ocean floors. Some of the ridges even expand into submarine plateaus, or are surmounted by islands (*e.g.* Malagasy, New Zealand) whose composition or structure are continental.

On earth data, the honours clearly lie with du Toit rather than the computer.

Notwithstanding, newer geological researches require revision of du Toit's Gondwanaland reconstruction.

Necessary Revisions of the Gondwanaland Reconstruction

The correlations between South America and Africa have been tested by many geologists. Independently, Leinz, Maack and Martin have each furnished reports that show the correspondences to be even better than du Toit had understood them to be. But the refits between the other southern continents have on re-examination proved to be less satisfactory.

Thus Malagasy has been refitted to Africa by Flores (1970) (Fig. 4) far to the south of Kenya where du Toit placed it; and this apparently trifling readjustment had the consequence that Antarctica can no longer be refitted into the reassembly where du Toit placed it. A more southerly placing of Antarctica had, indeed, been suggested by King in 1956. The reassembly then shows signs of 'opening up' in the eastern half. But there is no help for this. Acceptance of Flores's work inevitably entails further readjustments in the reassembly.

(*a*) *The position of India*

So, with the necessity for revision before us, we examine the position assigned to India.

In their reconstructions, both du Toit and later Smith and Hallam assigned a position to India alongside the Indian Ocean sector of Antarctica. Both these regions consist principally of Basement gneisses, but no clear or detailed correlations have been established even by the extensive post-war explorations of the relevant areas of Antarctica. On the contrary, charnockites of Madras and of Antarctica show marked disagreements of geochronology.

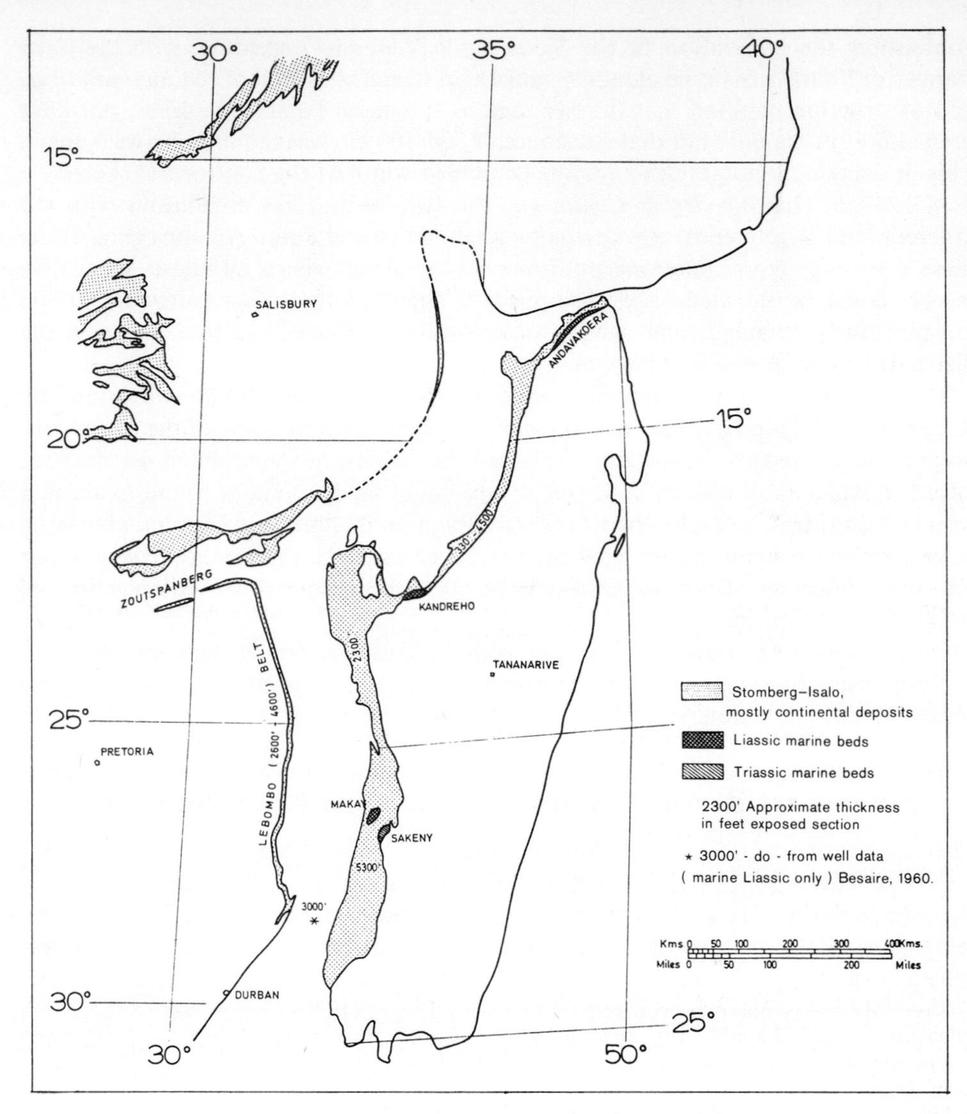

FIGURE 4 The revised refit of Malagasy and East Africa by Flores (1970), based on matching of geological features.

The position of India was reviewed by Ahmad (1960, 1961) who related it not to Antarctica as du Toit had done, but to Western Australia (Fig. 5). According to Ahmad, whose reconstruction was followed by Krishnan (1969), there are better geological correspondences in the basement and in the Gondwana rock sequences with Western Australia than with Antarctica. These statements have been regarded with disfavour by some authors; but there is a further line of argument that I find compelling. On du Toit's reconstruction the eastern end of the Himalayan mountain system has no logical continuation. It is left pointing in the direction of the southwest corner of Australia—where there are *no* Tertiary fold mountains. Clearly, the true connection

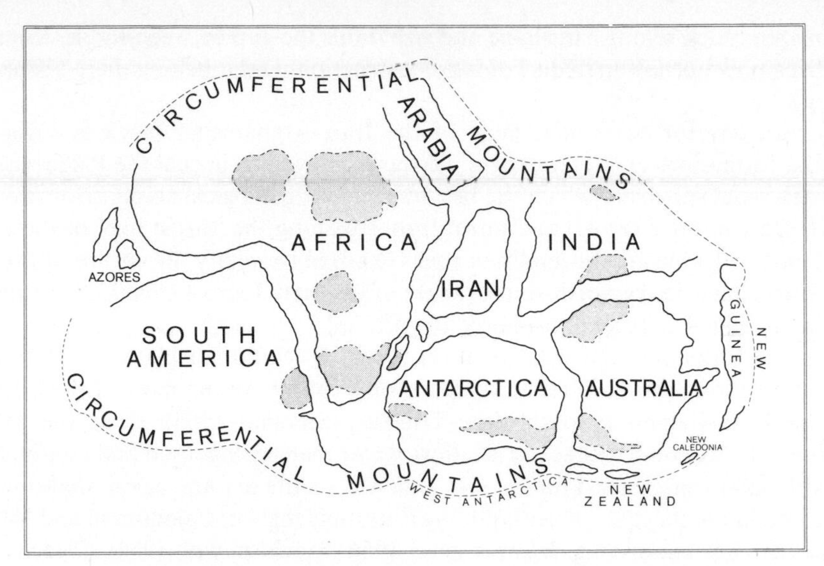

FIGURE 5 The revised reassembly of Gondwanaland. Boundaries of basins of Gondwana sedimentation (stippled) approximate only, in the light of modern areas of outcrop. Note the space left between the continents to allow for suboceanic ridge systems.

must be with New Guinea, and Ahmad's reconstruction achieves this correlation. This topic is therefore taken up again under a later section of this paper.

If Ahmad's placing of India is correct, the western side of the peninsula can no longer be apposed to Somaliland and Arabia in the reconstruction; and the readjustments required by the evidence from Malagasy and from India inevitably leave a great gap in the outline of Gondwanaland. A new element is needed to fill this gap. That element, we suggest, is the block of Iran–Afghanistan.

(*b*) *The inclusion of Iran–Afghanistan* (Fig. 5)

The mass of Iran–Afghanistan stands as a 'median maas' (Stahl 1911) between the mountain ranges of Laurasia (Elburz–Pamir) and of Gondwana (Zagros–Baluchistan). This position, apparently as part of neither parent super-continent, is anomalous if not unique.

But the economy of continental matter renders such a supposition highly improbable. Both Laurasia and Gondwana existed as primitive entities since way back in Precambrian time—when they may have originated as opposed polar earth masses. So far as can be ascertained, these two super-continental masses included *all* the continental matter of the globe. Each parent super-continent, moreover, was completely ringed about by folded mountain ranges (King 1967), and the remains of these structures can still be found along those aspects of the modern continents that were originally coastlands of Laurasia and Gondwana. All other continental coasts are of fractured type and originated at the Mesozoic break-up of the two supercontinents. How does it happen therefore that the mass of Iran-Afghanistan is situated in modern times between the confines of Laurasia and Gondwana?

Stocklin (1965) reviewed and rejected the hypothesis of Iran as a 'median mass' on the basis of geological studies there. If it was not a 'median mass', where did the Iran–

Afghanistan block belong? In shape and size it fills the gap between India, Antarctica and Africa now opened up in du Toit's reconstruction. Did it belong there? What is its geology?

Unfortunately for our study, most of the Iran–Afghanistan block is covered by terrestrial formations of Cretaceous or Tertiary age which conceal the Palaeozoic and older rock sequences from which the best evidence of geological comparisons could be derived. Only in three areas, (a) Central Iran, (b) along the 'thrust line' of the Zagros and (c) east of Kabul are Precambrian rocks exposed to signify the nature of the basement. Particularly in the north-south ranges of Kerman-Tabas-Ozbakkuh the late Precambrian is folded and metamorphosed to gneisses.

Then 'the formation and consolidation of the Precambrian platform of Iran was followed by a long period of remarkable tectonic calm covering most of the time span from the Infra-Cambrian to the late Triassic' (Stocklin 1965). Over the interior, sequences of Palaeozoic rocks are shallow water marine, lagoonal and continental in type with many hiatuses. Triassic and Jurassic sediments are again shallow water marine or continental, the whole signifying a cratonic regime. Caledonian and Variscan tectonics are not known (e.g. Ruttner *et al.* 1966; Stocklin *et al.* 1964). Clearly, therefore, the craton belonged to Gondwana and not to Laurasia. In other words, the Iran–Afghanistan block appears admirably qualified by its pre-Jurassic geology to fill the gap in our Gondwana reconstruction between India and East Africa (Fig. 5).

With Iran so placed, the Permian–Triassic marine transgressions of northern Madagascar and East Africa appear to be related with the Iranian sequence. Datings of the marine Mesozoic rocks further suggest an early breaking out from the parent

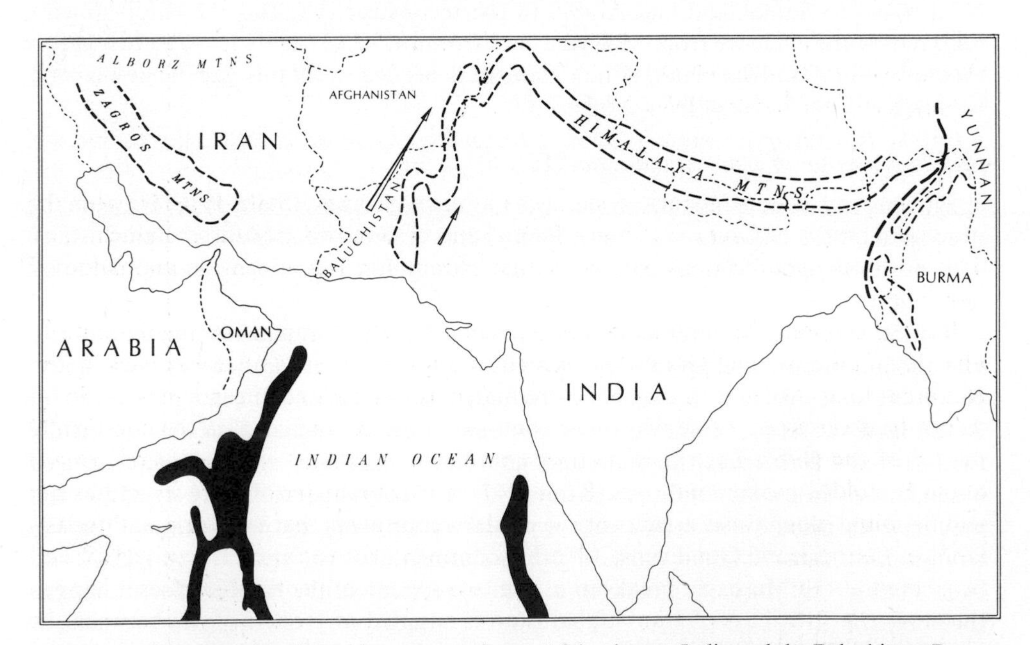

FIGURE 6 Relationships of Arabia—Oman—Iran—Afghanistan—India and the Baluchistan Ranges following the break up of this region of Gondwanaland during the Jurassic or earlier. Both Iran Afghanistan and India have moved northwards but the former relatively more so. Submarine ridges black.

land mass, not later than the end of the Triassic, which would make it the first part of Gondwanaland to break away from the parent.

So the continental-type mass of Iran–Afghanistan could have assumed station between Laurasia and Gondwana by the early Mesozoic; distinctly earlier than the Cretaceous fragmentation leading to formation of the present continents.

As the Iranian block moved northward, a relative movement developed between it and India, which was also moving northward though more slowly. As a result of such relative movement may have developed the tectonic pattern later followed by the Tertiary mountain arcs in Baluchistan and Mekran (Fig. 6).

After the break-up of Gondwanaland, before the late Cretaceous, Central Iran in its present relative position partook of violent orogenesis of Himalayan type which has continued intermittently through the Tertiary until the Plio-Pleistocene.

As these ideas develop, we note that Takin (1971) has also recently pointed out the necessity for thinking in terms of continental drift to explain the geology of Iran.

When the above readjustments are made, a new reconstruction emerges with Gondwanaland *as a simple ovoid body surrounded by a girdle of tectonic mountain ranges* (Fig. 5).

Testing the new Reconstruction

Three lines of testing to assess the validity of the proposed reconstruction may be developed: (*a*) palaeomagnetic studies, (*b*) studies of the pattern of submarine ridges in the Indian Ocean basin, and (*c*) studies of the mountain girdle about Gondwanaland.

(*a*) *Palaeomagnetic studies.* Despite enquiry, I have not been able to obtain palaeomagnetic measurements from Iran to compare with those from other countries within Gondwanaland, and can offer no data under this heading at present.

(*b*) *The Pattern of Submarine Ridges in the Indian Ocean Basin.* The main submarine ridges are often assumed to be sited over the late Mesozoic fracture zones of the disintegrating supercontinents Laurasia and Gondwana. The most elementary demonstration is on the mid-Atlantic Ridge, with simple withdrawal of the continents east or west on either side. Moreover, a single date, early Cretaceous, is accepted for this separation.

But the pattern of submarine ridges within the Indian Ocean is more complex. The mid-Indian Ridge is supplemented by the Carlsberg and Mascarene Ridges, the Ninety East Ridge and the volcanic Maldive-Chagos Ridge. The main mid-Indian Ridge itself bifurcates towards the south. Continental masses involved are Africa (with Arabia), Iran, India, Australia and Antarctica, and the separation of each may be individual or in concert with others, and the relative positions of each to the others may have changed considerably with time during the break up of Gondwanaland. On stratigraphic data, different parts of the dismemberment took place at various times from Permian till the early Cretaceous, with movement of the emerging continents in the interim.

Analysis of the break-up of Gondwanaland in the Indian Ocean sector requires more data on the composition and structure of the various sub-oceanic ridges than are at present available, and no detailed analysis upon this basis is attempted here.

But stratigraphic data from the various landmasses, especially the coastal sectors, suggest the following movements:

(i) the earliest opening-up of Gondwanaland was heralded by a shallow north-south gulf which stretched across Iran and as far south as Malgache during Permian time. In this gulf flourished Permian marine fauna of Tethyan derivation. The second phase including the exodus of the Iran-Afghanistan mass from Gondwanaland may be dated from the late Triassic tectonic events of Iran (Stocklin 1965). It then assumed a position in the Tethys between Laurasia and Gondwana, and from that time forth behaved in conformity with the Himalayan tectonics.

(ii) The east coast of Africa and Cutch in India preserve extensive developments of mid-Jurassic marine strata showing that this sector was again actively breaking up. Progressing southward, the split opened up the whole of the east coat of Africa by the lowest Cretaceous (Uitenhage fauna) and the eastern part of Gondwana (India, Australia and Antarctica) drifted away to the east while Africa and South America drifted westward as described by King (1958, 1967).

(iii) Fragmentation of the eastern half of Gondwana, with opening up of seaways between the present continents, was accomplished by the Albian (mid-Cretaceous (King 1958, 1967)).

(iv) Opening of the South Atlantic with separation of Africa and South America was of similar age, the earliest sediments relating to the present coasts being lagoonal Aptian followed by marine Albian.

(*c*) *Completion of the Circumvallation of Gondwanaland*

All the modern fragments of Gondwanaland are bordered upon one side by fold mountain ranges; the other seaboards are fractured or monoclinal coasts with geological strikes often at variance with the modern coastline. The mountainous margins: Andes, Atlas, Himalaya, eastern Australia with New Zealand and the ranges of West Antarctica, are rejuvenated structures inherited from Gondwanaland, and the whole circumference of that parent super-continent must be deemed (King 1967) to have been defended by a mountain rampart: like a Roman camp, *vallum fossaque*. (With a mound and a ditch.)

In all sectors the direction of folding is inward, towards the centre of Gondwanaland; and a similar disposition holds for the Laurasian circumvallation. Any portion of the girdle detached from one parent continent by grazing impact with the other retains its structural import and, if attached by the other super-continent, appears in startling structural contrast with its new environment (see below).

Any valid reconstruction of Gondwanaland must therefore show unbroken circumvallation, and this du Toit's model, and the Smith & Hallam version, fail to do. Four sectors require our attention: (*a*) the loss of the mountain girdle between Turkey and the Atlas Mountains, (*b*) the eastward continuation of the girdle beyond the Himalaya, (*c*) the South American–New Zealand linkage, and (*d*) the South America–North Africa link.

(i) The loss of the girdle between Turkey and Tunis has been explained by Carey (1955) (Fig. 7), and was referred to by King (1967) in relation to an initial northward movement of Africa during the mid- to late Mesozoic disruption of Gondwanaland. The resulting collision with southern Europe, and subsequent detachment, left part of

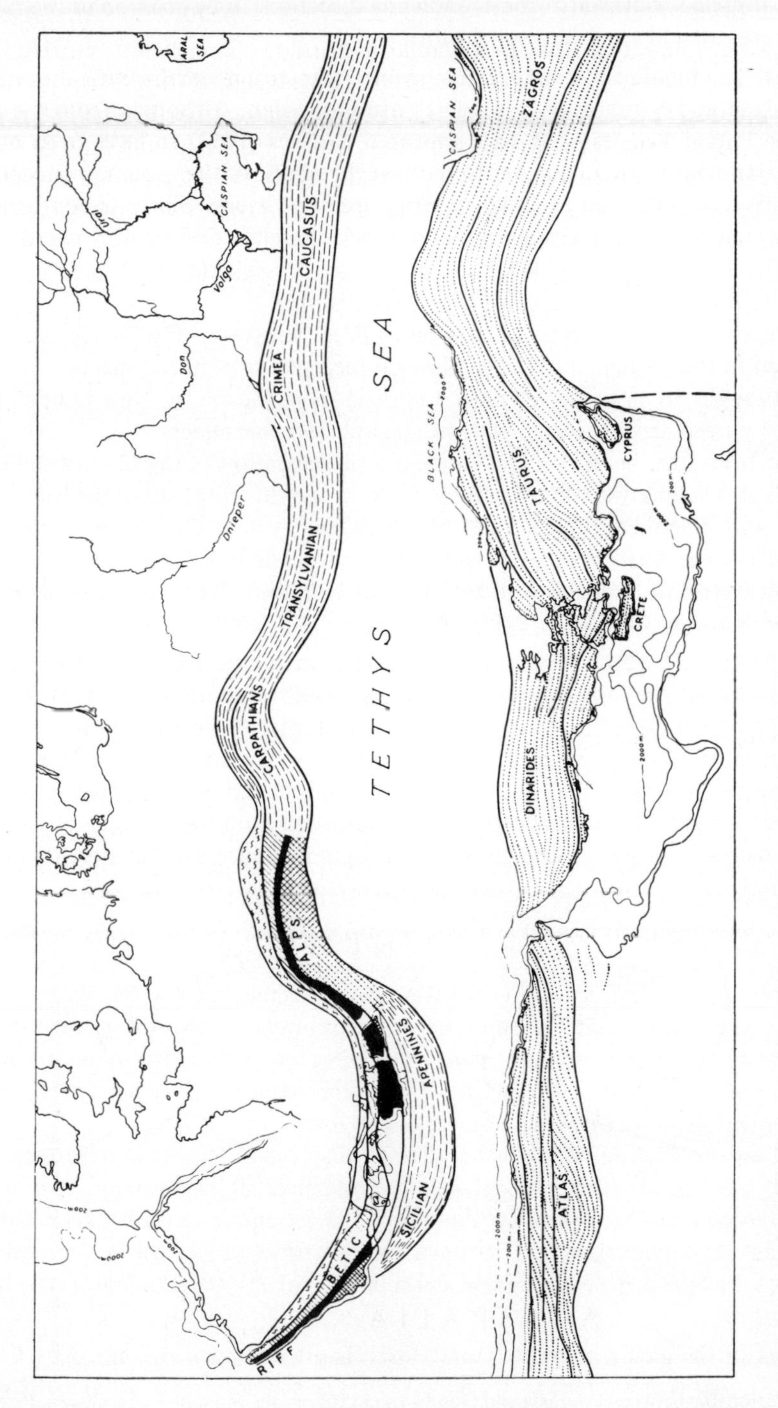

FIGURE 7 Restoration of the Gondwana circumvallation in the region of North Africa by S. W. Carey. Following collision with Laurasia, the Dinaric Alps (of Gondwana structure) have been attached to Europe.

the Gondwana girdle attached to Europe as the Dinaric Alps, the folded structure of which is still directed southward (the Gondwana direction) in opposition to the Alps, which are Laurasian and are folded northward.

(ii) The eastern Himalayan Ranges abut against the strong north-south mountain ranges of Burma and Yunnan (Fig. 6) which are Laurasian. Although, from the air, certain of the Patkai Ranges of Burma appeared to Lees (1952) to have been over-ridden by the Himalayan trend, farther to the east the whole of the mountain structure is Asian (north-south) and the Himalayan structure is no more. This is in conformity with the view that India is a Gondwanaland block that has drifted northward and driven into the Laurasian mass. Where then are we to seek the next sector of the Gondwana circumvallation? It can only be in New Guinea.

Ahmad's reconstruction of India in relation to West Australia (Fig. 5) achieves this and is adopted in the present study. The New Guinea sector is indeed spectacular, for as India is thought to have driven into Laurasia from the south, New Guinea has collided with Laurasia from the east, with remarkable tectonic effects.

From West Irian (the 'head' of New Guinea) a prolongation of the Gondwana fold girdle extends as a submarine ridge far to the west. Upon this ridge stand the islands of Misool, Obi and Kapulauan Sula, each of which has typical Gondwana structure. Westward drift of the Australian Gondwana segment has driven this prolongation into the island arc system of Indonesia, indenting it and causing the spectacular intussusception of the Sulawesi (Celebes) arc (Fig. 8).

(iii) A much broken sector of the Gondwana circumvallation may be traced from New Zealand through the ranges of West Antarctica and round the Scotia Arc to South America. When these far-strung-out fragments are replaced into a continuous line (King & Downard 1964), they span unbroken upon our reconstruction from New Zealand to South America. The rocks are appropriately Palaeozoic geosynclinal in origin and are always folded towards the original centre of Gondwanaland,

(iv) The most difficult gap to bridge on the meagre evidence available is that of the

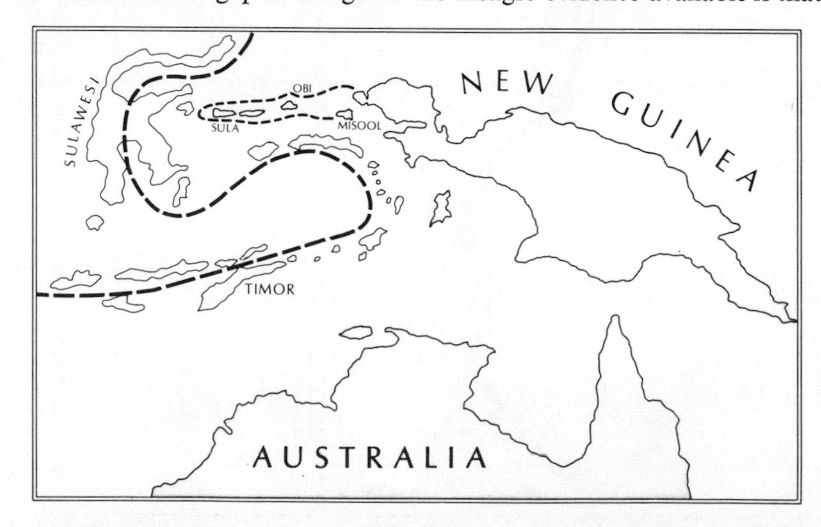

FIGURE 8 Relationship between Laurasia and Gondwana in Indonesia, indicative of collision. Regions of Gondwanaland stippled, Laurasia blank. Dotted line follows between outer and inner island arcs of Indonesia. Note how this has been turned inside out at Sulawezi (Celebes).

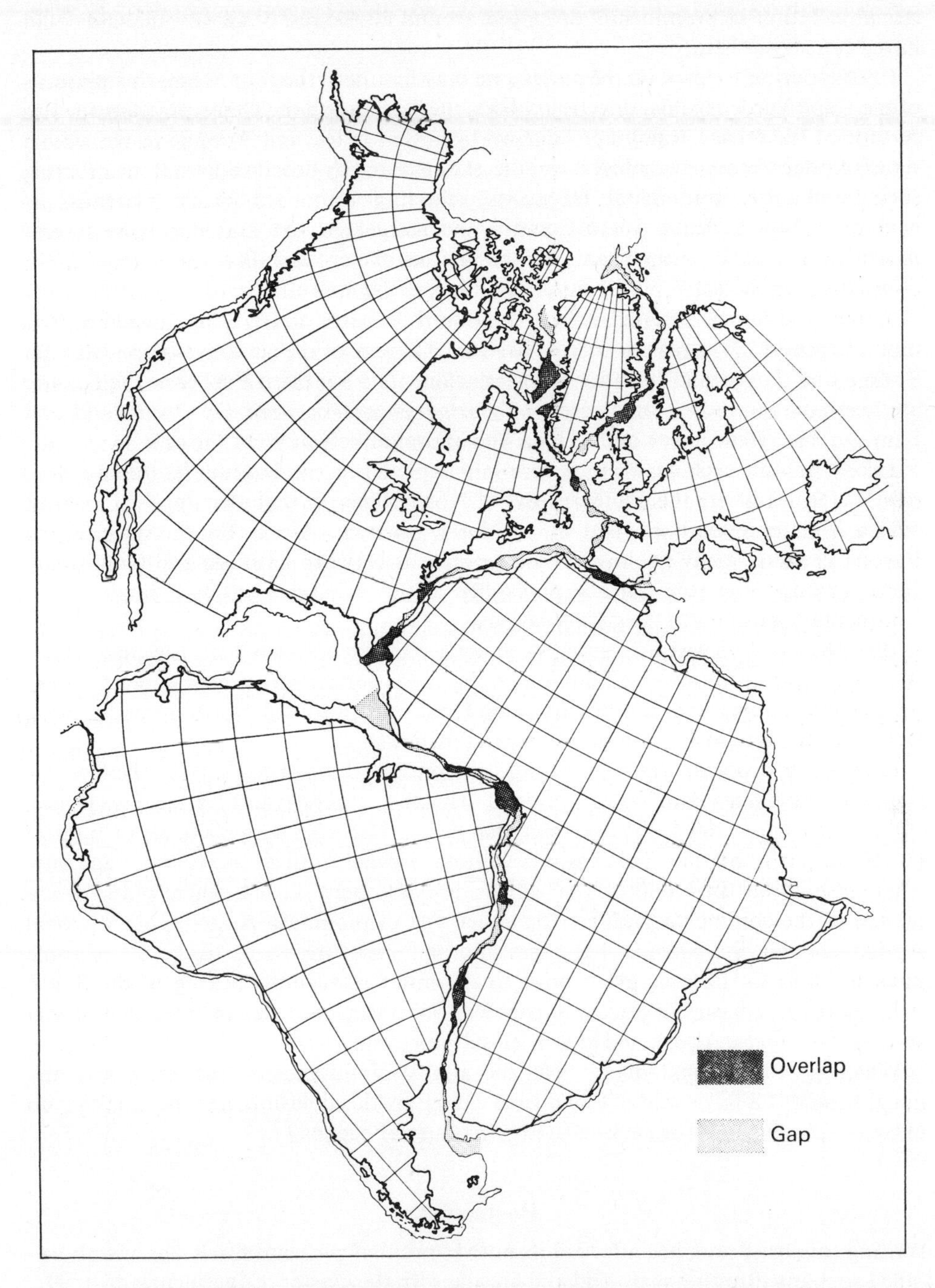

FIGURE 9 Computerized reassembly of North America, Europe, South America and Africa by Bullard, Everett & Smith, obtained by matching the outlines of the continents at the 500 fathom isobath. Discussion in the text suggests that this is not a geological entity but a 'grazing impact' of Laurasia and Gondwana.

Atlantic from Trinidad to the Atlas of West Africa, yet it is inconceivable that a circumvallation so prominently displayed around all the rest of Gondwanaland could have been absent here.

Endeavouring to 'pick up the pieces', we may first note the four submarine plateaus of the Cape Verde Islands, the Canary Isles, the Azores and the Demerara plateau. The Azores, at least, has a topography distinct from that of the mid-Atlantic Ridge, with a series of ridges trending northwest–southeast, separated by broad valleys. Remembering such indubitably continental fragments, standing upon submarine plateaus, as Malagasy, New Zealand, South Georgia, and Kerguelen, one may find these several Atlantic plateaus to be significant; but they are insufficient in bulk to fill the gap.

What made the gap? Was it simple opening of the Atlantic basin by contrary drift of Africa and South America, or was some further factor active in this area about the time of rupture? The loss to North Africa of that part of the circumvallation between Turkey and Tunis (and its subsequent attachment to southern Europe as the Dinaric arc, example (i) above), as a result of grazing impact between Gondwanaland and Laurasia, has already been quoted. Is a similar event likely in West Africa?

Likely or not, such an event is certainly depicted in the diagram by Bullard *et al.* (1965) (Fig. 9) where the southern part of North America is shown apposed to West Africa. This computerized 'best fit' of North America, Africa, South America and Europe presents many geological incongruities as between Laurasia and Gondwana. These incongruities rule out any possibility of the refit representing a single super-continental land-mass in the geological past.

But when it is viewed as temporary grazing contact between Laurasia and Gondwana shortly before major disruption of the super-continents, it seems by no means improbable. If that is so, loss of part of the Gondwana girdle to North America, in the vicinity of the Bahamas Plateau, becomes a possibility.

Geological proof of this supposition is at present difficult to find, for the foundation of the Bahamas Bank is deeply buried beneath Cretaceous and Tertiary limestones. But on the African side is the deep marginal basin of Senegal with 3000 feet of Jurassic (!) strata, partly marine. This basin is without parallel in position, form or age anywhere else about the coastlands of western and southern Africa, where oldest strata related to the continental outlines (on either side of the South Atlantic) are lagoonal Aptian and marine Albian. The basin appears, therefore, to be the result of some unusual, Jurassic tectonic event prior to the mid-Cretaceous opening of the South Atlantic oceanic basin. A grazing impact with Laurasia, recorded in other phenomena such as the Dinaric Alps, could have been such an event.

Which poses the final query: was the almost simultaneous fracturing and disintegration of Laurasia and of Gondwana caused by this collision, like the breaking up of two impinging fields of pack ice in the circumpolar oceans?

Postscript

Writing this brief note has afforded its author peculiar pleasure for it has vividly recalled so many discussions by field and fireside with the master of continental drift, 'Dr. Alex.' L. du Toit, who first proved *the fact* of drift—on visible geological data—in 1927! This achievement, without sophisticated equipment or funds, came naturally to him; for he walked—prodigiously—in humility and love of the earth.

References

Ahmad, F. (1960). Glaciation and Gondwanaland, *Rec. Geol. Surv. India*, **86,** pp. 637–674.

Ahmad, F., 1961. Palaeogeography of Gondwana period in Gondwanaland with special reference to India, *Mem. Geol. Surv. India*, **90,** 142.

Bullard, Sir E. C., Everett, J. E. and Smith, A. G., 1965. The fit of the continents around the Atlantic, *Symo. Continental Drift. Phi. Trans. R. Soc.* **258A,** 41–51.

Carey, S. W., 1955. The Orocline Concept in Geotectonics, *Proc. Roy. Soc. Tasmania.* **89,** 255–88.

du Toit, A. L., 1927. A Geological Comparison of South Africa with South America, *Carnegie Inst. Wash.* 318, 1–157,

du Toit, A. L., 1937. *Our Wandering Continents.* Oliver & Boyd, 361 pp.

Flores, G., 1970. Suggested Origin of the Moçambique Channel, *Tr. Geol. Soc. S. Africa*, **73,** 1–16.

King, L. C. 1958. The Origin and Significance of the great sub-Oceanic Ridges, *Continental Drift. Sympos. Tasmania*, 1956, 62–102.

King, L. C., 1967. *The Morphology of the Earth*, Oliver & Boyd.

King, L. C., and Downard, T. W., 1964. The Importance of Antarctica in the Hypothesis of Continental Drift, *Antarctic Geology*, ed. R. J. Adie, 727–35.

Krishnan, M. S., 1969. Continental Drift, *J. Indian Geophys. Union*, Vol. VI, 1–34.

Lees, G. M., 1952. Foreland Folding, *Quart. Journ. Geol. Soc. Lond.*, vol. 108, 1–34.

Ruttner, A., Nabavi, M. H. and Jajian, J., 1966. Geology of the Shirgesth Area (Tabas Area, East Iran), *Geol. Surv. Iran*, Rept. No. 4, 133.

Smith, A. G. and Hallam, A., 1970. The fit of the Southern Continents, *Nature*, 225, 139–44.

Stahl. F., 1911. Persien: *Handb. der Regionalen Geologie*, **5,** Heft. 8.

Stocklin, J., 1965. A Review of the Structural History & Tectonics of Iran, *Geol. Surv. of Iran*, 23 pp.

Stocklin, J., Ruttner, A., and Nabwi, M., 1954. Data on the lower Palaeozoic and Precambrian of North Iran, *Geol. Surv. of Iran*, Rept. 1, pp, 29.

Takin, M., 1971. Iranian Geology & Continental Drift, some new interpretations, *Geol. Surv. Iran.* Address.

Comment

N. SPJELDNAES

Geological Institute, The University,
Aarhus, Denmark.

We have seen a number of more or less computer based geometrical fits of continents. It may be necessary to give a warning that the fossils and sediments on these continents put certain constraints on the amount of movement you can construct. If you get coral reefs at the poles or widespread glaciations at the equator, this is a sign that there is something seriously wrong with the reconstruction.

To put the record straight—the Bullard fit for North America and Africa is absolutely incompatible with the biological and sedimentological evidence from the Lower Palaeozoic. In order to reconstruct the Palaeozoic continent, it will be necessary to change their shape, and to rotate them so much that it will be a waste of manpower and machine time to try to make a reconstruction based on their present geometrical shape. The Lower Palaeozoic stratigraphy and fossils of Iran is also incompatible with the position suggested by Professor King, and suggest that the region belonged to the north of Gondwanaland.

8.3

ROBERT S. DIETZ
NOAA, Atlantic Oceanographic &
Meteorological Laboratories
15 Rickenbacker Causeway
Miami, Florida 33149, USA

Morphologic Fits of North America/ Africa and Gondwana: A Review

Introduction

The purpose of this paper is to review morphologic fits as applied to continental drift reconstructions and especially the reconstruction of Gondwana. The appearance of a fit between continents has, of course, been the classic inspiration for drifters since the time of Wegener and even before.

Morphologic fits should be based upon the matching 2000 m or 1000 fm isobaths, as this contour is approximately equal to one half of the isostitic freeboard of continents (Carey 1958). It is understandable, however, that Bullard *et al.* (1965) obtained a better fit by matching the 500 fm (~1000 m) isobaths, as the 2000 m isobath is more apt to be displaced by sedimentary and volcano-tectonic excrescences. In those cases where the 2000 m isobath falls beyond the continental slope, my method has been to use an extrapolated or inferred former position of the 2000 m isobath by projection of the slope of the continental slope.

One can surmise many possible reasons for misfits in continental drift reconstructions. A few are indicated here, but the list is not intended to be complete. It is remarkable that the continents did break apart rather cleanly as evidenced by the close congruency between Africa and South America. Both Wegener and du Toit were satisfied with the mere semblance of a fit between these two continents and did not bother to test the fit precisely. They argued that the exigencies of geologic history were such that the margins did undergo major changes of outline. Wegener also supposed that the Mid-Atlantic Ridge and certain island groups such as the Cape Verde Islands were sialic. If this were true, they would need to be incorporated into the jigsaw-puzzle fit.

Gaps or underlaps generally are more difficult to account for than overlaps. Possibly there is some logic for making computerized fits based on some criterion which minimizes gaps at the expense of permitting considerable overlap. Basaltification or 'oceanization,' whereby the sialic craton is engulfed, foundered, and converted to sima, is a commonly offered explanation for such things as the disappearance of hypothetical Appalachia. While basins such as the North Sea basin and the Michigan basin commonly form within cratons, I doubt that the complete assimilation and disappearance of a craton is a real geologic process.

Tectonic translation by strike slip, but without detachment of the shifted block from the craton, may be one explanation for a gap. In this case the gap should be compensated by an equal area of overlap nearby. Tectonic detachment of a marginal sialic block through sea floor spreading associated with continental drift is another possible way of producing gaps. In this case the microcontinental block should be preserved in the new ocean basin. The Seychelles Islands are presumably an example.

So far as overlaps are concerned, sedimentation on the continental slope causes a post-rift accretion. This is especially true if a delta pile is imposed. The Niger delta may be regarded as the type example, for it forms a large overlap in the fit between Africa and South America. The continental slope off Louisiana and Texas offers another example. There the continental margin has been extensively built outward by the emplacement of a series of depocenters, creating a smeared deltaic pile which has migrated eastward along Texas to the present debouchment of the Mississippi River off Louisiana. Even in the absence of deltas, some sediment may be carried across the shelf and prograde the margin over oceanic crust, but such accretionary elements must be of small scale causing only minor irregularities in the fit.

Special problems and complexities are associated with Y-junctions, or triple points. The development of the Afar triangle in the Red Sea at a Y-junction between three crustal plates is a case in point. A large area of new ground apparently has been created with a complex volcano-tectonic history. It seems likely that ancient equivalents of such regions are now preserved as marginal plateaus. The Naturaliste Plateau off the southwest point of Australia may be an example, and many others could be cited.

Africa to North America Fit

The Africa to North America fit is critical to drift reconstructions, as it forms the principal join between Gondwana and Laurasia. The failure of this fit would mean (to drifters) that the continents were once configured into the supercontinents of Laurasia and Gondwana, but not into a Pangaea. The congruency as, for example, achieved by the Bullard fit at first was unconvincing to many drifters, including myself, but a re-examination of this fit supports its validity.

Our fit (Dietz *et al.* 1970), as shown in Fig. 1, is similar to the Bullard fit, but with less total mismatch; also it is somewhat looser in that the Guinea Nose (near the southern end) is placed about 200 km west of the Bullard position which overlaps about one third of Florida. Also unlike the Bullard fit, a snug congruency is attained between Morocco and North America which argues that this portion of the African craton has remained rigid and has not been appreciably crumpled or translated by Cenozoic orogeny in the Atlas mountains. A gap in the fit appears off the Ifni enclave, but this may be accounted for by the displacement of eastern Canaries island group (Lanzarote, Fuerteventura, and Concepcion Bank) of the Canary Islands as a microcontinental fragment (Dietz & Sproll 1970a). Unlike the fully volcanic western Canaries, this island group may well have a sialic foundation (e.g., Dash & Bosshard 1968). A more detailed analysis of the overlaps and underlaps in the Africa-North America fit has been presented elsewhere (Dietz & Sproll 1970b).

Bahama Platform Overlap

The major region of misfit of Fig. 1 is the Bahama platform overlap, an enormous area half the size of Texas, which must be accounted for as 'new ground' if the North

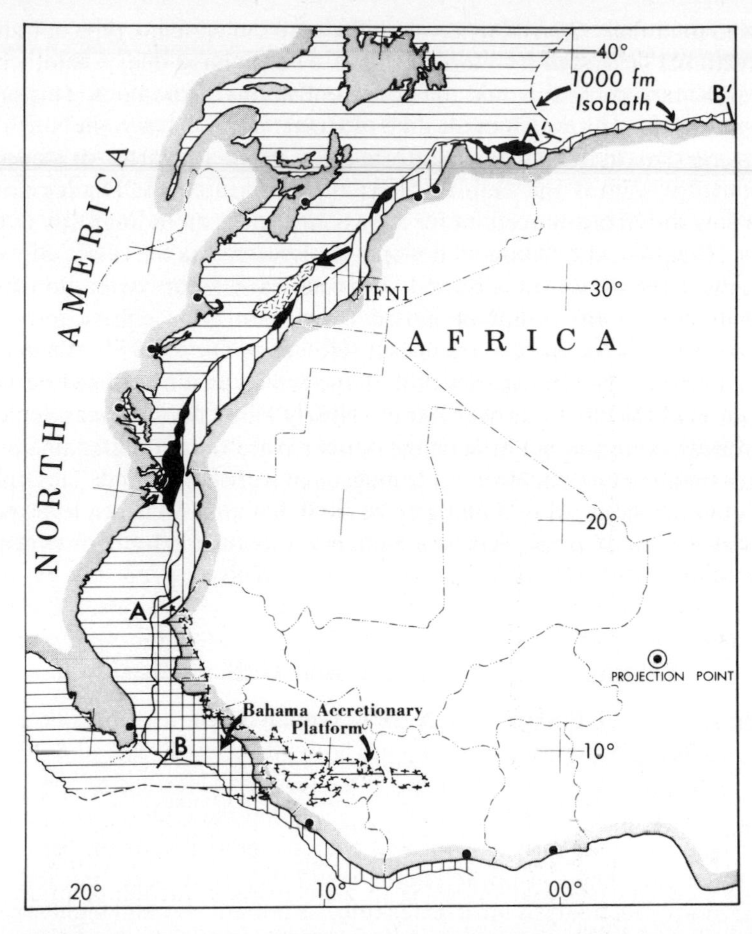

FIGURE 1 A computerized continental drift reconstruction fit between Africa and North America matching the 1000 fm isobaths and based on the criterion of smallest average misfit. The match was made by comparing the North American margin between points A to A′ with the African margin between points B to B′. The North American shelf is shown by horizontally ruled lines and the African shelf by vertically ruled lines. The overlaps are shown as opaque areas except for the Bahama platform which is cross-hatched. The clear areas are gaps or underlaps in the fit. The West Canaries block, which may have fitted into the Ifni gap, is shown in random dash pattern. (From Dietz *et al.*, 1970.)

America–Africa fit is valid. A 5 km thick section of Cretaceous and Cenozoic shallow water carbonates caps the platform, implying a history of great subsidence. Dietz *et al.* (1970) have suggested that this platform is not underlain by the usual sialic basement complex, but by a thick layer of Jurassic clastics beneath the carbonate which rests directly on oceanic crust. We supposed that a small ocean basin was filled with these clastics when North America initially split away from Africa about 190 m.y. Reef corals then attached themselves and flourished, creating upbuilding sufficient to offset subsidence so that the platform has maintained a sea level freeboard. This interpretation obviates the overlap problem.

A modification of this evolution now seems in order whereby it is proposed that an Iceland-like 'hot spot' rose 195 m.y. as a lava plume from the deep mantle at the Bahamas site. It served to initially drive North America away from Africa, a type of motive force for plate tectonics suggested by Morgan (1971). Thus, a large pile of alkalic basalts as well as sedimentary detritus would underlie the Bahama carbonates. This hot spot is inferred by the strike of lower Jurassic dike swarms (180 to 193 m.y.) in North America which converge toward the Bahama platform (May 1971). In North Africa similar dike swarms converge toward the Guinea Nose, the conjugate point under drift reconstruction. This would nicely account for the asymmetrical spreading early in the opening history of the Atlantic Ocean. A hot spot apparently has the effect of 'overprinting' the symmetrical splitting of injected dikes as occurs with normal sea-floor spreading. Thus, the mid-ocean rift is not permitted to migrate. This effect can be observed today for Iceland and for the Galapagos rift (Holden & Dietz, 1972). When the North Atlantic first opened, North America moved away from an almost fixed Africa so that the hot spot held the rift fixed near Africa. Nearly all of the lava was spilled on the North American plate and but little on the African plate. Thus the Bahama platform is much larger than the Guinea Nose, its conjugate area of new ground. This history also would account for the very rapid filling up of the Bahama basin to sea level, permitting the early attachment of coral reefs where up-growth would offset subsequent subsidence.

Closing of the Caribbean and Gulf of Mexico

Another difficulty of the Bullard fit is the large overlap of southern Mexico and Central America onto South America. A solution has been offered by Freeland & Dietz (1971), whereby this overlap is obviated by using plate tectonic rotations of the overlapping areas so as to close the Gulf of Mexico. We suggest that the Gulf of Mexico was rather quickly blocked out in the early Jurassic with the Luanne salt being the basal formation as a deep-water salt. The Caribbean region would then have been blocked out in the lower Cretaceous. Our analysis must be regarded as preliminary and highly speculative; however, it may offer the proper type of scenario to explain the geotectonic evolution of the region. The Caribbean and Gulf of Mexico would then be new ocean basins created after the breakup of Pangaea.

South America/Africa Fit

This fit remains the most convincing of all fits and it is supported, for example, by the matching of Precambrian provinces of similar radiometric age (Hurley 1971). Overlaps caused by the Niger delta and volcanic excrescences off the Walvis and Rio Grande ridges serve only to support the fit, as these are clearly areas of new ground. Local areas of salt domes along both Africa and South America are presumably also strips of new ground so that future studies of the fit should probably subtract these areas. The Benue Trough, trending northeast through Nigeria, was presumably an incipient, but abortive, rift which opened in the Lower Cretaceous and closed in the Upper Cretaceous. About 5 km of closure is indicated by the folding within the trough which is much less than the initial spreading. It would seem, therefore, that some small eastward rotation of the stem of Africa may have occurred with respect to the congruent margin of South America.

Reconstructing the Indian Ocean

Unlike the Atlantic, almost a pure rift ocean, the Indian Ocean is a mixed ocean basin —both a rift ocean and a subduction ocean. And, rather than being created by simply splitting a craton, the re-arrangement of Tethys is involved. The reconstruction, or 'closing,' of the Indian Ocean remains far from being solved. However, Smith and Hallam (1970) have made an admirable attempt, and a similar solution has been offered by Dietz & Holden (1970). Dietz & Sproll (1970c) have presented a fit of Antarctica against Africa which avoids the problem of positioning the horn of Antarctica (the Antarctic peninsula) under the assumption that it did not exist in the Triassic, at least not in its present position. The fit achieved is good in the sense that the total mismatch is small, but uncritical in that the congruent margins are both nearly straight. It also assumes that Madagascar was initially located in the Tanzania position, while apparently there was growing geologic evidence that it should be placed in the Mozambique position. From a morphologic viewpoint, Madagascar fits poorly in both positions. The adoption of the 2000 m isobath solves a major problem in the Smith & Hallam reconstruction, as it reveals that the Weddell Sea, which they show as

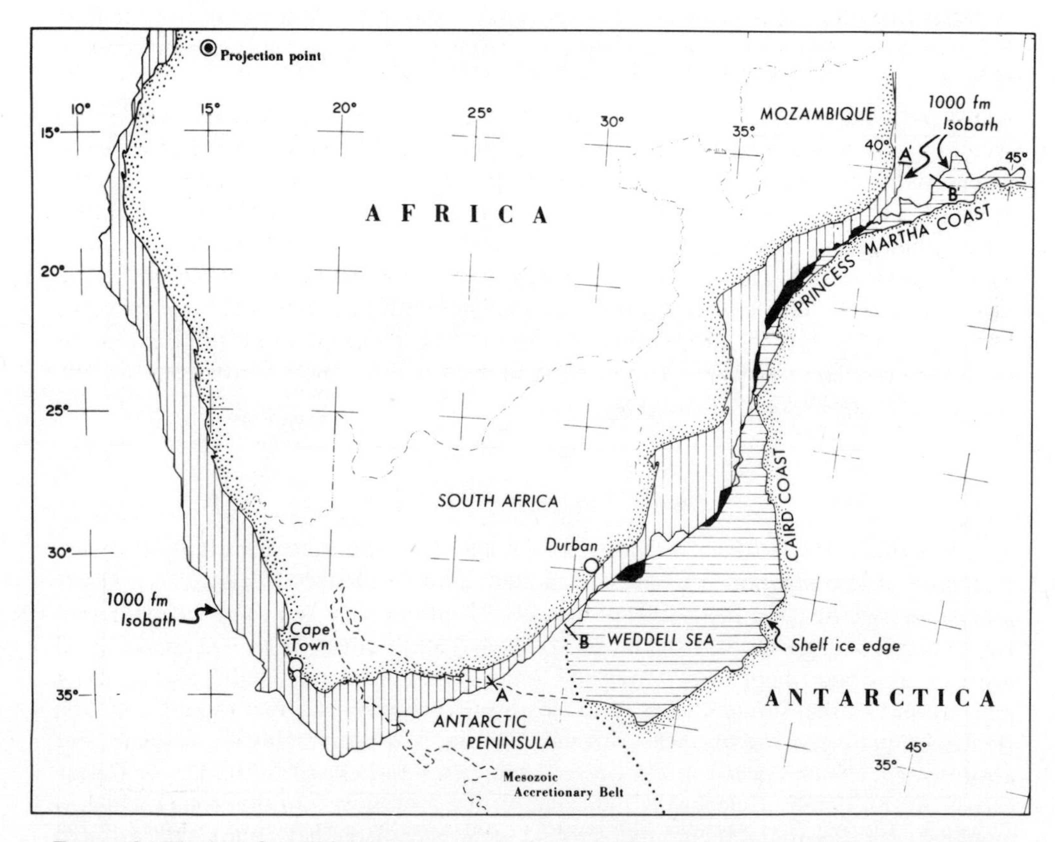

FIGURE 2 The best fit position between Antarctica and Africa, a continental drift reconstruction. Departures from congruency are shown as overlaps (opaque areas) or as underlaps (clear areas). (From Dietz & Sproll, 1970c.)

a 'gap', can be considered to be a deep shelf-sea and therefore part of the Antarctica cratonic block.

India fits equally badly in several positions (Dietz, in press) against Antarctica, Australia or Madagascar, but all of these fits can be accommodated by 'major surgery', e.g., removing the Exmouth Rise for the fit of Australia and India or moving Ceylon for the fit of Antarctica against India. My view is that the east coast of India belongs against Antarctica, but this is based upon interpreting the strike of major tectonic features on the ocean floor as flow lines for the drift of India. The idea (Dietz & Holden, 1971) that the Wharton Basin may be a remnant of Sinus Australia (a southern bay in ancient Tethys) and hence remnant Paleozoic ocean crust must now be abandoned, as it is not supported by the recent JOIDES drilling which suggests that this ocean floor is Cretaceous.

Australia/Antarctica and New Zealand

There has always been a general agreement on the fit of Australia with respect to Antarctica which can be assigned almost by inspection. This, however, has been given more precision by Sproll & Dietz (1969), who found that the best fit position is also blocked in the three marginal plateaus, the Iselin, Bruce and Naturaliste plateaus, which suggests that effusion of lavas over hot spots like today in the Afar triangle of the Red Sea may have played a role in the creation of triple junctions and subsequent breakup (Fig. 3).

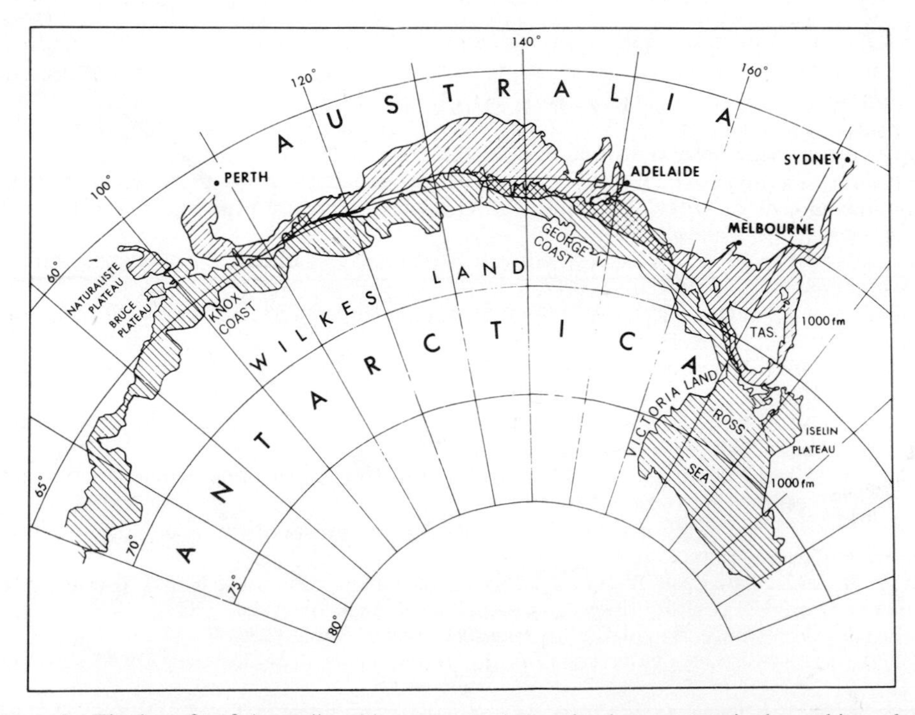

FIGURE 3 The best fit of Australia with respect to Antarctica by computerized matching of the 1000 fm isobaths. Ruled line pattern indicates continental shelf and slope from shoreline (inner contour) to the 1000 fm isobath (outer contour). Overlap areas are cross-ruled and underlap areas are blank. (From Sproll & Dietz, 1969.)

The New Zealand plateau appears to have been positioned against eastern Australia and West Antarctica. The Tasman Sea opened up first, commencing 85 m.y. and ending 60 m.y. The cessation of opening was caused by the rifting of Australia away from stationary Antarctica beginning 60 m.y. and continuing to the present day at about 5.4 cm/yr (Griffiths, 1971, Griffiths & Varne, 1972, and Weissel & Hayes, 1971).

It is interesting to note that, if the Tasman Sea continued to open while Australia moved away from Antarctica, a quadruple junction would have been involved. It would seem that in plate tectonics such junctions are prohibited and only triple junctions are permitted. It appears as if the Earth adjusts itself to maintain only six main spreading axes and six *major* plates with eight triple junctions at any one time—and thus to adopt the symmetry of a 'cube' or, more accurately, a spherical hexahedron.

References

Bullard, E. C., Everett, J. E., and Smith, A. G., 1965. The fit of the continents around the Atlantic. *In* Symposium on continental drift, *Phil. Trans., R. Soc. A*, **258,** (1088), 41–51.

Carey, S. W., 1958. A tectonic approach to continental drift. *In* Continental Drift, a Symposium, S. W. Carey, ed., Univ. of Tasmania Press, Hobart, 177–355.

Dash, B. P., and Bosshard, E., 1968. Crustal studies around the Canary Islands, Rep. 23rd Int. Geol. Congr., Prague, **1,** 249–260.

Dietz, R. S. (in press). Antarctica and continental drift; *In* Proc. SCAR Symposium on Antarctica Geology and Solid Earth Geophysics, Oslo 1970.

Dietz, R. S., and Holden, J. C., 1970. Reconstruction of Pangaea: Breakup and dispersion of continents, Permian to Present, *J. Geophys. Res.*, **75,** (26), 4939–4956.

Dietz, R. S., and Holden, J. C., 1971. Pre-Mesozoic oceanic crust in the eastern Indian Ocean (Wharton Basin)? *Nature*, **229,** (5283), 309–313.

Dietz, R. S., Holden, J. C., and Sproll, W. P., 1970. Geotectonic evolution and subsidence of Bahama platform, *Bull, Geol. Soc. Amer.*, **81,** (7) 1915–1928.

Dietz, R. S., and Sproll, W. P., 1970a. East Canary Islands as a microcontinent within Africa-North America continental drift fit, *Nature*, **226,** 1043–1045.

Dietz, R. S., and Sproll, W. P., 1970b. Overlaps and underlaps in North America to Africa continental drift fit. *In* vol. 1 Geology of Atlantic Continental Margins, 143–151, H. M. Stationery Office, London.

Dietz, R. and Sproll, W., 1970c. Fit between Africa and Antarctica: a continental drift reconstruction, *Science*, **197,** 1612–1614.

Freeland, G. L., and Dietz, R. S., 1971. Plate tectonic evolution of Caribbean-Gulf of Mexico region, *Nature*, **232,** 20–23.

Griffiths, J., 1971. Reconstruction of the south-west Pacific margin of Gondwana, *Nature*, **234,** 203–207.

Griffiths, J., and Varne, R., 1972. Plate tectonic model for evolution of Tasman Sea, Macquarie Ridge and Alpine fault, *Nature*, **235,** 83–86.

Holden, J. C., and Dietz, R. S., 1972. Galapagos gore, NazCoPac triple junction and Carnegie/Cocos ridges, *Nature*, **235,** 266–269.

Hurley, P., 1971. The confirmation of continental drift. *In* Continents Adrift, 57–67, Freeman and Co., San Francisco.

May, P. R., 1971. Pattern of Triassic-Jurassic diabase dikes around the North Atlantic in the context of predrift position of the continents, *Geol. Soc. Amer. Bull.*, **82** (5), 1285–1291.

Morgan, J., 1971. Convection plumes in the lower mantle, *Nature*, **230,** 42–43.

Smith, A., and Hallam, A., 1970. The fit of the southern continents, *Nature*, **235,** (5228), 139–144.

Sproll, W. P., and Dietz, R. S., 1969. Morphological continental drift fit of Australia and Antarctica, *Nature*, **222,** 345–348.

Weissel, J., and Hayes, D., 1971. Asymmetric sea floor spreading south of Australia, *Nature*, **231,** 513–521.

Comment

J. V. AVIAS
Laboratoire de Geologie
Université de Montpellier, France

I would like to point out the close agreement of palaeofaunal evidence with Dietz's reconstruction of land and sea for the Caribbean and Mexican areas in early Mesozoic times.

(1) Liassic ammonoids (family Psiloceratidae) are well known in Pacific side of both North and South America but are unknown in early Jurassic times (Hettangian) showing that this fauna, coming from the west, had not yet reached the Caribbean region, because of the existence of a 'Panamean' barrier (Avias 1966).

(2) In Sinemurian times (Avias 1966, Freneix 1956, Guerin-Franiatte 1966) Arietitidae faunas are known only on the east side, not on the Pacific side, of the same region. This, together with the close affinities with European faunas, shows that:

(*a*) the sea communication and faunal migration was from the proto-Atlantic ocean sea to the east.

(*b*) the same 'Panamean' barrier was still in existence.

(3) Pacific-Atlantic sea communication (2–3) occurred in this area in post Lower Liassic times.

References

Avias, J. V., 1966. Sur le probleme des relations entre les faunes du Lias inferieur de Nouvelle-Caledonie, d'Amerique et d'Europe. C. R. XX° *Congr. Geol. Intern. Mexico* 1957, 1–5 1966.

Freneix, S., 1956. Contribution a l'etude des Lamellibranches du Cretace de Nouvelle-Caledonie. Sc. de la Terre. T. IV, 153–208.

Guerin-Franiatte, S., 1966. Ammonites du Lias inferieur de France. Psilocerataceae: Arietitidae. C. N. R. S. Paris 1966.

8.4

P. E. KENT
British Petroleum Co. Ltd.

East African Evidence of the Palaeoposition of Madagascar

The former position of Madagascar was much discussed at the 1972 NATO Conference, and the evidence from the East African coast is directly relevant. It is the writer's view that a former position alongside Kenya (or Tanzania) is very unlikely, but that the island is closely related to Mozambique—i.e. that it has not moved very far in the course of crustal adjustments. A short note on the history of the western Indian Ocean was submitted to *Nature* before the Conference (Kent, 1972); the more detailed basis for the belief in Madagascar's near-autochthoneity is now given here.

1. *The Indian Ocean Coast in Tanzania and Kenya*

It is known from surface exposures and from deep borings that early in the Jurassic a marine regime was instituted on the East African coast, the evidence extending from 5° N in Somalia to 9° S in Tanzania. Locally at least this invasion took place by the early Lower Jurassic (Lias); it was general by the Middle Jurassic (Bajocian–Bathonian) and persisted from then onwards.

In Tanzania the detailed evidence from the Jurassic, Cretaceous and Tertiary is consistent with this being an open coast, facing the Indian Ocean or its precursor, rather than a narrow seaway (Kent *et al.*, 1971).

For the Kenya coast the evidence can be listed chronologically as follows:

(*a*) Bajocian shales with ammonites (fully marine) are known south of Mombasa.

(*b*) Callovian and later Jurassic beds are normally marine and well exposed in Mombasa harbour and in railway cuttings nearby; the earlier beds are relatively shallow water and show collapse/slump structures related to basin edge conditions—possibly to a faulted coast. Later beds (Miritini beds, etc.) are marine ammonite-bearing shales lacking a shallow water fauna; their thickness is 30–70 m or more.

(*c*) A deeper water facies of Upper Jurassic (Kimmeridgian) is known 50 km further north at Kilifi; this shows large scale contemporary sliding which suggests deposition on the upper continental slope.

(*d*) Shelf facies Lower Cretaceous is exposed at Freretown on the southern Kenya coast.

(*e*) In the Lamu embayment further north a normal Upper Cretaceous development is followed by a Tertiary succession some 3000 m thick; both Cretaceous and Tertiary attenuate inland and become fully marine towards the coast.

(*f*) Structures which controlled Cretaceous and later stratigraphy are known to extend 100 km offshore, to the edge of the continental shelf near Malindi.

The picture is thus of a east-facing coast in the Middle Jurassic with a deeper water facies of Jurassic succeeding this, and with an embayment deepening seawards well documented from the Cretaceous and Tertiary (Walters & Linton, 1972). In Kenya, as in Tanzania, there is a consistency of history from the Middle Jurassic onwards, with no reason to postulate any major change in the regime of a kind which might be associated with removal of Madagascar from contiguity with this coast.

2. *The Palaeoposition of Madagascar*

The original evidence for restoring Madagascar to a position close to the Kenya coast was based on the broad similarity of the Karroo successions as then known. On the evidence of the outcrops this was a valid standpoint; it became less strong when one considered that the sequence of coarse arkosic lower beds, a non-marine to paralic middle series and massive quartzose sandstone upper beds, is one found in several of the Karroo basins; it was probably climatically controlled and does not necessarily imply propinquity. The case for a fit of the Karroo of Madagascar with that of any part of the mainland was however greatly weakened, if not eliminated, by the documentation of transition of the continental Karroo of Madagascar into a marine facies at several levels, in both western and northern basins (Besairie, 1952). This is a feature without parallel in East Africa.

There are several places where the twin Majunga/Morondava Karroo-Mesozoic basins, with their dividing *schwelle* at Bemolanga–Cap St. André, could fit neatly against a comparable southeast–northwest feature on the African coast. Wherever this is done, however, there is the problem of the absence on the mainland of any representative of marine facies of the Permian, particularly the rich marine beds of the northern basin, with their fusulinids and ammonites. This development is perhaps most comfortably placed within range of Australia, and it independently favours a former southerly position.

In the Triassic and Jurassic the evidence is equivocal. The indicated marine connection from Arabia through Somalia and Kenya to Madagascar in the Lias would be shortened by a more northerly position for the island, but the extension of the Bajocian/Bathonian limestone south through Tanzania shows that a seaway did in fact extend a long way in this direction soon afterwards.

The most specific evidence on palaeolatitude comes from the Lower Cretaceous. The westerly basins of Madagascar were blanketed by basalts, which vary in date slightly from place to place but are broadly Cenomanian to Turonian (Brasseur *et al.*, 1959). No representative of such a series is known in Somalia, Kenya or Tanzania, but there was a similar closely contemporaneous episode in Mozambique. It is logical to regard the two areas of extrusion as closely related. They may be linked with the minor plate separation in the Mozambique channel which Green has recently postulated (1972).

As has been commented elsewhere, it seems illogical to move Madagascar south when India was moving north. It is hard to conceive a major Cretaceous movement at a time when the East African coast was particularly quiescent and free from tectonic disturbances. It seems unsound to separate the only two major Cretaceous lava fields on this coast—Madagascar and Mozambique—in favour of a questionable correlation with the Karroo of Kenya.

The weight of modern evidence is thus much on the side of a position for Madagascar near Mozambique from the Permian onwards, with a modest degree of shift towards the east-northeast on the evidence of tensional movements contemporary with the basalt extrusion in Middle Cretaceous times.

References

Besairie, H., 1952. Les Formations du Karroo a Madagascar, Sympos. sur les Ser. de Gondwana, XIX Cong. Geol. Int., Alger.

Brasseur, R. and others, 1959. Préreconnaissance Pétrolière du Bassin du Majunga (Madagascar), Inst. Franc. de Pet. XIV.

Kent, P. E., 1972. Mesozoic History of the East Coast of Africa. *Nature*, **238,** 147–8.

Kent, P. E., Hunt, J. A. and Johnstone, D. W., 1971. The Geology and Geophysics of Coastal Tanzania, Geophys. Paper 6, Inst. Geol. Sci. London.

Walters, R. and Linton, R. E., 1972. The Sedimentary Basin of Coastal Kenya, 24th Int. Geol. Cong. (Montreal).

Comment

D. H. TARLING
The University, Newcastle upon Tyne, England

The evidence outlined by Kent (Section 8.4 and 8.10) confirms the difficulty, if not impossibility, of a northerly position for Madagascar prior to the break-up of Gondwanaland. Most of the evidence previously used for a northerly origin (du Toit, 1937), such as the critical matching of Madagascan Jurassic shorelines with those in Somalia, has now been shown to be of dubious value as Jurassic shorelines clearly extended the length of eastern Africa (Dingle & Klinger, 1971). However, the fossil biota during most of the Phanerozoic clearly requires a close connection between Africa and Madagascar and there is increasing evidence for a southerly derivation of Madagascar (Flores, 1970; Green, 1972), probably from the Mozambique region. The original location of Madagascar is, however, of more than local interest as it is critical to any reconstruction of Eastern Gondwanaland (Tarling, 1971) and a southerly origin rules out several widely accepted models, particularly those of du Toit (1937), and Smith & Hallam (1970).

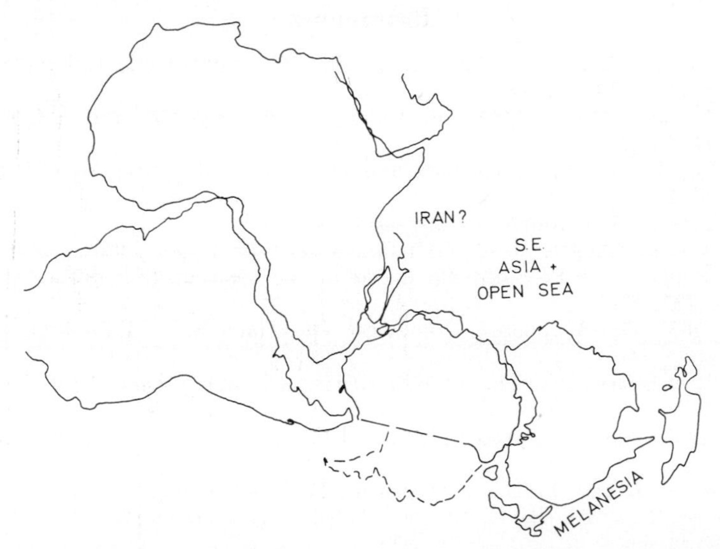

FIGURE 1 An alternate model of Gondwanaland in the Late Palaeozoic–Early Mesozoic.

Testing of any reconstruction of the continents is extremely difficult and the restricted fossil evidence from the Gondwanan continents prevents any unique interpretation being given on this data. The use of sea-floor spreading and magnetic anomaly matching allows reconstructions to be determined with reasonable precision for the last 75 m.y., i.e. for the period when reversals of the geomagnetic field were frequent, but the paucity of reversals in the Cretaceous restricts the construction of earlier models for the Indian Ocean (McKenzie & Sclater, 1971; Laughton *et al.*, 1973). At this time, 75 m.y. ago, Australia and Antarctica were clearly united but

Madagascar still had its present relationship to Africa. Palaeomagnetic tests have shown that the du Toit model is a very much more realistic one than that showing the continents in their present position (Creer, 1964, 1965), but much of the data is still of uncertain reliability and apparent discrepancies in the more reliable data have been explained in terms of multipole components of the geomagnetic field (Briden *et al.*, 1971; Creer, 1972), at least in the late Palaeozoic–Early Mesozoic. In view of the increasing evidence for a southerly origin of Madagascar, a new geometric reconstruction (Fig. 1) was attempted and tested against the palaeomagnetic evidence (Tarling, 1972). This showed that, for Mesozoic observations, the scatter of palaeomagnetic poles was significantly less for this model than for the Smith & Hal am/ du Toit model, and the very much less reliable Palaeozoic data was only slightly, and not statistically significantly, less scattered on the Smith & Hallam model.

Applying Occam's razor, this suggests that the new model for Gondwanaland is more realistic, in general terms, although this is certainly not proven and, at best, only affords a basis for further testing. It does, however, show that there is still no 'unique' Gondwanaland (McElhinny & Luck, 1970) and that it is premature to evaluate the magnitude of past geomagnetic components when the basic continental reconstruction has not yet been established.

References

Briden, J. C., Smith, A. G. and Sallomy, J. T., 1971. The geomagnetic field in Permo-Triassic time, *Geophys. J.*, **23,** 101–17.

Creer, K. M., 1964. Palaeomagnetic data and du Toit's reconstruction of Gondwanaland, *Nature*, **204,** 369–70.

Creer, K. M., 1965. Palaeomagnetism and the time of the onset of continental drift. *Nature*, **207,** 51.

Creer, K. M., 1972. This conference but not published.

Dingle, R. V. and Klinger, H. C., 1971. Significance of Upper Jurassic sediments in the Knysna Outlier (Cape Province) for timing of the break-up of Gondwanaland, *Nature Phys. Sci.*, **232,** 37–8.

du Toit, A., 1937. Our Wandering Continents. Oliver and Boyd, Edinburgh and London, 366pp.

Flores, G., 1970. Suggested origin of the Mozambique Channel, *Trans. Geol. Soc. S. Afr.*, **73,** 1–16.

Green, A., 1972. Sea-floor spreading in the Mozambique Channel, *Nature Phys. Sci.*, **236,** 19–21.

Laughton, A. S., Sclater, J. G. and McKenzie, D. P., 1973. Section 2.5, Vol. **I,** 203-12.

McElhinny, M. W. and Luck, G. R., 1970. Palaeomagnetism and Gondwanaland, *Science*, **168,** 830–2.

McKenzie, D. P. and Sclater, J. G., 1971. The evolution of the Indian Ocean since the Late Cretaceous, *Geophys. J.*, **24,** 437–528.

Smith, A. G. and Hallam, A., 1970. The tilt of the southern Continents, *Nature*, **225,** 139–44.

Tarling, D. H., 1971. Gondwanaland, palaeomagnetism and continental drift, *Nature*, **229,** 17–21.

Tarling, D. H., 1972. Another Gondwanaland, *Nature*, **238,** 92–3.

8.5

W. HAMILTON
US Geological Survey
Denver, Colorado

Tectonics of the Indonesian Region*

The exceedingly active present seismicity, volcanism, and tectonism of the Indonesian region are products of motions between lithospheric plates. Throughout most of Cenozoic time, the Australian–Indian Ocean plate has been moving northward relative to the Asian plate, and the Pacific plate has been moving relatively west-northwestward (e.g. Heirtzler *et al.*, 1968). The Indonesian–Melanesian region represents a broad 'soft' boundary zone by which the obliquely convergent motion of the three megaplates has been accommodated. The behaviour of the region has been characterized by constantly changing arrays of platelets, subduction zones, transform faults, migrating arcs, and oroclines.

The continuous, curving Andaman–Sumatra–Java–Timor–outer Banda–Seram subduction system now bounds the Indonesian region against the Indian Ocean–Australian plate. The Benioff zone of mantle earthquakes dipping under Indonesia from the trench has gentle dips at shallow depths but steepens downward in most sectors (Fig. 1). Volcanoes form a magmatic arc lying mostly above that part of the Benioff zone which is 100 to 200 km deep. There is a continuous outer-arc ridge between magmatic arc and trench. This ridge is wholly submarine in the Java sector, where only thin pelagic sediments veneer the Late Cretaceous and early Tertiary oceanic plate being subducted, but it rises to produce many islands to both east and west where voluminous sediments are being subducted. These islands consist largely of subduction melanges probably formed mostly in Miocene time. The distance from ridge to trench represents the amount of post-middle Miocene accretion to the hanging wall of the subduction system as materials were scraped off against it from the underflowing oceanic plate. In the Sumatra–Nicobar–Andaman sector, the terrigenous clastic sediments of the eastern lobe of the Bengal abyssal fan are being subducted. Around the Banda arc sector, shelf sediments of Australia and New Guinea are being subducted. Thus, in both eastern and western sectors of the

* Since early 1970, I have been making a tectonic analysis of the Indonesian region as part of a programme of assistance, supported in part by the US Agency for International Development, for the Government of Indonesia and the Economic Commission for Asia and the Far East. The work has included the study of most of the published primary geologic and geophysical literature on the region, the integration of data given generously by many petroleum companies, and the study of marine seismic-reflection profiles, particularly the voluminous records obtained by the Lamont-Doherty Geological Observatory and kindly made available to me by Maurice Ewing. A preliminary version of my in-progress tectonic map was released recently for limited distribution (Hamilton, 1972). The present note summarizes some of the conclusions reached in the course of the study. The descriptive basis for this summary is contained in more than a thousand published papers; in this note, I have cited only a few papers that develop conclusions particularly important for plate tectonics. Publication authorized by the Director, US Geological Survey.

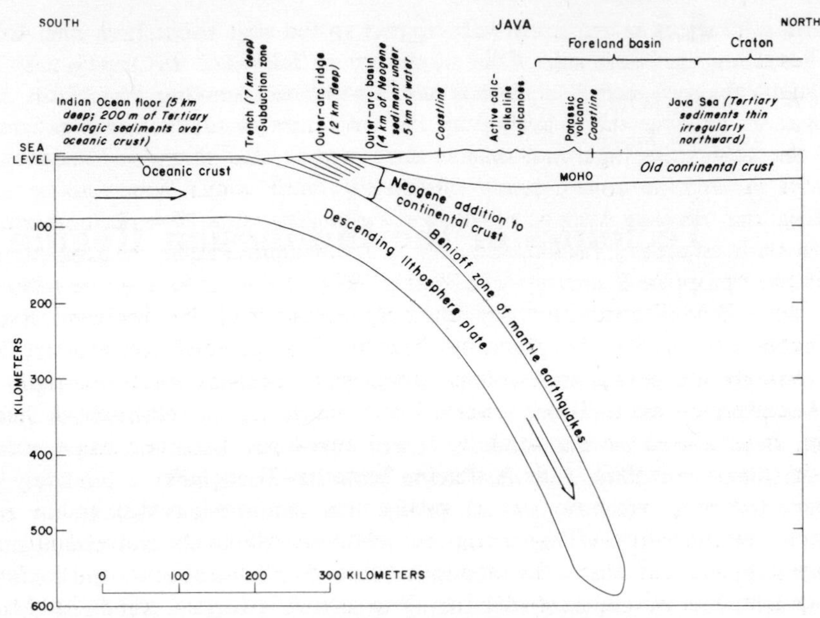

FIGURE 1 Generalized section through central Java, along the meridian of 111° E, showing the relationship of tectonic and magmatic zones to the Benioff seismic zone (which is located from data by Fitch, 1970).

subduction system, voluminous sediments yield voluminous low-density melanges, which buoy up the ridge. Behind the ridge is a broad submarine basin, mostly about 2 km deeper than the outer-arc ridge and filled with Neogene sediments which are generally 3–7 km thick along the basin axis.

The Holocene volcanic rocks erupted above the present Benioff zone show the usual correlation of ratios of potassium to silicon with depth to the seismic zone (Hatherton & Dickinson, 1969). However, bulk compositions of the volcanic rocks are controlled by the type of crust through which the magmas rise from the Benioff zone. The rocks are mostly silicic and intermediate in Sumatra, where the magmas rise through old continental crust; intermediate in Java, where the crust is continental but relatively thin, mafic, and young; and intermediate and mafic around the rest of the arc, which is closely flanked on both sides by oceanic crust. There is no apparent correlation between the thickness of sediments being subducted and the intensity of magmatism above the Benioff zone; very thick abyssal clastic sediments go into the north Sumatra–Nicobar–Andaman sector of the subduction system, where volcanism is of minor volume, whereas only thin pelagic sediments enter the Java sector, where volcanism is intense.

Other subduction systems are now active farther north in the Indonesian region. The Mindanao trench extends south to an abrupt end off northeastern Halmahera, probably at a left-lateral transform fault trending east-southeastward toward New Guinea. A short trench trends southward along the west side of central Halmahera. Another trench lies along the east side of northeasternmost Sulawesi and the Sangir island arc that trends from there to south-central Mindanao. This Sangir subduction system is imbricate in the north with respect to the Mindanao–northeast Halmahera Benioff zone, and convergent in the south with respect to the west Halmahera one. Another trench lies along the north side of the north arm of Sulawesi; this trench ends at the west against the very active left-lateral north-northwest trending Palu fault, and at the east at an orocline that is rolling up the north arm as subduction flips from one side to the other. Geologic and seismic data suggest that along northeastern New Guinea, a subduction zone dipping moderately south under New Guinea is now overriding one dipping steeply north under the Schouten–New Britain arc. The Mariana arc is migrating Pacificward even as it consumes Pacific crust, opening new ocean in the Philippine Sea in its wake (Karig, 1971). Most of these active subduction systems have Benioff zones that dip relatively steeply, and the landward slopes of the trenches are marked by platforms bearing thin, perched sedimentary basins rather than by the broad and strongly developed outer-arc ridge-and-basin pairs of the Andaman–Sumatra–Timor system. The trench along the north side of Sulawesi, however, is paralleled on the south by a well-developed outer-arc basin, although this basin is narrower than is the Andaman–Sumatra–Timor one.

Inactive trenches, not now sites of subduction, are also plentiful in the region. A recently inactivated trench lies along the northwest side of the Sulu Islands, and a more ancient one lies along the northwest side of Palawan and nearby Borneo. The Yap trench is now almost inactive—Yap Island is formed mostly of Miocene melange—and the Mariana festoon may be sliding southward past the Yap trench as it expands and migrates Pacificward. The Yap ridge bears a similar festoon relationship to the earlier Tertiary Palau ridge and trench. Yap and Palau systems continue northward, past the festoon junctions, as remnant arcs representing old positions of the Mariana system (Karig, 1971).

The positions and polarities of older subduction systems can be deduced from onshore geology. Many subduction zones are recorded by tectonic melanges—vast shear zones which typically contain such indicators as ophiolites (fragments of oceanic crust and mantle), abyssal sediments (carried on the conveyor-belt plates), and glaucophane schist and other high-pressure metamorphic rocks. Terrigenous clastic sediments, where available in quantity, are however the dominant component of some of these melanges. Magmatic arcs formed above ancient Benioff zones are now represented by andesitic and basaltic rocks where formed in island arcs, or by dacitic, rhyolitic, and granitic rocks where formed in continental plates.

The present subduction and magmatic-arc system trending eastward from Java has only recently reached its present position, probably by migrating away from Sulawesi. The Late Cretaceous and very early Tertiary ancestral subduction zones are exposed in several areas in the west half of Java, and can be traced northeastward across the Java Sea to southeastern Borneo. The subduction melanges of the southwest arm of Sulawesi likely formed as part of this same system, before the opening by rifting of Makassar Strait. Subsequent early and middle Tertiary subduction built the melanges of the east arms of Sulawesi. The magmatic rocks formed above the Benioff zone of this southeast-migrating subduction system lie in southern Sumatra(?), beneath the northwestern half of the Java Sea, in south-central Borneo(?), and in the southwest and north arms of Sulawesi.

The south half of New Guinea has been part of the Australian landmass since Precambrian time. The Tasman orogenic terrain—itself largely a product of subduction of oceanic material relatively westward beneath Australia during Palaeozoic time (Solomon & Griffiths, 1972)—trends northward in the subsurface and in infrequent outcrops to medial New Guinea, where it was apparently truncated by continental rifting early in Mesozoic time. During the Jurassic, Cretaceous, and early Tertiary, southern New Guinea formed the continental shelf bounding Australia on the north; the continental margin was mostly stable, and open ocean lay to the north. About the middle of the Miocene, however, the shelf was compressed as imbricate thrust faults drove southward, and a foreland basin formed along the southern margin of the deformed terrain, receiving sediments from the newly uplifted region. Calc-alkaline volcanic and granitic rocks were erupted through the old shelf materials later in the Neogene. The old New Guinea shelf sediments are now bounded on the north by highly contorted deep-water terrigenous sediments, and those in turn are bounded by a broad belt of melange that includes such indicators of subduction as glaucophane schist. North of the melange are great masses of north-topping ophiolite (Davies, 1971). Still farther north, projecting through voluminous upper Neogene fill, are Upper Cretaceous, Palaeogene, and lower Miocene andesites, basalts, diorites, volcanogenic and subordinate carbonate sediments, and, locally, their metamorphosed equivalents.

This geology is interpreted to indicate that, as Australia and New Guinea moved northward in early Tertiary time, their northern margin was a stable continental shelf, beyond which lay ocean floor which was being subducted beneath a south-facing island arc. The arc migrated southward as the continental plate moved northward, and they collided in Miocene time. The great ophiolite masses are the hanging wall of the arc; the voluminous melanges south of the ophiolites formed mostly as the subduction zone consumed the continental rise-and-slope sediments of New Guinea; and the volcanic rocks of northern New Guinea represent the volcanic island arc itself. After the collision, which inaugurated the deformation of the old shelf, the site of subduction shifted, and subsequent subduction was southward beneath the continent as enlarged by the accretion of the island arc. The late Cenozoic calc-alkaline volcanic rocks of interior New Guinea formed above this new subduction zone, a short sector of which is now exposed in the melange beneath south-topping ophiolites of the north-coast Cycloops Mountains.

The Caroline Sea floor presumably opened behind this southward-migrating island arc. Appropriately, the basaltic crust of the central part of that sea has been dated by JOIDES drilling as of early or middle Oligocene age. The boundary between the Caroline Sea plate and the older Pacific floor farther east may be the Mussau system of wholly submarine troughs and ridges, which, if so, represents an extinct island arc.

The various arc and collision belts of New Guinea trend westward to the Geelvink Bay region, where they swing northwestward to truncations against the left-lateral, west-trending Sorong fault system of northern New Guinea. Far to the west, the Sula Islands consist of rocks like those truncated in northwestern New Guinea, including Mesozoic shelf sediments, from continental sources, lying unconformably upon basement quartz monzonite and schist (which are dated as middle Palaeozoic age in northwestern New Guinea). The Sula Islands apparently stand on a sliver of continental crust, carried 1000 km westward during late Tertiary time by a strand of the Sorong as the oceanic plate to the north vanished into the eastern Sulawesi subduction system.

The broad 'eugeosynclinal' crescent of the 'Northwest Borneo geosyncline' is interpreted to be a melange of Eocene age, formed by subduction southeastward beneath Borneo of voluminous abyssal, terrigenous quartzose sediments of Asian source. The southern and eastern belt of the crescent contains abundant indicators of subduction, including ophiolite, galucophane schist, and polymict melange; fossils are of late Late Cretaceous and Palaeocene or early Eocene age. Most of the remainder of the crescent, however, is formed of abyssal Eocene sediments alone, although broad terrains of these sediments are sheared into monoformational melange. Lying unconformably upon this melange(?) in the northwest is an Oligocene and Neogene section which was deposited in gradually shoaling water, and which is interpreted as the fill of an outer-arc basin formed on a melange basement.

Three Tertiary arc systems trend northeastward from northern Borneo into the southwestern Philippine Islands. The melanges of all three systems contain abundant quartzose sandstones, which must have had sources in Borneo or southeast Asia inasmuch as richly quartzose materials are otherwise lacking in the Philippines. The eastern Sula arc, only recently inactivated, comes ashore in the northeast as the Zamboanga Peninsula of Mindanao, and in the southwest as eastern Sabah, both terrains consisting of Miocene melanges flanked on the southeast by appropriate andesites. The Palawan arc, also at least partly of Miocene age, spans from the north tip of Borneo to Mindoro. An arc midway between Sulu and Palawan systems is represented by a submarine ridge, which comes ashore in the northeast as western Panay Island. The products of these three arcs and of the Eocene melange terrain of northwestern Borneo all come together in northern Borneo, where spatial relationships are not obvious from available reconnaissance mapping.

The Philippine Islands consist of subduction melanges, calc-alkaline volcanic rocks of Benioff-zone source, and volcanogenic sediments, arrayed in patterns of baffling complexity. The melanges locally include rocks as old as Permian, but no data known to me require that any of the melanges themselves, or any of the calc-alkaline volcanic rocks, have formed before Late Cretaceous or very early Tertiary time. The bulk of the southern Philippines appears to represent the products of the three northeast-trending arcs noted in the previous paragraph, of the Sangir arc system trending northward to southern Mindanao, and of the Mindanao trench system. Complications are provided by minor trench systems, possibly still active, along the southwest margins of Mindanao and Negros. The great Mindanao trench system ends northward east of the central Philippines, and the Manila trench continues northward on the opposite side of the island group. Between the north end of the Mindanao trench and the south end of the Manila trench, a broad, sigmoidal double orocline serves to transform lateral motion across the Philippines.

Southeast Asia, Sumatra, and westernmost Borneo were part of Asia by Cretaceous time, but their earlier history included Palaeozoic and Triassic subduction and aggregation of continental fragments, and middle Mesozoic rifting. Indochina is bounded against the old platform of southern China by the Red River terrain, which records late Palaeozoic and Triassic subduction beneath the Indochinese side until the two plates collided. The melange and magmatic-arc belts of this terrain trend southeastward to a truncation, presumably by rifting, against the South China Sea. A Cretaceous magmatic arc is overprinted, with northeastward trend, across the truncated end of the older system in southeastern Indochina. Another late Palaeozoic or

Triassic suture is indicated by polymict melanges exposed discontinuously from northern Thailand through southeastern Thailand to southwestern Cambodia. Ambiguous data suggest that this latter suture may also trend southward through medial Malaya to central Sumatra, where it is truncated obliquely against the Indian Ocean; in any case, west-central Sumatra was part of a stable continental platform in Permian and Triassic time. In Burma, westward growth of the continental mass by accretion of subduction melanges is displayed by the westward-younging Late Mesozoic and Early and Middle Tertiary complexes. The youngest Burmese melanges, those of the Arakan coast, trend southward into the outer-arc ridge of the Andaman–Nicobar–Sumatra subduction system.

References

Davies, H. L., 1971. Peridotite-gabbro-basalt complex of eastern Papua—an overthrust plate of oceanic mantle and crust, *Australia Bureau Mineral Resources, Geol. and Geophys. Bull.*, **128,** 48.

Fitch, T. J., 1970. Earthquake mechanisms and island arc tectonics in the Indonesian–Philippine region, *Seismol. Soc. America Bull.*, **60,** 565–91.

Hamilton, W., 1972. Preliminary tectonic map of the Indonesian region, 1:5,000,000, U.S. Geol. Survey Open File Report, 3 sheets.

Hatherton, T. and Dickinson, W. R., 1969. The relationship between andesitic volcanism and seismicity in Indonesia, the Lesser Antilles, and other island arcs, *J. Geophys. Res.* **74,** 5301–10.

Heirtzler, J. R., Dickson, G. O., Herron, E. M., Pitman, W. C. III, and Le Pichon, X., 1968. Marine magnetic anomalies, geomagnetic field reversals, and motions of the ocean floor and continents, *J. Geophys. Res.*, **73,** 2119–36.

Karig, D. E., 1971. Origin and development of marginal basins in the western Pacific, *J. Geophys. Res.*, **76,** 2542–61.

Solomon, M. and Griffiths, J. R., 1972. Tectonic evolution of the Tasman orogenic zone eastern Australia, *Nature Phys. Sci.*, **237,** 3–6.

8.6

N. S. HAILE
Department of Geology
University of Malaya
Kuala Lumpur, Malaysia

The Recognition of Former Subduction Zones in Southeast Asia

'Those who have handled sciences have been either men of experiment or men of dogmas. The men of experiment are like the ant; they only collect and use; the reasoners resemble spiders who make cobwebs out of their own substance. But the bee takes a middle course; it gathers material from the flowers of the garden but transforms and digests it, by a power of its own. . . .'

Frances Bacon, 'Novum Organum', 1620.

The theory of plate tectonics, moving out from the mid-ocean ridges as inexorably but far more rapidly than the plates themselves, is now impinging with full force upon the continental margins. In parts of the circum-Pacific area, where comparatively simple chains of island arcs and arcuate trenches occur, it has been possible to adapt the theory to the models of trench-arc systems already erected by geologists and geophysicists without a great deal of modification, although even there many uncertainties remain.

When attempting to apply the ideas to complex areas such as Southeast Asia where several plates, two continents, and several microcontinents are involved, and to extend the models back in time it is inevitable that complications and uncertainties exist. Factual data are patchy, as the region has been for the most part only surveyed on a reconnaissance basis, and the quality of these surveys varies. Geochronological data, in particular, is lacking for many important areas, notably for the very important West Kalimantan (Indonesian Borneo) area for which even now not a single fossil or radiometric age determination is available. Geochronological data for the granite batholiths of the Malay Peninsula is only now becoming available (Bignell & Snelling, 1972) and the first palaeomagnetic work to be attempted in Southeast Asia is currently in progress (Haile & McElhinny, 1972).

Malaya, the Sunda Shelf, and Borneo were divided into a number of structural zones by Bemmelen (1949), who also explained the history of the Sunda Mountain System (e.g. Sumatra, Malay Peninsula, West Borneo, northern Sunda Shelf) in terms of orogenies centred on the Anambas Zone, and becoming progressively younger to northeast and southwest. In recent analyses, the writer (1972, in press, a and b) has attempted to show that although, especially in view of the almost non-existent stratigraphic control from the Sunda Shelf and the Anambas Zone in particular, Bemmelen's zonation was very percipient, it requires revision, and should not be accepted as an established scheme (as was done, e.g. by Holmes, 1965). A suggested geotectonic zonation is shown in Fig. 1, and in more detail for West Borneo in Fig. 2. The zonation is based on the broad regional structure and geotectonics, as far as is known, and is present as a preliminary hypothesis for subsequent modification and improvement. As far as the Malay Peninsula is concerned, the zonation is derived from published data and in particular the forthcoming compilation on the geology of the Malay Peninsula (Gobbett & Hutchison, in press).

The oldest known rocks are in the *West Malaya Zone*, which consists of Lower Palaeozoic carbonate/clastic shelf sedimentary rocks in the west, grading to deeper water strata in the east, and intruded by the Main Range granite batholith, a complex which seems mainly of Late Carboniferous and Triassic age. The *Central Malaya Zone* is characterized by folded clastic Triassic–Jurassic carbonate and continental deposits, with, along the junction with the West Malaya Zone, a north–south belt of Palaeozoic marine sediments showing deep water 'eugeosynclinal' characteristics —the 'Bentong Line' described below. The *East Malaya Zone* is composed essentially of Upper Palaeozoic, predominantly deep-water marine clastic rocks, with patches

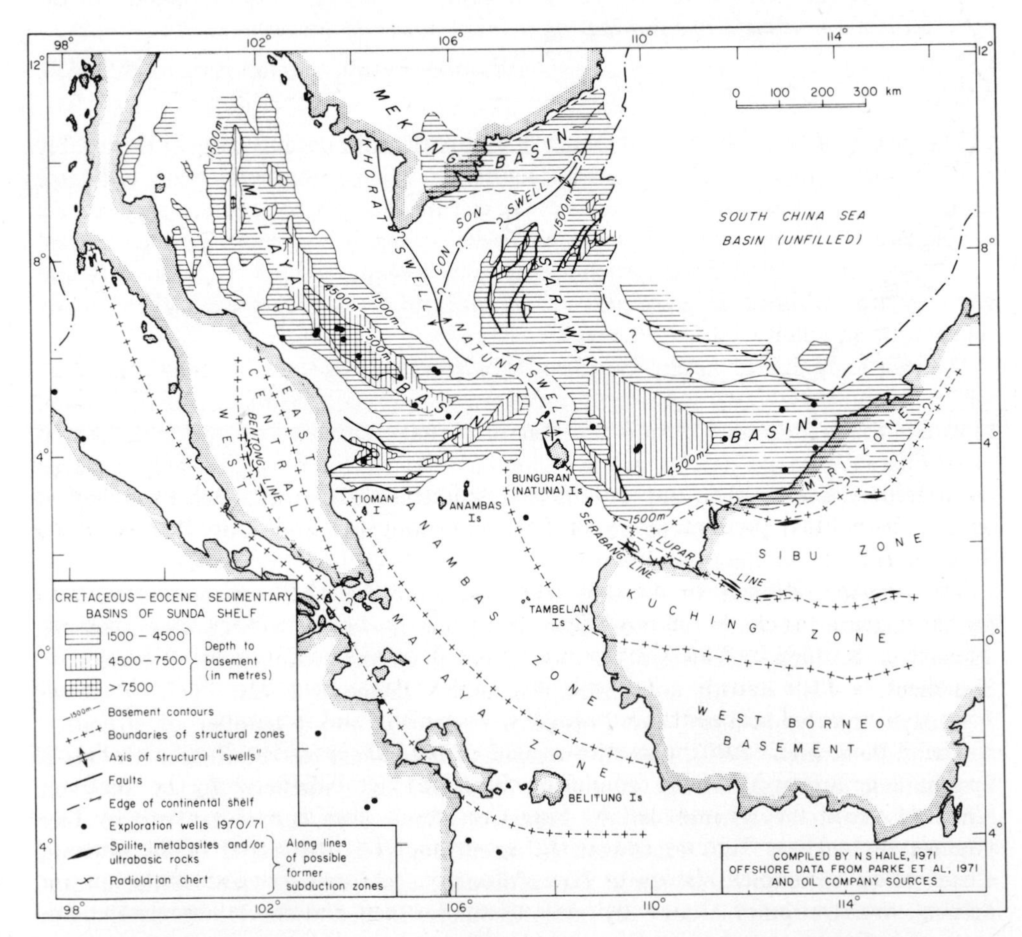

FIGURE 1

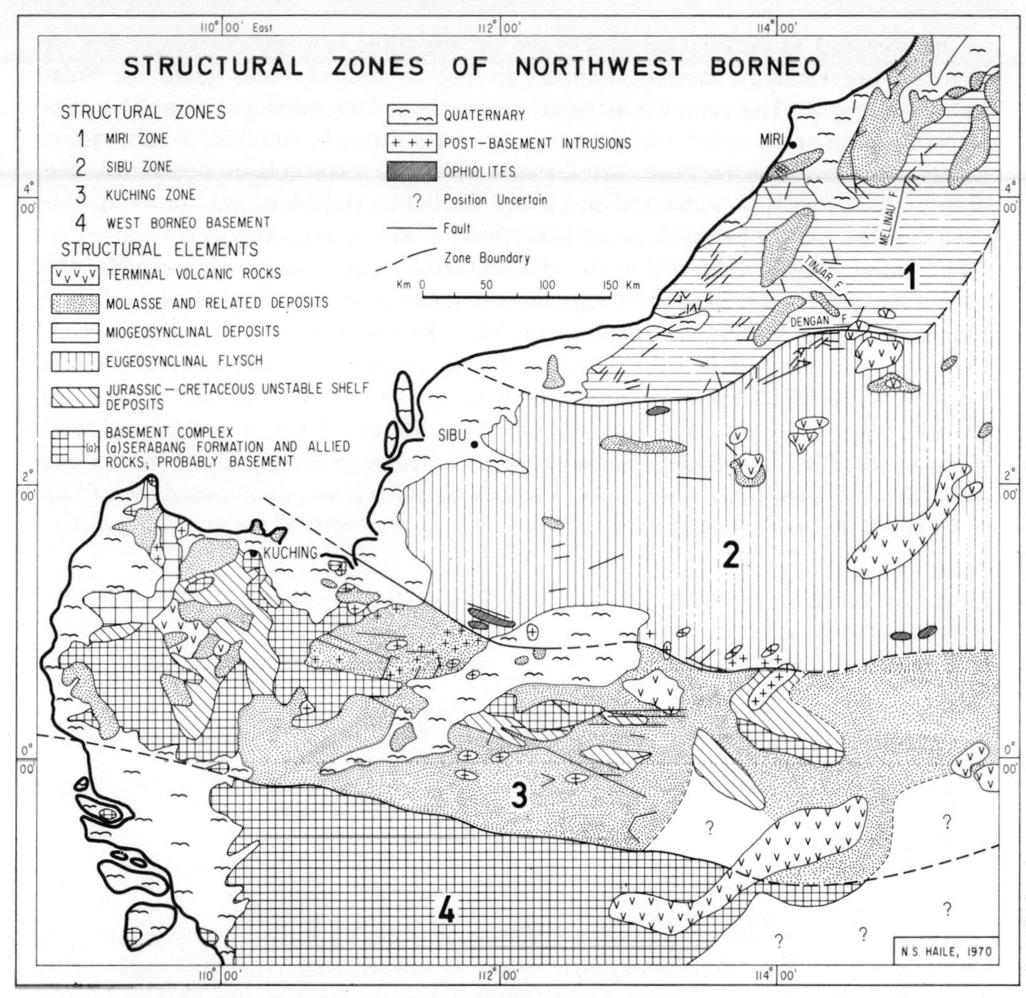

FIGURE 2

of unfolded Upper Mesozoic continental sediments. It is intruded by extensive granite batholiths, predominantly of Late Carboniferous and Late Permian/Early Triassic age.

The *Anambas Zone* is known only from islands on the shelf and is characterized by basic igneous rocks of unknown age, overlain by metasedimentary rocks, possibly Mesozoic, intruded by Late Cretaceous granite. It passes east into the West Borneo Basement, a little-known zone with a complex Palaeozoic and Early Mesozoic history of repeated sedimentation, orogeny, volcanism, and intrusion.

The *Natuna Swell* and the northwest end of the *Kuching Zone* (see also Fig. 2) are characterized by Mesozoic radiolarian cherts and metabasites (along the 'Serabang Line' described below) intruded by Late Cretaceous granite and overlain by Late Cretaceous to Early Tertiary continental sandstone. The main part of the Kuching Zone shows a complex history of Late Palaeozoic to Cainozoic sedimentation and folding, with mobility shown by very thick Tertiary continental sandstone and numerous Tertiary stocks and intrusive sheets.

The *Sibu* and *Miri Zones* are part of the Cretaceous–Cainozoic Northwest Borneo Geosyncline. The *Sibu Zone* is characterized by enormously thick intensely folded Late Cretaceous to Eocene flysch, with chert, spilite and dolerite along the border with the Kuching Zone (the 'Lupar Line' described below). The *Miri Zone* contains Upper Tertiary neritic and paralic sediments, with fold structures broader than those of the Sibu Zone. It continues off-shore into the Sarawak Basin.

Within the area considered, three linear belts which show some of the characters supposed to be indicative of former subduction zones are known: two, in Borneo, named here the Lupar and Serabang Lines, and the third, the Bentong Line, in the Malay Peninsula. The line of the Lupar River in Sarawak and its continuation east-southeast into the Danau (Lake) District of Kalimantan represents a very abrupt major facies boundary as is shown in Table 1.

TABLE 1 Facies contrast in the Lupar Valley, West Sarawak

Southwest of Lupar Valley (part of Kuching Zone)	Northeast of Lupar Valley (part of Sibu Zone)
Cainozoic stocks and intrusive sheets extremely common	Cainozoic intrusions absent
Late Cretaceous to Cainozoic molasse (Plateau Group) very thick and wide-spread	Molasse deposits absent
No Cenomanian–Eocene flysch	Extremely thick Cenomanian–Eocene flysch occupies entire zone
No radiolarite or spilite known	Cretaceous radiolarite/spilite belts along north and south borders
Mesozoic granites	No Mesozoic granites

These differences are seen in the drainage basin of the Lupar River. Taking the whole Sibu and Kuching Zones, the contrast is still striking. The differences listed apply, with the exceptions that some thin patches of molasse are known unconformably overlying the flysch in the northern part of the Rajang Zone, and a flysch and radiolarite/spilite association is known in the extreme east of the Kuching Zone.

The Kuching Zone contains thick Jurassic–Cretaceous rocks including biohermal limestones, marine shales, tuffs, and lavas and littoral conglomerates, facies entirely lacking in the Rajang Zone. The presence of Triassic volcanic and sedimentary rocks, a patch of Carbo-Permian Limestone and shale, and inliers of pre-Muscovian schist in the Kuching Zone complete the contrast with the Rajang Zone.

The concept of a convergent plate boundary—presumably a southward-dipping subduction zone—between the Rajang and Kuching Zones explains the sharp facies boundary, and the lithology of the strata of the Rajang Zones. The 'dynamic polarity' observed in the geology of the Northwest Borneo geosyncline, by which stages of sedimentation, orogeny, and volcanism become progressively younger, not only northwards across the strike, but eastwards along the strike, also can be explained in terms of an oceanic microplate travelling south impinging upon the Borneo microcontinent, in such a way that the flysch sediments carried by it met the western end of the Borneo microcontinent first, in about Cenomanian times. The younging northwards of the flysch deposition and other processes would, on such a model, be due to progressive northward migration of the resulting subduction zone, of the sort which has been described from other areas.

However, the Sibu Zone possesses a number of characters not typical of subduction zones, including:

(1) It does not face an ocean, nor could it have faced an ocean during its formation unless it has undergone major rotational and translational movements. It faces a marginal sea, and is concave to that sea.

(2) The ophiolite-chert belt (which is the only unit which could be regarded as melange) is very narrow, and confined to the southern rim, with a few isolated fault slices along the northern rim.

(3) The main flysch development (Upper Cretaceous to Upper Eocene) contains no chert or ophiolite, and is not a melange (except locally in Sabah). For the major part of its outcrop it shows extremely regular dips.

(4) No glaucophane-bearing rocks are known, and most of the rocks show only slight dynamic metamorphism—the shales being argillites, and locally, slates and low-grade graphitic schist.

The Serabang Line. In extreme west Borneo an argillite/greywacke/radiolarite complex, which contains greenstone metabasites in which both meta-volcanics and metagabbros have been identified, the Serabang Formation possesses several of the 'ear marks' of a former subduction zone. It contains some polymict bouldery conglomerate which could be regarded as melange, though it has not been proved whether this is a sedimentary or a tectonic feature (Wolfenden & Haile, 1963). In the Bunguran (Natuna) Islands some 230 km to the northwest of Serabang, the little-studied Bunguran beds, composed of shales and sandstone, also contain radiolarian chert, radiolarian tuffs, serpentinite, and gabbro, diorite and norite (Haile, 1972). The Bunguran Islands and Serabang District are here referred to as the Serabang 'Line'. The structural high through the Bunguran Islands to South Vietnam is known as the Natuna Swell (see Fig. 1).

Both the Bunguran beds and the Serabang Formation have been intruded by granite which have yielded K/Ar ages of 70/80 m.y. BP (Late Cretaceous). The Serabang and Bunguran Formations are thus older than 70 m.y. but their age is not certain. Radiolaria in the cherts in the Serabang Formation are of Jurassic–Cretaceous and Cretaceous age, according to G. F. Elliot (in Wolfenden & Haile, 1963, p. 19). A Cretaceous age for the Serabang Formation has been questioned, because in the Sarawak valley a complete sequence of Late Jurassic to Late Cretaceous (at least as young as Turonian) in an unstable shelf environment (reef limestone and clastic sediments, tuffs) is exposed (Wilford & Kho, 1965). These strata (the Kedadom, Bau Limestone, and Pedawan Formations) are within 15 km of the Serabang Formation, the contact being obscured, in Sarawak, by Late Cretaceous to Lower Tertiary sandstone; in Indonesia the formations are presumably in contact, but the area of the contact has not been mapped. Such a rapid facies change could only be explained by major faulting, such as occurs at a subduction zone. Nevertheless, the possibility that the Serabang Formation is older than the Late Jurassic–Cretaceous of the Bau District cannot be excluded, in which case it is unnecessary to postulate faulting.

In any case, it appears that the northwest tip of Borneo (the area of the Serabang Formation, Fig. 2) is more closely related structurally to the Natuna Swell than it is to the rest of the Kuching Zone, with which, however, it shares the common feature of Tertiary mobility as exemplified by the presence of molasse sandstone of the Plateau Group. Bemmelen (1949) recognized the anomalous position of this area by excluding it from his Kuching Zone and showing it as a subzone of his Basement Complex (which otherwise was equivalent to the west Borneo Basement of Fig. 2). However, the Natuna Swell, Serabang area and the main part of the Kuching Zone are unified by showing effects of Tertiary mobility exemplified by the presence of molasse-type sediments of the Plateau Group. Perhaps the relationship, as presently known, would be best expressed by regarding the Natuna Swell and the northwest tip of Borneo (the Serabang area) as together forming a subzone of the Kuching Zone.

Malay Peninsula

In the west central part of the Malay Peninsula, the so-called 'Bentong Group', extending north–south, has been suggested to represent a former Palaeozoic subduction zone (Warren Hamilton, personal communication, 1972; Hutchison, 1972). The 'Bentong Group' includes abundant radiolarian chert, some serpentinite masses, metabasites, and dolerite sills; its outcrop approximately separates the West Malaya Zone (of Fig. 1), where Lower Palaeozoic is developed in limestone facies, becoming more clastic to the east, and including acid volcanic rocks, from the Central and East Malaya Zones, where no Lower Palaeozoic is exposed. The geology is complicated by the large granite batholiths, concentrated in the West and East Malaya zones, representing intrusion in Late Cretaceous, Early Triassic, and Late Triassic to Early Jurassic (Bignell & Snelling, 1972).

Conclusions

The three belts briefly described, named here Lupar, Serabang, and Bentong Lines, all show some of the features considered typical of former subduction zones, but none can be regarded as proved to be such. The most convincing is the Lupar Line, the least, the Bentong Line.

It would be a mistake to regard these ideas as any more than working hypotheses which could stimulate collecting of relevant data. If Southeast Asia is, in fact, a mosaic of small plates which have moved hundreds or thousands of kilometres relative to each other, as has been suggested (e.g. by Hamilton, 1972), the unravelling of the geological history of this region will be a formidable task, and will require more facts than are at present available. The main value of attempts to apply plate tectonic theory is in stimulating the search for relevant data.

Although any new data bearing on the geologic history of the region may well be significant, the type of studies most needed at present include:

(1) Geochronologic studies, mainly radiometric dating. In West Malaysia an intensive geochronologic investigation of the main granite batholiths is nearing completion, preliminary results (see Bignell & Snelling, 1972) indicate three

major periods of intrusion between Late Carboniferous and Early Jurassic. Data on Palaeozoic and Mesozoic volcanic rocks in West Malaysia are now required, and work on this is being done by Dr. R. Crawford of the Australian National University. West Kalimantan (West Borneo) is a very important area of Palaeozoic and Mesozoic rocks from which no single radiometric or fossil age determination is available.

(2) Palaeomagnetic studies. The first palaeomagnetic data from Southeast Asia became available only this year (Haile & McElhinny, 1972). Many more determinations on rocks of all ages in the Southeast Asia region will provide information about the palaeolatitude and former relative positions of the structural elements, if these have in fact moved relative to each other. In view of the low angle of inclination of the magnetic field, palaeomagnetic studies of rocks which formed at low latitudes show subsequent rotational or translational movements very clearly.

(3) Stratigraphic and sedimentologic studies. Reinvestigation of the so-called 'Bentong Group' indicates that it is probably composed of three lithologically distinct formations, separated by major unconformities. Obviously it will be necessary to sort this out before basing any conclusions on the geotectonic significance of the Bentong Line. Likewise, the source of the enormous thickness of flysch in the Sibu Zone to the north of the Lupar Line is unknown, and a study of the sedimentary structures and petrology of the rocks would be likely to be very productive.

If the theory of plate tectonics can be applied hand-in-hand with, and controlled by, the collection of new and relevant data, it is likely to make a rapid contribution to the understanding of the geology of the region. It is important, particularly as so many sub-disciplines are involved, that the terminology used is clearly defined. For example, the term ophiolite appears to have been used for everything from a full fledged assemblage of spilite, gabbro and ultrabasic rocks (with associated radiolarian cherts) to isolated occurrences of basic lavas. Similarly, different authors seem to have different conceptions of the term melange, and confusion would be avoided if a definition (such as that given in the International Tectonic Dictionary (Dennis, 1967)) could be agreed.

Finally, whereas in any work aimed at investigating the validity of plate tectonic concepts in the region, it is natural that those features which appear to be indicative of former subduction zones, such as ophiolite-radiolarite belts, glaucophane schist, melange, major fault zones truncating orogenic belts, will be sought and featured, it is important not to exaggerate minor features which happen to be in a convenient location for the current theory. The author has in mind some recently postulated faults in Southeast Asia, which, starting from a few minor observations in the field, or on air-photographs, have been extrapolated to an extent by which the very coherence of the planet might seem to be threatened. Apart from the evidence for subduction, evidence for stability, continuity, and awkward facts which do not fit the current theory should not only not be suppressed, but should be emphasized, since it is from these anomalous facts that progress is likely to ensue. *Pace* Francis Bacon, whose illuminating analogy is quoted at the beginning of this paper, geologists cannot afford to be as selective in their collections as is the bee, otherwise they might well conclude that the Earth is made of pollen.

Acknowledgements

The writer has benefited from discussions with many colleagues, in particular Drs. C. S. Hutchison, P. H. Stauffer, and Warren Hamilton.

References

Bignell, J. D. and Snelling, N. J., 1972. The geochronology of the Thai–Malay Peninsula (abstract), *Annex to Geol. Soc. Malaysia Newsletter 34.*

Dennis, J. G. (ed.), 1967. International Tectonic Dictionary (IGC/CGMW), *American Assoc. Petrol. Geol.*, Memoir, 7.

Gobbett, D. J. and Hutchison, C. S. (eds.), in press. The Geology of the Malaya Peninsula: West Malaysia and Singapore. John Wiley, Interscience, New York.

Haile, N. S., 1970. Notes on the geology of the Tambelan Anambas, and Bunguran Islands, Sunda Shelf, Indonesia: United Nations ECAFE/CCOP Technical Bulletin, 3, 55–75.

Haile, N. S., 1972. The Natuna Swell and adjacent Cainozoic basins on the Sunda Shelf (abstract), *Annex to Geol. Soc. Malaysia Newsletter 34.*

Haile, N. S. and McElhinny, M. W., 1972. The potential value of palaeomagnetic studies in restraining romantic speculation about the geological history of Southeast Asia (abstract), *Annex to Geol. Soc. Malaysia Newsletter 34.*

Haile, N. S. (in press, a). Northwest Borneo. *In:* Data for Orogenic Studies. Geological Society, London, special publication.

Haile, N. S. (in press, b). The geomorphology and geology of the northern part of the Sunda Shelf and its place in the Sunda Mountain System, *Pacific Geology*, No. 6.

Hamilton, W., 1972. Plate tectonics of Southeast Asia and Indonesia (abstract), *Annex to Geol. Soc. Malaysia Newsletter 35*, April 1972.

Hutchison, C. S., 1972. Tectonic evolution of the Malay Peninsula and Sumatra—a personal view (abstract), *Annex to Geol. Soc. Malaysia Newsletter 35*, April 1972.

Holmes, A., 1965. Principles of Physical Geology. Oliver & Boyd, Edin., p. 1288.

Parke, M. L., Emery, K. O., Szymankiewicz, R. and Reynolds, L. M., 1971. Structural framework of continental margin in South China Sea: *American Assoc. Petrol. Geol. Bull.*, **55**, **5**, 723–51.

Wilford, G. E. and Kho, C. H., 1965. The Geology and Mineral Resources of the Penrissen Area, West Sarawak. *Geol. Survey Malaysia*, *Rept*. 2.

Wolfenden, E. B. and Haile, N. S., 1963. Sematan and Lundu area, west Sarawak: *Brit. Borneo Geol. Survey Rept.* 1.

8.7

W. ALVAREZ

*Lamont-Doherty Geological Observatory**
of Columbia University
Palisades, N.Y. 10964

The Application of Plate Tectonics to the Mediterranean Region

Introduction

Difficulties in applying plate tectonics to the Mediterranean result from the scarcity of usable evidence, the complexity of the region, and possibly from the effects of processes differing from standard plate tectonics mechanisms. Although many different plate motions can be proposed, only very rarely is it possible to test these hypotheses. This is illustrated by a proposed episode of Jurassic rifting between Sicily and Peninsular Italy.

Even when evidence is available it can be difficult to make precise determinations of plate motions. For example many lines of evidence support a counterclockwise rotation of Corsica and Sardinia in the Middle to Late Miocene, with the leading edge of the plate descending in a former Benioff zone beneath Italy. But the most probable location of the former trench requires a complicated motion of the small plate that cannot yet be reconstructed in a satisfactory manner.

The stratigraphy and structure of mountain belts may preserve a record of the geologic history of plate margins. However some Mediterranean orogenic belts may have formed by uplift and gravity tectonics alone; they may not mark plate margins. For example, it is difficult to explain the structure of the Apennines by the subduction process. The partly chaotic allochthon of the Apennines seems to be the product of submarine gravity sliding and not of deformation in a trench. Thus the nature of Mediterranean orogenic belts must be examined very carefully before they can be used as the basis of a plate tectonic interpretation of this region.

Although the theory of plate tectonics has been very successful in explaining the geological and geophysical features of much of the earth's surface, it has not yet clarified the pattern of extremely complicated areas such as the Mediterranean, the Alpine-Himalayan orogenic belt, and Indonesia. Why is this so? Is it simply that evidence bearing on the plate tectonic behaviour of these regions is scarce and difficult to interpret? Or is a different kind of tectonic behavior partially or entirely responsible for their development?

It seems likely that both factors are involved. In this paper we shall first consider the kinds of evidence that have been used to analyze plate motions elsewhere and see whether they are applicable in the Mediterranean. Then we shall review two examples

*Contribution no. 1793

of the kinds of problems that arise when one tries to make detailed plate tectonic interpretations of this region.

The Problem of Evidence

Many kinds of evidence have been used to work out the movement histories of major lithosphere plates. These include the map patterns of submarine topographic features such as mid-ocean ridges, transform faults, trenches, seamount chains, and bathymetric contours on continental margins. Geophysical measurements of magnetic anomaly stripes, paleomagnetism, and seismicity have been particularly useful, as have geological data on continental blocks that were once adjacent, on cores from deep sea bore holes, on volcanic belts, and on the stratigraphy and structure of mountain systems. For one reason or another most of these kinds of evidence are not usable in the study of the Mediterranean.

The sea-floor morphology of the Mediterranean is of little help. Although the Mediterranean Ridge was once thought to be a small mid-ocean ridge, Ryan *et al.* (1971) have shown that it is probably a flexural swell related to the Hellenic Trough. The Balearic Basin has the crustal structure typical of mid-ocean ridges, but its geography does not allow such an interpretation (Alvarez, 1972c). No transform faults are known in the Mediterranean Sea; perhaps this is because of the thick cover of sediment that hides the morphology of the oceanic basement. One good trench exists in the Mediterranean; this is the Hellenic Trough, flanking Greece and the Greek islands from the southeastern tip of Italy to the Turkish coast. Deep sea drilling into the inner wall of this trench showed older material resting on younger (Hsü & Ryan, 1972), but this probably does not indicate that all of the African plate is underthrusting all of the European plate. As discussed below, it seems probable that the Aegean microplate, and nothing else, is overriding the Eastern Mediterranean crust.

There are no linear seamount chains in the Mediterranean, but it is worth pointing out that the volcanic center of Mount Etna has undergone a slow displacement northward from a Mesozoic position in southeastern Sicily to its present location 50 to 100 km to the north (Pichler, 1970).

Matching of bathymetric contours in the Mediterranean is generally difficult because sedimentation and tectonic deformation have altered the geometry of the continental margins, and because the continental fragments, if such they are, are so small that a unique fit is seldom obtained. Reconstructions of Mesozoic paleogeography based on the shape of today's continental fragments are very hazardous. Only in the case of very recent movements, such as the rotation of Corsica and Sardinia (Nairn & Westphal, 1968; Alvarez, 1972a; Ryan, Hsü *et al.*, 1972) does the matching of continental margins give reliable reconstructions.

Of the various geophysical techniques, mapping and correlation of marine magnetic anomaly stripes have provided the most precise information on the history of sea-floor spreading and continental drift, but this technique has been totally unsuccessful in the Mediterranean. Most of the Eastern Mediterranean is magnetically undisturbed (Vogt & Higgs, 1969), but it is not clear whether this is crust formed by sea-floor spreading with magnetic anomaly stripes never formed, or perhaps formed and

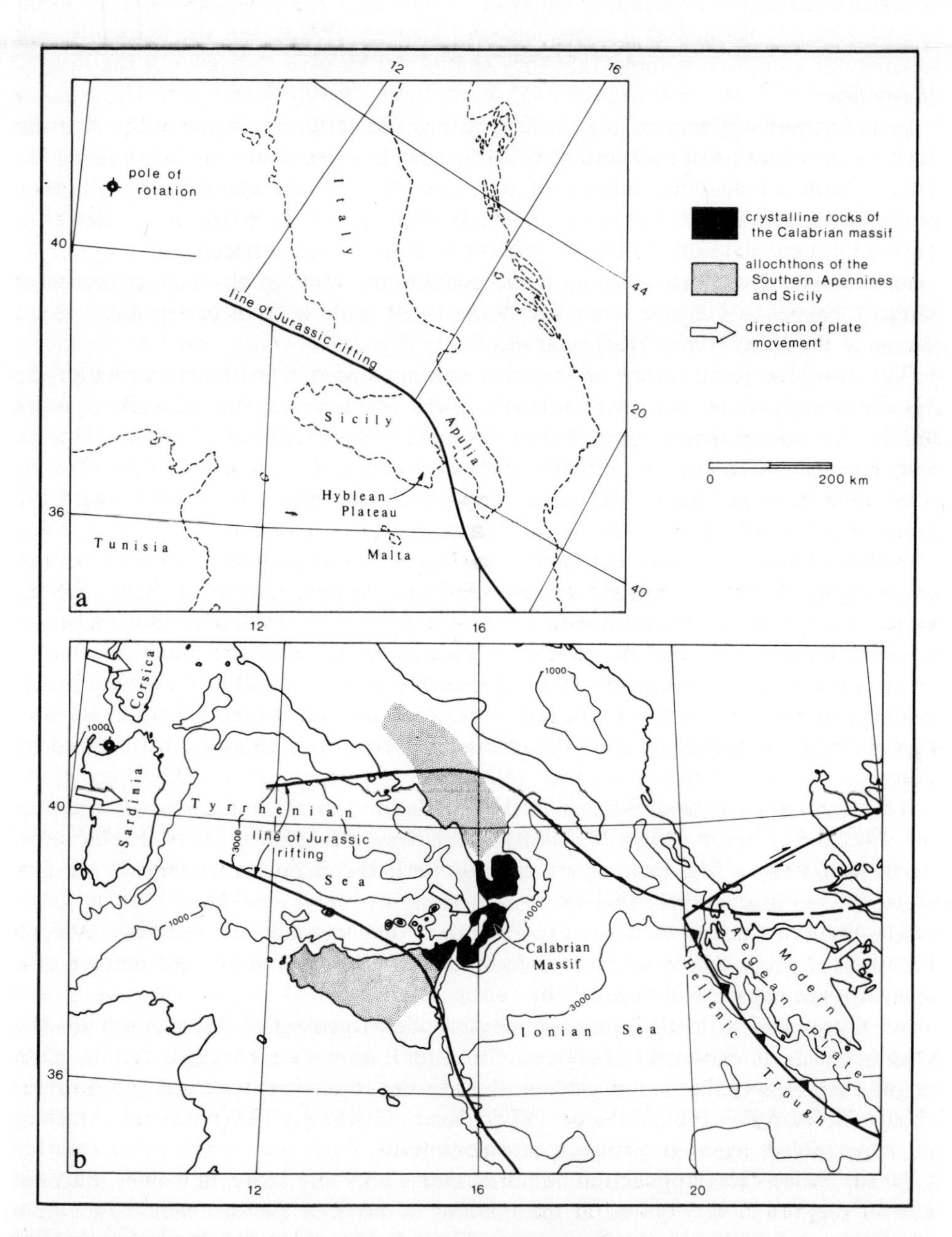

FIGURE 1 A possible episode of Jurassic rifting between southern Italy and Sicily. *a*. Prior to the rifting, which would have opened a small ocean basin roughly corresponding to the present Tyrrhenian, Ionian, and eastern Mediterranean Seas. *b*. Present situation. *This hypothesis has not been proven; it is presented only as an example of the difficulties involved in testing plate tectonic hypotheses in the Mediterranean.*

subsequently removed, or whether it was formed by a different mechanism altogether. Magnetic anomalies in the Tyrrhenian Sea appear to be associated with volcanic seamounts only (Morelli, 1970). Anomaly stripes are present in the Ligurian Sea between Corsica and Provence (Vogt *et al.*, 1971; LeBorge *et al.*, 1971). This basin almost certainly formed as a sphenochasm (Alvarez, 1972a), but the anomalies are so irregular and discontinuous that they cannot yet be used to work out the history of rotation.

In an indirect way magnetic anomaly patterns are useful, for those in the Atlantic permit determination of the motions of Africa and Europe relative to North America. This in turn specifies the motion of Africa relative to Europe since the Jurassic (unpublished work of J. F. Dewey, W. C. Pitman, W. B. F. Ryan, and J. Bonnin). This is fundamental information for the study of the Mediterranean.

Paleomagnetic work has been most successful in the study of small plates that have rotated, particularly Spain (Van der Voo, 1969) and the Corsica-Sardinia block (Nairn & Westphal, 1968; DeJong *et al.*, 1969; Zijderveld *et al.*, 1970). It cannot be expected to give useful results when plates have separated a small distance with little change in orientation. A further difficulty in the Mediterranean is the lack of rocks suitable for paleomagnetic work. For example, in Peninsular Italy the autochthonous sequences have virtually no units that can be studied paleomagnetically; formations with stable remnant magnetism occur only in allochthonous sequences where the results cannot be trusted.

Studies of seismicity and earthquake mechanisms have provided much important information about the present tectonics of the Mediterranean (Ritsema, 1969; McKenzie, 1970). The problem here is that earthquake studies tell nothing of earlier patterns of movement, and the motions appear to have changed frequently.

The most striking seismic features of the Mediterranean are the Benioff zones dipping northward under the Aegean (Papazachos and Comninakis, 1971) and Tyrrhenian Seas (Caputo *et al.*, 1970; Ritsema, 1969), both of which are marked by active or recently active volcanoes (Ninkovich & Hays, 1972). Although these features resemble the Benioff zones of the Pacific in vertical section, their island arcs are shorter and more sharply curved in map view than most of the Pacific arcs. Furthermore there is a pronounced gap between them, both in seismicity and in volcanism, so it is unlikely that they result from a general descent of African lithosphere beneath Europe. A better explanation is that they are due to southeastward motion of a small Calabrian plate (Ritsema, 1969) and southwestward motion of a small Aegean plate (McKenzie, 1970).

In comparison with the scanty geophysical data, geological information on the Mediterranean is extremely abundant. One application of geological data in plate tectonic analysis is the use of geological features for testing proposed pre-drift reconstructions. An example is the demonstration that, when Africa and South America are reassembled, age date provinces are continuous from one to the other (Hurley & Rand, 1968). This application requires that rocks and structures older than the time of separation be exposed at the margins of both of the continental blocks, a situation uncommon in the Mediterranean. The only example of this kind of geological testing in the Mediterranean is the demonstration by Nairn & Westphal (1968) and by Ryan *et al.* (1972) that the geology of Corsica and Sardinia continues into France when the islands are placed back against the coast of Provence. In most other places in the Mediterranean region sedimentary cover postdating the probable times of separation precludes the use of this technique.

Deep-sea drilling has produced much information bearing on the plate tectonics of oceanic regions. In the Mediterranean, Leg 13 of the Deep Sea Drilling Project has raised the interesting possibility that complete desiccation of the deep Mediterranean basins may be responsible for the presence of thick Upper Miocene evaporites (Hsü & Cita, 1972). It was not possible, however, to penetrate below the evaporites with the drill, and since most important plate movements would have preceded deposition of the evaporites, the results of the drilling have not added greatly to our understanding of the plate tectonics of the region.

It is now well established that volcanic arcs occur above descending lithosphere slabs, and plots of potash *vs* silica can be used to establish the geometry of Benioff zones that are no longer active (Ninkovich & Hays, 1972; Dickinson, 1970). This method promises to be useful in the Mediterranean, but it may be that the technique is not precise enough to reconstruct plate motions accurately when the plates are as small as they are in this region. An example of the application of this method is given in the next section.

Much of the history of convergence between adjacent plates should be recorded in the stratigraphy and structure of orogenic belts. Classifications of the types of interactions and the types of plates involved have been proposed by Mitchell & Reading (1969), Dewey & Bird (1970), and Dewey & Horsfield (1970). Thus the geology of the Alpine-Mediterranean mountain ranges may hold the key to the history of plate movements in this region. On the other hand some of the mountain ranges may have formed largely as a result of vertical uplift and gravity sliding; they may not be due to interactions along plate boundaries. This question is considered in detail in the last section of this paper.

It is possible to propose any number of small plate motions to account for various geological and geophysical features of the Mediterranean. The problem is that in most cases there is no way to test such hypotheses. As an example, let us consider the arcuate bend of Calabria, linking the northwest-southeast trending Apennines with the east-west ranges of Sicily. The Hercynian massif of Calabria has been shown on geological grounds to have moved toward the southeast to reach its present position (Caire, 1970), and seismological work (Ritsema, 1969) has shown that this motion is still continuing. If the Calabrian massif were removed far to the northwest a gap would remain between Peninsular Italy and Sicily (Fig. 1). The gap might have been even wider than the width of the Calabrian massif, for much of Lucania and northern Sicily is occupied by allochthonous units that originated in the area of the Tyrrhenian Sea, and whose basement is unknown (Ogniben, 1969; Alvarez & Gohrbandt, 1970). This gap would link the Tyrrhenian and Ionian Seas into a continuous oceanic belt extending into the eastern Mediterranean. It seems quite possible that this oceanic belt was formed by Jurassic sea-floor spreading, and that this is the origin of the Jurassic ophiolites found in the allochthonous terranes of the Apennines. The fit of the continental margins on either side of the oceanic belt is rather good, if one ignores the recent modification imposed by southwestward movement of the Aegean plate.

This is a typical Mediterranean plate tectonic hypothesis; can it be tested? Sea floor morphology is of no help. The Mediterranean Ridge, even if it were a ridge of mid-ocean type, could not be responsible for the spreading in question, for bathymetric features of Jurassic age would be buried by a great thickness of sediment in the Mediterranean. As mentioned above the magnetic field of the eastern Mediterranean is quite undisturbed; is this because of spreading during a time of no reversals in the Jurassic, or because the hypothesis is not correct? We have noted previously that there are no suitable rocks for paleomagnetic measurements in peninsular Italy. Geological information is equally incapable of testing the hypothesis. In the reconstruction Apulia and the Hyblean Plateau of southeastern Sicily lie next to each other. There is unfortunately no possibility of applying geological matching here, for Jurassic and older rocks are known only from bore holes in the Hyblean Plateau, while a single deep well in southern Apulia was abandoned after passing through more than 4500 m of monotonous Cretaceous and possibly Upper Jurassic carbonates (King, 1960, p. 1089). The deep sea drill holes, which have not reached deeper than the Upper Miocene, clearly tell us nothing about Jurassic plate movements. Finally, since this hypothesis postulates minor separation of the trailing margins of

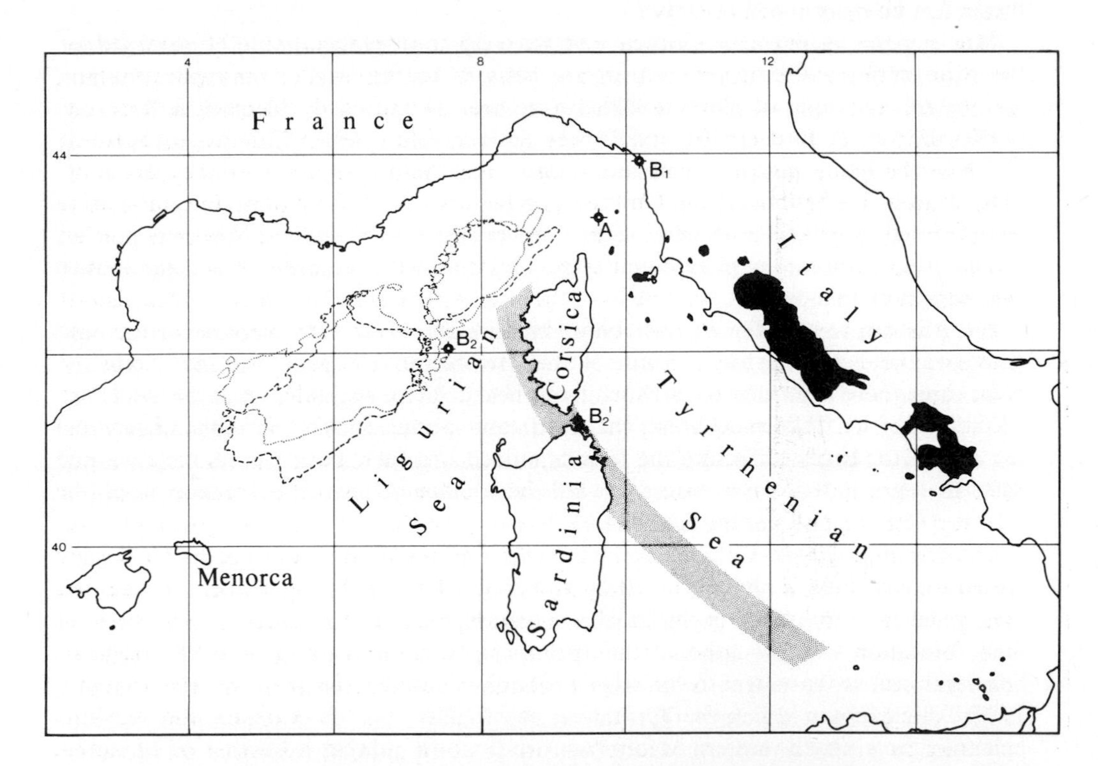

FIGURE 2 The rotation of Corsica and Sardinia. Solid lines show the present coasts. Dashed lines show the best fit of Corsica and Sardinia against France if the island block moved as a single unit (pole at A: 43°22′ N, 9°38′ E, rotation 49°). Dotted lines show the improved fit if Corsica and Sardinia have rotated relative to one another, opening a small sphenochasm to the west of the Straits of Bonifacio. In this case the Corsica–France pole is at B_1 (43°56′ N, 10°11′ E, rotation 32°), and the Sardinia–Corsica pole is at B_2 (position in the reconstruction) and B_2' (present position: 41°17′ N, 9°20′ E, rotation 33°). Dotted area shows the location of the trench inferred from a hypothetical Benioff zone reconstructed on the basis of potash content of volcanic and plutonic rocks of central Italy (black).

two large plates, it is difficult to know where we should go to look for related leading-edge features, such as volcanic arcs or orogenic belts.

Thus, like so many other proposed Mediterranean plate movements, this remains an attractive hypothesis that is, for the moment, untestable. In fact, the only Mediterranean plate movements that have been established so far are the rotations of the Iberian Peninsula (Van der Voo, 1969; Le Pichon & Sibuet, 1971) and of Corsica-Sardinia (Alvarez, 1972a).

Corsica–Sardinia and the Problem of Precise Reconstructions

Let us now consider the rotation of Corsica and Sardinia and see how precisely the tectonics of this microplate can be reconstructed (Fig. 2). In a complete reconstruction we should be able to specify the original position of the continental blocks, the extent of the plate that moved, the pole of rotation, any changes in the plate or the pole with time, the chronology of the movement, effects of the leading edge such as volcanism and deformation, and the probable reason for the movement. Which of these can be determined reliably?

The general position of Corsica and Sardinia along the coast of France before rotation is quite well established on the basis of the fit of the continental margin, geological matching of Corsica with the Esterel Massif, and paleomagnetic results (Alvarez, 1972a). In detail, however, there is a question whether Corsica and Sardinia were in the same position relative to each other before rotation as they are now. The islands are separated only by the narrow Straits of Bonifacio, but submarine morphology west of the straits suggests the presence of a small sphenochasm at an angle of 33°, opening westward and largely filled in with sediment. When the islands are bent back to close this gap, the fit of the western margin of the island block against France is improved (Fig. 2 and Ryan, Hsü *et al.*, 1972). This reconstruction also satisfies the palaeomagnetic results somewhat better, although still not perfectly. Palaeomagnetic measurements show a rotation relative to stable Europe of 25° to 30° for Corsica (Nairn & Westphal, 1968) and about 50° for Sardinia (Zijderveld *et al.*, 1970). In the fit where the islands moved together, both would have rotated 49°. If the fit is based on closing the Bonifacio sphenochasm the rotation would be 32° for Corsica and 65° for Sardinia.

As for the extent of the plate, the trailing edges of the continental blocks are obvious, but there is nothing in the bathymetry, seismic profiling results, or magnetic anomalies to suggest the presence of a spreading ridge in the Ligurian Sea; perhaps the formation of new crust in the sphenochasm took place in a more irregular, disorganized way. A transform edge probably bounded the plate on the south; a small circle about the pole of rotation at 43°22′N, 9°38′E (Corsica and Sardinia rotating as a single unit) fits nicely with the cut-off margin northeast of Menorca, the south end of the Corsica-Sardinia block, and the southeast end of associated volcanic areas in Italy (Alvarez, 1972a, Fig. 2). These relationships hold even if the islands have rotated with respect to each other, although the exact location of the transform edge would be slightly different. The most difficult question is the location of the trench; this is dealt with below.

The pole of rotation depends of course on whether Corsica and Sardinia moved together or independently. Poles for each of these cases are shown in Fig. 2.

If the area west of the Straits of Bonifacio did open as a sphenochasm, there must have been at least one change in the pole of rotation. Whether bending of the island block came early or late in the general rotation is not yet clear, but consideration of the pattern of magnetic anomalies of probable volcanic origin suggests that bending of Sardinia relative to Corsica was the first event in the history of movement (Ryan *et al.*, 1972).

Since magnetic anomalies, the most important guides to chronology in oceanic regions, are not usable here, indirect evidence must be found to date the rotation (Alvarez, 1972a). Stratigraphic and geomorphological events in Provence and the beginning of volcanism in the islands west of Tuscany combine to indicate the beginning of movement in the Tortonian, about 11·5 m.y. ago. Rotation must have ceased by about 6 m.y. ago, for the Ligurian sphenochasm is filled with Messinian evaporites, but volcanism above the leading edge has continued into the Pleistocene or even the Recent; the reason for this is not clear. (See note added in proof on p. 905.)

Volcanism in west-central Italy is apparently a result of subduction of the leading edge of the Corsica-Sardinia plate, but there is no sign of deformed trench sediments; these might, however, be hidden beneath the Tyrrhenian Sea. Age dates on the volcanics show an eastward movement, probably reflecting progressive subduction of the leading edge of the rotating plate (Alvarez, 1972a), and the potash contents of the volcanoes show that the former subduction zone dipped about 45° eastward beneath Italy, as expected (Alvarez, in progress). When the subduction zone thus determined is projected up to the surface, however, this line—an approximation to the site of the former trench—is rather difficult to explain, for it passes in a northwest-southeast direction *between* Corsica and Sardinia. If the location of the trench has been correctly determined by this method, it may be that Corsica has somehow overridden the trench, although this is difficult to visualize. At any rate, it is unlikely that the present bathymetric depression immediately east of Corsica represents the former trench as previously suggested (Alvarez, 1972a), for igneous rocks on the islands only 10 or 20 km east of this depression indicate depths of more than 150 km to the former Benioff zone. Perhaps part of the problem is that the rotating plate is shaped like a slice of pie with the pole at the apex; this requires a continuous change in the rate and perhaps the angle of subduction along the length of the trench. It is also possible that the location of the trench changed during the rotation.

Why did the Corsica-Sardinia plate rotate? It is probably not the result of Africa and Europe shearing past each other in a left lateral sense, as suggested by Carey (1958, p. 261), for the magnetic anomalies in the Atlantic imply no east-west relative movement between Africa and Europe since the Eocene (W. C. Pitman, personal communication, 1972). A small rising plume of mantle material could be responsible for the motion; rotation might have resulted from greater drag at the northeast end of the Corsica-Sardinia block where it had to move along the edge of the Italian continental crust, instead of into an oceanic region as in the southwest. Clearly any hypothesis on the cause of the rotation is little more than speculation.

In conclusion, we can say that the rotation of the Corsica-Sardinia plate is quite well established, but that there is a lack of precision regarding the details of the movement. It has been suggested (Anonymous, 1972) that one reason for the success of marine geophysics in the development of plate tectonic theory is the inherently fuzzy nature of geophysical measurements, and that the most successful geologists in this field are those who know where to draw the line in worrying about details. Perhaps this accurately describes scientific developments during the past few years, but it would be foolish to take it as a guide to the future. Surely discrepancies between theory and observation are opportunities to improve the theory, not embarrassments to be glossed over. It is to be expected that as we learn more about the Mediterranean the reasons for the problems discussed above will become clear.

The Apennines and the Problem of Orogenic Belts

Dewey & Bird (1970) have shown how the stratigraphy and structure of mountain belts can be used to interpret the history of plate boundaries. Figure 14 of that paper shows the fold belts of the Alpine-Mediterranean system and the motions at plate boundaries that are inferred from the locations of the deformed belts. In that map the Apennines with their eastward structural polarity (direction of overturning, overthrusting, and migration of orogenic phenomena) are interpreted as the result of Mesozoic-Tertiary consumption of Italian lithosphere in a subduction zone dipping west beneath the Tyrrhenian and Ligurian Seas. Boccaletti *et al.* (1971) have shown in a series of cross-sections how this might produce the observed structure of the Apennines.

The crustal consumption hypothesis for the Apennines is actually quite new; since 1951 the Apennines have been considered a classic example of gravity tectonics resulting from vertical uplift (Merla, 1951; Maxwell, 1953, 1959). The new interpretation is in part due to the evolution of geologic thought concerning the origin of chaotic strata. In the 1950s much of the allochthonous terrane which overlies the Apennine autochthon was mapped as *argille scagliose* ('scaly clays': chaotic assemblages of blocks of all sizes and of varied lithologies in a thoroughly sheared and disrupted clay matrix). In 1964 a group of American geologists visited the Apennines (Wise & Bird, 1964; Alvarez, 1972b). They were struck by the similarity between the *argille scagliose* and the chaotic Franciscan terrane of California, and this led to interpretation of the Franciscan as a product of gravity sliding (Hsü, 1965; Hsü & Ohrbom, 1969). In the later 1960s, however, as the theory of plate tectonics was gaining wide acceptance, it was suggested that underthrusting of oceanic lithosphere at a trench produces crumpling and thickening of the overlying sediments as they are scraped off and plastered against the inner wall of the trench (Chase and Bunce, 1969). Meanwhile, the view that late Mesozoic underthrusting of the Pacific plate beneath California was responsible for chaotic deformation and high pressure-low temperature metamorphism of the Franciscan was becoming increasingly popular (Davis, 1968, 1969; Hamilton, 1969; Blake *et al.*, 1969; Ernst, 1970, 1971; Dickinson, 1970; Barbat, 1971). By this time the term *mélange*, indicating chaotically deformed sedimentary units, had gained acceptance in the English literature. This term was applied both to the Franciscan (Hsü, 1968; Hsü & Ohrbom, 1969) and to the Apennine allochthon (Bailey & McCallien, 1963; Hsü, 1968). Thus by a gradual reversal of analogy the *argille scagliose* of the Apennines could be considered a product of deformation on the inner wall of a trench, and deformed trench sediments were labeled *argille scagliose* in the drawings of Dewey & Bird (1970, Figs. 8, 10).

Since the existence of a former Apennine trench would have an important bearing on the tectonics of the Mediterranean, it is worth examining the geology of the

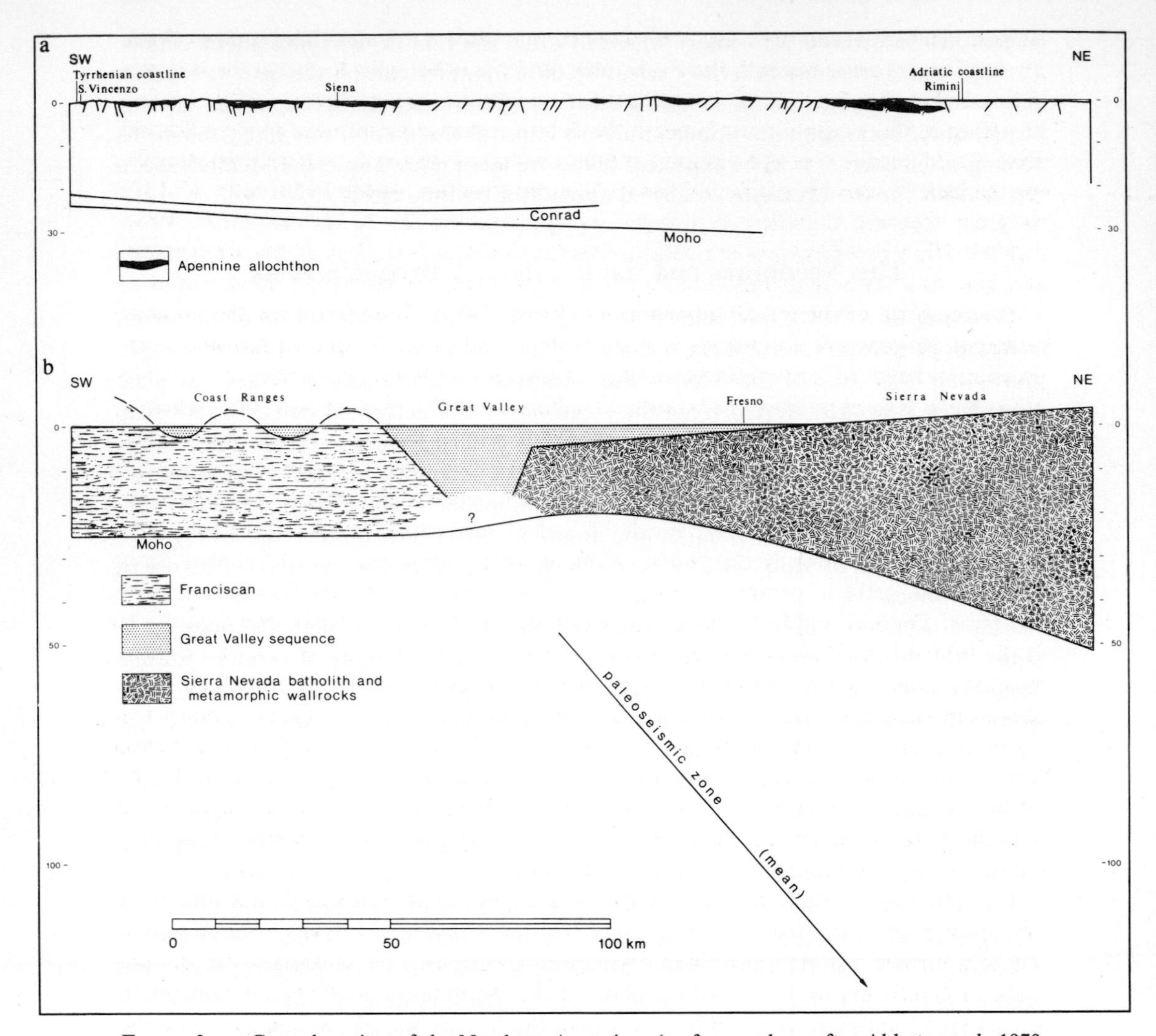

FIGURE 3 *a*. Crustal section of the Northern Apennines (surface geology after Abbate *et al.*, 1970, Fig. 5; deep structure after Morelli *et al.*, 1969, Fig. 10). *b*. Crustal section of California (after Dickinson, 1970, Fig. 6; palaeoseismic zone based on chemistry of the Sierra Nevada batholith). Both sections to the same scale, with no vertical exaggeration. This figure illustrates the difference in gross geological relationships between the Apennine allochthon and the Franciscan.

Apennines to see if such a trench is reasonable. First we shall review the features in California that suggest subduction of the Franciscan, and then we shall see whether similar conditions hold in the Apennines (Fig. 3).

In California a long thrust fault system separates the Franciscan on the west from the Great Valley sequence farther east (Ernst, 1970). The Franciscan is largely a chaotic mélange while the Great Valley sequence is a more orderly succession of miogeosynclinal rocks; the eastern part of the Great Valley sequence rests on continental crust, while the western part rests on oceanic crust (Bailey *et al.*, 1970). Although the two units cover almost exactly the same time interval there is no transitional facies, and the contact between them is always a fault. There is no evidence for continental crust beneath the Franciscan, and the lithologies found in the mélange indicate that the Franciscan was deposited on oceanic crust (Ernst, 1970). Underthrusting of this oceanic crust beneath the North American continent along a Benioff zone would account for the ubiquitous fault contacts with the Great Valley sequence, for the lack of transitional facies, for the chaotic structure of the Franciscan, and for its great tectonic thickness (estimated as approaching 25–30 km by Ernst, 1970, p. 888). High pressure–low temperature metamorphism is typical of the Franciscan, and here as elsewhere around the Pacific it seems that this blueschist metamorphism is characteristic of the trench environment (Ernst, 1970). The reason for the peculiar temperature–pressure conditions is not yet clear; they may be due to tectonic overpressure related to compression in the trench, or to lower temperatures as cold lithosphere descends into the mantle. Dickinson (1970) has shown that when a paleoseismic zone is constructed on the basis of potash values in the Sierra Nevada batholith, the zone projects up westward into the fault contact between the Franciscan and the Great Valley sequences, and he goes on to suggest that it may even be possible to correlate the principal episodes of blueschist metamorphism in the Franciscan with plutonic activity in the Sierra Nevada area. Some late modifications have affected the pattern produced by late Mesozoic–early Tertiary lithosphere consumption. Diapiric uplift of the Franciscan through its Great Valley sequence cover in the Miocene to Pleistocene has changed the east-dipping thrust contact to vertical in some places (e.g. the Diablo antiform). Meanwhile development of the San Andreas strike-slip fault (Atwater, 1970) moved a portion of coastal California toward the north and resulted in an east-west repetition of the Franciscan–Great Valley sequence contact in the area between San Francisco and Los Angeles (King, 1959; Ernst, 1970). Despite these complications the evidence seems very strongly to support the hypothesis that the chaotic deformation of the Franciscan took place during subduction of Pacific lithosphere beneath California.

The situation is very different in the Apennines. Although California and Italy are roughly the same size and both look westward to the sea and to the supposed site of a former trench, lithosphere consumption is said to have affected the western plate in California and the eastern plate in the Apennines. In fact it is difficult to fit a subduction model to any of the features of the Apennines. (We are here considering the tectonic movements that formed the major Apennine structures between the Cretaceous and the Middle Miocene [Sestini, 1970], not the Late Miocene to Quaternary movement of the Corsica-Sardinia plate, which does seem to have descended beneath Italy.)

At present allochthonous terranes cover the Apennine autochthon over a 200 km width (Fig. 3). Although some material would probably slump into an active trench, at least 150 km of the eastern plate would have had to go down the Benioff zone, if the allochthon owes its position over the autochthon to subduction. The basement of the autochthon is nowhere exposed, but there is strong indirect evidence that it consists of continental crust: gravity work shows a 30 km crust containing a Conrad discontinuity (Morelli *et al.*, 1969); the Triassic basal clastics (Verrucano) contain much conglomerate composed mainly of pebbles of quartz, quartzite, and schist, while footprints, mudcracks, and raindrop impressions show deposition in a shallow-water environment, probably lagoonal or deltaic (Bortolotti *et al.*, 1970, p. 347–351);

and igneous rocks generated in this area immediately after the Apennine orogeny are of sialic anatectic origin (Marinelli, 1967). If somehow 150 km of continental crust had been able to descend along the subduction zone, with the leading edge probably reaching a depth of 100 km, extreme metamorphism would have occurred; no metamorphism is found, however, except locally (greenschist facies) beneath parautochthonous thrust sheets (Giglia & Radicati di Brozolo, 1970). Clearly subduction is not the way in which allochthonous material moved 200 km out over the autochthon. A far better hypothesis is Merla's (1951) theory of gravity sliding off a series of submarine ridges rising one after the other toward the east.

On the other hand, some features of the Apennine allochthon do suggest subduction—or obduction—particularly the abundant occurrence of Jurassic ophiolites. Perhaps some compromise solution will be necessary, possibly involving Eocene subduction with a trench west of Italy, followed by Miocene gravity sliding eastward over the present Italian Peninsula. Nevertheless, as discussed below, it is very difficult to draw a satisfactory pattern of trenches to accommodate subduction in this part of the Mediterranean. At any rate the *present* position of the Apennine allochthon must be due to gravity sliding alone.

This conclusion is supported by the nature of the allochthonous material itself. Recent work in the Apennines (summarized in Sestini, 1970) has shown that most of the allochthon consists of very large slabs whose stratigraphy can be worked out, arranged in nappes apparently propelled by gravity. The allochthon is not a chaotic mélange, although some of its sequences are surrounded by *argille scagliose*, now called the 'Chaotic Complex' (Abbate & Sagri, 1970, p. 315). In contrast to what would be expected in the case of subduction, the proportion of chaotic material increases eastward (Abbate *et al.*, 1970, p. 243), *away* from the presumed site of the trench. Furthermore the Apennine allochthon is apparently unaffected by blueschist metamorphism.

In California there are east-dipping thrust faults marked by sheets of serpentinite which clearly mark the site of the ancient Benioff zone. In the Apennines no such features are found. Possibly they exist west of Italy under the Tyrrhenian Sea, but in that case it is very difficult to draw a reasonable pattern of trenches and plate motions. This is because structures apparently produced by gravity sliding continue from the Northern Apennines into the Southern Apennines (Ogniben, 1969), then around a sharp bend into the east-west ranges of Sicily (Alvarez & Gohrbandt, 1970) and the Atlas Mountains (Durand-Delga, 1967). The Apennines and the Sicily-Atlas Mountains make an angle of about 45° and face each other across the Tyrrhenian Sea, with the structural polarity outward in each case. The structural similarity all along this mountain chain is remarkable (Caire, 1970). If the Apennines were produced by subduction, then surely the rest of the system was too. But this would require a tightly curving trench with the Tyrrhenian Sea overriding as it moved roughly east-southeast, and the transform component of motion at the trenches would be much greater than the compressional component. If the Atlas and Italy are considered part of the African plate and the Tyrrhenian a prong of the European plate, the suggested motion conflicts with predominantly north-south relative movement between Africa and Europe since the Eocene based on the Atlantic magnetic anomalies (W. C. Pitman, personal communication, 1972). It is also difficult to see why the overriding Tyrrhenian plate should have subsided to oceanic depths almost immediately after the movement ceased, as it seems to have done (Pannekoek, 1969; De Booy, 1967).

We can conclude that no convincing evidence has yet been presented for a subduction origin for the Apennine allochthon. Mere lithologic analogy with the Franciscan is not enough, and for the moment uplift and gravity sliding, with little or no crustal shortening, is a more attractive hypothesis. It seems likely that chaotic 'mélange' units can form in more than one way.

If, as this suggests, orogenic belts can form at sites other than plate boundaries, it will be necessary to examine a mountain belt very carefully before it is cited as evidence for plate movements. In fact, of the mountain belts surrounding the western Mediterranean, only the Alps *s.s.* show convincing evidence of significant crustal shortening.

Acknowledgements

The ideas and reasoning presented in this paper result from an effort to apply the principles of plate tectonics to the Mediterranean area. During this time I have been helped greatly by discussions with M. Boccaletti, T. Cocozza, S. Connary, M. Ewing, G. Guazzone, R. L. Larson, D. Ninkovich, W. B. F. Ryan and W. B. Travers, not all of whom agree with everything in the paper. The work was supported by a NATO Postdoctoral Fellowship, and by the Industrial Associates of Lamont-Doherty Geological Observatory of Columbia University.

Note added in proof: Recent evidence on metamorphism in the Northern Apennines (Radicati di Brozolo & Giglia, 1973) and on the palaeomagnetism of Sardinian volcanics (De Jong *et al.*, in press) supports rotation of the Corsica–Sardinia microplate in the Early to Middle Miocene. Thus the previous suggestion of rotation during the interval 11·5–6·0 m.y. (p. 892; Alvarez, 1972a) may be slightly too young.

References

Abbate, E., Bortolotti, V., Passerini, P. and Sagri, M., 1970. Introduction to the geology of the Northern Apennines, p. 207–49. *In:* Development of the Northern Apennines geosyncline, edited by G. Sestini. *Sediment. Geol.*, **4**, 203–647.

Abbate, E. and Sagri, M., 1970. The eugeosynclinal sequences, p. 251–340. *In:* Development of the Northern Apennines geosyncline, edited by G. Sestini. *Sediment. Geol.*, **4**, 203–647.

Alvarez, W., 1972a. Rotation of the Corsica–Sardinia microplate, *Nature Phys. Sci.*, **235,** 103–5.

Alvarez, W., 1972b. Recent Italian Research, *Geotimes*, **17** (2), 14–7.

Alvarez, W., 1972c. Uncoupled convection and subcrustal current ripples in the western Mediterranean. *In:* The H. H. Hess Volume: Studies in Earth and Space Science, edited by R. Shagam. Geol. Soc. America, Memoir 132, 119–132.

Alvarez, W. Igneous evidence for a fossil seismic zone beneath Italy; in progress.

Alvarez, W. and Gohrbandt, K. H. A., eds., 1970. Geology and History of Sicily. Petr. Expl. Soc. Libya, Tripoli, 291 pp.

Anonymous, 1972. News and views: Afar and Iran. *Nature*, **235,** 126–7.

Atwater, T., 1970. Implications of plate tectonics for the Cenozoic tectonic evolution of western North America, *Geol. Soc. Amer. Bull.*, **81** (12), 3513–36.

Bailey, E. B. and McCallien, W. J., 1963. Liguria Nappe: Northern Apennines. *Trans. Roy. Soc. Edin.*, **65,** 315–33.

Bailey, E. H., Blake, M. C., Jr., and Jones, D. L., 1970. On-land Mesozoic oceanic crust in California Coast Ranges. U.S. Geol. Survey, Prof. Paper 700-C, pp. C70–C81.

Barbat, W. F., 1971. Megatectonics of the Coast Ranges, California, *Geol. Soc. Amer. Bull.*, **82** (6), 1541–62.

Blake, M. C., Jr., Irwin, W. P. and Coleman, R. G., 1969. Blueschist facies metamorphism related to regional thrust faulting, *Tectonophys.*, **8** (3), 237–46.

Boccaletti, M., Elter, P. and Guazzone, G., 1971. Plate tectonic models for the development of the western Alps and northern Apennines, *Nature Phys. Sci.*, **234,** 108–11.

Bortolotti, V., Passerini, P., Sagri, M. and Sestini, G., 1970. The miogeosynclinal sequences, pp. 341–444. *In* Development of the Northern Apennines geosyncline, edited by G. Sestini. *Sediment. Geol.*, **4**, 203–647.

Caire, A., 1970. Sicily in its Mediterranean setting, *In:* Geology and History of Sicily, edited by W. Alvarez and K. H. A. Gohrbandt, pp. 145–70. Petr. Expl. Soc. Libya, Tripoli, 291 pp.

Caputo, M., Panza, G. F. and Postpischl, D., 1970. Deep structure of the Mediterranean Basin, *J. Geophys. Res.*, **75,** 4919–23.

Carey, S. W., 1958. A tectonic approach to continental drift. *In:* Continental Drift: a Symposium, pp. 177–363. Geol. Dept., Univ. of Tasmania, Hobart, 363 pp.

Chase, R. L. and Bunce, E. T., 1969. Underthrusting of the eastern margin of the Antilles by the floor of the western North Atlantic Ocean, and origin of the Barbados Ridge, *J. Geophys. Res.*, **74,** 1413–20.

Davis, G. A., 1968. Westward thrust faulting in the south-central Klamath Mountains, California, *Geol. Soc. Amer. Bull.*, **79,** 911–34.

Davis, G. A., 1969. Tectonic correlations, Klamath Mountains and western Sierra Nevada, California, *Geol. Soc. Amer. Bull.*, **80,** 1095–1108.

De Booy, T., 1967. Repeated disappearance of continental crust during the geological development of the western Mediterranean area, *Verhandelingen Kon. Ned. Geol. Mijnbouwk. Gen.*, **26,** 79–103.

De Jong, K. A., Manzoni, M. and Zijderveld, J. D. A., 1969. Palaeomagnetism of the Alghero trachyandesites, *Nature*, **224,** 67–9.

De Jong, K. A., Manzoni, M., Stavenga, T., Van der Voo, R., Van Dijk, F. and Zijderveld, J. D. A. Rotation of Sardinia: Palaeomagnetic evidence for rotation during the Early Miocene, *Nature*, in press.

Dewey, J. F. and Bird, J. M., 1970. Mountain belts and the new global tectonics, *J. Geophys. Res.*, **75,** 2625–47.

Dewey, J. F. and Horsfield, B., 1970. Plate tectonics, orogeny, and continental growth, *Nature*, **255,** 521–5.

Dickinson, W. R., 1970. Relations of andesites, granites, and derivative sandstones to arc-trench tectonics, *Rev. Geophys. Space Phys.*, **8,** no. 4, 813–60.

Durand-Delga, M., 1967. Structure and geology of the northeast Atlas Mountains. *In:* Guidebook to the Geology and History of Tunisia, edited by L. Martin, pp. 59–83. Petr. Expl. Soc. Libya, Tripoli, 293 pp.

Ernst, W. G., 1970. Tectonic contact between the Franciscan Mélange and the Great Valley Sequence—crustal expression of a Late Mesozoic Benioff zone, *J. Geophys. Res.*, **75,** 886–901.

Ernst, W. G., 1971. Petrologic reconnaissance of Franciscan metagraywackes from the Diablo Range, Central California Coast Ranges, *J. Petrol.*, **12,** 413–37.

Giglia, G. and Radicati di Brozolo, F., 1970. K/Ar age of metamorphism in the Apuane Alps (Northern Tuscany), *Boll. Soc. Geol. Ital.*, **89,** 485–97.

Hamilton, W., 1969. Mesozoic California and the underflow of Pacific mantle, *Geol. Soc. Amer. Bull.*, **80,** 2409–30.

Hsü, K. J., 1965. Franciscan rocks of Santa Lucia Range, California, and the *argille scagliose* of the Apennines, Italy: a comparison in style of deformation. Geol. Soc. America, Abstracts for 1965, p. 210.

Hsü, K. J., 1968, Principles of mélanges and their bearing on the Franciscan-Knoxville paradox, *Geol. Soc. Amer. Bull.*, **79,** 1063–74.

Hsü, K. J. and Ohrbom, R., 1969. Mélanges of San Francisco Peninsula: geologic reinterpretation of type Franciscan, *Am. Assoc. Petrol. Geol. Bull.*, **53**, 1348–67.

Hsü, K. J. and Cita, M. B., 1972. The origin of the Mediterranean evaporite. *In:* Initial Reports of the Deep Sea Drilling Project, vol. 13, ch. 43. U.S. Govt. Printing Office, 1203–1231.

Hsü, K. J. and Ryan, W. B. F., 1972. Summary of the evidence for extensional and compressional tectonics in the Mediterranean. *In:* Initial Reports of the Deep Sea Drilling Project, vol. 13, ch. 37. U.S. Govt. Printing Office, 1011–1019.

Hurley, P. M. and Rand, J. R., 1968. Review of age data in West Africa and South America relative to a test of continental drift. *In*: The History of the Earth's Crust, edited by R. A. Phinney, pp. 153–60. Princeton Univ. Press, 244 pp.

King, P. B., 1959. The Evolution of North America. Princeton University Press, 190 pp.

King, R. E., 1960. Petroleum exploration and production in Europe in 1959, *Bull. Amer. Ass. Petrol. Geol.*, **44,** 1058–1101.

Le Borge, E., Le Mouël, J. L. and Le Pichon, X., 1971. Aeromagnetic survey of Southwestern Europe, *Earth Planet. Sci. Lett.*, **12,** 287–99.

Le Pichon, X. and Sibuet, J. C., 1971. Comments on the evolution of the north-east Atlantic, *Nature*, **233,** 257–8.

Marinelli, G., 1967. Genèse des magmas du volcanisme plio-quaternaire des Apennins, *Geol. Rundsch.*, **57,** 127–41.

Maxwell, J. C., 1953. Review: "Geologia dell'Appennino settentrionale", by Giovanni Merla; I cunei composti nell'orogenesi, by C. I. Migliorini, *Am. Assoc. Petrol. Geol. Bull.*, **37,** no. 9, 2196–2202.

Maxwell, J. C., 1959. Turbidite, tectonic and gravity transport, northern Apennine Mountains, Italy, *Am. Assoc. Petrol. Geol., Bull.*, **43,** 2701–19.

McKenzie, D. P., 1970. Plate tectonics of the Mediterranean region, *Nature*, **226,** 239–43.

Merla, G., 1951. Geologia dell'Appennino settentrionale, *Boll. Soc. Geol. Ital.*, **70,** 95–382.

Mitchell, A. H. and Reading, H. G., 1969. Continental margins, geosynclines, and ocean floor spreading, *J. Geol.*, **77,** 629–43.

Morelli, C., 1970. Physiography, gravity and magnetism of the Tyrrhenian Sea, *Boll. Geofis. Teor. Appl.* (*Trieste*), **12,** 275–309.

Morelli, C., Carrozzo, M. T., Ceccherini, P., Finetti, I., Gantar, C., Pisani, M. and Schmidt di Friedberg, P., 1969. Regional geophysical study of the Adriatic Sea, *Boll. Geophys. Teor. Appl.* (*Trieste*), **11,** no. 41–42, 3–56.

Nairn, A. E. M. and Westphal, M., 1968. Possible implications of the palaeomagnetic study of Late Palaeozoic igneous rocks of northwestern Corsica, *Palaeogeography, Palaeoclimatol., Palaeoecol.*, **5,** 179–204.

Ninkovich, D. and Hays, J. D., 1972. Mediterranean island areas and origin of high potash volcanoes, *Earth and Plan. Sci. Lett.*, **16,** 331–345.

Ogniben, L., 1969. Schema introduttivo alla geologia del confine calabro-lucano, *Mem. Soc. Geol. Ital.*, **8,** 453–763.

Pannekoek, A. J., 1969. Uplift and subsidence in and around the western Mediterranean since the Oligocene: a review, *Verh. Kon. Ned. Geol. Mijnbouwk. Gen.*, **26,** 53–77.

Papazachos, B. C. and Comninakis, P. E., 1971. Geophysical and tectonic features of the Aegean Arc. *J. Geophys. Res.* **76,** 8517–33.

Pichler, H., 1970. Volcanism in eastern Sicily and the Aeolian Islands. *In:* Geology and and History of Sicily, edited by W. Alvarez and K. H. A. Gohrbandt, pp. 261–81. Petr. Expl. Soc. Libya, Tripoli, 291 pp.

Radicati di Brozolo, F. and Giglia, G., 1973. Further data on the Corsica–Sardinia rotation, *Nature*, **241,** 389–391.

Ritsema, A. R., 1969. Seismic data of the west Mediterranean and the problem of oceanization, *Verh. Kon. Ned. Geol. Mijnbouw. Gen.*, **26,** 105–20.

Ryan, W. B. F., Stanley, D. J., Hersey, J. B., Fahlquist, D. A. and Allan, T. D., 1971. The tectonics and geology of the Mediterranean Sea, *The Sea*, **4** (2), 387–492.

Ryan, W. B. F., Hsü, K. J. *et al.*, 1972. Boundary of Sardinia slope with Balearic abyssal plain—Sites 133 and 134. *In:* Initial Reports of the Deep Sea Drilling Project, vol. 13, ch. 14, U.S. Govt. Printing Office, 465–514.

Sestini, G., ed., 1970. Development of the Northern Apennines geosyncline, *Sediment. Geol.*, **4**, 203–647.

Van der Voo, R., 1969. Palaeomagnetic evidence for the rotation of the Iberian Peninsula, *Tectonophys.*, **7**, 5–56.

Vogt, P. R. and Higgs, R. H., 1969. An aeromagnetic survey of the eastern Mediterranean Sea and its interpretation: *Earth Planet Sci. Lett.*, **5**, 439–448.

Vogt, P. R., Higgs, R. H. and Johnson, G. L., 1971. Hypotheses on the origin of the Mediterranean Basin: magnetic data, *J. Geophys. Res.*, **76**, 3207–28.

Wise, D. and Bird, J. M., 1964. International field institute, Italy, 1964, *Geotimes*, **9**, no. 5, 12–16.

Zijderveld, J. D. A., De Jong, K. A. and Van der Voo, R., 1970, Rotation of Sardinia: Palaeomagnetic evidence from Permian rocks, *Nature*, **226**, 933–934.

8.8

E. STUMP

Institute of Polar Studies
and Department of Geology and Mineralogy
The Ohio State University
Columbus, Ohio 43210, U.S.A.

Earth Evolution in the Transantarctic Mountains and West Antarctica*

Introduction

In the last decade many Antarctic workers have offered stratigraphic syntheses and tectonic schemes as the knowledge of the area expanded (see Appendix). This paper traces the evolution of the active portion of the continent from Late Precambrian through Cretaceous by the use of palaeogeographic maps, and suggests some explanations for certain patterns. Much of the information presented must necessarily be speculative due to the great area covered by ice and the paucity of data in some exposed areas, particularly portions of West Antarctica where even reconnaissance coverage is thin. The base map (Fig. 1) is drawn following Elliot's reconstruction (1972) in which West Antarctica is shifted to the west to accommodate straightening of the Scotia Arc (Dalziel & Elliot, 1971). In addition, the Thurston Island area has been distinguished and the Ellsworth Mountains included in their present

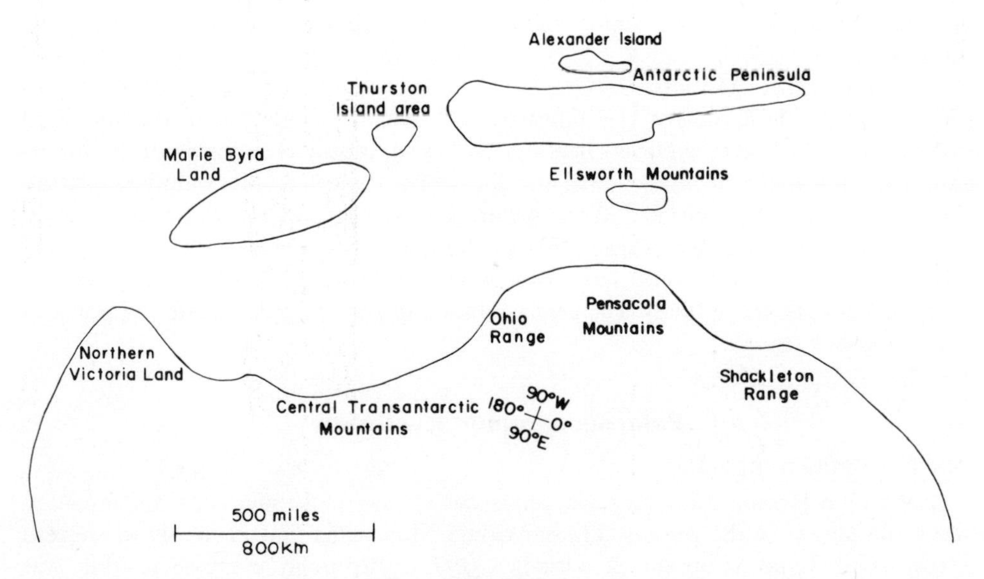

FIGURE 1

* Contribution No. 242 of the Institute of Polar Studies.

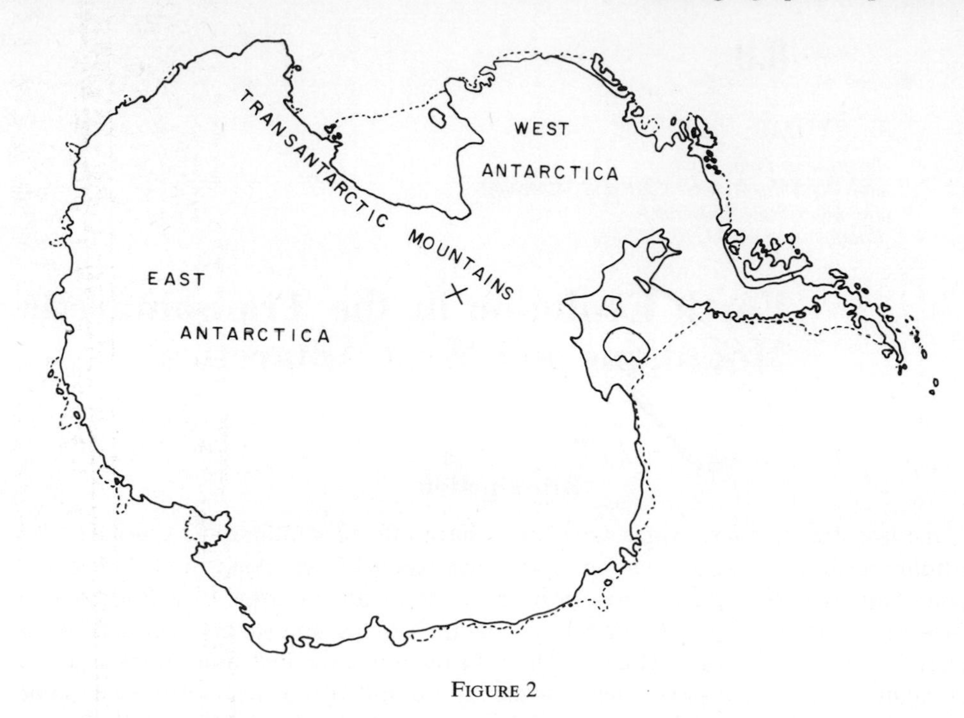

FIGURE 2

location, though rotated into structural parallelism with surrounding areas as suggested by others (Hamilton, 1967; Frakes & Crowell, 1968; Schopf, 1969; Elliot, 1972).

The continent can be subdivided into the provinces of East and West Antarctica on geologic, structural and topographic grounds (see Fig. 2). East Antarctica is primarily a Precambrian shield bounded on one side by younger orogenic belts of the Transantarctic Mountains. Crustal thickness is of the order of 40+ km (Woollard, 1962). If the ice was removed and glacial rebound occurred, the land surface would be above sea level (Bentley, 1965) and, on the craton side of the Transantarctic Mountains, it would have the form of a dissected, block-faulted plateau (Drewry, 1972). By contrast, West Antarctica is a group of continental crustal blocks, 30+ km thick (Woollard, 1962), containing no proven old cratonic material. Without the ice there, several large islands would be found bounded in places by deep trenches (Bentley, 1965). The areas outlined in Fig. 1 approximate the portions that are above sea level.

Palaeogeographic Evolution

Late Precambrian (Fig. 3)

In the Late Precambrian, turbidite sequences of graywacke and shale accumulated along the length of the present Transantarctic Mountains and probably in western Marie Byrd Land (Lopatin & Orlenko, 1972). Although everywhere this was the first depositional phase at the margin of the crystalline craton, the timing was quite different in various areas. For example, deposition began in the Pensacola

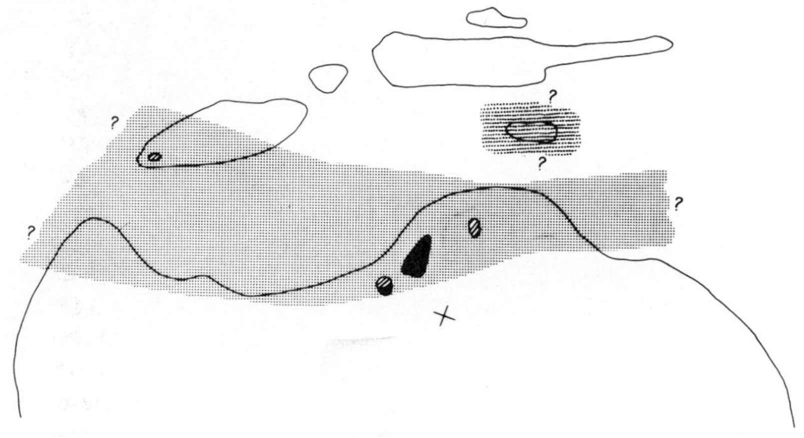

FIGURE 3

Mountains area prior to minor felsic extrusion dated at about 1210 m.y. (Eastin, 1970), whereas fossil spores found in northern Victoria Land indicate that deposition there continued into the Cambrian (Gair *et al.*, 1969). Silicic, pyroclastic volcanics of this period (Minshew, 1967; Ford & Sumsion, 1971) overlie the sediments in the central part of the range. Subsequent calc-alkaline intrusion in the same area (Murtaugh, 1969) occurred at the time of the Beardmore Orogeny (Grindley & McDougall, 1969), marked by isoclinal folding throughout the range, with the possible exception of northern Victoria Land (Gair *et al.*, 1969).

Considerable thicknesses of shallow-marine limestone accumulated in the area of the Ellsworth Mountains prior to known Late Cambrian deposition and are inferred to be Precambrian in age (Craddock, 1969). If the inference is correct, it would indicate a shallowing of the depositional basin at this time toward some area away from the craton.

Cambrian (Fig. 4)

Fossil-dated Cambrian limestones are found throughout the Transantarctic Mountains. Contact with the underlying rocks is unconformable wherever seen (Schmidt *et al.*, 1965; Laird *et al.*, 1971). However, in the Ellsworth Mountains the Late Precambrian(?) limestones pass conformably into Cambrian shallow-marine clastic rocks (Craddock *et al.*, 1964). As in the previous depositional sequence, pyroclastic volcanics mark the final deposits, being interbedded with the limestones in the central part of the range (Minshew, 1967) and overlying them in the Pensacola Mountains (Schmidt *et al.*, 1965). Turbidites of northern Victoria Land were possibly deposited at this time (as mentioned above) in an offshore position from limestones of the region.

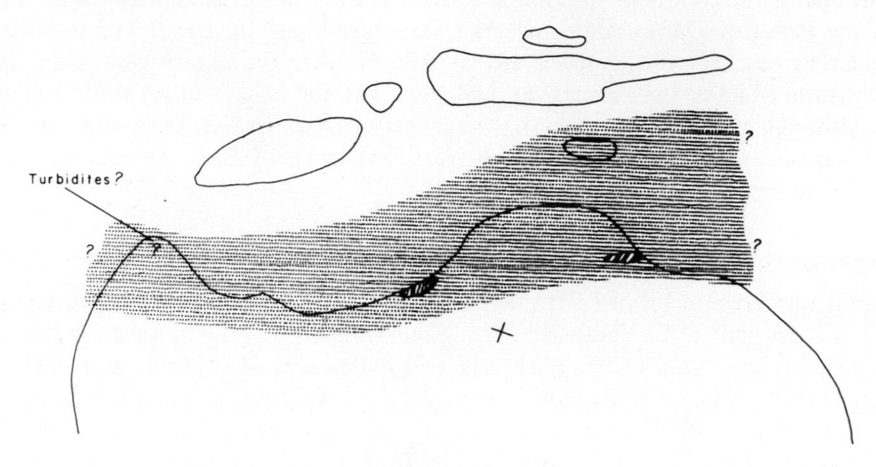

FIGURE 4

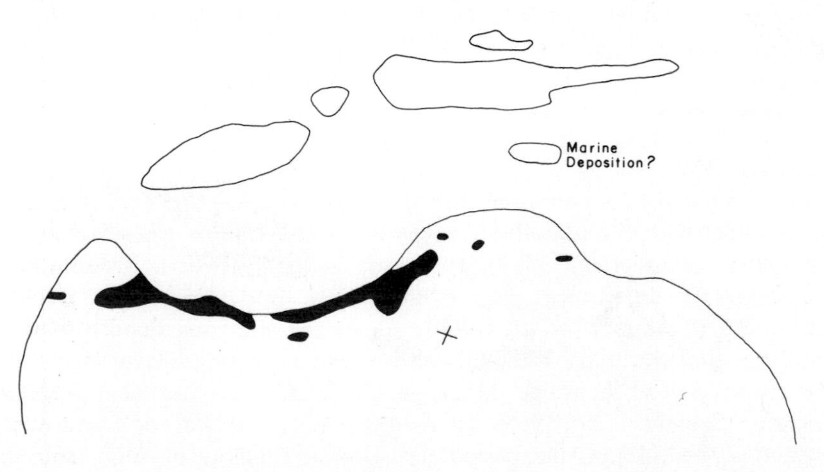

FIGURE 5

Ordovician (Fig. 5)

Volcanics of the Late Cambrian were precursors of major calc-alkaline batholiths intruded throughout the Transantarctic Mountains. Preceding the intrusion in part, intense deformation and metamorphism along the mobile axis culminated in the Ross Orogeny (Gunn & Warren, 1962). Dates of 500± m.y. record a thermal event throughout the continent (Angino & Turner, 1964). The intrusion was less pronounced in the Pensacola Mountains area than elsewhere along the range and is not merely apparent because erosion levels are comparable with the rest of the range, and the structural trend of the Pensacolas is aligned with the axis of intrusion to the west.

Although no Ordovician fossils have been found in the Ellsworth Mountains, the sequence is thick and conformable from the Cambrian to the Devonian (Craddock, 1969), so deposition at this time was likely.

Devonian (Fig. 6)

Uplift and erosion followed the Ross Orogeny and the sea did not again return to the western half of the Transantarctic Mountains area. Fossil-dated shallow-marine deposition continued in the Ellsworth Mountains area (Craddock & Webers, 1964) and the Ohio Range (Doumani *et al.*, 1965). Some volcanic detritus is documented near the top of the Devonian section in the Ellsworths (Craddock, 1969). In the Pensacola Mountains area coarse gravel from local mountains was followed by a probable marine transgression and regression (Schmidt & Ford, 1969), while local basins in the central Transantarctic Mountains received clastic sediments (Barrett *et al.*, 1972).

In northern Victoria Land intrusion occurred in a zone shifted away from the craton with respect to the Ordovician intrusions (Gair *et al.*, 1969) (cf. Figs. 5 and 6). Rhyolite extrusion accompanied this event (Sturm & Carryer, 1970). The regional trend of fold axes in the turbidite sequence is approximately 15° counterclockwise to the Ross trend in the central and eastern portions of the range. The time of folding is interpreted as Silurian on the basis of K-Ar whole rock dates on a shale (see Gair *et al.*, 1969). While this must be viewed with caution, post-Ross orogenic

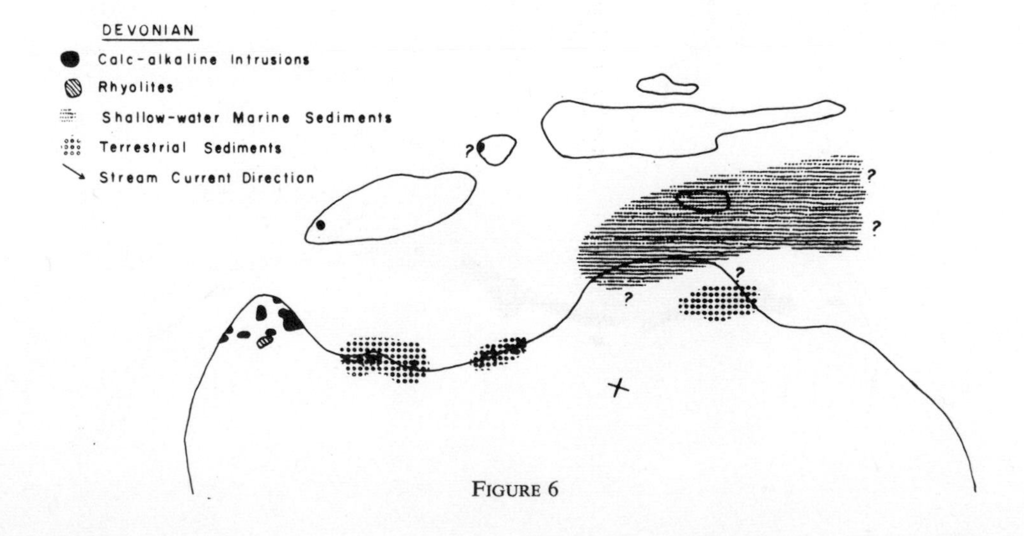

FIGURE 6

activity is nevertheless well established by the intrusive event. One Devonian granodiorite body is known to exist in western Marie Byrd Land (Halpern, 1968) and several Devonian ages are reported for rocks on Thurston Island (Craddock, 1970b). No conclusive evidence from the Antarctic Peninsula has yet been found for orogenic activity at this time.

Permo-Carboniferous (Fig. 7)

The basin configuration established in the Devonian existed during the Late Carboniferous and Early Permian when a major ice sheet was active in the area of the Transantarctic and Ellsworth Mountains. Frakes *et al.* (1971) outline three major basins of deposition, the first being active in the eastern sector. In general, the Ellsworths continued to receive thick sediments in a marine depression followed by continental glaciers depositing tills in the central Transantarctics. The Pensacola Mountains area was transitional between these two facies. Ice flow directions are parallel to the range, with most movement indicated from west to east.

Turbidites mark the first definite deposition in the Antarctic Peninsula area (e.g. Aitkenhead, 1965; Elliot, 1965). Similar rocks are reported from Alexander Island (Grikurov, 1972). It is not known whether these basins were connected to each other or to the glaciomarine basin in the Ellsworth Mountains area.

Post-glacial Permian (Fig. 8)

Following retreat of the glaciers fluvial and lacustrine deposition predominated in both the Transantarctic (Barrett *et al.*, 1972) and Ellsworth Mountains areas (Craddock *et al.*, 1964). Deposition is known to have extended onto the craton in

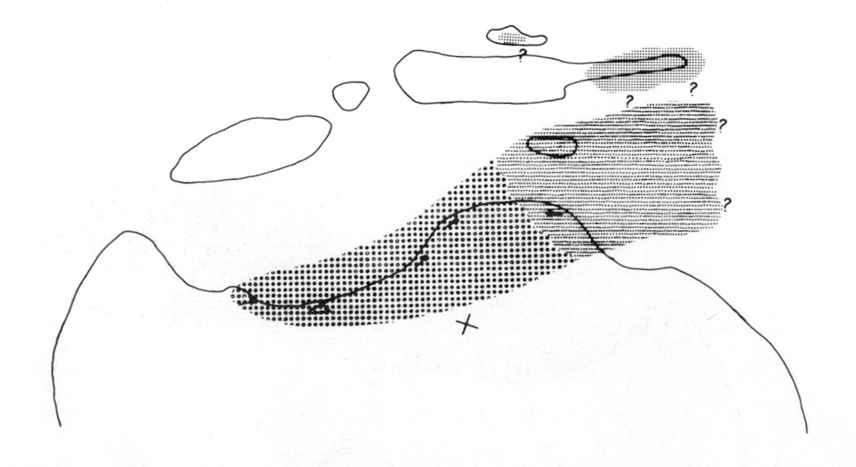

FIGURE 7

the area of the Shackleton Range (Stephenson, 1966) and western Queen Maud Land (Aucamp *et al.*, 1972). Northern Victoria Land remained a source area for most of the period but thin Permian beds are present there in an isolated exposure (Dow & Neall, 1972). Currents flowed predominantly from the west to the east along the axis of the present range (Barrett, 1970). Widespread coal deposition occurred during this interval and an intermediate-acid volcanic source supplied a major clastic fraction to the rock in the central portion of the range (Minshew, 1967; Barrett, 1969) and in the Shackleton Range area (Stephenson, 1966).

Triassic (Fig. 9)

At the beginning of the Triassic, uplift in the eastern sector blocked the drainage outlet and caused a reversal of the regional palaeoslope (Barrett, 1970). This climaxed during the Triassic in the Gondwanide Orogeny which was predicted to have occurred in Antarctica by Du Toit in 1937. Rocks in the Pensacola and Ellsworth Mountains were strongly folded (Craddock, 1969; Ford, 1972) but no calc-alkaline intrusion accompanied deformation at either location. Fluvial deposition continued through the period in the central and western portions of the range, with an ever-increasing fraction of epiclastic volcanics being recorded (Barrett, 1969). By the end of the Triassic(?) the volcanic source reached the central Transantarctic Mountains area as tuffs and associated pyroclastic debris were erupted there (Barrett & Elliot, 1972).

A 200 m.y. whole rock Rb-Sr date on an orthogneiss from the base of the Antarctic Peninsula (Halpern, 1972) probably indicates orogenic activity in that area. In addition, Triassic dates have been determined for some rocks in the Thurston Island area (Craddock, 1970b).

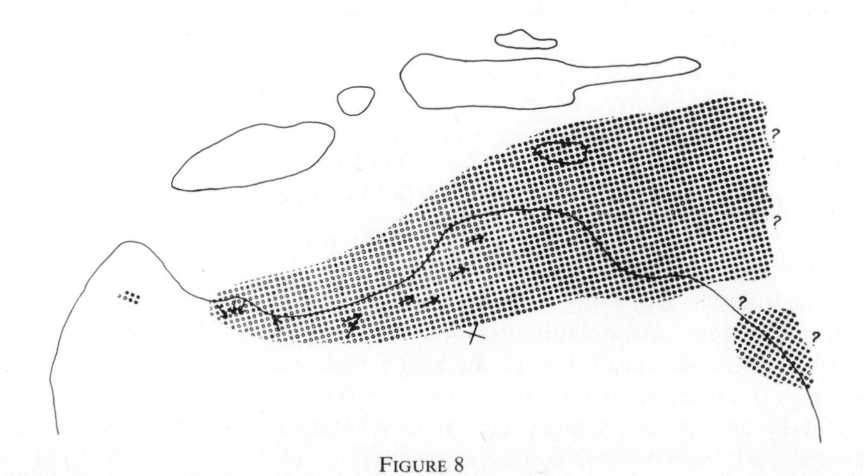

FIGURE 8

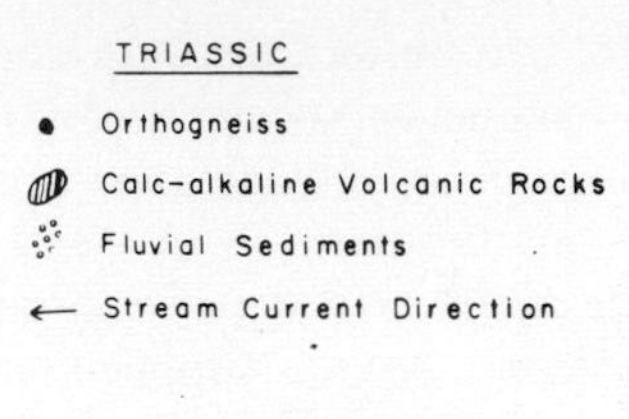

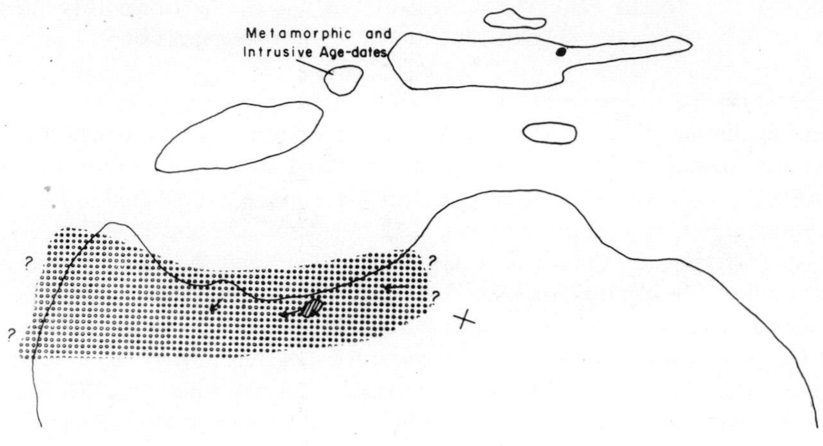

FIGURE 9

Jurassic (Fig. 10)

Shallow-marine basins are indicated in parts of the Antarctic Peninsula during the mid-Jurassic where shales and volcaniclastic sediments were deposited (Bibby, 1966; Laudon, 1972; Williams *et al.*, 1972). However, the extent of these basins is unknown. The Late Jurassic was a time of widespread calc-alkaline volcanism throughout much of the Peninsula area (e.g. Adie, 1964; Curtis, 1966) and to an apparently limited extent in western Marie Byrd Land (Craddock, 1970c). Early to mid-Jurassic plutonism occurred in the area between the Ellsworth and Pensacola Mountains (Halpern, 1966; Craddock, 1970b), and also to a limited extent in western Marie Byrd Land and the Antarctic Peninsula (Boudette *et al.*, 1966; Rex, 1972).

Diabase sills and tholeiitic flood basalts of this period are found throughout the Transantarctic Mountains (Hamilton, 1965; Gunn, 1966; Stephenson, 1966; Elliot, 1970). These are interpreted as precursors to transform faulting between East and West Antarctica accompanying the separation of Africa and Antarctica (Elliot, 1972). This faulting may account for the rotation of the Ellsworth Mountains and the easterly shift of the rest of the West Antarctic blocks in Fig. 11.

Cretaceous (Fig. 11)

An Early Cretaceous trough existed in the area of Alexander Island with shallow-marine sediments accumulating toward the craton and deeper-water sediments farther off shore (Horne, 1969). In the Late Cretaceous a shallow-marine basin was located at the northern end of the Peninsula (Bibby, 1966). The extent of both of these basins was probably much greater than indicated by exposures and included large areas presently covered by ocean.

JURASSIC

- Calc alkaline Intrusions
- Calc alkaline Volcanic Rocks
- Tholeitic Basalts and Diabase Sills
- Shallow-water Marine Sediments

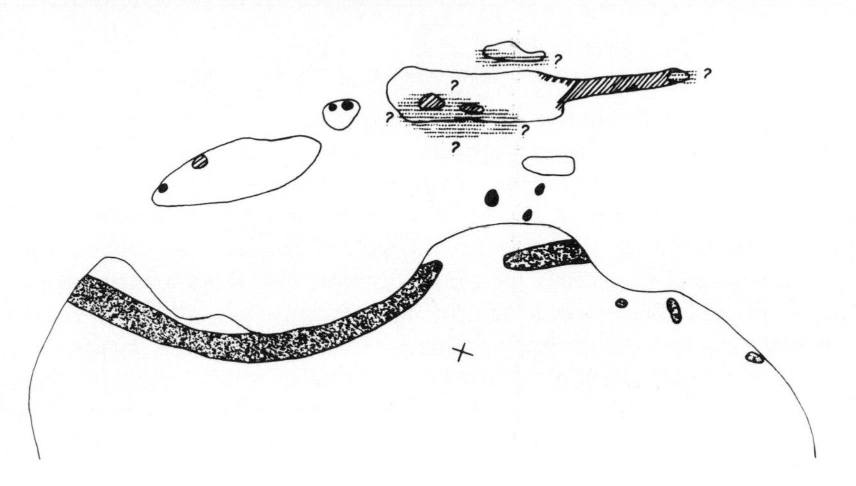

FIGURE 10

CRETACEOUS

- Calc-alkaline Intrusions
- Turbidite Sediments
- Shallow-water Marine Sediments

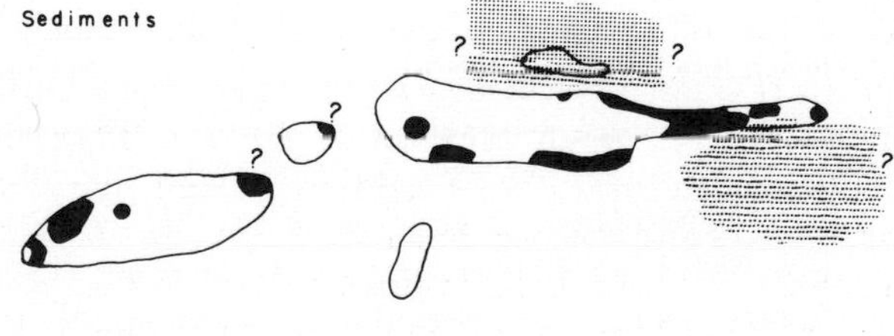

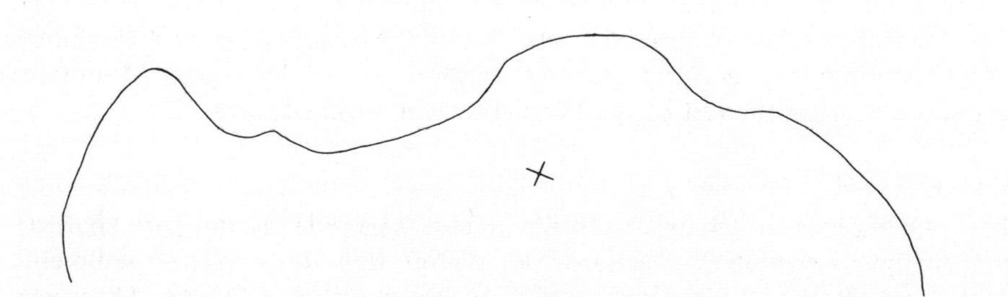

FIGURE 11

The predominant event of the Cretaceous, however, was the Andean Orogeny which produced large volumes of calc-alkaline intrusions from the Antarctic Peninsula to Marie Byrd Land (Adie, 1955, 1969a, b; Wade, 1969).

The subsequent Tertiary evolution of the Peninsula and Scotia Arc areas will not be discussed here, but are given by Dalziel & Elliot (in press).

Discussion

Recapitulating the information from the foregoing: a compressive margin to Antarctica is indicated by (1) calc-alkaline intrusion which occurred in the Late Precambrian, Ordovician, Devonian, Triassic(?), Jurassic and Cretaceous; (2) by calc-alkaline volcanism which is substantiated by deposits or implied by clastic fractions in sediments in the Late Precambrian, Cambrian, Devonian, Permian, Triassic, Jurassic and Cretaceous; and (3) by compressional deformation which was recorded in the Late Precambrian, Ordovician, Silurian(?), Triassic and Cretaceous. If it is correct that episodic orogeny is a manifestation of persistent subduction, then nearly continuous subduction is indicated for the area of Antarctica under consideration from the Late Precambrian through the Cretaceous. This is not to imply, however, that significant adjustments within an evolving plate system did not occur.

Continental Accretion

Many authors have recognized a progression of orogenic belts away from the craton through time in Antarctica (see Tectonic reference in Appendix). A horizontal accretion of crustal material is best appreciated by observing the relative locations of the Ross and Andean orogenic belts. Activity in the intervening period apparently did not produce such through-going effects. These will be discussed in the next section.

A vertical accretion may be inferred by considering the depth of the depositional basin through time along the axis of the present Transantarctic Mountains. Initially, deep-water turbidite deposition occurred. Following the Beardmore Orogeny, shallow-marine limestones were deposited and, following the Ross Orogeny, terrestrial sediments accumulated. If it is assumed that erosion established 'stable' isostatic conditions following each orogeny, then the successively shallower basin at the onset of each new depositional period would indicate that during the previous orogeny the generation of new continental material resulted in a higher buoying of the crust. In the eastern sector, where substantially smaller volumes of magma were generated during the Ross Orogeny, the basin remained deeper.

Lateral Variation

Barrett *et al.* (1972) postulate a geanticlinal ridge confining a through-going intracratonic geosyncline in this area during the Late Palaeozoic. Such an elevated area seems necessary to account for the ice movement and stream current data, and offers a plausible source for the volcanic detritus found in the terrestrial sediments. However, consideration of the palaeogeographic maps presented in this paper would imply that there was a basin of deposition widening and deepening to the east but not extant in the area of northern Victoria Land. Frakes & Crowell (1968) postulate such a basin. I visualize a configuration for Antarctica at this time which was similar to the present area where the Aleutian Arc meets the North American continent; that is, subduction occurring beneath a region which transits from continent to island arc and marginal sea.

Devonian igneous activity is established on the craton in northern Victoria Land and western Marie Byrd Land. No definite extension of this has been documented in the Peninsula area (but see Craddock, 1970d) although volcanic detritus in the Devonian sediments of the Ellsworth Mountains possibly indicates a contiguous source. The Carboniferous was apparently a period of inactivity, however, in the Permian volcaniclastics accumulated throughout the Transantarctic Mountains area.

A major tectonic re-orientation following the Ross Orogeny may be indicated by the regional trend of the folds in northern Victoria Land. If such a trend were followed as far as the Peninsula area, the zone of subduction there would be considerably removed from its location during the Ross Orogeny.

Parallelism of the Ross and post-Ross subduction zones would, however, only mean that the Late Palaeozoic marine basin to the east was narrower and not that a given depositional environment was necessarily continuous. That a deeper basin existed to the east during the Late Palaeozoic and that orogenic activity followed in the Triassic may be related to the apparently small volume of magma generated in that area during the Ross Orogeny. Following arguments of the 'Accretion' section, I suggest that the Silurian(?) deformation of the turbidites and the Devonian intrusion in northern Victoria Land occurred due to subduction beneath a thicker and more rigid continental crustal margin, while deposition was occurring at the same time to the east where the crust was thinner, more flexible, and thus prone to downwarping when submitted to the same regional stresses.

In the Triassic, the thickening accumulations in the eastern marine basin were deformed, possibly by lateral pressure from an expanding mobile core (Dewey & Bird, 1970) in the area of the suggested geanticlinal ridge (Barrett *et al.*, 1972). The lack of intrusion at this time in the Ellsworth and Pensacola Mountains supports this hypothesis. Synchronous folding is not seen in the western sector where the crust is postulated to have been more rigid.

Acknowledgements

David H. Elliot first turned me on to Antarctica's desolate beauty. He and Gunter Faure have engaged me in many hours of stimulating discussion regarding my work. Charles E. Corbató, Helmut E. Ehrenspeck, David H. Elliot, Gunter Faure and John Splettstoesser kindly reviewed the manuscript. Support has come from NSF Grant No. GV–26652.

Appendix

This compilation is hoped to be an aid to those unfamiliar with the literature of the Transantarctic Mountains and West Antarctica. I have listed references in the following manner:

(1) A.G.S. Folio. The American Geographical Society Antarctic Map Folio 12 (1969, 1970) contains regional maps with accompanying texts and bibliographies. A good, visual starting place.
(2) Whole Treatments. Review papers, each covering the geology of the entire area.
(3) Tectonics. Primarily concerned with orogenic evolution.
(4) Drift. Global tectonic considerations, Gondwanaland recontructions.
(5) Specific Areas. Broad approaches to particular geographic regions or rock groups.

A complete compilation of radiometric age dates through 1969 exists in Craddock, 1970b. Likewise, all known fossil localities through 1969 are compiled in Craddock, 1970a.

A.G.S. Folio	Whole Treatments	Tectonics
Adie, 1969a	Adie, 1962	Angino & Turner, 1964
Adie, 1969b	Adie, 1964	Craddock, 1970d
Craddock, 1969	Anderson, 1965	Grikurov, 1972
Gair *et al.*, 1969	Craddock, 1970c	Grikurov *et al.*, 1972
Grindley & Laird, 1969	Ford, 1964	Hamilton, 1964
Laudon *et al.*, 1969	Gunn, 1963	Hamilton, 1967
McGregor & Wade, 1969	Harrington, 1965	Voronov, 1964
Mirsky, 1969	Klimov *et al.*, 1964	
Schmidt & Ford, 1969	Warren, 1965	
Wade, 1969		
Warren, 1969		

Drift	Specific Areas
Dietz & Holden, 1970	Barrett, 1971
Dietz *et al.*, 1972	Barrett *et al.*, 1972
Dietz & Sproll, 1970	Dalziel & Elliot, in press
Du Toit, 1937	Frakes & Crowell, 1968
Elliot, 1972	Grindley & McDougall, 1969
Scharnberger & Scharon, 1972	Grindley & Warren, 1964
Schopf, 1969	Lopatin & Orlenko, 1970
Smith & Hallam, 1970	Wade & Wilbanks, 1972
Van der Linden, 1969	

References

Adie, R. J., 1955. The petrology of Graham Land: II. The Andean granite-gabbro intrusive suite. *Falkland Islands Dependencies Survey Sci. Rpt.*, *12*, 39 pp.

Adie, R. J., 1962. The geology of Antarctica, *Geophys. Monograph 7*, *Am. Geophys. Union*, 26–39.

Adie, R. J., 1964. Geological history. *In:* Antarctic Research, edited by R. Priestley, R. J. Adie and G. DeQ. Robin, pp. 118–62. Butterworths, London.

Adie, R. J., 1969a. Northern Antarctic Peninsula, *Am. Geog. Soc. Map Folio Series*, **12,** Plate I.

Adie, R. J., 1969b. Southern Antarctic Peninsula, *Am. Geog. Soc. Map Folio Series*, **12,** Plate II.

Aitkenhead, N., 1965. The geology of the Duse Bay–Larsen Inlet area, northeast Graham Land (with particular reference to the Trinity Peninsula Series), *Brit. Ant. Surv. Sci. Rpt.*, *51*, 62 pp.

Anderson, J. J., 1965. Bedrock geology of Antarctica: a summary of exploration: 1831–1962, *Ant. Res. Ser.*, *6*, *Am. Geophys. Union*, 1–70.

Angino, E. E. and Turner, M. D., 1964. Antarctic orogenic belts as delineated by absolute age dates. *In:* Antarctic Geology, edited by R. J. Adie, pp. 551–6. North-Holland, Amsterdam.

Aucamp, A. P. H., Wolmarans, L. G. and Neethling, D. C., 1972. The Urfjell Group, a deformed (?)Early Palaeozoic sedimentary sequence, Kirwanveggen, Western Dronning Maud Land. *In:* Antarctic Geology and Geophysics, edited by R. J. Adie, pp. 557–62. Universitetsforlaget, Oslo.

Barrett, P. J., 1969. Stratigraphy and petrology of the mainly fluviatile Permian and Triassic rocks from Beardmore Glacier area, Antarctica, *Inst. Polar Studies Rpt. 34*, *Ohio State Univ. Research Foundation*, *Columbus*, *Ohio*, 132 pp.

Barrett, P. J., 1970. Paleocurrent analysis of the mainly fluviatile Permian and Triassic Beacon rocks, Beardmore Glacier area, Transantarctic Mountains, Antarctica, *J. Sed. Pet.*, **40,** 395–411.

Barrett, P. J., 1971. Stratigraphy and paleogeography of the Beacon Supergroup in the Transantarctic Mountains, Antarctica. Second Gondwana Symposium, South Africa, pp. 249–56. Council for Sci. and Industrial Res., Pretoria.

Barrett, P. J. and Elliot, D. H., 1972. The early Mesozoic volcaniclastic Prebble Formation, Beardmore Glacier. *In:* Antarctic Geology and Geophysics, edited by R. J. Adie, pp. 403–9. Universitetsforlaget, Oslo.

Barrett, P. J., Grindley, G. W. and Webb, P. N., 1972. The Beacon Supergroup of East Antarctica. *In:* Antarctic Geology and Geophysics, edited by R. J. Adie, pp. 319–22. Universitetsforlaget, Oslo.

Bentley, C. R., 1965. The land beneath the ice. *In:* Antarctica, edited by T. Hatherton, pp. 259–77. Reed, Wellington.

Bibby, J. S., 1966. The stratigraphy of part of north-west Graham Land and the James Ross Island Group, *Brit. Ant. Surv. Sci. Rpt.*, *44*, 33 pp.

Boudette, E. L., Marvin, R. F. and Hedge, C. E., 1966. Biotite, potassium-feldspar, and whole-rock ages from adamellite, Clark Mountains, West Antarctica, *U.S. Geol. Surv. Prof. Paper 550–D*, D190–D194.

Craddock, C., 1969. Geology of the Ellsworth Mountains, *Am. Geog. Soc. Map Folio Series*, **12,** Plate IV.

Craddock, C., 1970a. Fossil map of Antarctica, *Am. Geog. Soc. Map Folio Series*, **12,** Plate XVIII.

Craddock, C., 1970b. Radiometric map of Antarctica, *Am. Geog. Soc. Map Folio Series*, **12,** Plate XIX.

Craddock, C., 1970c. Geologic map of Antarctica, *Am. Geog. Soc. Map Folio Series*, **12,** Plate XX.

Craddock, C., 1970d. Tectonic map of Antarctica, *Am. Geog. Soc. Map Folio Series*, **12,** Plate XXI.

Craddock, C., Anderson, J. J. and Webers, G. F., 1964. Geologic outline of the Ellsworth Mountains. *In:* Antarctic Geology, edited by R. J. Adie, pp. 155–70. North-Holland, Amsterdam.

Craddock, C. and Webers, G. F., 1964. Fossils from the Ellsworth Mountains, Antarctica, *Nature*, **201,** 174–5.

Curtis, R., 1966. The petrology of the Graham Coast, Graham Land, *Brit. Ant. Surv. Rpt.*, *50*, 51 pp.

Dalziel, I. W. D. and Elliot, D. H., 1971. Evolution of the Scotia Arc, *Nature*, **233,** 246–52.

Dalziel, I. W. D. and Elliot, D. H., in press. The Scotia Arc and Antarctic Margin. *In:* The Ocean Basins and Continental Margins: 1. The South Atlantic, edited by F. G. Stehli and A. E. M. Nairn. Plenum Pub. Corp., New York.

Dewey, J. F. and Bird, B. J. M., 1970. Mountains and belts and the new global tectonics, *J. Geophys. Res.*, **75,** 2625–47.

Dietz, R. S. and Holden, J. C., 1970. Reconstruction of Pangea, its breakup and dispersion of continents—Permian to Present, *J. Geophys. Res.*, **75,** 4939–56.

Dietz, R. S., Holden, J. C. and Sproll, W. P., 1972. Antarctica and continental drift. *In:* Antarctic Geology and Geophysics, edited by R. J. Adie, pp. 837–42. Universitetsforlaget, Oslo.

Dietz, R. S. and Sproll, W. P., 1970. Fit between Africa and Antarctica: a continental drift reconstruction, *Science*, **167,** 1612–14.

Doumani, G. A., Boardman, R. S., Rowell, A. J., Boucot, A. J., Johnson, J. G., McAlester, A. L., Saul, J., Fisher, D. W. and Miles, R. S., 1965. Lower Devonian fauna of the Horlick Formation, Ohio Range, Antarctica, *Ant. Res. Ser.*, *6*, *Am. Geophys. Union*, 241–81.

Dow, J. A. S. and Neall, V. E., 1972. Summary of the geology of lower Rennick Glacier, northern Victoria Land. *In:* Antarctic Geology and Geophysics, edited by R. J. Adie, pp. 339–44. Universitetsforlaget, Oslo.

Drewry, D. J., 1972. Subglacial morphology between the Transantarctic mountains and the South Pole. *In:* Antarctic Geology and Geophysics, edited by R. J. Adie, pp. 693–703. Universitetsforlaget, Oslo.

Du Toit, A. L., 1937. Our Wandering Continents. Oliver and Boyd, Edinburgh, 336 pp.

Eastin, René, 1970. Geochronology of the basement rocks of the central Transantarctic Mountains, Antarctica. Unpub. Ph.D. dissertation, The Ohio State University, Columbus, Ohio.

Elliot, D. H., 1965. Geology of the north-west Trinity Peninsula, Graham Land, *Brit. Ant. Surv. Bull.*, **7**, 1–24.

Elliot, D. H., 1970. Jurassic tholeiitic basalts of the central Transantarctic Mountains, Antarctica. Second Columbia River Basalt Symposium, Cheney, Washington, 301–25.

Elliot, D. H., 1972. Aspects of Antarctic geology and drift reconstructions. *In:* Antarctic Geology and Geophysics, edited by R. J. Adie, pp. 849–58. Universitetsforlaget, Oslo.

Ford, A. B., 1964. Review of Antarctic geology, *Trans. Am. Geophys. Union*, **45,** 363–81.

Ford, A. B., 1972. Weddell Orogeny—latest Permian to early Mesozoic deformation at the Weddell Sea margin of the Transantarctic Mountains. *In:* Antarctic Geology and Geophysics, edited by R. J. Adie, pp. 419–25. Universitetsforlaget, Oslo.

Ford, A. B. and Sumsion, B. S., 1971. Late Precambrian silicic pyroclastic volcanism in the Thiel Mountains, *Ant. J. U.S.*, **6,** 185–6.

Frakes, L. A. and Crowell, J. C., 1968. Late Palaeozoic glacial facies and the origin of the south Atlantic basin, *Nature*, **217,** 837–8.

Frakes, L. A., Matthews, J. L. and Crowell, J. C., 1971. Late Paleozoic glaciation: part III, Antarctica, *Geol. Soc. Am. Bull.*, **82,** 1581–604.

Gair, H. S., Sturm, A., Carryer, S. J. and Grindley, G. W., 1969. The geology of northern Victoria Land, *Am. Geog. Soc. Map Folio Series*, **12,** Plate XII.

Grikurov, G. E., 1972. Tectonics of the Antarctandes. *In:* Antarctic Geology and Geophysics, edited by R. J. Adie, pp. 163–7. Universitetsforlaget, Oslo.

Grikurov, G. E., Ravich, M. G. and Soloviev, D. S., 1972. Tectonics of Antarctica. *In:* Antarctic Geology and Geophysics, edited by R. J. Adie, pp. 457–68. Universitetsforlaget, Oslo.

Grindley, G. W. and Laird, M. G., 1969. Geology of the Shackleton Coast, *Am. Geog. Soc. Map Folio Series*, **12,** Plate XIV.

Grindley, G. W. and McDougall, I., 1969. Age correlation of the Nimrod group and other Precambrian rock units in the central Transantarctic Mountains, Antarctica, *New Zealand J. Geol. Geophys.*, **12,** 391–411.

Grindley, G. W. and Warren, G., 1964. Stratigraphic nomenclature and correlation in the western Ross Sea region. *In:* Antarctic Geology, edited by R. J. Adie, pp. 314–33. North-Holland, Amsterdam.

Gunn, B. M., 1963. Geological structure and stratigraphic correlation in Antarctica, *New Zealand J. Geol. Geophys.*, **6,** 423–33.

Gunn, B. M., 1966. Modal and element variation in Antarctic tholeiites, *Geochim. et Cosmochim. Acta*, **30,** 881–920.

Gunn, B. M. and Warren, G., 1962. Geology of Victoria Land between Mawson and Mulock Glaciers, Antarctica, *New Zealand Geol. Surv. Bull.*, **71,** 157 pp.

Halpern, M., 1966. Rubidium-strontium date from Mt. Byerly, West Antarctica, *Earth Planet. Sci. Letters*, **1,** 455–7.

Halpern, M., 1968. Ages of Antarctic and Argentine rocks bearing on continental drift, *Earth Planet. Sci. Letters*, **5,** 159–67.

Halpern, M., 1972. Rb-Sr Total-rock and mineral ages from the Marguerite Bay area, Kohler Range, and Fosdick Mountains. *In:* Antarctic Geology and Geophysics, edited by R. J. Adie, pp. 197–204. Universitetsforlaget, Oslo.

Hamilton, W., 1964. Tectonic map of Antarctica—a progress report. *In:* Antarctic Geology, edited by R. J. Adie, pp. 676–9. North-Holland, Amsterdam.

Hamilton, W., 1965. Diabase sheets of the Taylor Glacier region, Victoria Land, Antarctica, *U.S. Geol. Surv. Prof. Paper 456–B*, 71 pp.

Hamilton, W., 1967. Tectonics of Antarctica, *Tectonophysics*, **4,** 555–68.

Harrington, H. J., 1965. Geology and morphology of Antarctica. *In:* Biogeography and Ecology in Antarctica—Monographiae Biologicae, edited by J. van Mieghem and P. van Oye, 15, pp. 1–71.

Horne, R. R., 1969. Sedimentology and palaeogeography of the lower Cretaceous depositional trough of south-eastern Alexander Island, *Brit. Ant. Surv. Bull.*, **22,** 60–76.

Klimov, L. V., Ravich, M. G. and Soloviev, D. S., 1964. Geology of the Antarctic platform. *In:* Antarctic Geology, edited by R. J. Adie, pp. 681–91. North-Holland, Amsterdam.

Laird, M. G., Mansergh, G. D. and Chappell, J. M. A., 1971. Geology of the central Nimrod Glacier area, Antarctica, *New Zealand J. Geol. Geophys.*, **14,** 427–68.

Laudon, T. S., 1972. Stratigraphy of eastern Ellsworth Land. *In:* Antarctic Geology and Geophysics, edited by R. J. Adie, pp. 215–23. Universitetsforlaget, Oslo.

Laudon, T. S., Lackey, L. L., Quilty, P. G. and Otway, P. M., 1969. Geology of eastern Ellsworth Land, *Am. Geog. Soc. Map Folio Series*, **12,** Plate III.

Lopatin, B. G. and Orlenko, E. M., 1972. Outline of the geology of Marie Byrd Land and the Eights Coast. *In:* Antarctic Geology and Geophysics, edited by R. J. Adie, pp. 245–50. Universitetsforlaget, Oslo.

McGregor, V. R. and Wade, F. A., 1969. Geology of the western Queen Maud Mountains, *Am. Geog. Soc. Map Folio Series*, **12,** Plate XV.

Minshew, V. H., 1967. Geology of the Scott Glacier and Wisconsin Range areas, central Transantarctic Mountains, Antarctica. Unpub. Ph.D. dissertation, The Ohio State University, Columbus, Ohio.

Mirsky, A., 1969. Geology of the Ohio Range–Liv Glacier area, *Am. Geog. Soc. Map Folio Series*, **12,** Plate XVI.

Murtaugh, J. G., 1969. Geology of the Wisconsin Range batholith, Transantarctic Mountains, *New Zealand J. Geol. Geophys.*, **12,** 526–50.

Rex, D. C., 1972. K-Ar age determinations on volcanic and associated rocks from the Antarctic Peninsula and Dronning Maud Land. *In:* Antarctic Geology and Geophysics, edited by R. J. Adie, pp. 133–6. Universitetsforlaget, Oslo.

Scharnberger, C. K. and Scharon, L., 1972. Palaeomagnetism and plate tectonics of Antarctica. *In:* Antarctic Geology and Geophysics, edited by R. J. Adie, pp. 843–7. Universitetsforlaget, Oslo.

Schmidt, D. L. and Ford, A. B., 1969. Geology of the Pensacola and Thiel Mountains, *Am. Geog. Soc. Map Folio Series*, **12,** Plate V.

Schmidt, D. L., Williams, P. L., Nelson, W. H. and Ege, J. R., 1965. Upper Precambrian and Paleozoic stratigraphy and structure of the Neptune Range, Antarctica, *U.S. Geol. Surv. Prof. Paper 525–D*, D112–D119.

Schopf, J. M., 1969. Ellsworth Mountains: position in West Antarctica due to sea-floor spreading, *Science*, **164,** 62–66.

Smith, A. and Hallam, A., 1970. The fit of the southern continents, *Nature*, **223,** 139–44.

Stephenson, P. J., 1966. Theron Mountains, Shackleton Range and Whichaway Nunataks with a section on palaeomagnetism of the dolerite intrusions by D. J. Blundell, *Trans-Antarctic Expedition 1955–58, Sci. Rpt. 8, Geology*, 79 pp.

Sturm, A. G. and Carryer, S., 1970. Geology of the region between Matusevich and Tucker Glaciers, northern Victoria Land, Antarctica, *New Zealand J. Geol. Geophys*, **13**, 408–35.

Van der Linden, W. J. M., 1969. Rotation of the Melanesian complex and of West Antarctica —a key to the configuration of Gondwana?, *Palaeog., Palaeocl., Palaeoec.*, **6**, 37–44.

Voronov, P. S., 1964. Tectonics and neotectonics of Antarctica. *In:* Antarctic Geology, edited by R. J. Adie, pp. 692–700. North-Holland, Amsterdam.

Wade, F. A., 1969. Geology of Marie Byrd Land, *Am. Geog. Soc. Map Folio Series*, **12,** Plate XVII.

Wade, F. A. and Wilbanks, J. R., 1972. Geology of Marie Byrd and Ellsworth Lands. *In:* Antarctic Geology and Geophysics, edited by R. J. Adie, pp. 207–14. Universitetsforlaget, Oslo.

Warren, Guyon, 1965. Geology of Antarctica. *In:* Antarctica, edited by T. Hatherton, pp. 279–320. Reed, Wellington.

Warren, Guyon, 1969. Geology of the Terra Nova Bay–McMurdo Sound area, Victoria Land, *Am. Geog. Soc. Map Folio Series*, **12,** Plate XIII.

Williams, P. W., Schmidt, D. L., Plummer, C. C. and Brown, L. E., 1972. Geology of the Lassiter Coast area, Antarctic Peninsula: preliminary report. *In:* Antarctic Geology and Geophysics, edited by R. J. Adie, pp. 143–8. Universitetsforlaget, Oslo.

Woollard, G. P., 1962. Crustal structure in Antarctica, *Geophys. Monograph 7, Am. Geophys. Union*, 53–73.

8.9

P. R. DAWES
Geological Survey of Greenland,
Copenhagen, Denmark

The North Greenland Fold Belt: A Clue to the History of the Arctic Ocean Basin and the Nares Strait Lineament

Introduction

In continental drift studies of the North Atlantic and Arctic regions, there has been a tendency for Greenland to be almost nonchalantly accommodated into hypotheses or models explaining continent-ocean relationships. This is illustrated by the fact that the fold belt flanking Greenland's northern coast has at various times been branded as Caledonian, Hercynian and Alpine in age and, as such, fitted into the regional tectonics. During the past few years investigations by the Geological Survey of Greenland and others have clarified many aspects of the bedrock geology of the extreme northern part of Greenland and papers reporting our present knowledge of this region have been presented (Dawes, 1971; Dawes & Soper, in press).

The purpose of the present article is to discuss some features of the geology of North Greenland with respect to the structure of the adjacent part of the Arctic Ocean Basin and the Nares Strait lineament. It is far from intended as a comprehensive treatment of the evolution of the Arctic and North Atlantic Ocean system but rather as an initial work on which it is intended larger syntheses will be based.

The Geological Setting

The Greenland crystalline shield, which is exposed at the margins of the Inland Ice, passes to the north under a platform cover extending from the west to east coasts, and composed of thick unmetamorphosed Late Precambrian and Lower Palaeozoic strata. To the north these rocks give way to a thick suite of Lower Palaeozoic sediments of the North Greenland geosyncline, which in the southern part are undeformed and dip shallowly to the north as a continuation of the stable platform block. In the north these sediements become involved in a roughly east–west trending belt of folding and metamorphism (the North Greenland fold belt) parallel to the north coast. This belt, which contains the world's most northerly exposures of crustal rocks, forms a border to the Arctic Ocean and is separated by the Nares Strait lineament from the larger Canadian part of the orogenic system (see Fig. 1).

The structural nature of the Arctic Ocean Basin has been under steady review

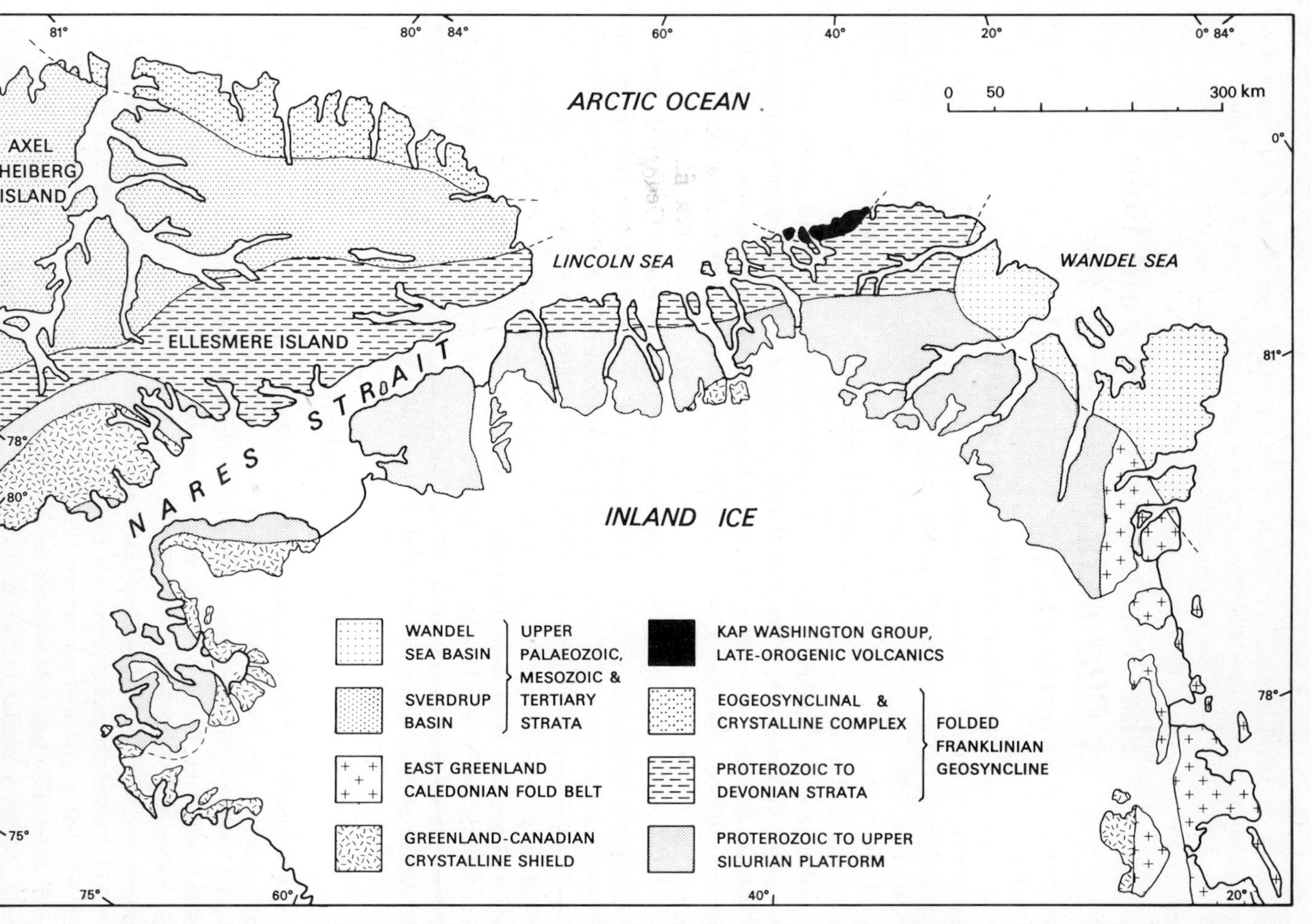

FIGURE 1 Simplified sketch map of the northern part of Greenland and adjacent Canada showing the main structural-stratigraphical units.

for many years (Oliver *et al.*, 1955; Hope, 1959; Eardley, 1961; Ostenso, 1962; Demenitskaya *et al.*, 1962; King *et al.*, 1966; Gakkel & Dibner, 1967; Churkin, 1969; Demenitskaya & Hunkins, 1970; Ostenso & Wold, 1971). Recently, with the progress in deep sea research and with the advent of the theories of ocean-floor spreading and plate tectonics (Dietz, 1961; Hess, 1962; Vine & Matthews, 1963; Vogt & Ostenso, 1967; McKenzie & Parker, 1967; Le Pichon, 1968), interest in the evolution of the ocean floor and ocean-continent relations has never been greater.

Morphology of the Arctic Ocean Basin

Since the initial soundings by Fridtjof Nansen (1904) proved that the Arctic Ocean is a deep water basin, the main morphological features of the basin have been revealed during the last two decades through various submarine traverses, ice-island programmes and airborne surveys (Webster, 1954; Dietz & Shumway, 1961; Hunkins, 1961; Heezen & Ewing, 1961; Ostenso, 1962; Atlasov *et al.*, 1964; Dibner *et al.*, 1965; Kutschale, 1966). It is now clear that the basin is not a simple deep structure with uniform character but that it is made up of individual basins separated by submarine ridges of contrasting morphology.

The Lomonosov Ridge, a flat-topped relatively narrow structure, which reaches to within 954 m of the ocean surface, splits the basin into two unequal parts; the Eurasian and the Amerasian Basins. This ridge extends for 2000 km from the continental shelf off Ellesmere Island to the shelf off the New Siberian Islands of Russia (see Fig. 2).

The Eurasian Basin is divided by the seismically active Nansen Cordillera (known also as the Arctic Mid-Oceanic Ridge, the Nansen Ridge and the Gakkel Ridge), which as a northerly extension of the Atlantic Mid-Oceanic Ridge reaches the Arctic Basin via the Lena Trough between Spitsbergen and Greenland. Major fracture zones, which characteristically offset the ridge in the Greenland Sea (Johnson & Heezen, 1967b) are apparently absent or poorly developed in the Eurasian Basin.

The Amerasian Basin is crossed by the Alpha Cordillera (known also as the Alpha Rise, Alpha Range, Alpha Ridge, or the Mendeleyev Ridge), a broad fractured ridge of rugged topography and complex morphology which reaches to about 1800 m below the ocean surface and which divides the basin into two; the Canada Basin and the much smaller Marakov or Central Arctic Basin (see Fig. 2).

Origin of the Ridge System

The three celebrated ridges of the Arctic Ocean are parallel. The aseismic Lomonosov Ridge, first discovered in 1948 (Webster, 1954) has been interpreted in a number of ways. Early Soviet reports indicate the ridge as a range of Mesozoic or Tertiary tectonics (Panov, 1955; Saks *et al.*, 1955; Gakkel, 1961) with some evidence of recent submarine volcanicity (Gakkel, 1958). More recently Gakkel & Dibner (1967), Egiazarov *et al.* (1969) and Meyerhoff (1970) treat the ridge as a Caledonian fold structure. Carey (1958) described the ridge as a 'nematath' and Vogt *et al.* (1969) regard it as a possible 'microcontinent'. Whatever its age, the Lomonosov Ridge is now generally considered to be formed of continental material and many workers (e.g. Harland, 1966; Johnson & Heezen, 1967a; Demenitskaya & Karasik,

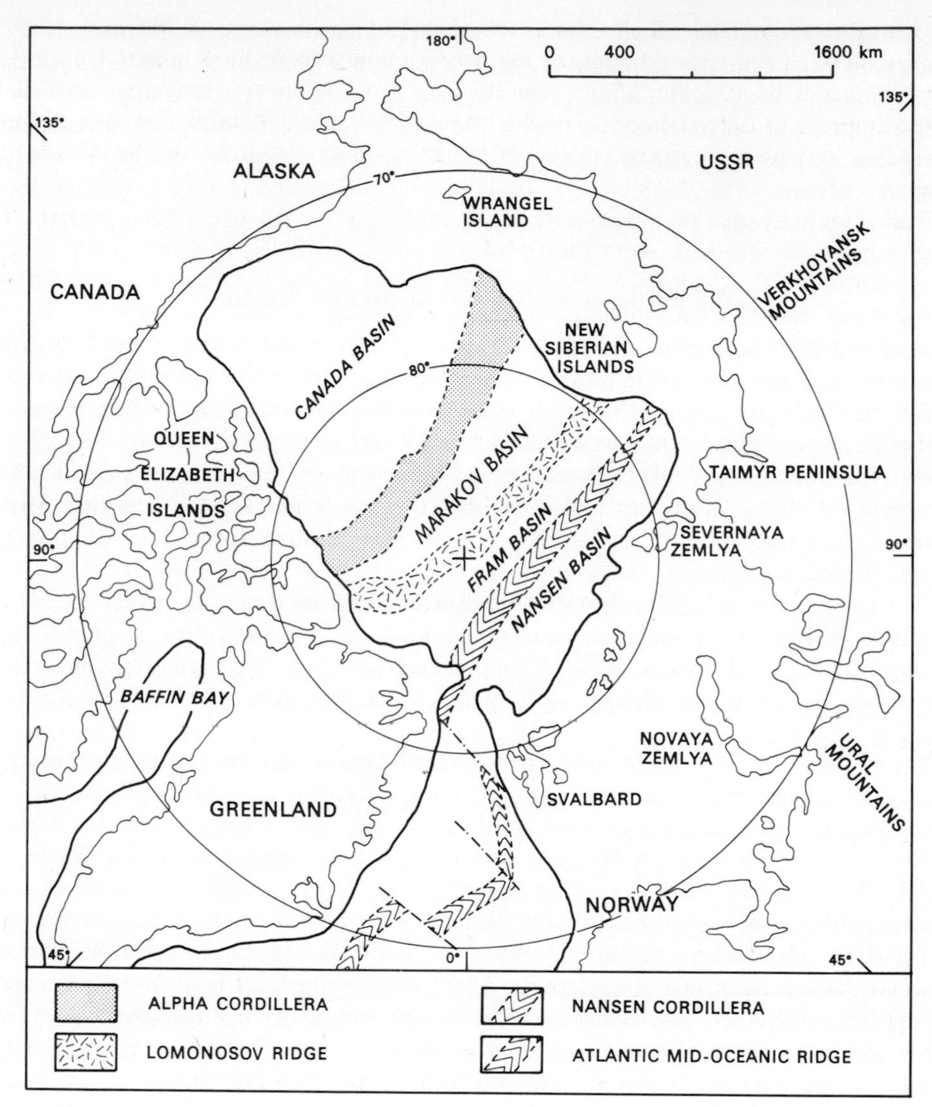

FIGURE 2 Physiographic sketch map of the Arctic Ocean and adjacent North Atlantic showing the major ocean ridges. The boundary of the continental shelf is indicated in a full heavy line; main fracture zones are shown by the conventional fault symbol.

1969; Vogt & Ostenso, 1970; Ostenso & Wold, 1971) support Wilson's (1963a) view that the ridge represents part of the Eurasian continent fractured off during the opening of the Eurasian Basin.

The Nansen Cordillera is an axis of present-day sea-floor spreading (Heezen & Ewing, 1961; Johnson & Heezen, 1967a) being associated with a well-defined pattern of earthquake epicentres (Sykes, 1965). Magnetic anomalies over the ridge (King *et al.*, 1966; Ostenso, 1968) have rather small amplitudes compared with the magnetic anomalies normal over oceanic ridges (Vine & Mathews, 1963), but nevertheless a linear pattern of magnetic anomalies parallel to the ridge is definable (Demenitskaya

& Karasik, 1966, 1969; Rassokho *et al.*, 1967). Demenitskaya & Karasik (1969) suggest as a preliminary estimate of the average rate of spreading about 1·1 cm/yr for the last 8 to 10 m.y., while Vogt *et al.* (1970) tentatively conclude that the spreading rate in the axial region of the ridge is less than 0·9 cm/yr. The age of this spreading is generally thought to have commenced at the close of the Mesozoic era.

The Alpha Cordillera, discovered in 1957, has an associated linear pattern of high amplitude magnetic anomalies (Ostenso, 1962, 1968; King *et al.*, 1966; Vogt & Ostenso, 1970; Ostenso & Wold, 1971) which along with gravity and other geophysical data (Lachenbruch & Marshall, 1966) suggest that the ridge is a fossil site of sea-floor spreading. Vogt & Ostenso (1970) consider this spreading was active 60 m.y. ago and that it ended abruptly 40 m.y. ago. However, Vogt & Ostenso (1970, p. 4935) also remark that 'it is, of course, possible that the feature represents something entirely different'; an important point to be borne in mind in discussions about the evolution of the Arctic Ocean Basin. The same authors suggest a link between the Alpha Cordillera and the Mid-Labrador Sea Ridge through the Nares Strait.

The North Greenland Fold Belt

The North Greenland fold belt, forming a border to the Eurasian Basin, is the easterly extension of the Innuitian orogenic system of Arctic Canada, being composed of a complex of highly folded, mainly Lower Palaeozoic sediments overlain by less-severely deformed Upper Palaeozoic and Mesozoic–Tertiary rocks. In Greenland the fold belt is traceable for 600 km from Hall Land on the Robeson Channel to the east coast of Greenland in Peary Land. The trend of the structural elements is approximately east–west parallel to the northern coast (see Fig. 3).

Sediments

The platform, overlying the crystalline basement, is composed of at least 1000 m of Proterozoic quartzites and sandstones and an Eocambrian to Silurian section at least 3500 m thick. The Eocambrian and Cambrian section is a mixed sequence of clastics, shales and limestones; the Ordovician and Silurian rocks are dominated by limestones. These platform rocks dip shallowly north and pass into the folded Lower Palaeozoic sediments of the North Greenland (Franklinian) geosyncline, where two main facies exist; (1) a southern, dominantly carbonate facies of the shelf zone characterized by regionally developed reefs of mainly Silurian age with some clastics and (2) a northern clastic facies of the trough zone containing subsidiary carbonates and cherts. These clastics form a thick monotonous, mainly unfossiliferous sequence, traceable from Ellesmere Island and throughout the fold belt. The upper part of the section reaches at least lower Ludlovian age. The total Lower Palaeozoic section in the trough probably exceeds a thickness of 5000 m.

In North Greenland, no clear distinction between mio- and eugeosynclinal rocks can be made. There is a gradual transition from platform rocks to the carbonates and argillaceous limestones of the reef complex which pass directly into gently folded clastics which farther to the north have turbidite structures indicating a steepening of slope. No well-defined boundary is present between platform and geosyncline and no geosynclinal volcanic rocks are exposed.

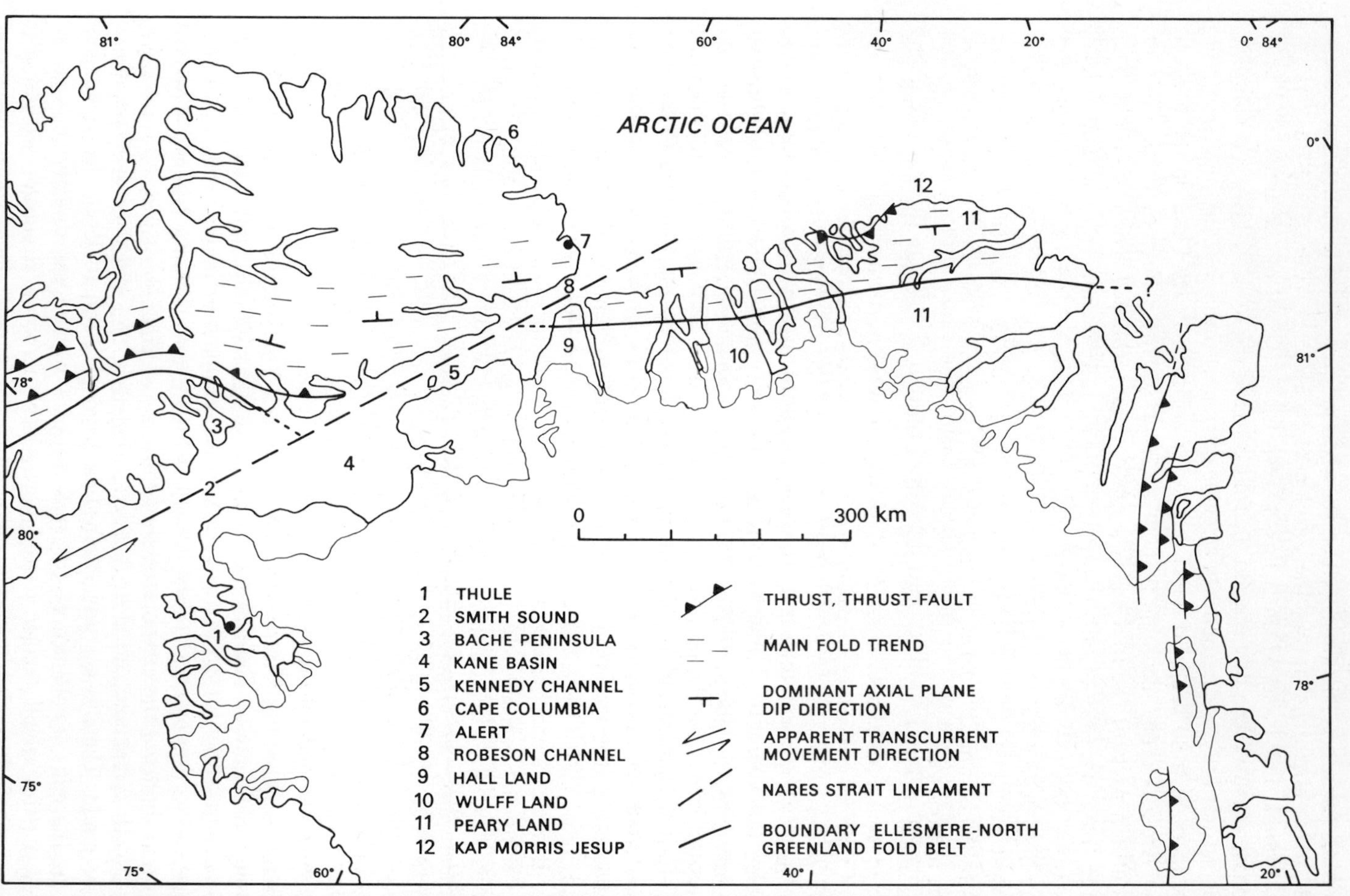

FIGURE 3 Generalized structural map to show the main tectonic elements of the folded rocks of the Franklinian geosyncline either side of the Nares Strait. Main direction of tectonic transport in the East Greenland fold belt is included for comparison. Important place names are indicated.

Structure and metamorphism

A striking feature of the fold belt is its asymmetrical tectonic pattern, a feature noted in certain areas during the early geological work (Ellitsgaard-Rasmussen, 1955; Fränkl, 1955) and which is now known to be typical of the entire fold belt. Polyphase diastrophism has affected the fold belt and the tectonic history is complex and as yet not fully understood. On a regional scale deformational and metamorphic effects increase from south to north so that in northern Peary Land on the borders of the Arctic Ocean, medium- to high-grade crystalline schists outcrop, while in the southern parts of the fold belt low-grade metamorphosed sediments occur. The passage from unmetamorphosed platform rocks into metamorphosed and folded rocks is transitional and the southern boundary of the fold belt is not complicated to any high degree by dislocations or thrust tectonics (see Fig. 3). Six, more or less east–west trending metamorphic-tectonic zones can be mapped from platform rocks to the complexly folded, amphibolite facies rocks of the north coast (Dawes & Soper, in press).

The main direction of overturning of fold structure is towards the north and the assumed centre of the orogen. Axial plane surfaces, particularly in the northern part of the fold belt where close and isoclinal folds exist, characteristically dip south and in places at low angles. The unfolded rocks of the platform when traced northwards pass into a gently folded zone characterized by open symmetrical folds, north of which the majority of folds become asymmetrical and overturned to the north. Accompanying this northerly change in fold style is an intensification of the cleavage. The northerly overturned folds are of at least two ages and some post-date at least one episode of deformation which produced southerly-facing structures.

The folded rocks reach the waters of the Arctic Ocean. How far the metasediments continue out under the Arctic Ocean is not specifically known although according to King *et al.* (1966) thick Palaeozoic geosynclinal sediments exist on the continental slope. If such Palaeozoic strata show the same northerly increase in metamorphic grade as do the exposed rocks, the occurrence of granite and other plutonic rocks towards what must be considered the centre of the orogen, is not unlikely. Earlier statements however, (Koch, 1934, 1935), that gneisses with 'eruptions of granite' occur in Peary Land are now disproved, although crystalline metamorphic erratics on the north coast could indicate the metamorphic conditions in the unexposed or removed part of the fold belt (Ellitsgaard-Rasmussen, 1955; Dawes, 1970).

Fracturing and magmatic activity

On the edge of the Arctic Ocean in northern Peary Land outcrops the Kap Washington Group of bedded, essentially rhyolitic, lavas and tuffs (see Fig. 1), a volcanic pile at least 1500 m thick. A distinct positive magnetic anomaly off the northern coast (Haines *et al.*, 1970; Miss E. R. King, pers. comm.; Ostenso & Wold, 1971) may indicate the offshore extent of these volcanics, provided a rather high intensity of magnetization of the rocks is accepted. The volcanics are non-metamorphic, slightly deformed and cleaved and are calc-alkaline in type, being of the andesite-rhyolite orogenic kindred. SiO_2 content in the volcanics analyzed so far is high and in most acid types it reaches over 75%. The metamorphic nature and geological setting of the volcanics suggest a post-Lower Palaeozoic age. If the K/Ar whole rock age determinations (Dawes & Soper, 1970) reveal the true age of extrusion and not the effect of later deformation, then the volcanics are of Late Phanerozoic (Late Cretaceous and/or Tertiary) age.

The extrusives are bordered on the south by the Kap Cannon Thrust, a shallow southerly-dipping thrust over which the Lower Palaeozoic metasediments of the fold belt have been transported. The major thrusts and thrust-faults of the fold belt trend more or less parallel to the strike of the bedrocks and the majority have southerly-dipping planes.

The main direction of regional high-angled faults is also approximately east–west parallel to the strike of the folded bedrock. Subsidiary directions are northwest–southeast and northeast–southwest. Two of the main fractures—the Harder Fjord and the Kap Bridgman Faults—flank the mountainous uplifted block of northern Peary Land.

Cross-cutting basic intrusives occur in the eastern part of the fold belt where, in Peary Land dykes of three main directions occur; north–south (varying to northwest–southeast in southern Peary Land), northeast–southwest and approximately east–west. The dykes post-date the main phases of folding; most are fresh and undeformed, some however show evidence of metamorphism and alteration. Basic igneous material also occupies the planes of major east–west faults. Two K/Ar dates of $66{\cdot}0 \pm 6{\cdot}6$ m.y. (Henriksen & Jepsen, 1970) and $72{\cdot}9 \pm 9{\cdot}0$ m.y. (Dawes & Soper, 1971) on north-west–southeast and east–west dykes respectively support the idea that all dykes are members of a Late Phanerozoic suite.

Age of the diastrophism

The present character of the belt is a product of both Palaeozoic and Late Phanerozoic orogenesis. The one unconformity preserved is between folded Lower Palaeozoic rocks and a less-severely deformed Pennsylvanian to Tertiary sequence of the Wandel Sea Basin (Dawes & Soper, in press). This stratigraphical relationship simulates that in Arctic Canada between the folded Palaeozoic rocks of the Franklinian geosyncline and the younger deposits of the Sverdrup Basin. The main diastrophism is assumed to have affected the North Greenland geosyncline between Late Silurian and Late Devonian time. This orogenic activity embraced metamorphism and associated polyphase deformation resulting in a northerly overturning of fold structures.

Tertiary orogenesis is indicated by the folding and faulting of the Wandel Sea Basin strata, the slight deformation and metamorphism of Late Cretaceous basic dykes, the thrusting of metasediments over the Kap Washington Group volcanics and by mylonitization and low-grade metamorphism associated with the Kap Cannon Thrust. K/Ar isotopic age dates between 84 m.y. and 42 m.y. on the Lower Palaeozoic metasediments indicate a significant regional thermal event (or events) in Late Phanerozoic time.

Geological Links with the North Greenland Fold Belt

The tectonic strike of the North Greenland fold belt is continuous with the main structural trends of the Lower Palaeozoic sediments of the Franklinian geosyncline in northeastern Ellesmere Island. The sedimentary history of the Lower Palaeozoic strata in the Hall Land–Wulff Land area of Greenland adjacent to the Robeson Channel (Allaart, 1965; Dawes, 1966, 1971) has clear similarities to that in Ellesmere

Island (Christie, 1964; Kerr, 1967a; Trettin, 1970, 1971) and there is no doubt about the sedimentary unity of the two regions. The same relationships exist between platform, shelf and trough deposits on both Greenland and Ellesmere Island and the development of Palaeozoic reefs associated with major facies variations is typical on both sides of the Nares Strait (Kerr, 1967a; Dawes, 1971; Norford, 1972). The Franklinian geosyncline can be traced continually from Melville Island on the Beaufort Sea across the Queen Elizabeth Islands to eastern North Greenland (Schuchert, 1923; Fortier *et al.*, 1954).

The projected intersection of the North Greenland fold belt with that of East Greenland occurs in the Wandel Sea. Early ideas favoured the North Greenland fold belt as a harmonious continuation of the Caledonian East Greenland fold belt (Schuchert, 1923; Koch, 1934, 1935; Nielsen, 1941), an idea still entertained by some (Meyerhoff, 1970). Since the differences in sedimentary and tectonic histories of the two orogenic belts have become known in more detail, the East Greenland fold belt is considered part of the North Atlantic (Scandinavian) Caledonian orogenic system, whereas the North Greenland fold belt appears to show more affinities to the orogenic systems of western Russia. The discovery of significant Late Phanerozoic diastrophism in North Greenland associated with regional reheating of the fold belt to give persistent Late Cretaceous-Tertiary isotopic dates (Dawes & Soper, 1971) is evidence supporting this view.

Although there has been much discussion about the age of the Lomonosov Ridge, most Soviet attempts at geological correlation across the Arctic Ocean Basin have regarded the Ridge as a fixed structure in relation to the Russian and Ellesmere Island–Greenland landmasses. Saks *et al.* (1955) and Panov (1955) regarded the Lomonosov Ridge as a zone of Mesozoic tectonics, linking the northern part of the Canadian Arctic with the New Siberian Islands and the Verkhoyansk Range of Siberia. Earlier E. Suess had drawn attention to similarities between the geology of Ellesmere Island and the folded rocks of this part of Russia (see Hope, 1959). Meyerhoff (1970) regards the Lomonosov Ridge to have formed in Late Proterozoic or earlier time and to form part of an orogenic belt referred to as the 'Lomonosovides' which borders the Amerasian Basin and stretches from Alaska and the northern part of Ellesmere Island across the Arctic Ocean to the New Siberian Islands and the Chukotsk Range of Siberia. Although the age of the Lomonosovides is said to be unknown 'evidently they were deformed several times during the late Proterozoic(?)–early Palaeozoic', and are classified by Meyerhoff as Caledonian. Some Russian workers (Saks *et al.*, 1955; Gakkel, 1958) have postulated a Hercynian (Upper Palaeozoic) belt of folding across the Eurasian Basin connecting the Innuitian fold belt of Ellesmere Island and North Greenland to the Severnaya Zemlya–Novaya Zemlya–Taimyr Peninsula–Urals orogenic system (Uralides) of Russia, a correlation favoured by others (e.g. Carey, 1958; Hamilton, 1970). Part of the evidence for the earlier Soviet proposals was the fact that the tectonic trends of the Russian fold belts, particularly the New Siberian Islands–Verkhoyansk belt, strike, albeit at various angles, towards the Arctic Ocean.

On the other hand, the structures of the Hazen Trough of Ellesmere Island and of the North Greenland fold belt strike parallel to the continental slope, a point which influenced Christie (1964) to note that the main tectonic features of northeast Ellesmere and adjacent North Greenland do not invite direct correlation with suboceanic tectonic features of the Arctic Ocean Basin. More recently, however, Trettin (1969a), from the northern coast of Ellesmere Island, reports the presence of a north–south trending fold belt aligned with the Lomonosov Ridge which was deformed during the Caledonian (Mid-Ordovician or earlier), Hercynian (Late Silurian to Middle Pennsylvanian) and Tertiary. If the reason for this alignment is other than coincidental, then support is provided for the earlier concepts of geological connection via the Lomonosov Ridge. Reports describing the 'fold basement' of the Lomonosov Ridge as being composed of crystalline limestones, quartzites, double mica and orthogneisses similar to the Archean–Proterozoic and Riphean metamorphic rocks of Ellesmere Island, overlain by Mesozoic sediments (Gakkel & Dibner, 1967), are clear statements of support for the Ellesmere–Lomonosov–Russia hypothesis.

However, many workers now believe that the juxtaposition of the Lomonosov Ridge adjacent to northern Ellesmere Island may well be a consequence of continental drift. If the Ridge represents a strip of the Eurasian continent (Wilson, 1963a) fractured off during the Late Phanerozoic continental break-up, the alignment of the Ellesmere Island structures with the Lomonosov Ridge is most probably coincidental (see p. 929). The important point which arises in this case is that the Lomonosov Ridge could nevertheless theoretically represent a link between Ellesmere Island and Russia in the sense of a displaced portion of the Innuitian–Uralides orogenic system.

Evolution of the Arctic Ocean Basin

The origin and age of the Arctic Ocean Basin are questions which have been continually discussed in recent years. They are by no means finally solved.

Present consensus

Early support for the Arctic Ocean being a sunken part of the continental crust (Shatskiy, 1935; Eardley, 1948) has been superseded by a general consensus that the Basin is all, or in part, oceanic in character being floored by a relatively thin crust above the Mohorovičić discontinuity (Oliver *et al.*, 1955; Demenitskaya, 1958). Differences in character between the Eurasian and Amerasian Basins have gradually come to light and some authors have proposed that the nature of the floor differs on either side of the Lomonosov Ridge. King *et al.* (1966), for example, favoured the presence of oceanic crust on the Eurasian side, continental sunken rocks on the Amerasian side. Others (Churkin, 1969; Vogt & Ostenso, 1970; Ostenso & Wold, 1971) regard the Amerasian side as being floored primarily by oceanic rocks.

The age of the Arctic Ocean Basin is clearly dependent on the type of model favoured to explain the various structures involved, but essentially it involves a choice between an old permanent ocean (proto-Arctic Ocean) dating from the Proterozoic or Early Palaeozoic (Harland, 1966; Churkin, 1969; Meyerhoff, 1970; Dietz & Holden, 1970), or a relatively young Late Phanerozoic evolution of the Basin (Carey, 1958; Vogt & Ostenso, 1970; Ostenso & Wold, 1971; Hamilton, 1970; Vogt *et al.*, 1970). Acceptance of the existence of a proto-Arctic Ocean does not necessarily exclude support for a significant Late Phanerozoic development of part of the present-day Arctic Ocean (Harland, 1969; Churkin, 1969) although in some cases it does (Meyerhoff, 1970). There is a tendency for advocators of the Late Phanerozoic evolution to underestimate or shun the importance of any proto-Arctic Ocean.

To summarize, the present consensus appears to be that many of the features of the present-day Arctic Ocean Basin developed in Late Phanerozoic time, by either continental wandering induced by rifting or by sea-floor spreading (Carey, 1958; Vogt & Ostenso, 1970; Vogt *et al.*, 1970; Hamilton, 1970) or by sinking of the surrounding continental crust (Dibner, 1970). Although parts of the Basin may well represent old features, at least part, if not the majority or all, of the ocean floor of the Basin is much younger, the product of volcanicity associated with Late Phanerozoic sea-floor spreading.

Evidence from North Greenland

Since it is now known that the Arctic Ocean represents a complex feature of the Earth's crust containing a number of individual structural units whose origin must differ, a simple discussion between an 'old' or 'young' basin or between 'continental' or 'oceanic' crust is no longer appropriate, and discussion must be directed at which parts of the basin represent 'old' or 'young' features, which parts are floored by continental, which by oceanic crust. And finally, what is the age of any oceanic crust involved?

North Greenland has a role to play since it forms a border of the present-day Eurasian Basin, being situated between the Lomonosov Ridge and the Nansen Cordillera (Fig. 2). The east–west trend of the North Greenland fold belt is more or less at right angles to the trend of the Lomonosov Ridge. North–south trending structures in the Ellesmere part of the Franklinian geosyncline parallel with the Lomonosov Ridge (Trettin, 1969a) are totally absent in Greenland and the strike of the fold belt is consistently parallel to the coast and the continental slope.

The sense of tectonic transport in the fold belt is conspicuously northwards towards the Arctic Ocean (see p. 923). Hence the stable platform stretching from the west to east coasts south of the fold belt represents a hinterland or backland to the fold belt, not a foreland (Ellitsgaard-Rasmussen, 1950, 1955; Dawes, 1971). The northward sense of overturning seen throughout the 3000 m vertical exposure of the folded crust, involving in Peary Land a composite section of probably over 5000 m of strata, is considered due to a significant southerly-directed movement at depth beneath the fold belt. Earlier explanations considered the remarkable northward-facing nature of the fold belt as being bound up with superficial alpine-type nappe tectonics and intense overthrusting, with a sense of a 'push from the south' (Fränkl, 1955, 1956). This tectonic pattern has been refuted and although 'relative movement of higher crustal rocks over lower no doubt took place' it is found 'difficult to envisage how powerful tangential stress could have been transmitted through the undeformed platform sediments of the hinterland, nor is there evidence of a root zone from which northerly-directed nappes could have been derived' (Dawes & Soper, in press).

The site of the fold belt is at a continental margin flanking on the north a stable platform and crystalline block, the exposed parts of which show no effects of Palaeozoic tectonism. The southerly-directed movement of crustal material at depth is considered to be connected to the interaction of continental and oceanic structures, probably under the control of mantle forces, in a manner outlined by Dietz (1963, 1966). Therefore the model proposed here to account for the structure of the North Greenland fold belt proposes an underthrusting of a northern oceanic crust against and under the continental block of Greenland; the sense of tectonic transport in the trapped sedimentary prism being an indication of the main direction of underthrusting (see Fig. 4). The northerly overturning of the fold belt was initiated in Palaeozoic (Late Silurian to Late Devonian) time and hence the model also demands Palaeozoic oceanic crust. If this is accepted then there is evidence for the original existence of ancient oceanic floor at the site occupied by the present-day Eurasian Basin.

This proposal for the existence of a Palaeozoic Arctic Ocean to the north of Greenland conforms well with the findings of Churkin (1969), who from a study of geological and geophysical data in lands bordering the present-day Canada Basin, reached the conclusion that the Canada Basin is a true and very old ocean floored by oceanic crust. As evidence Churkin described the 'circumarctic geosynclinal belt' stretching from Wrangel Island and eastern Siberia, through the 'ancestral Brooks geosyncline' of Alaska to the Franklinian geosyncline, concluding that 'movement of the floor of the Arctic Ocean against the continental crust of North America' may explain the long Palaeozoic orogenic history of the belt.

However, the Greenland part of the Franklinian geosyncline is not mentioned or indicated in the maps by Churkin, conceivably because North Greenland forms part of the Eurasian and not the Canada Basin. The North Greenland geosyncline is in fact the eastern part of the 'circumarctic geosynclinal belt' and premises concerning the development of this part of the crust must also include Greenland. Hence it is probable that the oceanic crust which floored the Palaeozoic Canada Basin extended farther towards the Eurasian continent incorporating at least a substantial area north of Greenland.

Southerly-directed crustal movements in the area north of Greenland are not in harmony with the presence of the monolithic north–south trending Lomonosov

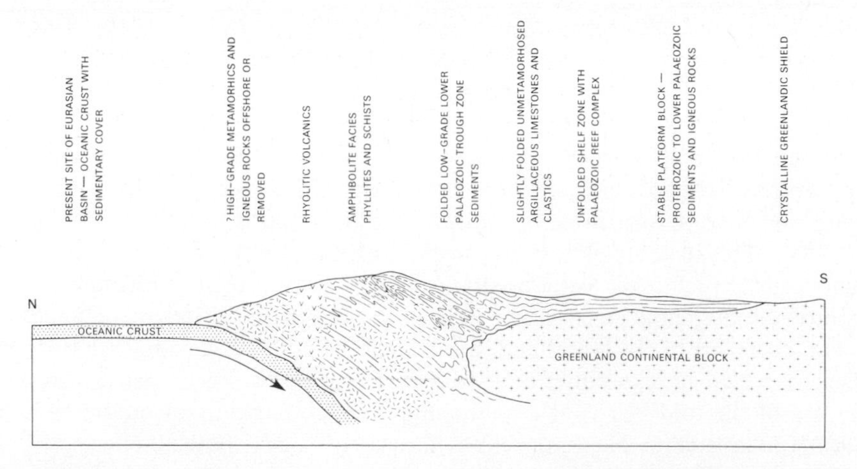

FIGURE 4 Schematic cross-section through North Greenland illustrating the proposed relationship between motion of the ocean floor and the tectonics of the North Greenland fold belt.

Ridge, unless the Ridge is part of the later evolution of the Arctic Ocean Basin. Hence the hypothesis is favoured that the Lomonosov Ridge arrived at, or developed in, its present position in post-Silurian time. Consequently any alignment of the Ridge with the north–south trending Lower Palaeozoic or older structures in Ellesmere Island (Trettin, 1969a) must be coincidental unless it is envisaged that the movement of the Ridge actually was the cause of the north–south trending structures in Ellesmere Island. Although Late Phanerozoic orogenesis is described from northernmost Ellesmere Island, this latter possibility is not explored by Trettin (1969a).

If the Lomonosov Ridge is indeed a part of the Eurasian continent fractured off during spreading from the Nansen Cordillera as many workers now believe, its drift path must have passed north of Greenland. In this case the movement of Greenland relative to the Lomonosov Ridge occurred along some sort of megashear north of Peary Land, more or less in the site of the morphodisjunction postulated by Gakkel & Dibner (1967) and on line with the Spitsbergen fracture zone.

Any crustal features originally to the north of Greenland which might have given evidence for the true nature of the southerly-directed Palaeozoic plate pattern, would be expected to exist, if at all, in the Amerasian Basin. The Alpha Cordillera, crossing this basin is a possible inactive site of ocean-floor spreading. If so, it will have produced new oceanic floor in zones parallel with the Lomonosov Ridge, a phase of crustal growth which could have essentially remodelled the Amerasian Basin and produced ocean floor which may have updated most or all of the Palaeozoic oceanic crust. If the Alpha Cordillera is not a fossil spreading center (see p. 921) then there may be a case for the presence of old oceanic crust in the present-day Amerasian Basin.

Origin of the Nares Strait Lineament

An extensive treatment of the origin of the Nares Strait is far from the scope of this paper. However, certain aspects of the geology of North Greenland have a direct bearing on its history.

Discussion about the Strait has a long history stemming from Taylor (1910) who drew attention to the striking physiography of the channel ('a passage so straight and persistent', p. 205) and suggested it was a major rift along which North America had moved southwest away from Greenland. Since then the Strait has been regarded by others as a sinistral transcurrent fault (Wegener, 1924; Du Toit, 1937; Wilson, 1963a), a megashear (Carey, 1958; Hilgenberg, 1966), a transform fault (Wilson, 1965), a rift valley (Kerr, 1967a, b) and a dextral transcurrent fault (Egiazarov *et al.*, 1969). Moreover, the Strait has been regarded as a northward expression of the Mid-Labrador Sea Ridge, joining either the Nansen Cordillera (Grachev *et al.*, 1967) or the Alpha Cordillera (Vogt & Ostenso, 1970). Whatever its nature the lineament has been accepted by all workers hitherto to be a relatively young feature generated during the Late Phanerozoic episode of continental break-up.

Age of the lineament

The geometry of the fold pattern in the Franklinian geosyncline differs on either side of the Nares Strait (Kerr, 1967a) and the northerly overturning so conspicuous throughout the North Greenland fold belt (Ellitsgaard-Rasmussen, 1955; Fränkl, 1955; Allaart, 1965; Dawes, 1966; Dawes & Soper, 1970) is not a striking feature in adjacent Ellesmere Island (Christie, 1964; Kerr, 1967a; Trettin, 1971). Kerr (1967a, p. 510) summarized the tectonic style in northeastern Ellesmere Island as asymmetrical being 'expressed as southward overriding (axial planes dip north)'. Christie (1964) originally described folds in northeastern Ellesmere Island with northerly-dipping axial planes and although not all folds are of this style, they are certainly predominant (Trettin, 1971). This clear divergence in structural pattern must be controlled in some way by the structure of the Strait (or by some predecessor), and whatever its true nature the Nares Strait must have been a significant structural site when the divergence in fold pattern emerged.

Kerr (1967a, p. 510) considered that the Palaeozoic structural pattern in North Greenland and Ellesmere Island was characterized by southerly overriding and that in Ellesmere Island 'it was reinforced along the same lines by the Tertiary orogeny', while in Greenland Tertiary compression resulted in northerly overriding. Evidence in Greenland for northerly-directed tectonic transport of Tertiary age exists in some localities (for example, the Kap Cannon Thrust) but the main northerly overturning of the folds is a result of Palaeozoic compression (Dawes & Soper, 1971). Thus, if the Palaeozoic deformation in Ellesmere Island truly resulted in southerly overriding, then Nares Strait most probably represented a significant structure in Palaeozoic (pre-Carboniferous) time. It is thus substantially older than hitherto thought.

Nature of the movement

Suggestions about sinistral movement along the Strait have varied from 'at the most a few kilometres of relative horizontal displacement' (Kerr, 1967a, p. 506), to estimates of 250 km (Wegener, 1924; Du Toit, 1937), 300 km (Hilgenberg, 1966), 320 km (Carey, 1958) and 528 km (Taylor, 1910). This disagreement is caused mainly by lack of reliable marker horizons, both structural and stratigraphical, on either side of the Strait.

In North Greenland boundaries proposed between mio- and eugeosynclinal rocks are arbitrary (see p. 921) and especially unsatisfactory if used in correlations with the so-called mio- and eugeosynclinal rocks on Ellesmere Island. There has been unlimited disagreement about the classification of the geosynclinal rocks in both Ellesmere Island and Greenland. For example, King (1965) regards the folded rocks of North Greenland as eugeosynclinal, Kerr (1967a) indicates the same rocks as miogeosynclinal. On the Canadian side, Christie (1964) regards the rocks bordering the Robeson Channel as miogeosynclinal, Thorsteinsson & Tozer (1960) and Kerr (1967a) regard the same rocks as eugeosynclinal. The boundary indicated on Greenland between so-called miogeosynclinal rocks and the 'central stable region' and which is used to correlate with a boundary in Ellesmere Island (Kerr, 1967a; Cowie, 1970), is purely arbitrary and thus misleading in drift studies. The boundary chosen by Egiazarov *et al.* (1969) between pericratonic 'Riphean–Middle Palaeozoic' and folded 'Middle Palaeozoic' in both Greenland and Canada, is also an arbitrary division. By means of the apparent displacement of this boundary Egiazarov *et al.* (op. cit.) interpret the Nares Strait as a dextral transcurrent fault showing a lateral displacement in the order of 60 km.

The boundary between folded and unfolded rocks of the Franklinian geosyncline is mappable on both sides of the Nares Strait and its position in northern Bache Peninsula on Ellesmere Island and in Hall Land on Greenland is generally agreed upon. Although structural trends cut the Strait at a small angle, projection of these boundaries to a median line in the Strait immediately suggests sinistral displacement of at least 200 km (see Fig. 3). However, the history of the Nares Strait is much more complicated than this suggests. The present fold belt boundary position is a product of a long history of orogenesis from Palaeozoic to Late Phanerozoic time, and there is evidence to suggest that the diastrophism affected Canada and Greenland differentially. For example, on the Canadian side, where Tertiary structures are apparently common (Thorsteinsson & Tozer, 1960; Kerr, 1967a), the Palaeozoic position of the tectonic boundary may have been significantly modified by the Tertiary orogenesis; the Palaeozoic position on the Greenland side may have been little affected. Furthermore, the time relation between the establishment of the present boundary position and any movement of Greenland relative to Canada, is not accurately known and strike-slip movement, sinistral and/or dextral, may have occurred prior to the formation of the present boundary position. Also, since the Palaeozoic structure in Nares Strait had such a marked influence on the structural pattern of the units either side of it, the boundary of the fold belt may not have been originally straight and of constant strike. This is also suggested by the far from constant trend of the boundary in Ellesmere Island. All these aspects lead to innumerable problems in the interpretation of the apparent displacement along the Strait.

In summary it can be said that our present knowledge of the geology on either side of the Nares Strait cannot exclude major transcurrent movement between Ellesmere Island and Greenland. The geology does, however, place serious restraints on an overall dextral strike-slip movement and an absolute sinistral strike-slip movement greater than about 250 km. The 500 km length and the striking linearity of the Strait certainly suggests a plane of movement; this becomes even more realistic when the Strait was in existence as a significant fracture of the crust in the Palaeozoic. It would be surprising if motion of one type or another had not occurred along this fracture during the earth movements which affected both Ellesmere Island and Greenland in Late Phanerozoic time, and which caused the relative motion of landmasses and the continental break-up of Laurasia. More detailed stratigraphical and structural data from the areas bordering the Nares Strait should give us a clue to the ages and relative parts played by any sinistral and dextral strike-slip or other movements in the history of the Strait.

Relation to Baffin Bay–Davis Strait

Nares Strait has played a central role in discussions about the drift of Greenland and North America. Early ideas of continental break-up in this region showed sinistral lateral slip along the Nares Strait as the cause of the opening of Baffin Bay and Davis Strait (Taylor, 1910; Wegener, 1924; Du Toit, 1937; Carey, 1958; Wilson, 1963a). Recently with the onset of the theory of sea-floor spreading many workers now believe that Nares Strait is more a consequence of the forces that opened Baffin Bay and Davis Strait (Drake *et al.*, 1963; Harland, 1966; Kerr, 1967a, b; Johnson & Pew, 1968) although about the nature of these forces there is still much discussion.

However, whether the Nares Strait is the cause of, or the result of, the opening of continental crust in the south, all workers seem to agree that formation of the Nares Strait and the opening of Baffin Bay and Davis Strait are contemporaneous events.

If this is so, then evidence from the Nares Strait presented here suggests that the Baffin Bay–Nares Strait structure may have started to form much earlier than the Late Phanerozoic episode of continental break-up with which it is generally held to be inherently associated. Precisely when the structure started to form must remain open, but it is particularly interesting to note that Fahrig *et al.* (1970) from entirely different evidence suggest that Baffin Bay and Davis Strait 'may have started to form as early as the late Hadrynian', i.e. latest Proterozoic.

Discussion

Like other orogenic regions surrounding the Arctic Ocean (U.S.S.R., Alaska, Canada and Spitsbergen), the North Greenland fold belt has been affected by different episodes of diastrophism, and the belt displays a long orogenic history from Palaeozoic to Tertiary time. Elucidation of this history helps in any understanding of the evolution of the Arctic Ocean Basin, which is inferred to have had an equally long history.

The model favoured to explain the essential features of the fold belt demands the existence of Palaeozoic oceanic crust, the southerly movement of which against the continental block of Greenland, provided the forces producing the tectonic pattern within the geosynclinal rocks. The North Greenland geosyncline developed on the active edge of the stable platform block and became the site of continued subsidence from the Early Palaeozoic (possibly Late Proterozoic) to at least Early Devonian time. The upper part of the section in the trough of the geosyncline is a thick clastic sequence similar to the graywacke and turbiditic strata of the Hazen Trough of Ellesmere Island (Christie, 1964; Trettin, 1970, 1971). However, Devonian rocks have not been proved in North Greenland, the youngest fossiliferous pre-Carboniferous sediments yet encountered being of Ludlow (possibly late Ludlow) age.

Schuchert (1923) postulated in the present area of the Arctic Ocean, north of Ellesmere Island and Greenland, the existence of an Archean landmass 'Pearya' from which he believed the sediments of the Franklinian geosyncline were derived. He cited the presence of metamorphic rocks in northern Ellesmere Island and Peary Land as evidence of this borderland. These metamorphics are now known not to be Archean. The Cape Columbia Complex (Blackadar, 1954; Christie, 1964) in Ellesmere Island is pre-Ordovician in age but is considered to contain Lower Cambrian and/or Late Proterozoic strata metamorphosed in Late Proterozoic and/or Early Cambrian times (Trettin, 1969b). The metamorphics in northern Peary Land have been derived at least in part from Lower Palaeozoic sediments and the metamorphism there is certainly post-Ordovician. Trettin (1971) has renamed 'Pearya', the 'Pearya Geanticline', interpreting it as a landmass uplifted out of the Franklinian geosyncline in Late Proterozoic or Early Palaeozoic time. In North Greenland no metamorphic complex equivalent in age to the Cape Columbia Complex exists; if present, it lies offshore or has been removed. However, some kind of 'Pearya' most probably existed offshore in Late Proterozoic or earliest Palaeozoic time from which the sedimentary fill of the North Greenland geosyncline has been wholly or partially derived. Whether this borderland remained a positive feature off North Greenland into the Devonian as it did in the Alaskan and Canadian part of the 'circumarctic geosynclinal belt' (Churkin, 1969), is yet unknown. However, at some time between the latest Silurian and Late Devonian the borderland ceased shedding deposits southwards and the thick geosyncline pile was deformed.

A subduction zone, dipping beneath the stable continental Greenland block, developed somewhere off North Greenland (continuous with that proposed at the margin of the Palaeozoic Canada Basin flanking the rest of the circumarctic geosynclinal belt). Underthrusting along this zone resulted in the consumption of ocean floor and a heat-flow for the deformation and metamorphism of the sedimentary prism above it. This subduction zone probably remained active until Early Carboniferous time, when, as inferred by the unconformity beneath the Wandel Sea Basin, the North Greenland fold belt had been regionally uplifted. In Pennsylvanian time parts of North Greenland and northern East Greenland became flooded by the Upper Palaeozoic sea which persisted into the Late Mesozoic and Tertiary.

No plutonic rocks are exposed in the North Greenland fold belt which might be interpreted as the expression of Palaeozoic magmatism from the subduction zone. The metamorphic grade of the fold belt, developed under a northerly-increasing temperature regime, suggests that plutonic rocks might well be present toward the interior of the orogenic zone, which at the present time lies beneath the Arctic Ocean or has subsequently been removed. In northern Ellesmere Island, where a much wider zone of the Franklinian geosyncline is exposed, a suite of magmatic rocks do occur (Frisch, 1967; Trettin, 1969a) which may be of this origin.

The silicic calc-alkaline volcanics of the Kap Washington Group outcropping on the present border of the Eurasian Basin occupy a position within the accreted geosynclinal part of the continental block, and as such are in a position to be classified as circumpacific in type being of the same setting as the major andesitic and rhyolitic occurrences associated with so many other active land margins. These rocks, which probably have a much larger presence offshore (see p. 923) represent late-orogenic volcanicity and it is tempting to consider them as products of melting and contamination of rocks in a subduction zone. However, as discussed earlier, the age of the extrusion is not known with certainty, and the volcanics may be connected with the post-Palaeozoic spreading evolution of the Arctic Ocean Basin.

Many consider that sometime before 60 m.y. ago the Alpha Cordillera became the site of sea-floor spreading (Vogt & Ostenso, 1970; Vogt *et al.*, 1970; Hall, 1970), which generated oceanic crust in the present-day Amerasian Basin. Crustal spreading from the ridge is said to have ended rather abruptly at about 40 m.y. when the Nansen Cordillera took over the major spreading role (Vogt & Ostenso, 1970). If this is so, the evolving oceanic crust from the Alpha Cordillera must have been consumed in trenches along the margins of the Amerasian Basin and also at the western margin of the Lomonosov Ridge (or at that part of the Eurasian continent destined to become the Lomonosov Ridge). If at this time the Lomonosov Ridge was situated to the east of North Greenland any movement of the sea floor generated from the Alpha Cordillera would also affect the North Greenland fold belt. Similarly, since about 40 m.y. ago sea-floor spreading is presumed to have opened the present

Eurasian Basin and to have split off the Lomonosov Ridge from the Eurasian continent. Consequently generation of sea floor parallel to the Nansen Cordillera would have affected the eastern part of the North Greenland fold belt.

Either or both of these spreading episodes could provide a mechanism for the generation of andesitic and rhyolitic rocks of the Kap Washington Group, providing at the same time an explanation for the regional reheating of the fold belt, indicated by the Late Phanerozoic isotopic ages on Palaeozoic metasediments. Later compressional features are indicated by the folding and thrusting of the volcanics and of the sediments of the Wandel Sea Basin. Basic magma reached the same level in the crust in Late Phanerozoic time as indicated by basic dykes, some of which are parallel to the coast, others at right angles to it. Tensional fractures are indicated by faults having approximately the same directions.

Although major transcurrent movements have not been recognized in North Greenland, the major east–west fault fractures in Peary Land are more or less on line with the Spitsbergen fracture zone (Johnson & Eckhoff, 1966; Johnson & Heezen, 1967b). Two of these faults, the Kap Bridgman and the Harder Fjord Faults have been interpreted as normal in type flanking the uplifted mountains of northern Peary Land (Fränkl, 1965). While the striking topographical expression marked by the faults suggests normal movements, strike-slip components may have occurred. If so, and even if not, the fractures may be onshore expressions of the Spitsbergen fracture zone which offsets the Atlantic Mid-Oceanic Ridge in the Greenland Sea.

An important point to arise from the Late Phanerozoic spreading pattern of the Arctic Ocean (Vogt & Ostenso, 1970) is that the Alpha Cordillera, which has been suggested to be linked to the Mid-Labrador Sea Ridge system via the Nares Strait (Vogt & Ostenso, 1970), may in fact be the original continuation or a branch of the Atlantic Mid-Oceanic Ridge, now inactive and displaced by dextral movement along a megashear north of Greenland. This idea, suggested earlier by Dr. E. F. Roots (pers. comm.) finds good support in the age of spreading from the Atlantic Mid-Oceanic Ridge south of the Spitsbergen fracture zone, where a major spreading phase began, according to Vogt *et al.* (1970), about 60 m.y. ago, i.e. contemporaneous with the proposed spreading from the Alpha Cordillera. Initiation of the Nansen Cordillera as a spreading center leading to the formation of the Lomonosov Ridge, may thus have caused the separation of the Alpha Cordillera from the North Atlantic Mid-Oceanic Ridge.

Conclusions

The structural features of the North Greenland fold belt can be correlated with stages in the evolution of the Arctic Ocean. The main tectonic character of the fold belt is Palaeozoic, orogenic overprinting having occurred in the Late Phanerozoic. The effects of these later events have not been severe enough to obliterate the Palaeozoic history of the fold belt. On the other hand, the Late Phanerozoic events connected with continental break-up and with the production of 'new' oceanic crust, have tended to overshadow the early history of the Arctic Ocean.

Evidence from North Greenland suggests that:

(1) the Arctic Ocean, floored by oceanic crust, existed in some form in Palaeozoic time, and
(2) the Nares Strait–Baffin Bay lineament represented a fundamental fracture of the crust as early as the Palaeozoic.

Both systems have been remodelled to varying degrees by the Late Phanerozoic episode of continental break-up. In the case of the Nares Strait itself, no radical changes need necessarily be envisaged since Palaeozoic time. Changes, however, along its southern extension have resulted in the final opening of Baffin Bay and Davis Strait.

In the case of the Arctic Ocean, the Late Phanerozoic events appear to have been important in controlling much of the present-day configuration of the floor. If the southerly-directed spreading of the Palaeozoic oceanic crust causing the orogenic activity in the Franklinian geosyncline resulted in complete closure of the proto-Arctic Ocean, then no Palaeozoic ocean floor is to be expected in the basin. On the other hand, if no collision occurred between the Greenland continental block with its accreted fold belt and a complimentary block on the Russian side of the proto-Arctic Ocean, Palaeozoic ocean floor probably existed right up to the onset of the Late Phanerozoic spreading episode. If this stage of crustal development has not updated the entire floor of the present-day Arctic Ocean, there may still be a case for the existence of Palaeozoic oceanic crust flooring part of the Basin. This becomes something of a real possibility if the Alpha Cordillera proves not to be a fossil-spreading center.

Acknowledgements

The field work in North Greenland which provided the basic thought for this paper was carried out for the Geological Survey of Greenland (GGU) between 1965 and 1971, in part in cooperation with the Geological Survey of Canada's 'Operation Grant Land' led by R. L. Christie and the 1969 British Joint Services Expedition led by J. D. C. Peacock. My thanks are due to R. L. Christie, H. P. Trettin and N. J. Soper of these expeditions for valuable discussions.

My thanks are also due to members of GGU for reading and criticizing earlier drafts of this paper, in particular A. E. Escher and T. C. R. Pulvertaft, who suggested important improvements. Publication is authorized by K. Ellitsgaard-Rasmussen, Director of GGU.

References

Allaart, J. H., 1965. The Lower Paleozoic sediments of Hall Land, North Greenland. Unpubl. rep., Grønlands geol. Unders., Copenhagen, 11 pp.

Atlasov, I. P., Yegiazarov, B. Kh., Dibner, V. D., Romanovich, B. S., Zimkin, A. V., Vakar, V. A., Demenitskaya, R. M., Levin, D. V., Karasik, A. M., Gakkel, Ya. Ya. and Litvin, V. M., 1964. Tectonic Map of the Arctic and Subarctic. Report XXII Int. Geol. Congr., Inst. Geol. Arctic., Leningrad.

Blackadar, R. G., 1954. Geological reconnaissance, north coast of Ellesmere Island, Arctic Archipelago, Northwest Territories, *Pap. geol. Surv. Can. 53–10*, 22 pp.

Carey, S. W., 1958. A tectonic approach to continental drift. *In:* Continental Drift, a symposium, pp. 177–355. Univ. Tasmania, Hobart.

Cowie, J. W., 1970. The Cambrian of the North American Arctic Regions. *In:* The Cambrian of the New World, edited by C. H. Holland, pp. 325–383. Interscience, London.

Christie, R. L., 1964. Geological reconnaissance of Northeastern Ellesmere Island, District of Franklin, *Mem. geol. Surv. Can. 331*, 79 pp.

Churkin, M., 1969. Paleozoic Tectonic History of the Arctic Basin North of Alaska, *Science*, **165**, 549–55.

Dawes, P. R., 1966. Lower Palaeozoic geology of the western part of the North Greenland fold belt, *Rapp. Grønlands geol. Unders.*, **11**, 11–5.

Dawes, P. R., 1970. Quaternary studies in northern Peary Land, *Rapp. Grønlands geol. Unders.*, **28**, 15–6.

Dawes, P. R., 1971. The North Greenland fold belt and environs, *Bull. geol. Soc. Denmark*, **20**, 197–239.

Dawes, P. R. and Soper, N. J., 1970. Geological investigations in northern Peary Land, *Rapp. Grønlands geol. Unders.*, **28**, 9–15.

Dawes, P. R. and Soper, N. J., 1971. Significance of K/Ar age determinations from northern Peary Land, *Rapp. Grønlands geol. Unders.*, **35**, 60–2.

Dawes, P. R. and Soper, N. J., in press. Pre-Quaternary history of North Greenland, *Bull. Amer. Ass. Petrol. Geol.*

Demenitskaya, R. M., 1958. Structure of the earth's crust in the Arctic, *Informatsionnyi Biulleten 7*, 42–49. Inst. Geol. Arctic., Leningrad. (In Russian.)

Demenitskaya, R. M. and Karasik, A. M., 1966. Magnetic data that confirm the Nansen–Amundsen Basin is of normal oceanic type. *In:* Continental Margins and Island Arcs, edited by W. H. Poole. *Pap. geol. Surv. Can. 66–15*, 191–6.

Demenitskaya, R. M. and Karasik, A. M., 1969. The active rift system of the Arctic Ocean, *Tectonophysics*, **8**, 345–51.

Demenitskaya, R. M., Karasik, A. M. and Kiselev, G. G., 1962. Results of studies of the geological structure of the Earth's crust in the Central Arctic by geophysical methods, *Problemy Arktiki i Antarktiki 2*. (In Russian.)

Demenitskaya, R. M. and Hunkins, K. L., 1970. Shape and structure of the Arctic Ocean. *In:* The Sea, edited by A. E. Maxwell, vol. 4, pt. 2, pp. 223–49. Interscience Publ., New York.

Dibner, V. D., 1970. Epi-Paleozoic horst-anticlinoria and shelf troughs of Arctic margin of Eurasia, *Bull. Amer. Ass. Petrol. Geol.* (*Abst.*), **54**, 2477.

Dibner, V. D., Gakkel, Ya. Ya., Litvin, V. M., Martynov, V. T. and Shurgayeva, 1965. The geomorphological map of the Arctic Ocean. *Trudy Nauchno-Issled., Instituta Geologii Arktiki*, **143**, 341–5. (In Russian.)

Dietz, R. S., 1961. Continent and ocean basin evolution by spreading of the sea floor, *Nature*, **190**, 854–7.

Dietz, R. S., 1963. Collapsing continental rises: An actualistic concept of geosynclines and mountain building, *Jour. Geol.*, **71**, 314–33.

Dietz, R. S., 1966. Passive continents, spreading sea floors, and collapsing continental rises, *Amer. J. Sci.*, **264**, 177–93.

Dietz, R. S. and Holden, J. C., 1970. Reconstruction of Pangaea: Breakup and Dispersion of continents, Permian to Present, *J. Geophys. Res.*, **75**, 4939–56.

Dietz, R. S. and Shumway, G., 1961. Arctic Basin geomorphology, *Bull. geol. Soc. Amer.*, **72**, 1319–30.

Drake, C. L., Campbell, N. J., Sanders, G. and Nafe, J. E. ,1963. Mid-Labrador Sea ridge, *Nature*, **200**, 1085–6.

Du Toit, A. L., 1937. Our Wandering Continents: a Hypothesis of Continental Drifting. Oliver and Boyd, Edinburgh, 366 pp.

Eardley, A. J., 1948. Ancient Arctica, *J. Geol.*, **56**, 409–30.

Eardley, A. J., 1961. History of geologic thought on the origin of the Arctic Basin. *In:* Geology of the Arctic, edited by G. O. Raasch, vol. 1, pp. 607–21. Univ. Toronto Press.

Egiazarov, B. Ch., Atlasov, I. P., Ravich, M. G., Demenitskaya, R. M., Dibner, V. D., Karasik, A. M., Kulakov, Ju. N., Puminov, A. P., Romanovich, B. S. and Tkachenko, B. V., 1969. Tectonic map of Polar regions of the Earth. Scale 1:10,000,000. Ministry of Geology of USSR. Inst. Geol. Arctic., Leningrad.

Ellitsgaard-Rasmussen, K., 1950. Preliminary report on the geological work carried out by the Danish Peary Land Expedition in the year 1949–50, *Meddr dansk geol. Foren.*, **11,** 589–95.

Ellitsgaard-Rasmussen, K., 1955. Features of the geology of the folding range of Peary Land, North Greenland, *Meddr Grønland,* **127** (7), 56 pp.

Fahrig, W. F., Irving, E. and Jackson, G. D., 1970. Franklin igneous events, tension faulting, and possible Hadrynian opening of Baffin Bay, Canada, *Bull. Amer. Ass. Petrol. Geol.* (*Abst.*), **54,** 2480.

Fortier, Y. O., McNair, A. H. and Thorsteinsson, R., 1954. Geology and petroleum possibilities in Canadian Arctic Islands, *Bull. Amer. Ass. Petrol. Geol.*, **38,** 2075–2109.

Fränkl, E., 1955. Rapport über die Durchquerung von Nord Peary Land (Nordgrönland) im Sommer 1953, *Meddr Grønland,* **103** (8), 61 pp.

Fränkl, E., 1956. Some general remarks on the Caledonian mountain chain of East Greenland, *Meddr Grønland,* **103** (11), 43 pp.

Frisch, T. O., 1967. Metamorphism and plutonism in northernmost Ellesmere Island, Canadian Arctic Archipelago. Unpubl. Ph.D. thesis, Univ. Calif. Santa Barb., 236 pp.

Gakkel, Ya. Ya., 1958. Signs of recent volcanic activity in the Lomonosov Range, *Priroda,* **4,** 87–90. (In Russian.)

Gakkel, Ya. Ya., 1961. Modern presentation of the Lomonosov Ridge. Materialy po Arktike i Antarktike 1, Leningrad. (In Russian.)

Gakkel, Ya. Ya. and Dibner, V. D., 1967. Bottom of the Arctic Ocean. *In:* International Dictionary of Geophysics, edited by S. K. Runcorn, vol. 1, pp. 152–65. Pergamon Press, London.

Grachev, A. F, Ivanov, S. S. and Karasik, A. M., 1967. Concerning the rift system of the Arctic, *Proc. Inst. Geol. Arctic, Reg. Geol.*, **10,** 65–70. Leningrad.

Haines, G. V., Hannaford, W. and Serson, P. H., 1970. Magnetic anomaly maps of the Nordic countries and the Greenland and Norwegian Seas, *Publ. Dominion Observ. Ottawa,* **39,** 119–49.

Hall, J. K., 1970. Geophysical evidence for ancient sea-floor spreading from Alpha Cordillera and Mendeleyev Ridge, *Bull. Amer. Ass. Petrol. Geol.* (*Abst.*), **54,** 2483–4.

Hamilton, W., 1970. The Uralides and the motion of the Russian and Siberian Platforms, *Bull. geol. Soc. Amer.*, **82,** 2553–76.

Harland, W. B., 1966. A hypothesis of continental drift tested against the history of Greenland and Spitsbergen, *Cambridge Research, no. 2*, pp. 18–22.

Harland, W. B., 1969. Contribution of Spitsbergen to understanding of tectonic evolution of North Atlantic region. *In:* North Atlantic—Geology and Continental Drift, a Symposium, edited by M. Kay. Mem. Amer. Petrol. Geol. 12, pp. 817–51.

Heezen, B. C. and Ewing, M., 1961. The Mid-Oceanic Ridge and its extension through the Arctic Basin. *In:* Geology of the Arctic, edited by G. O. Raasch, vol. 1, pp. 622–42. Univ. Toronto Press.

Henriksen, N. and Jepsen, H. F., 1970. K/Ar age determinations on dolerites from southern Peary Land, North Greenland, *Rapp. Grønlands geol. Unders.*, **28,** 55–8.

Hess, H. H., 1962. History of ocean basins. *In:* Petrologic Studies, edited by A. E. J. Engel, H. L. James and B. F. Leonard, pp. 599–620. Geol. Soc. Amer.

Hilgenberg, O. C., 1966. Bestätigung der Kennedy-Channel-Scherung durch die Bruchstruktur von Grönland und Nordost-Kanada. *Geotekt. Forsch.*, **22,** 1–74.

Hope, E. R., 1959. Geotectonics of the Arctic Ocean and the great Arctic Magnetic anomaly, *J. Geophys. Res.*, **64,** 407–27.

Hunkins, K., 1961. Seismic studies of the Arctic Ocean floor. *In:* Geology of the Arctic, edited by G. O. Raasch, vol. 1, pp. 645–65. Univ. Toronto Press.

Johnson, G. L. and Eckhoff, O. B., 1966. Bathymetry of the North Greenland Sea, *Deep-Sea Res.*, **13,** 1161–73.

Johnson, G. L. and Heezen, B. C., 1967a. The Arctic mid-oceanic Ridge, *Nature*, **215**, 724–5.

Johnson, G. L. and Heezen, B. C., 1967b. The morphology and evolution of the Norwegian-Greenland Sea, *Deep-Sea Res.*, **14**, 755–71.

Johnson, G. L. and Pew, J. A., 1968. Extension of the mid-Labrador Sea ridge, *Nature*, **217**, 1033–4.

Kerr, J. W., 1967a. Nares submarine rift valley and the relative rotation of North Greenland, *Bull. Can. Petrol. Geol.*, **15**, 483–510.

Kerr, J. W., 1967b. A submerged continental remnant beneath the Labrador Sea, *Earth Planet. Sci. Letters*, **2**, 283–9.

King, E. R., Zietz, I. and Alldredge, L. R., 1966. Magnetic data on the structure of the Central Arctic region, *Bull. geol. Soc. Amer.*, **77**, 619–46.

King, P. B., 1969. Tectonic Map of North America. Scale 1 : 5 000 000, U.S. Geol. Surv.

Koch, L., 1934. Some new main features in the geological development of Greenland. *In:* Zbiór Prac (collected papers), edited by H. Arctowski, pp. 149–59. Towarzystwo Geogr. Lwowie (Geogr. Soc. Lvov), Eugenjuszowi Vol.

Koch, L., 1935. A day in North Greenland, *Geogr. Ann. Sven Hedin Bd.*, 609–20.

Kutschale, H., 1966. Arctic Ocean geophysical studies: The southern half of the Siberia Basin, *Geophysics*, **31**, 683–710.

Lachenbruch, A. H. and Marshall, B. V., 1966. Heat flow through the Arctic Ocean floor: The Canada basin–Alpha rise boundary, *J. Geophys. Res.*, **71**, 1223–48.

Le Pichon, X., 1968. Sea-floor spreading and continental drift., *J. Geophys. Res.*, **73**, 3661–97.

McKenzie, D. P. and Parker, R. L., 1967. The north Pacific: an example of tectonics on a sphere, *Nature*, **216**, 1276–80.

Meyerhoff, A. A., 1970. Continental drift, II: High-latitude evaporite deposits and geologic history of Arctic and North Atlantic oceans, *Jour. Geol.*, **78**, 406–44.

Nansen, F., 1904. The bathymetrical features of the North Polar Seas, with a discussion of the continental shelves and previous oscillations of the shore-line. *In:* Norwegian North Polar Expedition, 1893–96. Scientific Results, edited by F. Nansen, vol. 4 (13), 231 pp. Jacob Dybwad, Christiania.

Nielsen, E., 1941. Remarks on the map and the geology of Kronprins Christians Land. *Meddr Grønland*, **126** (2), 34 pp.

Norford, B. S., 1972. Silurian stratigraphic sections at Kap Tyson, Offley Ø and Kap Schuchert, northwestern Greenland, *Meddr Grønland*, **195** (2), 40 pp.

Oliver, J., Ewing, M. and Press, F., 1955. Crustal structure of the Arctic regions from the Lg phase, *Bull. geol. Soc. Amer.*, **66**, 1063–74.

Ostenso, N. A., 1962. Geophysical investigation of the Arctic Ocean Basin. *Univ. Wisconsin Polar Res. Center, Res. Rep. 62–4*, 124 pp.

Ostenso, N. A., 1968. Geophysical studies in the Greenland Sea, *Bull. geol. Soc. Amer.*, **79**, 107–31.

Ostenso, N. A. and Wold, R. J., 1971. Aeromagnetic survey of the Arctic Ocean: Techniques and interpretations, *Marine Geophys. Res.*, **1**, 178–219.

Panov, D. G., 1955. Tectonics of the central Arctic. *Dokl. Akad. Nauk S.S.S.R.*, **105**, 339–42. (In Russian.)

Rassokho, A. I., Senchura, L. I., Demenitskaya, R. M., Karasik, A. M., Kicelev, Yu. G. and Timashenko, N. K., 1967. Podovodyni Shedinnyi arkticheskii khrebet i yego mesto v sisteme khrebtov severnogo ledovitogo okeana. *Dokl. Akad. Nauk S.S.S.R.*, **172**, 659–62.

Saks, V. N., Belov, N. A. and Lapina, N. N., 1955. Present concepts of the geology of the central Arctic, *Priroda*, **44** (7), 13–22. (In Russian.)

Shatskiy, N. S., 1935. Geologiya i poleznye iskopayemyi severa SSSR. 1st conf. geological explor. Moscow, 1935, Trudy 1, pp. 149–68.

Schuchert, C., 1923. Sites and nature of the North American geosynclines. *In:* Symposium on the structure and history of mountains and the causes of their development, *Bull. geol. Soc. Amer.*, **34**, 151–230.

Sykes, L. R., 1965. The seismicity of the Arctic, *Bull. Seismol. Soc. Amer.*, **55**, 519–36.

Taylor, F. B., 1910. Bearing of the Tertiary mountain belt on the origin of the earth's plan, *Bull. geol. Soc. Amer.*, **21**, 179–226.

Thorsteinsson, R. and Tozer, E. T., 1960. Summary account of structural history of the Canadian Arctic Archipelago since Precambrian time, *Pap. geol. Surv. Can. 60–7*, 25 pp.
Trettin, H. P., 1969a. A Paleozoic–Tertiary fold belt in northernmost Ellesmere Island aligned with the Lomonosov Ridge, *Bull. geol. Soc. Amer.*, **80,** 143–8.
Trettin, H. P., 1969b. Pre-Mississippian geology of northern Axel Heiberg and northwestern Ellesmere Islands, Arctic Archipelago, *Bull. geol. Surv. Can. 171*, 82 pp.
Trettin, H. P., 1970. Ordovician–Silurian flysch sedimentation in the axial trough of the Franklinian geosyncline, northeastern Ellesmere Island, Arctic Canada. *In:* Flysch sedimentology in North America, edited by J. Lapou, pp. 13–35. Geol. Ass. Can. Sp. Pap. 7.
Trettin, H. P., 1971. Geology of Lower Paleozoic Formations, Hazen plateau and southern Grant Land mountains, Ellesmere Island, Arctic Archipelago, *Bull. geol. Surv. Can. 203*, 134 pp.
Vine, F. J. and Matthews, D. H., 1963. Magnetic anomalies over oceanic ridges, *Nature*, **199,** 947–9.
Vogt, P. R. and Ostenso, N. A., 1967. Steady state crustal spreading, *Nature*, **215,** 810–7.
Vogt, P. R. and Ostenso, N. A., 1970. Magnetic and gravity profiles across the Alpha Cordillera and their relation to Arctic sea-floor spreading, *J. Geophys. Res.*, **75,** 4925–37.
Vogt, P. R., Ostenso, N. A. and Johnson, G. L., 1970. Magnetic and bathymetric data bearing on sea-floor spreading north of Iceland, *J. Geophys. Res.*, **75,** 903–20.
Vogt, P. R., Schneider, E. D. and Johnson, G. L., 1969. The crust and upper mantle beneath the sea. *In:* The Earth's Crust and Upper Mantle, Geophys. Monograph 13, edited by P. J. Hart, pp. 556–617. Amer. Geophys. Union, Washington.
Webster, C. J., 1954. The Soviet expedition to the central Arctic, 1954, *Arctic*, **7,** 59–80.
Wegener, A., 1924. The Origin of the Continents and Oceans. Methuen & Co., London, 212 pp.
Wilson, J. T., 1963a. Hypothesis of earth's behaviour, *Nature*, **198,** 925–9.
Wilson, J. T., 1965. A new class of faults and their bearing on continental drift, *Nature*, **207,** 345–7.

8.10

P. E. KENT
British Petroleum Co. Ltd.,
London, England

The Continental Margin of Tanzania

Exploration in coastal Tanzania has so far been unsuccessful in finding oil and gas but the operation spread over a number of years has provided a thorough knowledge of the structural development of this Indian Ocean margin. Surface mapping, magnetic, gravity and seismic surveys with fifty-two exploratory boreholes have provided knowledge of the distribution of sedimentary rocks from the crystalline basement contact to the edge of the continental shelf.

A memoir covering the new information has recently been issued by the Institute of Geological Sciences (Geophysical Paper No. 6, Geology and Geophysics of Coastal Tanzania, 1971).

In Permo-Triassic times (Karroo) this part of Africa showed a block and trough structure with almost completely continental sediments deposited in the troughs. The Indian Ocean, in the modern sense, came into existence in the Middle Jurassic and a marine transgression flooded across the older features. At this time marine sediments were laid down across pre-existing faults with throws in excess of 7000 m. Thence onwards sediments built out progressively from the continental margin, with non-marine and shallow water deposition in the coastal region and deeper water sediments further east.

The earlier Jurassic rocks are exposed on the inner edge of the basin; they are shelf facies oolites and sandy limestones, which give way to deeper water sediments down dip to the east. In northern Tanzania Bajocian/Bathonian reefal limestones transgressed across a major fault bounding a Karroo trough to rest on crystalline basement north of Ngerengere, and west of Dar-es-Salaam beds of early Jurassic date are coarsely conglomeratic, apparently derived from an active fault scarp. In southern Tanzania a combination of geophysical and drilling data shows that there again Middle Jurassic transgressed onto fault blocks, in that case bounding a Karroo evaporite basin.

In the main embayments (Dar-es-Salaam and Mandawa), the Upper Jurassic is represented by more than 1000 m of marine shales with some sandstone. Marginally (Matumbi Hills and Tendaguru) this facies interdigitates with a thinner estuarine development (coloured shales and sandstones), and there was regression of fully marine conditions in the Lower Cretaceous (Neocomian-Aptian) with an estuarine or paralic facies widespread. Marine shales in open water shelf conditions were deposited through the Upper Cretaceous and Palaeocene, followed in the Middle

Eocene by shallow water nummulite limestones which grade into a deeper water facies at depth towards the edge of the present continental shelf.

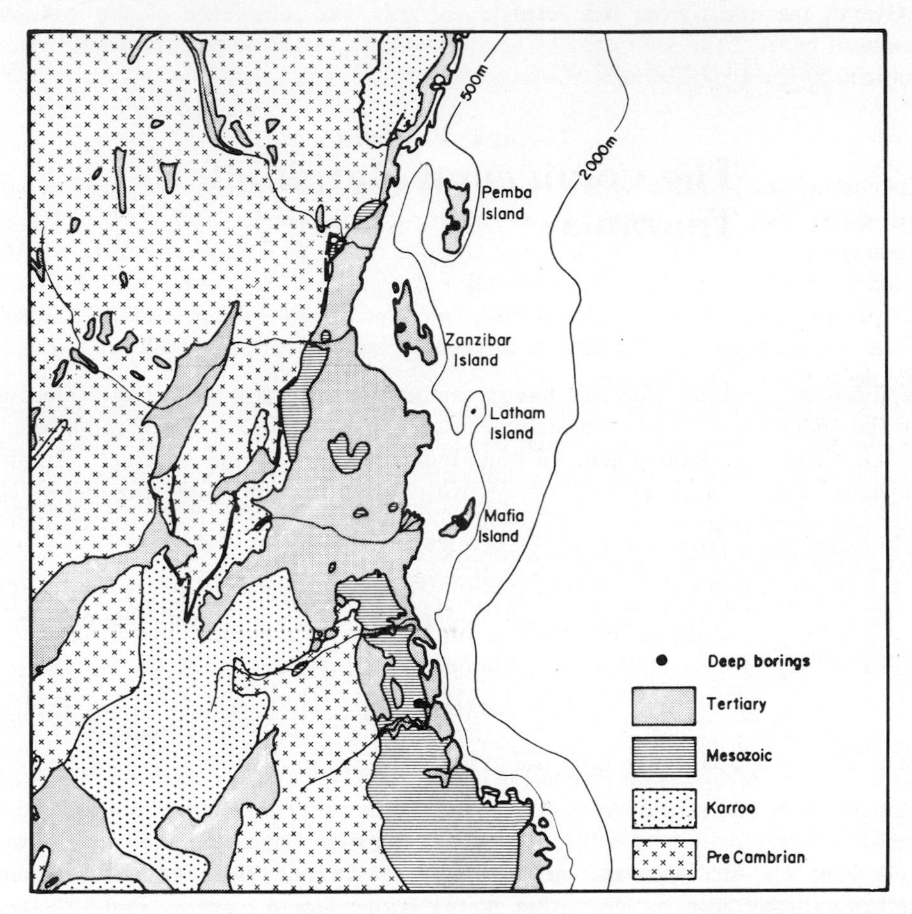

FIGURE 1 Coastal basin of Tanzania.

In line with occurrences elsewhere in the world there was a major regression in the Oligocene, locally associated with faulting, and Lower Miocene was widely transgressive. Thick later Miocene and Pliocene are known from seismic evidence in synclinal belts.

The measured sequences from the Jurassic through the Tertiary exceed 7000 m in total thickness; to this has to be added an unknown measurement for the Karroo beneath and perhaps 1000 m for the post-Middle Miocene sediments in the late downwarps. Not only are the exposed rocks mainly in shallow water facies—as might be expected near the inner basin edge—but also the long Tertiary sequences in the deep boreholes on the islands are predominantly shallow water. Thus the 2300 m of Miocene on Mafia Island is predominantly of shallow water shelf limestones; the equivalent on Zanzibar is a deltaic sequence. Deeper water sediments probably related to the upper continental slope are also recognized locally (e.g. the Eocene on Pemba Island). Essentially the sediments relate to progressive vertical subsidence continuing into and throughout Tertiary times—subsidence demonstrably more than 4500 m at Zanzibar since the Cretaceous.

Overall the effect from the Jurassic onwards was subsidence of the unknown basement beneath the sediments by some 10,000 m, and building out of the original continental coast by distances of up to 160 km.

Tectonic Features

The coastal belt is characterized by vertical tectonics; normal faulting is dominant, and shows two main trends, approximately north-northwest and north-northeast. These trends occur across Africa from coast to coast, as Furon demonstrated; they no doubt represent ancient structures liable to rejuvenation. The Karroo rifts follow this pattern, and a north-northeast fault belt which bounds the crystalline basement in the northern part of Tanzania developed a throw in excess of 7000 m before the Middle Jurassic transgression.

Although the Karroo faults of northern Tanzania and the Mandawa area were very large they were not reactivated in later adjustments, so that the Miocene and Plio-Pleistocene faulting and differential movements are developed further seawards, nearer the margin of the continental shelf. The large islands provide striking examples of progressive development; they are positive structures reflecting late (post-Pliocene) movement of blocks which were rising throughout Tertiary times.

The one normal anticline found in the coastal belt (Mandawa) proved to be due to a salt bulge, and the region shows a total absence of compressional features.

The history of fault movement is closely comparable to that of the Rift Valleys, with a major Karroo movement, a Lower Miocene phase and a Late Pliocene to Pleistocene phase. There is no good reason to suppose that contemporary stress conditions here were significantly different from those in the Rift Valleys themselves.

Summary

(1) Tanzania developed an open east-facing coast by Middle Jurassic times, and the history since then has been that of the Indian Ocean margin. Shoreline facies beds are known on the inner edge of the shelf; in many cases they can be traced eastwards into deeper water, and the continental edge has been progressively built out during the deposition of some tens of thousands of feet of sediments.

(2) The movements documented by geological and geophysical evidence were essentially vertical. The area is characterized by normal faults.

(3) The earlier faulting (7–10,000 m throw) occurred in Karroo to Early Jurassic times. Later movements have been smaller and located further seawards. The main blocking out of the new continental margin is ascribed to Early Jusassic.

(4) There is no evidence of low-angled faults or other features which would be associated with the plastic stretching of the crust which might be expected in an area of continental parting, but the fault movements differ from those inland in that subsidence of the downthrown sides does not appear to be associated with uplift of the upthrown block. The upthrown blocks remain near sea level and were subject to marine transgression.

(5) Collapse of an early broad regional dome could be postulated to fit the pattern of normal faulting (e.g. as shown in Tanzania Memoir, Fig. 44) but there is an

absence of stratigraphic evidence for this; known thinning is related to the smaller individual 'highs'; as east of Mandawa (Kiswere). Only the inland edge of the depositional shelf seems to have been actually emergent during the Jurassic, Cretaceous and Tertiary.

(6) Overall, the area shows a pattern of high blocks separated by downwarps, which may reflect and graben structures in the underlying crystalline Basement. The flanks tend to be faulted where the sediments are thin, but are monoclinal where the deep fractures are blanketed by thick sediments. The earlier downwarps were the site of salt basins in the south (Mandawa, Pindiro) and possibly in the Dar-es-Salaam embayment (Msanga). Others, known later in the regional history but possibly early origin, separate high blocks such as the large islands from mainland features. Pemba, one of these, lies separate from the continent, separated by a 700 m deep channel. Latham Bank, 100 km off Dar-es-Salaam, is analogous and is known from seismic survey to be anticlinal in a thick sedimentary series.

There is thus a series of homologous positive structures on the mainland, on the continental shelf and in the deeper ocean beyond, supporting the postulate of Loncarevic (1964) that the East African plate does not end sharply at the shoreline, but extends beneath the margin of the Indian Ocean as a continuous tectonic unit

References

Gregory, J. W., 1921. Rift Valleys and Geology of East Africa. London.

Kent, P. E., 1965. An evaporite basin in Southern Tanzania. *In:* Salt Basins Around Africa, pp. 41–54. Inst. Pet., London.

Kent, P. E., Hunt, J. A. and Johnstone, D. W., 1971. The geology and geophysics of coastal Tanzania, *Geophys. Paper 6*. Inst. Geol. Sci., London.

Loncarevic, B. D., 1964. Geophysical studies in the northwest Indian Ocean, *Endeavour*, **23,** 88, 43–47.

9

ANCIENT OCEANS

9.1

W. M. CADY*
U.S. Geological Survey,
Denver, Colorado 80225, U.S.A.

The Earmarks of Subduction

Introduction

The purpose of this paper is first to examine the kinds of evidence that may lead to the recognition of pre-Mesozoic subduction of both continental and oceanic lithosphere plates, and then to evaluate, if possible, whether or not plate tectonics operated continuously, cyclically, or at all in pre-Mesozoic time. If it operated in the pre-Mesozoic, when and how extensive was it? These items are to be discussed because 'a possible model' (Dewey, 1969, p. 124) for the evolution of the pre-Mesozoic Appalachian–Caledonian orogen, by subduction connected with the closing of the hypothetical "Proto-Atlantic Ocean' (Wilson, 1966), has been suggested and widely accepted, though not proved. And of related significance in the arguments that follow is Sutton's (1970) less familiar, later alternative that plate tectonics has indeed operated cyclically, but not continuously, during Earth history.

Subduction, though but one result of plate tectonics, may leave earmarks of plate convergence that are locally visible in old rocks. Evidence for subduction may be provided locally and also regionally by: (1) lithologic discontinuities across subduction zones, (2) blueschist within and (3) mélange terranes within and adjacent to subduction zones, (4) ophiolite assemblages immediately above subduction zones, and (5) increase of the potash content of superjacent igneous rocks in the down-dip direction of subduction zones. Other evidences of plate-tectonic activity, arrived at by interregional studies and more through conceptual tectonic restoration than by local observation, are chiefly: fit of continents, distribution of continental-shelf-facies faunal provinces, palaeomagnetic determinations of ancient geographical coordinates, spatial and temporal distribution of the effects of mantle melting or 'hot spots', and the record of magnetic polarity reversals shown by sea-floor stripes.

The earmarks of subduction become blurred with time and may be recognized only by those who believe the present to be an invariable key to the past. On the other hand, failure to recognize useful evidence of subduction in the ancient rocks may simply indicate that subduction did not take place prior to the Mesozoic. Instead, the proven and ongoing Mesozoic and Cenozoic subduction may have evolved less than 300 m.y. ago from a Precambrian and Palaeozoic setting that was equally significant but less mobile. Or perhaps sea-floor spreading was cyclic, marked by similar, albeit obviously more obscure, plate-tectonic activity related to the evolution of the first oceanic basins in the Early to Middle Precambrian time.

* Publication authorized by the Director, U.S. Geological Survey.

Extrapolation of Ideas of Subduction into the Past

The ideas regarding subduction have been extrapolated backward in time, to continental areas in which ocean floors have or appear to have become incorporated, on evidence much more speculative than that for the currently active sea-floor spreading and subduction.

'On-land' oceanic crust that contains a complete ophiolite sequence—ultramafites → gabbro → diabase → pillow basalt → chert → graywacke—raised up above a subduction (fossil Benioff) zone, seems very likely in the Mesozoic of the Coast Ranges of California (Bailey *et al.*, 1970). This sequence of rocks is comparable to that of existing ocean floors. The proposed subduction zone is sited along the Coast Range thrust. This thrust is marked by lithologic discontinuities in adjacent rocks (op. cit., Fig. 5, p. C78, Fig. 6, p. C79) and contains blueschist that identifies it appropriately as a site of high-pressure low-temperature metamorphism (op. cit., p. C77; Blake *et al.*, 1967; Ernst, 1971; *c.f.* Gresens, 1971; *see also* Barbat, 1971). Immediately west of and beneath the thrust is the ensimatic Franciscan mélange terrane considered to mark the former site of an oceanic trench (Ernst, 1970). And to the east of the trace of the east-dipping thrust is the parallel Mesozoic calc-alkaline plutonic belt of California (Dickinson, 1970, p. 841–844). This belt is reminiscent of the island arcs of the western Pacific, whose igneous rocks are characterized by increased potash content in the down-dip direction of subjacent and now-active subduction zones (Dickinson & Hatherton, 1967).

The blueschist-mélange terranes, as in the California Coast Ranges, and the adjacent calc-alkaline belts, have come to be considered key indicators of subduction zones (Dewey, 1971), and they may provide the only evidence that subduction occurred. Other criteria, which are less convincing but resorted to where these keys may or may not be available, include linear belts of alpine-type ultramafites and of mafic volcanic rocks and perhaps eclogites, which are also features of the Coast Ranges (Bailey *et al.*, 1964, p. 90; 1970, p. C71). These latter possible criteria of subduction are a last resort in continental terranes in which regional structural, stratigraphic, petrologic, palaeontologic, and palaeomagnetic discontinuities are obscure or appear to be obscure. The alpine-type ultramafites have long been recognized as features of the cores of orogens in both continental and oceanic settings, without any necessary implication of subduction (Hess, 1939).

Least convincing is evidence for very old subduction zones in Palaeozoic and Precambrian terranes. These terranes are commonly exposed far from now-converging plate margins such as those that adjoin the Pacific Ocean basin, or their structures are transected by now-diverging plate margins such as those of the Atlantic region. The earmarks and validity of these old subduction zones, especially in the Atlantic region, are the principal subjects of this discussion.

Proposed Palaeozoic Subduction Zones

Appalachian–Caledonian orogenic belt

Resort to suggestive but inconclusive evidence for subduction has been necessary in the Late Precambrian to Middle Palaeozoic rocks of the Appalachian–Caledonian belt, which trends northeastwards from eastern North America through the British Isles and Scandinavia (Fig. 1).

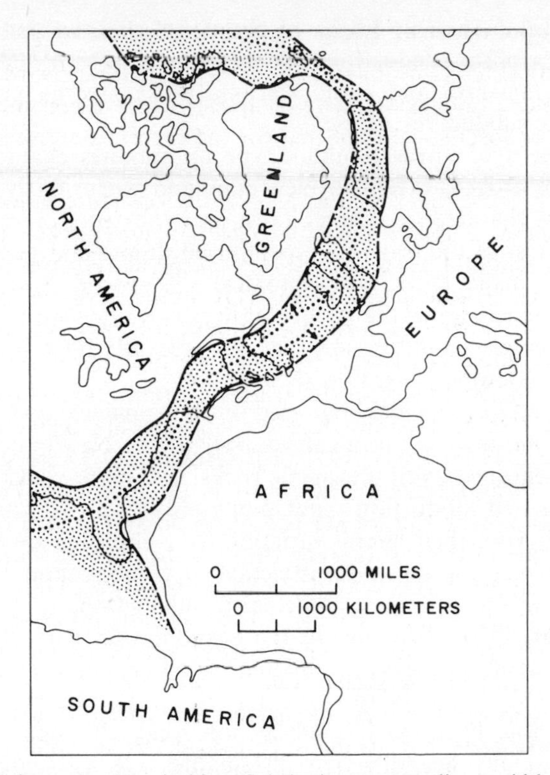

FIGURE 1 Lower Palaeozoic Appalachian–Caledonian geosyncline; width of the geosyncline before its cross section was halved during middle Palaeozoic orogeny is shown by the northwest–southeast attenuation of the outlines of the British Isles and Newfoundland; dotted lines bound an area possibly floored by oceanic crust at and northeast of Newfoundland and also suggest continuation beneath continental crust of an oceanic ridge system; arrows indicate possible crustal divergence during minimal early Palaeozoic sea-floor spreading.

Blueschist terranes are uncommon in this belt. Glaucophane schist, which may be loosely regarded of the blueschist facies (Bailey, 1961; Bailey *et al.*, 1964, Fig. 21, p. 110), and which reflects similar high-pressure and low-temperature conditions, is notable only in the Girvan-Ballantrae Complex in South Ayrshire, Scotland (Bloxam & Allen, 1960) and on the island of Anglesey, Wales (Greenly, 1919, p. 115–121). (*See also* Zwart, 1969, p. 11.) Mélange terranes are reported only in Anglesey (the type mélange, Greenly, 1919) and in northeastern Newfoundland (Horne, 1969). Granitic intrusives of all kinds are relatively less abundant in the Appalachian–Caledonian belt than in the nearby Middle-to-Late Palaeozoic Appalachian–Hercynian orogenic belt (Zwart, 1967, pp. 507, 513). An apparent exception is in the Maritime Provinces, New England and the Piedmont where the Appalachian–Hercynian belt encroaches westward from southwest England and western Europe. In the Maritime–New England region granitic plutons include elongate calc-alkaline batholiths (Cady, 1969, pp. 51–52) possibly reminiscent of the Sierra–Klamath calc-alkaline belt in California and Oregon. However, they are east of the trace of the the proposed subduction zone, which as presented (Bird & Dewey, 1970, Fig. 9B, p. 1047) dips westward away from instead of toward the plutons. A continent–continent collision, such as assumed by those who suggest subduction in the

Appalachian–Caledonian belt, might result in no effective subsequent control of the direction of dip of the subduction zone. Elsewhere in the Appalachian–Caledonian orogenic belt ultramafites and mafic volcanic rocks are the principal recourse in attempts to locate subduction zones.

Longitudinal belts of intrusive alpine-type ultramafites characterize the Appalachian–Caledonian orogen, except in Newfoundland where fairly complete ophiolite sequences such as characterize oceanic crust are reported (Church & Stevens, 1971) (see Fig. 1). In most of the orogen, serpentinized dunite and peridotite that are not in sequence with mafic rocks, are tectonically emplaced (Cady, 1969, pp. 47–9) much as in the Franciscan in the California Coast Ranges west of the Coast Range thrust. They are emplaced in terranes that have recently been suggested to lie above subduction zones (Dewey, 1969, Figs. 2, 3, 4).

Mafic volcanic rocks are widely reported in the Appalachian–Caledonian orogenic belt but, though intruded by ultramafites, only in the Newfoundland area are they in full ophiolite sequences with ultramafic rocks. The mafic rocks have been likewise attributed to suggested subduction zones, and more closely than have the ultramafites (op. cit., Dewey). Also, their lateral variation from tholeiitic through calc-alkalic to alkalic (presumably in the down-dip direction of a subduction zone) has been presented as evidence for otherwise undetected subduction, in England and Wales (Fitton & Hughes, 1970). Eclogites in the Scandinavian part of the Appalachian–Caledonian belt, which probably are mainly metamorphosed mafic rocks as well as ultramafites (Bryhni *et al.*, 1970), have recently been interpreted by one author (op. cit., Dewey, Fig. 1) to mark a subduction zone, although he 'gives no reason for supposing that any Scandinavian ultramafites are associated with big faults' (Nicholson, 1971, p. 2351).

The most obvious evidence against subduction in the Appalachian–Caledonian belt is in the absence of convincing discontinuities across the proposed traces of subduction zones. In the northern Appalachian belt alone mixed and transitional Pacific (western or American) and Atlantic (eastern or European-Baltic) shallow-water, shell-faunal elements occur together in single and tectonically unbroken rock units (Kay & Eldredge, 1968; Neuman & Whittington, 1964; Theokritoff, 1968, pp. 17–8). This seams to indicate locally continuous marine shelf conditions between the American and European and probably also the African cratons during the Early Palaeozoic, and it weakens the concept of a proto-Atlantic. Moreover, ubiquitous bimodal basalt-rhyolite volcanics without transitional andesitic facies (Cady, 1969, pp. 23, 105) bespeak extensional tectonics (Lipman, 1969) in the New England region during the Early-to-Middle Palaeozoic, as opposed to the compression implied by subduction. Also, evidence of lithologic discontinuities, such as that marked by the Coast Range thrust in California, is lacking.

Uralian orogenic belt

A Lower to Middle Palaeozoic blueschist terrane with associated eclogite occupies a nearly vertical shear zone ('Uralian fault') that trends northward through the length of the Ural Mountains in central Eurasia and is perhaps a subduction zone (Hamilton, 1970, p. 2563). Granitic and ultramafic plutons and basaltic to andesitic volcanic rocks are also found in the Urals, but their locations and directions of chemical variation relative to the shear zone are ambiguous (op. cit., pp. 2563–5). The granitic rocks are successively younger eastward, but episodically more potassic westward, which suggests west dip of successively newer subduction zones that originated progressively eastward. The ultramafites are east of the shear zone. The direction of compositional change of the volcanic rocks, if any, is not clear.

Palaeozoic subduction in the Uralian orogenic belt is suggested by palaeomagnetic and palaeoclimatic observations which show that the Russian and Siberian Platforms moved closer together during the Palaeozoic (Hamilton, 1970, p. 2555; Kropotkin, 1962, pp. 1220–5; 1967, p. 276; 1971, p. 265).

The Verdict of Geologic Mapping

Geologic mapping, along with regional stratigraphic and structural studies, is an obvious and probably practicable solution to many uncertainties about subduction. Such mapping will, it is hoped, prove or disprove continuity of stratigraphic sequences across postulated subduction zones. This approach has already shown the possibility of a subduction zone in the California Coast Ranges, where rocks of different facies but of the same age are juxtaposed through thrust faulting. In the Appalachians, where subduction zones have been suggested (Bird & Dewey, 1970; Dewey, 1969), stratigraphic, structural, and map studies fail to prove them because stratigraphic discontinuity cannot be clearly demonstrated. Where mapping is ambiguous, and subduction with lithosphere plate loss was great enough, palaeogeographic restorations based on palaeomagnetic studies may be decisive. Thus geologic mapping alone may or may not provide conclusive evidence against subduction, but should be supplemented by combinations of criteria of both local and interregional significance to prove subduction.

Another Viewpoint

Sutton (1969, pp. 240–3; 1970) has recently suggested an alternate, worldwide, approach that includes an evolutionary interpretation of plate tectonics instead of one that simply interprets the present as a strict key to the past. Sutton actually uses a reasonable combination of these approaches—an evolutionary interpretation that explains possible lack of evidence for subduction in the Late Precambrian and Palaeozoic and a uniformitarian interpretation that suggests a continuous cyclic process. These combined interpretations envision various plate movements, large and small, that took place during two previous episodes of about 200 m.y. each (in which ocean basins, notably the ancestral Pacific, first formed) and an ongoing Mesozoic–Cenozoic episode marked by the opening of the Atlantic and Indian oceans. Each episode evolved by tensional rifting of relatively stable shield areas that themselves had evolved—during approximately 1000 m.y. intervals—through tectonic, magmatic, and metamorphic consolidation of intraplate geosynclines to form networks of orogenic belts. Notable among the rifts formed 1000 m.y. ago, but which failed to open into an ocean basin, is the rift marked by the mafic Keweenawan lava flows and the related midcontinent gravity high of central United States (King & Zietz, 1971).

Conclusions

Evidence from the Appalachian–Caledonian orogenic belt seems to support the case for a Late Precambrian–Palaeozoic stable interval free of plate tectonism, but is opposed especially by the palaeomagnetic evidence relative to the Uralian belt. More investigation of these and other Precambrian and Palaeozoic orogenic belts is obviously required. Various alternate schemes are expectable and must be tested open-mindedly by those who are aware of but are not wholly engaged in 'the new global tectonics'.

Acknowledgements

The author is indebted to E. H. Bailey, G. P. Eaton, James Gilluly and P. W. Lipman for critical reading of drafts of this manuscript and to many others for profitable discussions of this and related subjects.

References

Bailey, E. H., 1961. Metamorphic facies of the Franciscan Formation of California and their geologic significance (abs.), *Geol. Soc. America Spec. Paper 68*, pp. 4–5.

Bailey, E. H., Irwin, W. P. and Jones, D. L., 1964. Franciscan and related rocks, and their significance in the geology of western California, *California Div. Mines and Geology Bull. 183*, 177 pp.

Bailey, E. H., Blake, M. C., Jr. and Jones, D. L., 1970. On-land Mesozoic oceanic crust in California Coast Ranges. *In:* Geological Survey Research 1970, pp. C70–C81. U.S. Geol. Survey Prof. Paper 700–C, 251 pp.

Barbat, W. F., 1971. Megatectonics of the Coast Ranges, California, *Geol. Soc. Amer. Bull.*, **82,** 1541–61.

Bird, J. M. and Dewey, J. F., 1970. Lithosphere plate—continental margin tectonics and the evolution of the Appalachian orogen, *Geol. Soc. Amer. Bull.*, **81,** 1031–59.

Blake, M. C., Jr., Irwin, W. P. and Coleman, R. G., 1967. Upside-down metamorphic zonation, blueschist facies, along a regional thrust in California and Oregon. *In:* Geological Survey Research, 1967, pp. C1–C9. U.S. Geol. Survey Prof. Paper 575–C, 251 pp.

Bloxam, T. W. and Allen, J. B., 1960. Glaucophane-schist, ecolgite, and associated rocks from Knockormal in the Girvan–Ballantrae Complex, South Ayrshire, *Royal Soc. Edinburgh Trans.*, **44,** 27 pp.

Bryhni, I., Green, D. H. and Heier, K. S., 1970. On the occurrence of eclogite in western Norway, *Contr. Mineral and Petrology*, **26,** no. 1, 12–19.

Cady, W. M., 1969. Regional tectonic synthesis of northwestern New England and adjacent Quebec, *Geol. Soc. America Mem. 120*, 181 pp.

Church, W. R. and Stevens, R. K., 1971. Early Paleozoic ophiolite complexes of the Newfoundland Appalachians as mantle—oceanic crust sequences, *J. Geophys. Res.*, **76,** 1460–6.

Dewey, J. F., 1969. Evolution of the Appalachian/Caledonian orogen, *Nature*, **222,** 124–9.

Dewey, J. F., 1971. Plate models for the evolution of the Alpine fold belt, *Geol. Soc. America Abstracts with Programs (Annual Meeting)*, **3,** no. 7, 543.

Dickinson, W. R., 1970. Relations of andesites, granites and derivative sandstones to arc-trench tectonics, *Rev. Geophys. Space Phys.*, **8,** 813–60.

Dickinson, W. R. and Hatherton, T., 1967. Andesitic volcanism and seismicity around the Pacific, *Science*, **157**, 801–3.

Ernst, W. G., 1970. Tectonic contact between the Franciscan mélange and the Great Valley sequence—crustal expression of a late Mesozoic Benioff zone, *J. Geophys. Res.*, **75,** 886–901.

Ernst, W. G., 1971. Do mineral parageneses reflect unusually high-pressure conditions of Franciscan metamorphism?, *Am. J. Sci.*, **270,** 81–108.

Fitton, J. G. and Hughes, D. J., 1970. Volcanism and plate tectonics in the British Ordovician, *Earth and Planet. Sci. Letters*, **8,** 223–8.

Greenly, E., 1919. The geology of Anglesey, *Great Britain Geol. Survey Mem.*, 980 pp.
Gresens, R. L., 1971. Discussion: Do mineral parageneses reflect unusually high-pressure conditions of Franciscan metamorphism?, *Amer. J. Sci.*, **271**, 311–8.
Hamilton, Warren, 1970. The Uralides and the motion of the Russian and Siberian Platforms, *Geol. Soc. Amer. Bull.*, **81**, 2553–76.
Hess, H. H., 1939. Island arcs, gravity anomalies and serpentinite intrusions, *Internat. Geol. Congr. Moscow 1937*, *Rept. 17*, **2**, 263–83.
Horne, G. S., 1969. Early Ordovician chaotic deposits in the central volcanic belt of northeastern Newfoundland, *Geol. Soc. Amer. Bull.*, **80**, no. 12, 2451–64.
Kay, M. and Eldredge, N., 1968. Cambrian trilobites in central Newfoundland volcanic belt, *Geol. Mag.*, **105**, 327–77.
King, E. R. and Zietz, Isodore, 1971. Aeromagnetic study of the mid-continent gravity high of central United States, *Geol. Soc. Amer. Bull.*, **82**, 2187–207.
Kropotkin, P. N., 1962. Paleomagnetism, paleoclimates and the problem of extensive horizontal movements of the earth's crust, *Internat. Geology Rev.*, **4**, 1214–34 (A.G.I. transl.).
Kropotkin, P. N., 1967. The mechanism of crustal movements, *Geotectonics*, 1967, 276–85 (A.G.U. transl.).
Kropotkin, P. N., 1971. Eurasia as a composite continent, *Tectonophysics*, **12**, 261–6.
Lipman, P. W., 1969. Relations between andesitic and rhyolitic volcanism, western interior United States, *Geol. Soc. London Proc.*, no. 1662, pp. 36–39.
Neuman, R. B. and Whittington, H. B., 1964. Fossils in Ordovician tuffs northeastern Maine, *U.S. Geol. Survey Bull. 1181–E*, 38 pp.
Nicholson, R., 1971. Faunal provinces and ancient continents in the Scandinavian Caledonides, *Geol. Soc. Amer. Bull.*, **82**, 2349–56.
Sutton, J., 1969. Rates of change within orogenic belts. *In:* Time and Place in Orogeny, edited by P. E. Kent, G. E. Satterthwaite and A. M. Spencer, 239–50. Geol. Soc. London Spec. Pub. 3, 311 pp.
Sutton, J., 1970. Migration of high temperature zones in the crust. *In:* Palaeogeophysics, edited by S. K. Runcorn. Academic Press, London and New York, 518 pp.
Theokritoff, G., 1968. Cambrian biogeography and biostratigraphy in New England. *In:* Studies of Appalachian Geology: Northern and Maritime, edited by E-an Zen, W. S. White, J. B. Hadley and J. B. Thompson, Jr., 9–22. Interscience Publishers, New York, 475 pp.
Wilson, J. T., 1966. Did the Atlantic close and then re-open?, *Nature*, **211**, 676–81.
Zwart, H. J., 1967. Orogenesis and metamorphic facies series in Europe, *Dansk. Geol. Fören. Medd.*, **17**, 504–16.
Zwart, H. J., 1969. Metamorphic facies series in the European orogenic belts and their bearing on the causes of orogeny, *Geol. Assoc. Canada Spec. Paper 5*, 7–16.

9.2

E. M. MOORES

Geology Department, University of California, Davis, California 95616, U.S.A.

Plate Tectonic Significance of Alpine Peridotite Types

Introduction

Ever since the classic papers of Steinmann (1905, 1926), Benson (1926) and Hess (1955), Alpine peridotite rocks have been a source of controversy, partly because of confusion about their mode of occurrence, but also because it was recognized rather early that they were mantle derived and therefore, if understood, would reveal a great deal about the nature of the mantle. Steinmann, Benson and Hess all recognized that these rocks were present in linear mountain belts, and Hess observed that they were 'intruded in the first great deformation of a mountain belt and probably do not recur in subsequent deformations of the same belt' (1955, p. 391). Despite these observations, much confusion has remained, partly because of widely varying experience of individual authors (see Jackson & Thayer, 1972, for short review).

In the light of Plate Tectonics Theory and much recent work on ultramafic rocks in general, a new prespective is possible on the occurrence of Alpine peridotites and their tectonic significance.

Terminology

Petrology in general, and particularly igneous petrology, suffers from a plethora of names, a hold-over from the days when it was thought that the way to understanding of rock occurrences lay through coining of new names. Though this tendency has diminished markedly in the last few years, still a confusing and disparate amount of terminology must be mastered if one is to be able to read the literature pertaining to these rocks. Therefore it is perhaps not inappropriate to review briefly some of the words involved; more complete glossaries are given by Wyllie (1967) and Williams *et al.* (1958). *Ultramafic* rocks contain more than 70% mafic minerals; *ultra-basic* rocks contain less than 45% SiO_2. Figure 1 shows a triangular diagram illustrating some commonly used terms for ultramafic rocks. *Peridotites* are olivine rich rocks, *dunite* is a rock containing more than 95% olivine. *Wehrlite* is a rock with 50% or more olivine and the rest dominantly calcium-rich pyroxene. *Harzburgite* is a rock containing more than 50% or more olivine plus calcium-poor pyroxene (generally enstatite or bronzite). *Lherzolites* are peridotites with approximately equal proportions of calcium-poor and calcium-rich pyroxene. *Pyroxenites* are generally called

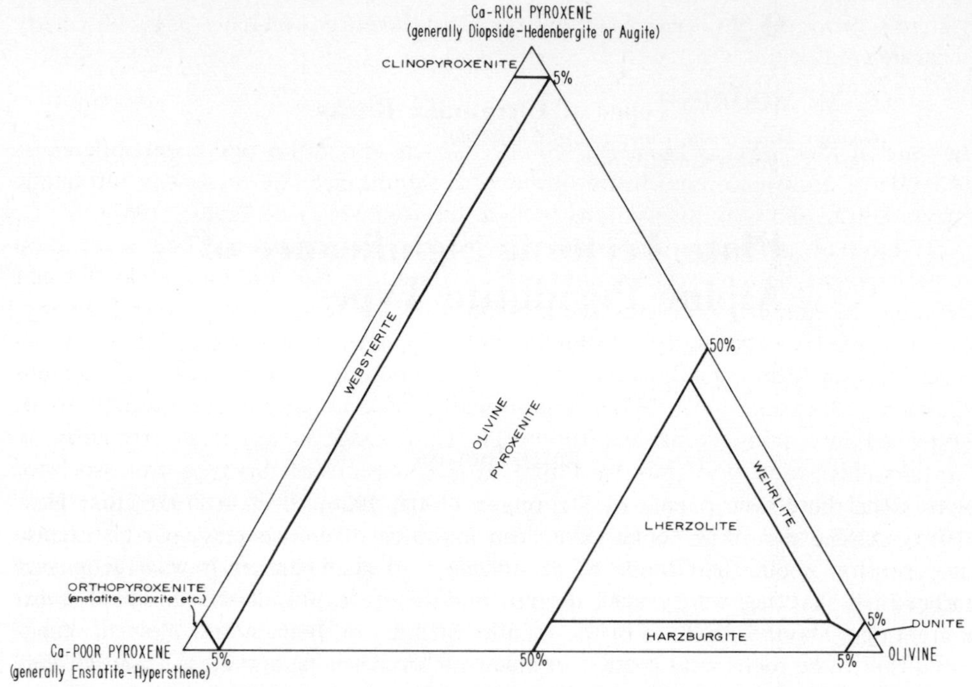

FIGURE 1 Ternary diagram illustrating generally the terminology of ultramafic rock types.

orthopyroxenite (or *enstatite*) if exclusively calcium-poor pyroxene, *websterite* if containing both calcium-poor and calcium-rich pyroxenes. *Eclogites* are rocks containing magnesian garnet plus omphacite (sodic calcium-pyroxene).

The above review is neither intended as exhaustive nor authoritative. Readers still curious are referred to Wyllie (1967, p. 3) or to Williams *et al.* (1958).

Ophiolites

Part of the problem of the significance of peridotites revolves about the question of the nature of the ophiolite sequence which has been well-known in European literature since Steinmann's (1905, 1926) original exposé. A great deal of recent work has been focused on the interpretation of these masses as fragments of oceanic crust and mantle (Hess, 1965; Dewey & Bird, 1971; Bailey *et al.*, 1970; Maxwell, 1969; Coleman, 1971; Brunn, 1960; Vuagnat, 1963; Gansser, 1966; Moores & Vine, 1971; Aumento *et al.*, 1971). The ideal ophiolite sequence consists from bottom to top of metamorphic (Alpine) ultramafics, cumulate textured ultramafic rock, grading up into gabbro, diorite and/or quartz diorite, and hypabyssal mafic-intermediate rock, commonly present as dike swarms (altered dolerite, keratophyre, epidosite, etc.), grading upward into pillow lava, and overlain by radiolarian chert or other deep sea pelagic sediment. If the term ophiolite is restricted in usage to this type of sequence, then the very powerful relationship follows that one is dealing with a slice of oceanic crust and mantle emplaced by plate tectonic processes. The presence of only one ophiolite body in the restricted sense in an intracontinental mountain system is proof that an ocean basin has been lost during convergence of two formerly separate continents.

Types of Ultramafic Rocks

In spite of their general mineralogic and chemical similarity, significant differences in mode of occurrence, and hence, in tectonic significance, are present in ultramafic rocks. Ultramafic rock associations include the following (see Wyllie, 1967):

(1) Layered gabbro–norite–peridotite associations of large area. This association is perhaps the best recognized ultramafic association, and includes rocks formed definitely by cumulate processes (see Jackson, 1971, for review, also Wager & Brown, 1967). These rocks generally are thought to be emplaced as magma and then crystallized by stratiform crystallization; hence they represent in a general way, igneous sediments (Jackson, 1967, 1971). Traditionally these intrusions are thought to be emplaced into non-tectonic environments. They range in age from Archaean to Tertiary. Like most flood basalts, Phanerozoic intrusions of this type are associated with rifted continental margins, or represent the remains of mantle 'Hot spots' (Morgan, in press). Whether the abundant Precambrian bodies presently also represent similar tectonic environments is unknown. Also, ultramafic rocks displaying cumulative textures are present in minor, but significant, amounts in ophiolite complexes (Davies, 1971; Moores & Vine, 1971; Jackson, 1971; Parrot, 1967), and also in oceanic crustal areas near ridges, where their presence has been revealed by dredging (Melson, 1970; Aumento *et al.*, 1971).

(2) Concentrically zoned bodies, the Alaska–Ural type of intrusion (Taylor & Noble, 1960; Jackson, 1971; Jackson & Thayer, 1972). These masses are small, concentrically zoned bodies with dunite cores, which display clear evidence of magmatic intrusion, and which are generally spatially restricted to the Pacific Margin and to the Urals. They are rich in Fe and Ca; hence they contain abundant iron oxides and Ca-rich pyroxene, and virtually no Ca-poor pyroxene. The age of a given belt of these intrusions is apparently restricted in a given mountain belt (Jackson & Thayer, 1972) but the tectonic significance of these bodies remains presently unknown.

(3) Ultramafic lavas and associated sills displaying quench-textured olivine (Naldrett & Mason, 1968). These bodies seem particularly common in Archaean Greenstone belt associations, but may also occur in very minor amounts in Phanerozoic ophiolites (Bear & Morel, 1960; Gass, 1958).

(4) Alpine peridotites generally occur in linear belts in deformed mountain regions. Their characteristics have been summarized by Wyllie (1967), Thayer (1967, 1968 and 1960) and Jackson & Thayer (in press). Characteristically they display tectonic fabrics (Raleigh, 1965; Ave Lallement, 1967; Jackson & Thayer, in press; Nicolas *et al.*, 1971). Their compositions are generally fairly magnesian, the olivines and pyroxenes generally displaying Mg/Fe ratios of approximately 9/1. Chromian spinal concentrations are generally fairly common in these rocks (Thayer, 1967). Most workers agree that they represent the exposure of mantle (lithosphere). Alpine peridotites are restricted in time to the Late Precambrian and Phanerozoic mountain systems (Jackson & Thayer, in press). In other words the oldest deformed mountain system known to contain Alpine peridotites is the Late Precambrian Pan-African-Baikalian system.

Types of Alpine Peridotites

Though generally similar, significant variations occur in Alpine peridotites which reflect different composition or history of mode of occurrence, or depth of equilibration. Jackson & Thayer (1972) have divided these rocks into the harzburgite and lherzolite subgroups, characterized by those peridotites respectively. Other bodies, however, display garnet peridotite assemblages, and raise the question whether Alpine masses can be separated easily into two groups of whether a continuum of types is present. Table 1 shows some characteristic features of different

TABLE 1. Some features of Alpine peridotites

Peridotite type	Assoc. Gabbroic composition	% Al_2O_3 OPx	Form	Size	Regional metaphorphic grade
Plagio-Spinel	Gabbro	0·5–1·5	Pseudostrat.	60–1000	Unmetamorphosed
Plagio-Spinel	Gabbro-Pyrox.	1–3	Stock-lense	10–60	Unmetamorphosed-Greenschist
Spinel	Pyroxenite	3–6	Stock-lense	20–30	Greenschist-Alm. Amphibolite
Garnet	Pyrox-Eclog.		Lense-Dome	5–25	Alm. Amphibolite-Granulite

types of alpine peridotites. Generally speaking one can distinguish these bodies on whether they contain plagioclase peridotite, spinel peridotite, or garnet peridotite assemblages. Some bodies, may show gradations between peridotite types from plagioclase to spinel, and the associated mafic compositions also show gradations from gabbroic to granulite to eclogite (Dickey, 1970). An interesting relationship indicated in Table 1 is that the size of the body decreases with increasing apparent depth of emplacement.

Moores (1970) and Moores & MacGregor (in press) have recently attempted a grouping of peridotite masses on the basis of their mode of occurrence, as follows:

(A) Mantle slabs. These rocks include allochthonous masses of suboceanic crust and mantle as well as subcontinental mantle. Suboceanic crust and mantle slabs are fairly common and consist principally of ophiolite complexes and eroded ophiolite complexes. Examples of these masses include the Troodos complex, Cyprus (Gass, 1968; Moores & Vine, 1971), other Mediterranean ophiolites, the Coast Range ultramafic sheet California (Bezore, 1969; Bailey *et al.*, 1970), perhaps the Trinity ultramafic pluton Klamath Mountains, California (Davis, 1968; Lipman, 1964; Lindsley-Griffin, oral communication, 1972), and the peridotite-gabbro complexes of New Guinea (Davies, 1971) and New Caledonia (Coleman, 1971).

Known subcontinental mantle slabs, on the other hand, are rare. Only one good example is known—the Ivrea-Lanzo body of the western Alps (Nicolas *et al.*, 1971).

(B) Disrupted Mantle or overthrust slabs presently incorporated into melange terranes, such as in the Franciscan terrane of California (Loney *et al.*, 1970; Hsu, 1969; Medaris, 1972) in the Tethyan belt (Gansser, 1959; Temple & Zimmerman,

1969). Many of these masses are discordant to surrounding rocks and display a wide variety of degree of serpentinization, of and in P-T condition of crystallization in respect to the surrounding rock.

(C) Sedimentary serpentinites and serpentine diapirs. These relatively common rocks recently have been the subject of a review by Lockwood (1971). They may form mobilization of serpentinized mantle material in convergent plate boundaries, but also may be present in mid-ocean ridges (Aumento *et al.*, 1971).

(D) Hot diapiric masses with thermal aureoles. These bodies are relatively small, rare, and include both the lherzolite and harzburgite subfacies of Jackson & Thayer (in press). Examples include the bodies at Mt. Albert, Quebec (Smith & MacGregor, 1960), Lizard, Cornwall (Green, 1964), Tinaquillo, Venezuela (MacKenzie, 1960), Beni Bouchera, Morocco (Kornprobst, 1969) and Ronda, Spain (Dickey, 1970). It seems clear that in each of these examples the peridotite was emplaced hot and thermally metamorphosed its surroundings. However, Thayer (1968) has recently questioned whether one of the principal examples of these rocks, the Lizard, really does not represent an ophiolitic mass. Most descriptions of these bodies were made before the full extent of oceanic metamorphism at mid-oceanic ridges became known (Miyashiro *et al.*, 1970). At the present time, therefore, it is not known whether these bodies represent a special class of occurrence or simply oceanic crust-mantle sections.

(E) Conformable bodies in regionally metamorphozed terranes. Masses included in this category include those in the western Sierra Nevada, California, where Hietanen (1951) has mapped lensoid bodies of ultramafic and mafic rock which are in a terrane intruded and deformed by plutons of the Sierra Nevada batholith. Similar bodies have been mapped by O'Hara & Mercy (1963) in the Ticino region of the Alps and in the Trondhjem region of Norway (O'Hara & Mercy, 1963; Battey, 1965). Whether all these masses should be classified as 'Alpine' *sensu stricto*, however, is questionable. Battey (1965) has shown that the Trondhjem bodies probably represent layers within a deformed stratiform intrusive complex. Many of these masses display garnet peridotite assemblages, and pronounced effects of deep-seated recrystallization and equilibration. Whether all masses represent deformed mafic-ultramafic complexes or whether some actually are the product of mantle diapirism into deep crustal levels is presently unknown.

Conditions of Equilibration

Using the phase relations worked out by MacGregor (1967), O'Hara (1967) and others (see Wyllie, 1970, for review), it is possible to obtain to a first approximation the conditions of equilibration of a given peridotite mass. Figure 2 shows a P-T plot with a number of bodies plotted upon it. Clearly a mass such as the Bay of Island complex (Church & Stevens, 1971) which represents a substantial amount of vertical section through the mantle, will display a range of P-T conditions of equilibration. If such a mass is disrupted by subsequent plate tectonic activity (e.g. mélange formation), then one will find high-pressure peridotites in a tectonic environment considerably out of equilibrium with the body (e.g. Medaris, 1972). It is clear that one must distinguish between the arrival of mantle material from the depths through diapiric rise and the emplacement of this material into the crust of the earth. The differences

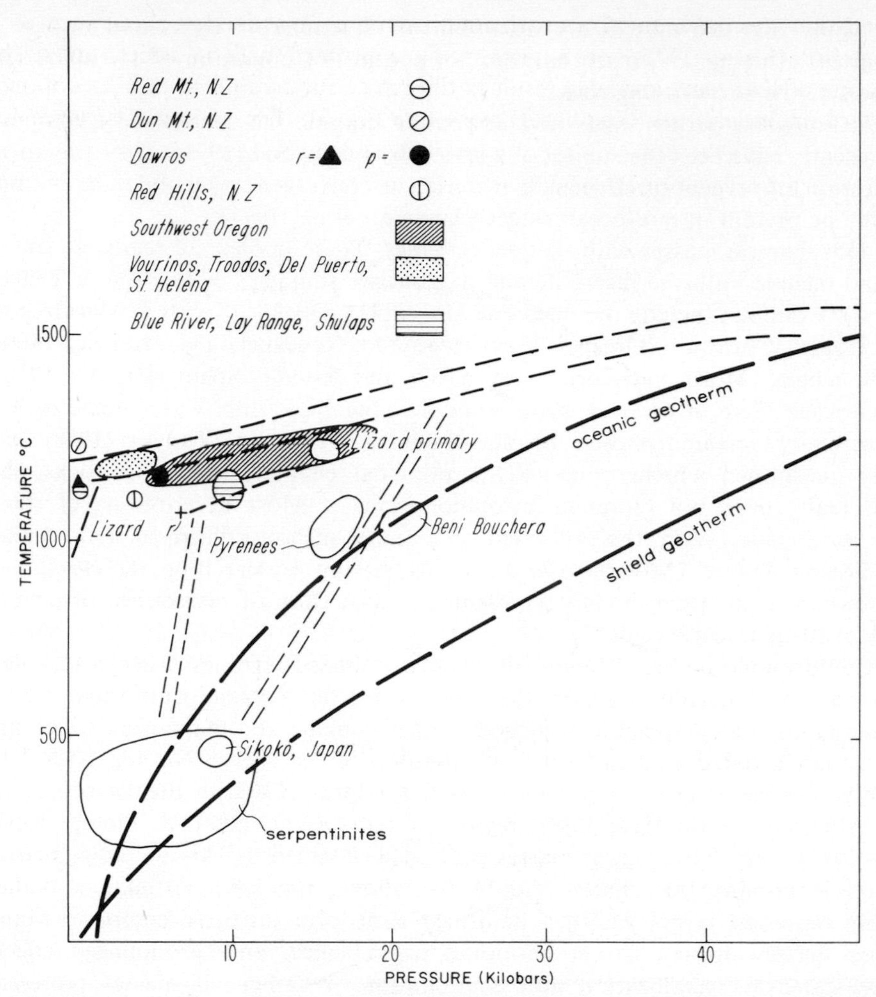

FIGURE 2 P–T diagram of peridotite occurrences, modified, after O'Hara (1967) and Moores & MacGregor (1972).

in thermal effects and disequilibrium relations which different bodies display may result from differences in the time of the action of these two processes. Hence a hot body with a thermal aureole represents a mass where the diapiric rise from the mantle, presumably at a spreading center, and its incorporation into the crust of the Earth were essentially simultaneous phenomena. Most ophiolite slabs, on the other hand, represent material which came to rest at a spreading center at some time before being tectonically emplaced at the margin of a continent. Hence the rocks were essentially cold at the time of the emplacement, as indicated by the general lack of thermal effects.

Emplacement of Ophiolites and Other Mantle Slabs

The emplacement of ophiolites has been a subject of controversy ever since the work of Steinmann, Hess and Benson, previously mentioned. MacKenzie (1969) observed that the presence of a continental (or other low density mass, such as a remnant arc (Karig, 1972)) on a plate being consumed will, in effect, arrest the subduction of that plate and may result in the flip of subduction so that it continues downward underneath the continental mass (see Fig. 3). The geologic evidence for

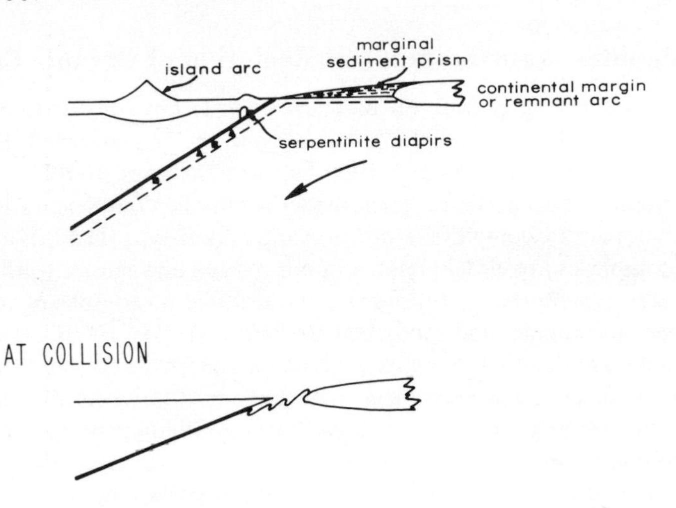

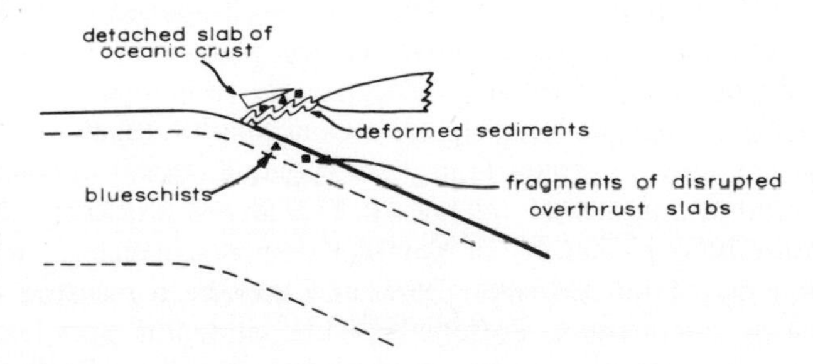

FIGURE 3 Plate tectonic emplacement of ophiolites, with one possible post-collision situation, that of 'flip' of subduction.

such a collision, as emphasized by the authors cited above, is the emplacement of an ophiolite slab. These relations are beautifully demonstrated by the Tertiary emplacement of a large ophiolitic body in New Guinea, resulting from collision of the continental mass of Australia with the Banda Arc (Davies, 1971; Hamilton, 1973). A corollary of this model is that consuming plate margins may not necessarily be originally generated at the margins of continents dipping under them. Subduction zones may start anywhere conditions are proper for their development (e.g. Sykes, 1970). These zones will consume lithosphere, without regard to whether these plates bear continents or not, until they intersect the continental margin. Though passive riders on the plates, the continents' inability to be subducted enable them to exert strong control on the location of consuming plate margins. There will be less tendency for such margins to be initiated at the continental margins than for them to collect there. The presence of ophiolite bodies in a mountains system, however, may record directly the collision of a continental margin and a subduction zone, and therefore, provide valuable evidence of the early stages of plate tectonic interaction prior to the convergence of two continental masses to form an intracratonic mountain system.

Ophiolites, Anorthosites, and Evolution of Oceanic Crust

Several recent workers (e.g. Jackson & Thayer, 1972) have noted that both ophiolite complexes, as well as true Alpine peridotite masses, are restricted in time to Late Precambrian to Recent time. Mobile belts between the ages of 1000 and 2000 m.y. display abundant evidence of stratiform mafic-ultramafic complexes and bodies, and two kinds of anorthosite complexes stratiform and massif-type (Herz, 1969). Stratiform anorthosite complexes are closely related in occurrence and nature to their stratiform mafic-ultramafic counterpart. Indeed many stratiform complexes contain both well-developed ultramafic and anorthositic zones (Hess, 1960). However, these ultramafic rocks are definitely cumulate, or magmatic, in origin, and do not display the abundant high pressure assemblages and tectonic fabrics coexisting contrasting side by side with cumulate fabrics which characterize Phanerozoic Alpine peridotites and ophiolite complexes.

A possible explanation for these observed relationships may be afforded by consideration of Phanerozoic exposures of mantle slabs. Nearly all Phanerozoic exposures of mantle slabs are ophiolites, or eroded remnants of ophiolites—that is they represent exposures of *suboceanic* mantle. The only well-defined subcontinental slab is the Ivrae-Lanzo mass described above. Hence, plate tectonic present inter-actions apparently do not provide enough vertical throw to expose subcontinental mantle, but do provide enough to bring up suboceanic mantle. Such vertical throw would therefore presumably be greater than 5 km (oceanic crustal thickness) and less than 20 km (minimum continental thickness). The largest thickness measured for any given ophiolite body is that of the Vourinos complex, which aggregates a thickness of approximately 11 km (Moores, 1969) and may be a measure of the maximum throw of such interactions. Presumably subcontinental mantle is not generally exposed because continents are too thick to expose their underlying mantle.

As mentioned above, most well-preserved ophiolite complexes show some cumulate rocks of ultramafic-mafic composition, rather similar to, but less well developed than, those displayed by stratiform intrusive complexes. These relations, all taken together, suggest the hypothesis that if plate tectonics applies to Proterozoic deformed belts, then during this time the amount of magmatic material generated at the equivalent of a mid-oceanic spreading center was greater than at present, so that the mantle never was exposed as a result of any plate tectonic interactions. This thickened hypothetical crust would be formed of mafic-ultramafic intrusive material and would be represented in the geologic literature as mafic-ultramafic or stratiform anorthositic stratiform complexes of Proterozoic age (see Fig. 4). Ultimately some of this crust would have become involved in plate marginal deformation. If the stratiform-type complexes are a Proterozoic equivalent of ophiolites, hence oceanic crust, then if that oceanic crust is taken down a subduction zone, perhaps the mafic and ultramafic

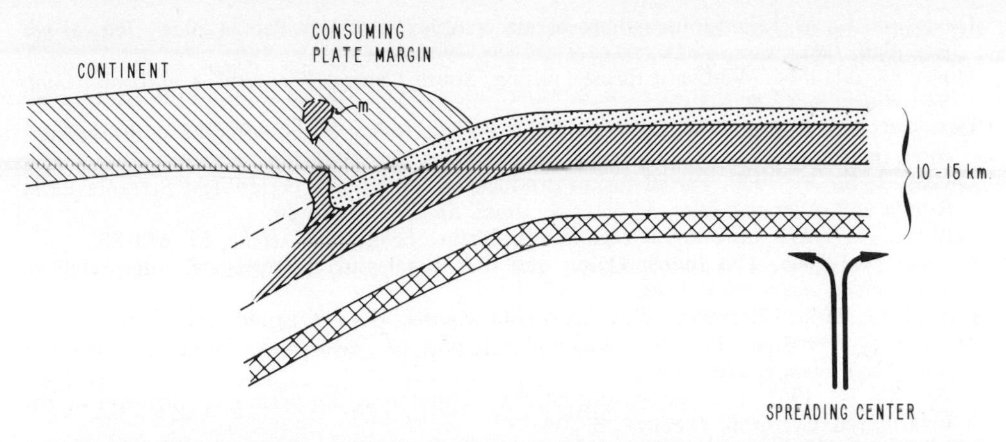

FIGURE 4. Hypothetical plate tectonic situation for middle Precambrian belts, involving thicker crust. Stippled—hypabyssal, extrusive or salic rocks; heavy diagonal-ruling—anorthositic layer of mafic crust; white—mafic layer; cross-pattern—ultramafic cumulate layer; light diagonal pattern—continental crust; m—massif-type-anorthosite. See text for explanation.

portions would have continued on down, possibly, ordered by phase changes to denser phase. However, any anorthositic layer may have detached and been reintruded as a massif-type body (Fig. 4). This model is speculative but is amenable to testing.

Acknowledgements

I thank F. J. Vine, J. C. Maxwell, J. L. Talbot, P. W. Lipman, I. D. MacGregor and E. D. Jackson for interesting discussions on these topics. This work was supported partially by N.S.F. Grants GA 1395 and GA 14298.

References

Aumento, F., Loncarevic, B. D. and Ross, D. I., 1971. Hudson Geotraverse, geology of the mid-Atlantic Ridge at 45° N., *Trans. Roy. Soc. Lond.*, **268,** 623–50.

Ave Lallement, G., 1967. Structure and petrofabric analysis of an 'Alpine type' peridotite: the lherzolite of the French Pyrenees, *Leidse Geol. Meded.*, **42,** 1–57.

Bailey, E. H., Jones, J. L. and Blake, M. C., 1970. Onland Mesozoic oceanic crust in California, Coast Ranges, U.S.G.S. *Prof. Pap. 700–C*, pp. C70–C81.

Battey, M. H., 1965. Layered structures in rocks of the Jotunheim complex, Norway, *Min. Mag.*, **34,** 35–51.

Bear, L. M. and Morel, S. W., 1960. Geology of the Agros–Akrotiri area, Cyprus, *Geol. Surv. of Cyprus, Mem. 7*, 88 pp.

Benson, W. N., 1926. The tectonic conditions accompanying the intrusion of basic and ultrabasic igneous rocks, *National Academy of Sciences Memoirs*, *19*, 1st Memoir, pp. 1–90.

Bezore, S. P., 1969. The Mt. St. Helena ultramafic-mafic complex of the northern California Coast Ranges, *Geol. Soc. Amer. Abstr. with Programs*, **3,** 5–6.

Brunn, J. H., 1960. Mise en place et differenciation de l'association pluto-volcanique du cortège ophiolitique, *Rev. Géog. Phys. et Géol. Dyn.*, **2,** 115–32.

Church, W. R. and Stevens, R. K., 1971. Early Paleozoic Ophiolite complexes of the Newfoundland Appalachians as mantle-oceanic crustal sequences, *J. Geophys. Res.*, **76,** 1460–6.

Coleman, R. G., 1971. Plate tectonic emplacement of upper mantle peridotites along continental edges, *J. Geophys. Res.*, **76,** 1212–22.

Davies, H. L., 1971. Peridotite-gabbro-basalt, complex in eastern Papua, *Austr. Bur. Mines Res. Bull. 128*.

Davis, G. A., 1968. Westward thrust faulting, South Central Klamaths, California, *Geol. Soc. Amer. Bull.*, **79**, 911–34.

Dewey, J. F. and Bird, J. M., 1971. Origin and emplacement of the ophiolite suite: Appalachian ophiolites in Newfoundland, *J. Geophys. Res.*, **76**, 3179–206.

Dickey, J. S., Jr., 1970. Partial fusion products in alpine-type peridotites; Serrania de la Ronda and other examples, *Miner. Soc. Amer. Sp. Pap.*, **3**, 33–50.

Gansser, A., 1959. Ausseralpine Ophiolithprobleme, *Eclog. Geol. Helv.*, **52**, 659–80.

Gansser, A., 1966. The Indian Ocean and the Himalayas: A geological interpretation, *Eclog. Geol. Helv.*, **59**, 831–48.

Gass, I. G., 1958. Ultramafic pillow lavas from Cyprus, *Geol. Mag.*, **95**, 241–251.

Gass, I. G., 1968. Is the Troodos complex a part of ocean floor formed by sea-floor spreading? *Nature*, **220**, 39–42.

Green, D. H., 1964. The petrogenesis of the high-temperature peridotite intrusion in the Lizard area, Cornwall, *J. Petrol.*, **5**, 134–88.

Hamilton, W., in press. Tectonics of Indonesia (this volume).

Hess, H. H., 1955. Serpentines, orogeny and epeirogeny, *Geol. Soc. America Special Paper 62*, pp. 391–408.

Hess, H. H., 1960. Stillwater igneous complex, Montana, a quantitative mineralogical study, *Geol. Soc. Amer. Mem.*, *80*, 230 pp.

Hess, H. H., 1965. Midoceanic ridges and tectonics of the sea floor, Submarine Geology and Geophysics. Proceedings of the Seventeenth Symposium of the Colston Research Society. Butterworth, London, pp. 317–34.

Herz, N., 1969. Anorthosite belts, continental drift and the anorthosite event, *Science*, **164**, 944–7.

Hietanen, A., 1951. Metamorphic and igneous rocks of the merrimac area, Plumas National Forest, California, *Geol. Soc. Amer. Bull.*, **62**, 565–607.

Hsü, K. J., 1969. Preliminary report and geologic guide to Franciscan mélanges of the Morro Bay–San Simeon area, California, *Calif. Div. Mines Pub. 35*.

Jackson, E. D., 1967. Ultramafic cumulates in the Stillwater, Great Dyke and Bushveld Intrusions. *In:* Ultramafic and Related Rocks, edited by P. J. Wyllie, pp. 20–38. Wiley, New York.

Jackson, E. D., 1971. Origin of ultramafic rocks by cumulus processes, *Fortschr. Miner.*, **48**, 128–74.

Jackson, E. D. and Thayer, T. P., 1972. Criteria for distinguishing between stratiform, zoned and Alpine peridotite–gabbro complexes. 24th Int. Geol. Congr., Montreal, sect. 2, pp. 289–296.

Karig, D. E., 1972. Remnant arcs, *Geol. Soc. Amer. Bull.*, **83**, 1057–68.

Kornprobst, J., 1969. Le massif ultrabasique des Beni Bouchera (Rif Interne Maroc), *Contr. Min. Pétrol.*, **23**, 283–322.

Lockwood, J. P., 1971. Sedimentary and gravity-slide emplacement of serpentinite. *Geol. Soc. Amer. Bull.*, **82**, 919–936.

Loney, R. A., Himmelberg, G. R. and Coleman, R. G., 1971. Structure and petrology of the alpine-type peridotite at Burro Mountain, Calif., U.S.A., *J. Petrol.*, **12**, 145–310.

Lipman, P. W., 1964. Structure and origin of an ultramafic pluton in the Klamath Mts., Calif., *Amer. J. Sci.*, **262**, 199–222.

MacGregor, I. D., 1967. Mineralogy of model mantle compositions. *In:* Ultramafic and Related Rocks, edited by P. J. Wyllie, pp. 381–92. Wiley, New York.

MacKenzie, D. B., 1960. High-temperature alpine-type peridotite from Venezuela, *Geol. Soc. Amer. Bull.*, **71**, 303–18.

MacKenzie, D. P., 1969. Speculations on the causes and consequences of plate motion, *Geophys. J.*, **18**, 1–32.

Maxwell, J. C., 1969. Alpine mafic and ultramafic rocks, the ophiolite suite; a contribution to the discussion of the paper: The origin of ultramafic and ultrabasic rocks, by P. J. Wyllie, *Tectonophysics*, **7**, 489–94.

Medaris, J., 1972. High-pressure peridotites in Southwestern Oregon, *Geol. Soc. Amer. Bull.*, **83,** 41–58.

Melson, W. G. and Thompson, J., 1970. Layered basic complex in oceanic crust, Romanche Fracture, Equatorial Atlantic Ocean, *Science*, **168,** 817–20.

Miyashiro, A., Shido, F. and Ewing, M., 1971. Metamorphism in the mid-Atlantic ridge near 24° and 30° N, *Phil. Trans. Roy. Soc., Lond.*, **A.268,** 589–60.

Moores, E. M., 1969. Petrology and structure of the Vourinos ophiolite complex, northern Greece, *Geol. Soc. Amer. Spec. Pap. 118*, 74 pp.

Moores, E. M., 1970. Ultramafics and orogeny, with models for the U.S. Cordillera and the Tethys, *Nature*, **228,** 837–42.

Moores, E. M. and MacGregor, 1972. Types of Alpine Ultramafics. Geol. Soc. Amer. Hess Volume in press.

Moores, E. M. and Vine, F. J., 1971. The Troodos Massif, Cyprus and other ophiolites as oceanic crust: evaluation and implications, *Trans. Roy. Soc., Lond.*, **268A,** 443–66.

Morgan, W. J., 1971. Plate motions and deep mantle connection. Geol. Soc. Amer. Hess Volume, in press.

Naldrett, A. J. and Mason, G. D. 1968. Contrasting Archean ultramafic igneous bodies in Dundonald and Clergur townships, Ontario, *Can. J. Earth Sci.*, **5,** 111.

Nicolas, A., Bouchez, J. L., Boudier, F. and Mercier, J., 1971. Textures, structures and fabrics due to solid state flow in some European lherzolites, *Tectonophysics*, **12,** 55–86.

O'Hara, M., 1967. Mineral facies in ultrabasic rocks. *In:* Ultramafic and Related Rocks, edited by P. J. Wyllie, pp. 7–18. Wiley, New York.

O'Hara, M. J. and Mercy, E. L. P., 1963. Petrology and petrogenesis of some garnetiferous peridotites, *Trans. Roy. Soc., Edinburgh*, **65,** 251–314.

Parrot, J. F., 1967. Le cortège ophiolitique du Pinde Septentrional (Grèce). OSTROM, Paris, 114 pp.

Raleigh, C. B., 1965. Structure and petrology of an Alpine peridotite on Cyprus Island, Wash., U.S.A., *Beit. Min. Petrog.*, **11,** 719–41.

Smith, C. H. and MacGregor, I. D., 1960. Ultrabasic intrusive conditions illustrated by the Mount Albert ultrabasic pluton, Gaspé, Quebec, *Geol. Soc. Amer.*, **71,** 1978 (abstr.).

Steinmann, G., 1905. Geologische Beobachtungen in den Alpen II. Die Schartt'sche Überfaltungstheorie und die geologische Bedeutung der Tiefseeabsätze und der ophiolithischen Massengesteine. *Ber. Nat. Ges. Freiburg*, 1 Bd. 16, pp. 44–65.

Steinmann, G., 1926. Die ophiolithischen Zonen in den mediterranean Kettengebirgen. 14th International Geol. Cong., Madrid, C.R. 2, pp. 638–67.

Sykes, L. R., 1970. Seismicity of the Indian Ocean, and a possible nascent island arc between Ceylon and Australia, *J. Geophys. Res.*, **75,** 5041.

Taylor, H. P., Jr. and Noble, J. A., 1960. Origin of the ultramafic complexes in Southeastern Alaska *Int. Geol. Congr., 21st Report 13*, 175–87.

Temple, P. G. and Zimmermann, J., 1969. Tectonic significance of Alpine ophiolites in Greece and Turkey, *Geol. Soc. Amer. Abs. with Programs, for 1969*, part 7, p. 221.

Thayer, T. P., 1960. Some critical differences between alpine-type and stratiform complexes, *Rept. Int. Geol. Congr., 21st, Copenhagen*, Sect. 13, pp. 175–87.

Thayer, T. P., 1967. Chemical and structural relations of ultramafic and feldspathic rocks in Alpine intrusive complexes. *In:* Ultramafic and Related Rocks, edited by P. J. Wyllie, pp. 232–9.

Thayer, T. P., 1968. Peridotite-gabbro complexes as keys to petrology of mid-ocean ridges, *Geol. Soc. Amer. Bull.*, **80,** 1515–22.

Vuagnat, M., 1963. Remarques sur la trilogie serpentinite-gabbro-diabases dans le bassin de la méditerranée occidental, *Geol. Rundsch.*, **53,** 336–58.

Wager, L. R. and Brown, G. M., 1967. Layered Igneous Rocks, 588 pp. W. H. Freeman & Co., San Francisco.

Williams, H., Turner, F. J. and Gilbert, C. M., 1958. Petrography, an Introduction to the Study of Rocks in Thin Section, 406 pp. W. H. Freeman and Co., San Francisco.

Wyllie, P. J. (ed.), 1967. Ultramafic and Related Rocks, 464 pp. Wiley, New York.

Wyllie, P. J., 1970. Ultramafic rocks and the upper mantle, *Min. Soc. Amer. Spec. Pap.*, **3,** 3–32.

Comments

J. V. AVIAS

Laboratoire de Géologie,
Université de Montpellier, France.

I would like to make a few remarks about the problem of the genesis and setting of ophiolites, based mainly on my experience with ultramafic rocks in New Caledonia (see bibliography).

(1) I do not think Moores' classification of shallow to deep ultramafic types is valid in all cases. Detailed field observations in New Caledonia, for example, show a *mixture* of all the rocks, cited as typical of each type, in close association.

(2) In the case of the Alpine peridotite type I would like to point out that:

(*a*) The grade of metamorphism of the associated rocks is always low (greenschist facies) which does not agree with postulated high temperature conditions.

(*b*) There is always a close association of vulcano-sedimentary rocks (including basic pillow lavas, radiolarites, etc.) which are more or less metasomatized and spilitized.

(*c*) Chloritization and serpentinization (serpentine being a magnesian chlorite) may occur through metasomatic differentiation of the vulcano-sedimentary rocks, as shown in ophispherite rock remanents which often occur in marginal parts of ultrabasic masses; furthermore, there is a 'geochemical complementarity' between the ultramafics and associated rocks.

(*d*) Serpentine may be very easily peridotitized in relatively low temperature conditions (around 400°C) such as those occurring in layer 3 of the oceanic crust (on the basis of thermodynamics and laboratory experimental results).

(*e*) In actual outcrops it is normal to find a 'secondary' (classical) serpentinization of peridotites.

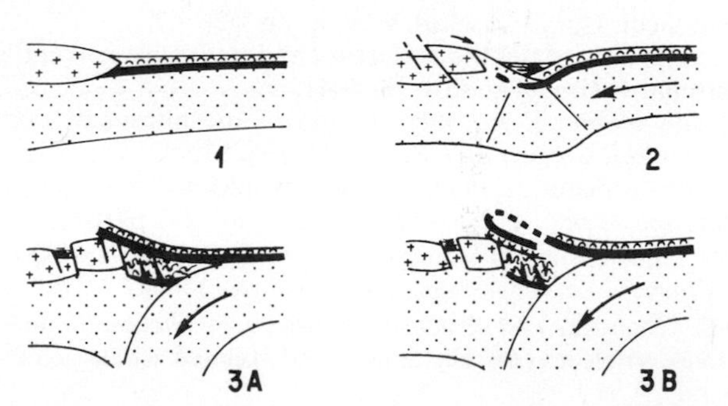

(3) I have, therefore, been led to propose the following hypothesis for the origin and setting of ultramafic overthrust masses (Fig. 1): the ultramafic overthrust sheets should be layer 3 (derived by metamorphism of layer 2 with help of connate sea water) overthrust, mainly in a solid state, in continental oceanic margin troughs at the end of the subduction which included sediments and even previous folded foreland sialic crust.

The overthrusting would be, on this basis, the consequence of a separation of layer 3 from the pyrolite surface of the lithosphere at the Mohorovičič level (which must be a mechanical discontinuity lubricated by greenschists and serpentinites); the forces being the consequence of the terminal phase of the arching of oceanic crust, following the consequences of sea floor spreading processes in sialic-oceanic crustal margins.

Bibliography

Avias, J. V., 1967. Overthrust structure of the main ultrabasic New Caledonian massives, *Tectonophysics*, **4**, 531–541.
Avias, J. V., 1971. *C.R. Acad. Sci., Paris*, **273**, 667–70.
Avias, J. V., in press. Major features of the New Guinea–Louisiade–New Caledonia–Norfolk Arc System. XII° PanPacific Science Congress, Canberra 1971, Proceedings. Univ. Press, The University of Western Australia.

Reply

by E. M. Moores to comments by J. V. Avias:

I concur with Professor Avias that a continuum of peridotite depth types is present. My separation into shallow to deep types is only an attempt to categorize some features which may be separable thus, but clearly a continuum and a *mixture* will be commonly present. I disagree somewhat with Professor Avias about the lack of thermal contacts of these bodies, to the extent of the occurrence of the high temperature types.

In his third point, Professor Avias has raised the interesting question of the nature of oceanic layer three, which may be partly mafic rock, and partly serpentinized peridotite as emphasized by Hess (1965).

9.3

A. GILBERT SMITH
Department of Geology, Sedgwick Museum,
Downing Street, Cambridge CB2 3EQ, England

The So-called Tethyan Ophiolites

Introduction

There are many ophiolites (see below) in the Alpine–Himalayan region (Fig. 1). What they represent has been a controversial question for many years, and will probably continue to be so for some time to come. For some geologists ophiolites are huge submarine lava flows erupted onto a pre-existing sea-floor (which need not be oceanic); for others they are diapirs of plastic mantle material that have been emplaced like salt domes into the overlying lighter crust; a third view is that ophiolites are slices of older ocean-floor which have been tectonically emplaced onto the edges of continents or island arcs. The third interpretation is well supported by much recent data on the nature of the ocean-floor in present-day oceans and also by detailed mapping of ophiolite terrains (Coleman, 1971; Dewey & Bird, 1971; Moores & Vine, 1971; Hynes *et al.*, 1972). The interpretation of ophiolites as ocean-floor slices is the view accepted in this paper.

In later Palaeozoic time the marine fauna and flora of what has since developed into the Alpine–Himalayan region appears to have belonged to a distinct province. The province is assumed to owe its existence and distinctive nature to the presence in the region of a former ocean which was named the Tethys (Suess, 1906, **3,** 19). Suess believed that the oceanic area in the Alpine–Himalayan region had been more extensive than it is today, and regarded the Mediterranean as a remnant of the Tethys. However, the extent, age, and even the nature of the Tethys have never been well defined (Sylvester-Bradley, 1967).

As the ophiolites in the Alpine–Himalayan region probably represent slices of ocean floor, and because the Tethys is thought to have existed in the same region, these ophiolites are commonly referred to as Tethyan ophiolites. Smith (1971) has argued that the so-called Tethyan ophiolites are remnants of oceans which, together with the Mediterranean and the Black Sea, came into existence in Jurassic or later time, long after the establishment of the Tethyan faunal province. In late Triassic time an ocean (or oceans) existed in the region (Fig. 2a), but during Jurassic, Cretaceous and possibly Tertiary time it gradually shrank to nothing as the new Jurassic or younger oceans came into being. New data and new interpretations (see below) suggest that disruption of the areas that were to develop into Jurassic and younger oceans began in mid- or late Triassic time, but it is not known whether ocean-floor spreading dates from this period.

Which of these mid-Triassic or younger oceanic areas should be known as the Tethys—the ocean(s) already in existence, or the ocean(s) that were just coming into

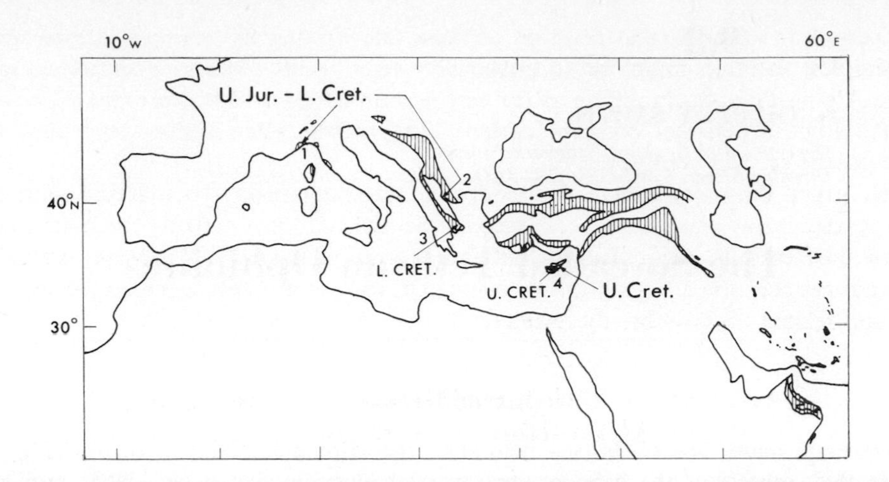

FIGURE 1 Sketch map showing the distribution of the principal areas of ophiolites and associated radiolarian-bearing sediments in the Mediterranean region. The crystallization ages of the ophiolites are believed to be Mesozoic and possibly lower Tertiary in parts of Turkey. Well-dated pelagic sediments lying near the top of the ophiolites are shown as: L. Cret., and so on. 1, Northern Apennines (Bortolotti and Passerini, 1970); 2, Vourinos complex, northern Greece (Moores, 1969); 4, Troodos massif, Cyprus (Moores & Vine, 1971). K-Ar ages are shown as: U. CRET., and so on. 3, Othris zone, eastern central Greece (Hynes *et al.*, 1972); 4, Troodos massif, Cyprus (Vine *et al.*, 1971). The ages imply the existence of several Jurassic and Cretaceous oceans that separated different parts of Yugoslavia, Greece and Turkey from one another at that time. The sizes and shapes of these oceans are not known. Mercator projection.

being in the same region? In the writer's view, the site of the older ocean(s) does not coincide with any part of the present-day Mediterranean or Black Sea, nor with any of the so-called Tethyan ophiolites. The shrinking of the older ocean(s) and the growth of the younger ocean(s) was a gradual process. It is most unlikely to have affected the fauna and flora to any marked extent. The fauna and flora which are characteristically 'Tethyan' probably flourished in and around the new ocean(s) just as well as they had in the old ocean(s). Thus a period of palaeobiological continuity spans a period of marked tectonic discontinuity.

Provisional reconstructions of Carboniferous and earlier time (Smith *et al.*, 1973) suggest that the oceanic area of Permo-Triassic time was a remnant of a much larger ocean, which at its greatest extent may have been of the same size and nature as the present-day Atlantic, Indian or Pacific oceans. In other words, it seems to have had an ancestry dating back at least to mid-Palaeozoic time. From a tectonic point of view it seems most logical to name this oceanic area the Tethys, and to give other names to the younger oceans.

Individual oceans are probably best named from the structural zone each of them presently occupies. For example, the western belt of ophiolites in Greece (location 3 in Fig. 1) lies structurally in the Othris (Othrys) zone, and the ocean it originally formed part of may be described as the Othris ocean. How to name the new oceans as a whole is not clear: one does not wish to confuse matters with a list of locally applicable names, but it is highly preferable not to refer to them as Tethyan. The

Mediterranean is the largest remnant of these late Triassic or younger oceans and the simplest solution might be to collectively refer to all of these younger oceans in the Alpine–Himalayan region as far east as Iran as 'proto-Mediterranean' oceans (C. P. Hughes, personal communication). The precise sizes, shapes and ages of these oceans are not yet known.

This paper does not repeat the evidence, arguments and speculations used to support the above views about the ages of the ocean(s), but extends the discussion to new problems. In particular, the maximum areas of the older (Tethyan) ocean(s) and younger ocean(s) are qualitatively estimated, and some recent work on the ophiolites and related rocks is briefly reviewed.

Areas of the Tethys and Mesozoic Oceans in the Alpine–Himalayan Region

The least-squares fit of the Atlantic continental edges (Bullard *et al.*, 1965) appears to be a good approximation to the relative positions of these continents before the Atlantic Ocean formed, probably in late Triassic or early Jurassic time (Le Pichon & Fox, 1971; Pitman & Talwani, 1972). By adding a least-squares fit of the southern continents to it (Smith & Hallam, 1970) one obtains an approximate outline of a Permo-Triassic Pangaea (Briden *et al.*, 1970). This re-assembly of the major continental areas of the world may be oriented relative to the mean palaeomagnetic pole to give a world map (Smith *et al.*, 1973). The map shows a wedge-shaped ocean opening to the east (Fig. 2a). It gives the maximum likely area of all the oceans that lay between the northern and southern continents in Permo-Triassic time.

By applying the rotations inferred from the Atlantic spreading data to the appropriate continents (Le Pichon & Fox, 1971; Laughton, 1971) and similar rotations, inferred from the Indian Ocean data, to India (Le Pichon, 1968), one may trace the evolution of this oceanic area between the northern and southern continents. The actual positions of the continental fragments between the northern and southern continents is not known during the interval considered (Triassic to Eocene time), and will be difficult to determine. But provided no more than a few small fragments have been transported into or out of the region, the total oceanic area at any given time is known from the maps (Fig. 2). The oceanic area between the northern and southern continents as far east as Iran progressively diminishes in size at a roughly uniform rate during the Triassic to present-day interval. If maintained, all of the ocean will have disappeared about 50 m.y. in the future. The maps could be refined by using the new Atlantic data of Pitman & Talwani (1972) and new Indian Ocean data of McKenzie & Sclater (1971).

The maps show only the total oceanic area available at a given time. They do not show how this area should be distributed among the Tethys and the postulated younger oceans in the region. A schematic distribution is easily made by using the information in Fig. 1. In the writer's view the younger Mesozoic and possibly Tertiary oceans, whose remnants are preserved as ophiolites, could never have been large oceans. The maps (Fig. 2) show that if modern-day analogues of these younger oceans exist, they must be sought in smaller areas believed to be underlain by ocean-floor, like the Red Sea, Gulf of California or Philippine Sea, and not in

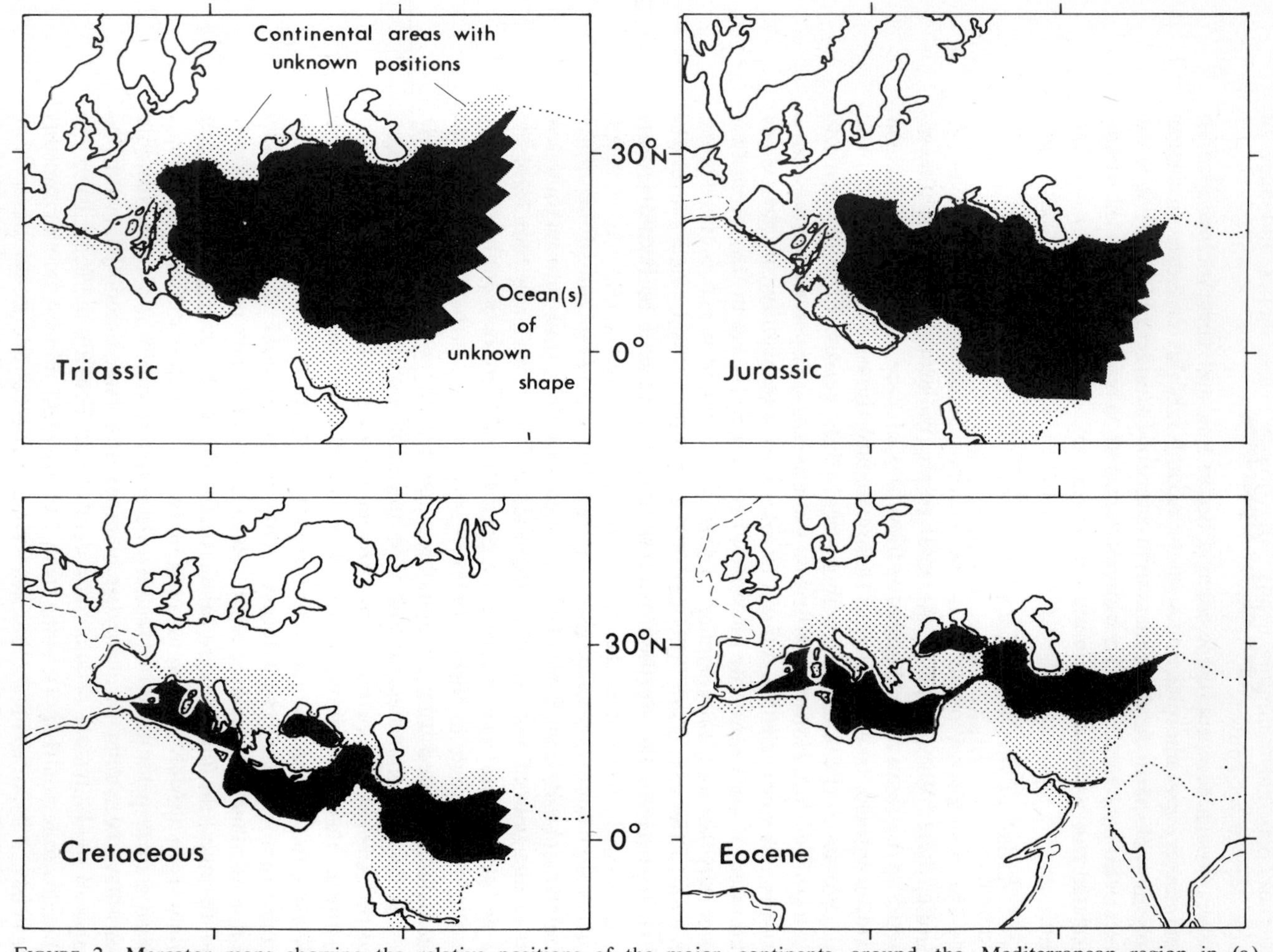

FIGURE 2 Mercator maps showing the relative positions of the major continents around the Mediterranean region in (a) Triassic; (b) Jurassic; (c) Cretaceous; (d) Eocene time. The continents have been repositioned by least-squares fitting methods and ocean-floor data. The latitude/longitude grid uses the mean palaeomagnetic pole as the palaeogeographic pole (Smith *et al.*, 1973). Stippled areas have been deformed during Triassic to present time and cannot yet be repositioned. The Triassic and Jurassic maps portray the continental fragments around the Mediterranean in the positions suggested by Smith (1971). The Cretaceous and Eocene maps arbitrarily show them in their present-day positions and split Iran along an arbitrary line. The dark areas show the total oceanic area that existed at any one time in the region. This area has to be distributed among an old Palaeozoic Tethys and young Late Triassic or younger oceans whose positions are now marked by the ophiolite belts (Fig. 1). No attempt has been made to show how such a distribution might be made.

the present-day Atlantic, Indian or Pacific Oceans. The positions of these younger, smaller oceans are marked by the ophiolite belts in parts of the Alpine–Himalayan chain.

Ages of the Ophiolites

The term ophiolite has been applied to many different kinds of rocks, including some kinds of sedimentary, metamorphic and igneous rocks. It is highly preferable to define them mainly as rocks that show a characteristic igneous stratigraphy, ranging from ultramafics at the base with a well-developed tectonite fabric (the so-called alpine peridotites), overlain by mafic units with a magmatic texture (mostly gabbros at the base grading up into dolerites and finally into basalts). Commonly the basalts are spilitized pillow lavas that pass down into innumerable vertical dykes that form 'sheeted complexes'.

Probably the best known ophiolite is the Troodos massif on Cyprus (Gass & Masson-Smith, 1963; Wilson, 1959) the stratigraphy of which has been directly related to that of the ocean floor (Moores & Vine 1971). It is remarkably undeformed and probably represents an early stage in ophiolite emplacement.

In many ophiolite terrains tectonic emplacement of the presumed ocean floor has chopped the igneous rocks into a number of thrust sheets, or even nappes, in which the original igneous stratigraphy is difficult to recognize or reconstruct. In the present state of knowledge, such areas are probably best reconstructed by assuming that their stratigraphy was originally similar to that of the Troodos massif. In a given area the tectonic slices may show only one of the several rock types of an ophiolite sequence, such as serpentine or pillow lava. Other units may have been sheared out, concealed or eroded. Sometimes the name ophiolite has been applied to isolated outcrops of pillow lava, gabbro, serpentine, and the like, even though all other members of the sequence are absent. This use is justified only if it can be shown that all the other members of the ophiolite sequence are likely to have been present. For example, geochemical evidence may suggest that the rocks concerned belong to an ophiolite suite (Bickle & Nisbet, 1972).

Two ages are associated with all ophiolites that are pieces of ocean floor; the spreading age and the emplacement age. The *spreading age* is the age of primary crystallization of the igneous rocks with a magmatic texture. It can in principle be determined from their isotopic ages. However, undeformed ocean-floor rocks are commonly tholeiites low in potassium and rubidium (Engel *et al.*, 1965); they may contain excess radiogenic argon (Dymond, 1970); potassium may be added to them by submarine weathering processes (Matthews, 1971); and they may be metamorphosed by the high heat-flow at ridge crests (Cann, 1971). The proper interpretation of potassium-argon ages of undeformed oceanic rocks is therefore difficult. Ophiolites have suffered a subsequent period of tectonic deformation during which the isotopic 'clocks' may have been reset by a number of other processes. For all these reasons isotopic ages of ophiolites are of uncertain significance. Few are available from the Alpine–Himalayan region (Fig. 1): some from Cyprus yield mid- to late Cretaceous ages (Vine *et al.*, 1971); others from the Othris zone in Greece yield Lower Cretaceous to lowest Tertiary ages (Hynes *et al.*, 1972).

A better estimate of the minimum spreading age of an ophiolite slice is probably provided by the age of the oldest conformable or interbedded sediments in it. Unfortunately these are commonly radiolarian cherts or unfossiliferous pelagic carbonates. It is only in the past few years that techniques have been developed for extracting radiolaria from siliceous sediments. The palaeontology of Cretaceous and older radiolaria is still in its infancy, but will undoubtedly improve considerably over the next few years and will allow much more precise ages to be assigned to Mesozoic radiolarian-bearing rocks (W. R. Riedel, personal communication). Nevertheless, some reasonably well-dated sediments are interbedded with pillow lavas at the top of ophiolite sequences or lie just above the pillows (Fig. 1). These include sediments in the northern Apennines (Bortolotti & Passerini, 1970); sediments on top of the Vourinos complex in Greece (Moores, 1969) and the Troodos massif in Cyprus (Moores & Vine, 1971). These sediements have ages ranging from Upper Jurassic to Upper Cretaceous age and suggest that the ophiolites crystallized a few million years earlier.

The writer knows of no large ophiolite bodies in the Mediterranean region, in the sense of a sequence with a well-defined igneous stratigraphy, which are as old as the Palaeozoic or as young as the Tertiary, except possibly in Turkey, where Eocene ophiolites may exist (Brinkmann, 1972). Palaeozoic ophiolites have been reported from the area, but either the ophiolites concerned do not have a well-documented igneous stratigraphy, or they lack the supporting isotopic or palaeontological data. Ophiolites have occasionally been dated as Palaeozoic because they are intermingled with Palaeozoic sediments. This may at times be a correct interpretation, but detailed studies of an area in Greece where Permian limestone blocks are intermingled with radiolarian cherts and associated with ophiolites suggests that the ophiolites are younger (Hynes *et al.*, 1972). The fossiliferous blocks are believed to have been picked up by the ophiolites and associated rocks from the underlying strata during tectonic emplacement onto the adjacent continental margin.

This does not mean that Palaeozoic ocean-floor did not exist in the region. As noted above, the Tethys of Permian time must have been floored by such ocean-floor, but none of it has so far been unambiguously identified in the Alpine–Himalayan belt. The plate margin(s) that eliminated the Tethys appear to have resembled the margins between the American and Antarctic plates in the Andes. Ocean-floor seems to disappear completely along such margins; none of it is emplaced onto the continent.

It could be supposed that the ophiolites in the Mediterranean region, which represent Jurassic and Cretaceous ocean-floor, are simply the youngest preserved remains of the oceans concerned. That is, the oceanic area concerned may at one time have been floored by much older ('Tethyan') ocean-floor, all of which has subsequently disappeared, leaving only the youngest to be emplaced onto the adjacent continents or island arcs. Whether or not this is a possibility can be determined from the former continental margins that bordered the former oceans. All the ocean-floor will obviously be younger than the time of break-up that led to the formation of the ocean and its margins. If it can be shown that the break-up or the margins which originally bordered the oceans that are now represented by ophiolites are of Jurassic or younger age, then these oceans cannot be part of an ocean that already existed in Triassic time. In the writer's view criteria for recognizing ancient continental margin sequences are not yet as well established as those for identifying old ocean-floor.

The emplacement ages are not particularly significant for determining the ages of the oceans concerned. They suggest that at the time of emplacement ocean-floor existed in the area.

Ages of the Continental Margins that Border the Ophiolites

Bernouilli (1972) and Bernouilli & Laubscher (1972) have interpreted the sedimentary history of the tectonic units in parts of Italy and Greece in terms of the evolution of a continental margin. They compare the early stages in the development of the inferred margin between northern Italy and the Ligurian Sea to the west, and between a continental area in western Greece and an ocean farther east with the early history of the Atlantic continental margin off the eastern United States. The inferred continental margins in Italy and Greece were probably initiated as rifts in mid- to late Triassic time, but probably did not develop into areas with characteristic submarine topography, sedimentation, and the like, until at least late Triassic or early Jurassic time. A very similar interpretation has been proposed for the inferred continental margin in Greece east of the ocean discussed by Bernouilli and Laubscher (Fig. 3, and Hynes *et al.*, 1972).

Stratigraphic sequences similar to those found in Italy and Greece exist in southern Turkey (Dumond *et al.*, 1972); in Iran (Ricou, 1971); and in Oman (Wilson, 1969). All three areas contain thick Mesozoic carbonate successions that could have been formed along continental margins that did not exist until Triassic or later time. All have subsequently been overridden by ophiolite nappes, and the thick carbonates could represent the marginal facies of the oceans from which the ophiolites were derived. If this is so, then the ophiolite belt from Greece, southern Turkey, Iran to Oman cannot be part of a Permo-Triassic ocean, and are not remnants of a Permo-Triassic Tethys. The ages of the ophiolite belts elsewhere in Turkey and Iran is less certain, but are likely to be of Mesozoic or younger age (Brinkmann, 1972).

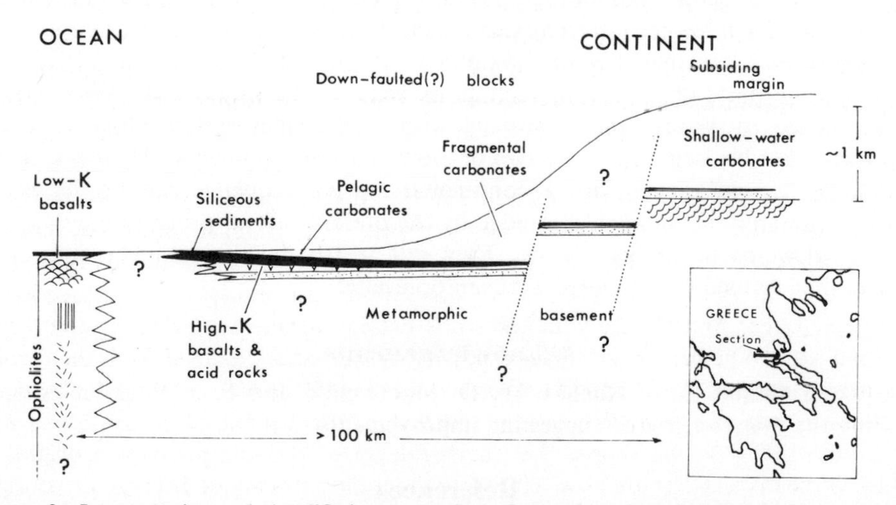

FIGURE 3 Interpretative and simplified cross-section of the edge of a continent and an ocean in eastern central Greece in Upper Jurassic time (Hynes *et al.*, 1972). The low-K basalts in the ophiolites give Lower Cretaceous K-Ar ages, but are likely to be older. The high-K basalts and acid igneous rocks are at least as old as Lower Jurassic and could be Triassic in age. They are not considered to be part of the ophiolite suite and are interpreted as igneous rocks associated with early rifting of the continent.

Despite the consistency of the ages of the inferred continental margins and the spreading ages of the ophiolites, it is by no means clear how ancient continental margins can be unambiguously distinguished from areas of prolonged subsidence and sedimentation that lie wholly within continents. The margins adjacent to ophiolites are always deformed. Pre-emplacement relationships between the sediments on the inferred margin and those deposited on the ophiolites are commonly obscure because the rocks concerned have become incorporated into a chaotic mélange (Ricou, 1971).

Detailed mapping may reveal the existence of systematic facies relationships prior to ophiolite emplacement even in highly deformed areas. For example, in eastern Greece it has been possible to interpret the field data in terms of a lateral change from a subsiding continental margin to an ocean (Fig. 3, and Hynes *et al.*, 1972). Such work may also reveal other relationships. For example, the mafic rocks in the ophiolites in eastern Greece are, as might be expected, low in potassium content. Before their formation there was an earlier (mid-Triassic to early Jurassic?) phase of vulcanism during which highly differentiated acid and basic igneous rocks were formed. The earlier mafic rocks have much higher potassium contents and are not part of the ophiolite suite, though they may be intimately associated with ophiolites in the field. Similar rocks occur in southern Greece (Bannert & Bender, 1968); in southern Turkey (Marcoux, 1970); and in Cyprus (Rocci & Lapierre, 1969). All are believed to be related to an early stage of rifting, probably in mid- to late Triassic time, prior to the formation of a continental margin proper (Hynes *et al.*, 1972). If generally present and correctly interpreted, their recognition and dating will greatly assist in pinning down the time that rifting began.

Conclusions

Although the structure and stratigraphy of the ocean floor and continental margins has become much clearer in recent years many problems remain. For example, what are ophiolites? Are they random samples of the ocean floor, or are they derived from specific parts of an ocean, such as its edge or the ridge crest? What criteria may be used to distinguish a continent with a rift valley system along it from a continent that has separated into two or more discrete fragments? How much of a separation is necessary before a continental margin sequence can be deposited? Though presently unanswered, to judge by the present rate of progress the questions will be answered in the near future. They will greatly aid the understanding of the oceans that formed the so-called Tethyan ophiolites.

Acknowledgements

The writer thanks C. P. Hughes, W. D. MacDonald and F. J. Vine for critically reading the manuscript and suggesting improvements to it.

References

Bannert, D. and Bender, H., 1968. Zur Geologie der Argolis-Halbinsel (Peloponnes, Griechenland), *Geologica et Palaeontologica*, **2**, 151–162.

Bernouilli, D., 1972. North Atlantic and Mediterranean Mesozoic facies: a comparison. *In:* Hollister, C. D., Ewing, J. I., *et al.*, Initial reports of the Deep Sea Drilling Project, **11**, 801–871, Washington.

Bernouilli, D. and Laubscher, H., 1972. The palinspastic problem of the Hellenides, *Eclogae Geol. Helvetiae* **65**, 107–118.
Bickle, M. J. and Nisbet, E. G., 1972. The oceanic affinities of some alpine mafic rocks based on their Ti–Zr–Y contents, *J. Geol. Soc., London*. **128**, 267–272.
Bortolotti, V. and Passerini, P., 1970. Magmatic activity. *In:* Development of the Northern Apennines Geosyncline, edited by G. Sestini. Spec. issue *Sedimentary Geology*, **4**, 599–624.
Briden, J. C., Smith, A. G. and Sallomy, J. T., 1970. The geomagnetic field in Permo-Triassic time, *Geophys. J. Roy. Astr. Soc.*, **23**, 101–17.
Brinkmann, R., 1972. Mesozoic troughs and crustal structure in Anatolia, *Geol. Soc. Amer. Bull.*, **83**, 819–26.
Bullard, E. C., Everett, J. E. and Smith, A. G., 1965. The fit of the continents around the Atlantic, *Phil. Trans. Roy. Soc.*, **A, 258**, 41–51.
Cann, J. R., 1971. Petrology of basement rocks from Palmer Ridge, NE Atlantic, *Phil. Trans. Roy. Soc.*, **A, 268**, 605–18.
Coleman, R. G., 1971. Plate tectonic emplacement of upper mantle peridotites along continental edges. *J. Geophys. Res.*, **76**, 1212–22.
Dewey, J. F. and Bird, J. M., 1971. Origin and emplacement of the ophiolite suite: Appalachian ophiolites in Newfoundland, *J. Geophys. Res.*, **76**, 3179–206.
Dumont, J. F., Gutnic, M., Marcoux, J., Monod, O., and Poisson, A., 1972. Le Trias des Taurides occidentales (Turquie). Définition du bassin Pamphylien: un nouveau domaine à ophiolithes à la marge externe de la chaîne Taurique. *Z. Deutsch. Geol. Ges.* **123**, pt. 2, 385–410.
Dymond, J., 1970. Excess argon in submarine basalt pillows, *Geol. Soc. Amer. Bull.*, **81**, 1229–32.
Engel, A. E. J., Engel, C. G. and Havens, R. G., 1965. Chemical characteristics of oceanic basalts and the upper mantle, *Geol. Soc. Amer. Bull.*, **76**, 719–34.
Gass, I. G. and Masson-Smith, D., 1963. The geology and gravity anomalies of the Troodos massif, Cyprus, *Phil. Trans. Roy. Soc. Soc.*, **A, 255**, 417–67.
Hynes, A. J., Nisbet, E. G., Smith, A. G., Welland, M. J. P. and Rex, D. C., 1973. Spreading and emplacement ages of some ophiolites in the Othris region, eastern central Greece. *Z. Deutsch. Geol. Ges.* **123** pt. 2, 455–468.
Laughton, A. S., 1971. South Labrador Sea and the evolution of the North Atlantic, *Nature*, **232.**
Le Pichon, X., 1968. Sea-floor spreading and continental drift, *J. Geophys. Res.*, **73**, 3661–97.
Le Pichon, X. and Fox, P. J., 1971. Marginal offsets, fracture zones, and the early opening of the North Atlantic, *J. Geophys. Res.*, **76**, 6294–6308.
Marcoux, J., 1970. Age carnien des termes effusifs du cortège ophiolitique des Nappes d'Antalya (Taurus lycien oriental, Turquie), *C. R. Acad. Sci. Paris*, **271**, 285–7.
Matthews, D. H., 1971. Altered basalts from Swallow Bank, an abyssal hill in the NE Atlantic and from a nearby seamount, *Phil. Trans. Roy. Soc.*, **A, 268**, 551–72.
McKenzie, D. P. and Sclater, J. G., 1971. The evolution of the Indian Ocean since the late Cretaceous, *Geophys. J. Roy. Astr. Soc.*, **25**, 437–528.
Moores, E. M., 1969. Petrology and structure of the Vourinos ophiolitic complex of northern Greece, *Geol. Soc. America Spec. Paper 118*, 74 pp.
Moores, E. M. and Vine, F. J., 1971. The Troodos massif, Cyprus and other ophiolites as oceanic crust: evaluation and implications, *Phil. Trans. Roy. Soc.*, **A, 268**, 443–66.
Pitman, W. C. and Talwani, M., 1972. Sea floor spreading in the North Atlantic, *Geol. Soc. Amer. Bull.*, **83**, 619–46.
Ricou, L.-E., 1971. Le croissant ophiolitique peri-Arabe, une ceinture de nappes mises en place au crétacé supérieur, *Rev. géogr. phys. géol. dyn.*, **13**, 327–50.
Rocci, G. and Lapierre, H., 1969. Etude comparative des diverses manifestations du volcanisme préorogénique au sud de Chypre, *Bull. suisse Min. Pét.*, **49**, 31–46.
Smith, A. G., 1971. Alpine deformation and the oceanic areas of the Tethys, Mediterranean and Atlantic, *Geol. Soc. Amer. Bull.*, **82**, 2039–70.
Smith, A. G. and Hallam, A., 1970. The fit of the southern continents, *Nature*, **225**, 139–44.
Smith, A. G., Briden, J. C. and Drewry, G. E., 1973. Phanerozoic world maps. *In:* Organisms and Continents through Time: Palaeontology, spec. Paper 12, 1–42.

Suess, E., 1904–1924. The Face of the Earth (English translation of German edition dating from 1885), 5 vols. Clarendon Press, Oxford.

Sylvester-Bradley, P. C., 1967. The concept of the Tethys. *In:* Aspects of Tethyan Biogeography, pp. 1–4. Systematics Assoc. Pub. 7.

Vine, F. J., Moores, E. M. and Gass, I. G., 1971. The Troodos Massif of Cyprus as a fragment of deep sea floor (Abs.). 1st European Earth and Planetary Physics Colloquium, Reading, p. 22.

Wilson, H. H., 1969. Late Cretaceous and eugeosynclinal sedimentation, gravity tectonics and ophiolite emplacement in Oman Mountains, southeast Arabia, *Amer. Ass. Petroleum Geologists Bull.*, **53**, 626–71.

Wilson, R. A. M., 1959. The geology of the Xeros-Troodos area, *Mem. Geol. Surv. Cyprus*, **1**, 1–135.

9.4

E. R. DEUTSCH and K. V. RAO
Physics Department, Memorial University of Newfoundland, St. John's, Newfoundland, Canada

Preliminary Magnetic Study of Newfoundland Ophiolites, Indicating the Presence of Native Iron

Introduction

The following is a brief account of some initial thermomagnetic measurements on Ordovician ophiolite samples from Newfoundland (Fig. 1). Our main finding, which was reported for the first time at this conference, is that some of these rocks have prominent Curie points compatible with pure or nearly pure iron. Occurrences of native ferromagnetic metals are rare in terrestrial rocks, and there appear to be no reported instances of their discovery through rock magnetism. These results will be published in greater detail elsewhere.

Our initial purpose in sampling the Newfoundland ophiolites for magnetic study was to assess their usefulness for palaeomagnetism, for example in testing the hypothesis of a closing proto-Atlantic Ocean (Wilson, 1966). Although magnetic results from ophiolites in Cyprus (Moores & Vine, 1971) and the Alps (Wagner, 1971) point to the potential palaeomagnetic value of these two rock complexes, it seems that no magnetic data on any pre-Mesozoic ophiolites have been previously published.

Most of our samples are from the well-exposed Early Ordovician ophiolite complex at Betts Cove, but we studied also a few samples from ophiolites of similar age at Bay of Islands and Hare Bay (Fig. 1). The Newfoundland ophiolites have been discussed in terms of plate tectonics by Dewey & Bird (1971), Church & Stevens (1971) and others. In the view of these authors the rocks are remnants of oceanic crust and uppermost mantle that developed in Early Palaeozoic time and were emplaced at a proto-Atlantic consuming plate margin. In the Betts Cove complex, a basal ultramafic member, consisting chiefly of variably serpentinized pyroxenite, harzburgite and dunite, is overlain in turn by gabbro, by a sheeted complex of basic dykes, by pillow lava and by sediments (Fig. 2(b)). Upadhyay *et al.* (1971) consider the pillow lavas to correspond to layer 2 and the gabbro and dykes to layer 3 of an Early Ordovician oceanic crust, while the ultramafics represent adjacent, depleted upper mantle.

Observations

The main property investigated at this preliminary stage was the temperature dependence of magnetic moment (M), measured in air in fields of about 1000 Oersteds.

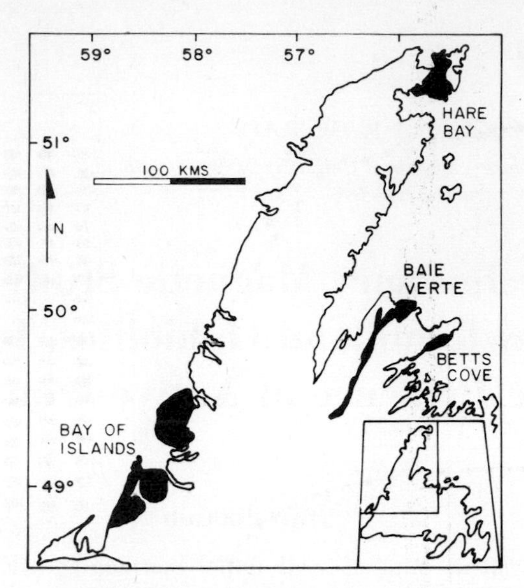

FIGURE 1 Early Palaeozoic ophiolite complexes in Newfoundland.

The specimens were coarse powders obtained by crushing the rock samples with non-magnetic tools to avoid ferromagnetic contamination. We used a thermomagnetic balance (Deutsch *et al.*, 1971) giving T_c for these rocks to $\pm 15°C$ or less, repeatable to $\pm 5°C$. A heating–cooling cycle to 800°C normally lasted 1–2 hours.

Figure 2(a) shows one representative *M–T* curve each for the four main rock layers at Betts Cove, out of some 30 powders measured from 10 samples. All four curves are highly irreversible on cooling. The pillow lava and dyke heating curves have dominant Curie points within ten degrees of that for pure iron (770°C), and they resemble magnetization curves for lunar fines obtained in vacuum (e.g. Strangway *et al.*, 1970).

Pillow lava

The upper curves in Fig. 2(a) are for the matrix of a large ($\sim$ 50 cm diameter) pillow, but Curie points indicating iron or iron with minor nickel, were also obtained from the glassy rim and core of that pillow and from four smaller pillows (mean $T_c = 775 \pm 15°C$, 15 specimens). However, the rock was very heterogeneously magnetized and the nature of the irreversibility was variable even over short distances in the same pillows, with the cooling curves sometimes rising above, and more often falling below, the heating curves. The relative prominence of smaller curve trends, such as inflexions suggesting magnetite (e.g. Fig. 2(a), pillow cooling curve), also varied greatly between specimens.

For these reasons an explanation of the detailed thermomagnetic behaviour of the pillows is premature, except for saying that probably the main cause of the irreversibility is oxidation of the high-T_c component (assuming it is iron) to magnetite and perhaps also hematite on heating in air. Oxidation of iron has also been invoked by Larochelle & Schwarz (1970) and Helsley (1970) to account for the irreversible curve trends of their lunar samples. Since these had been heated in a vacuum, it is

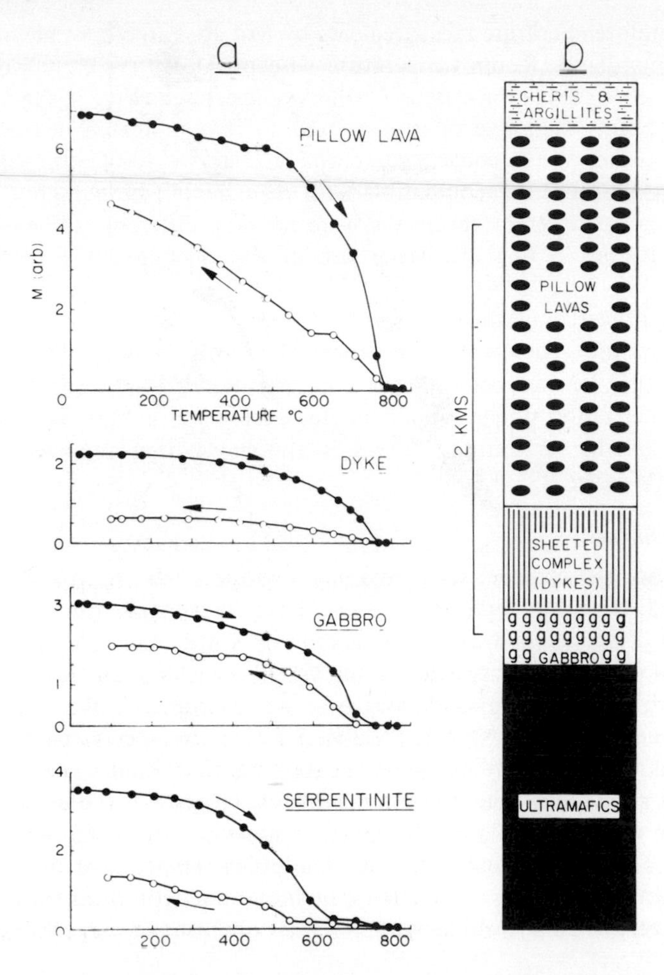

FIGURE 2 Temperature dependence of magnetic moment (M, in arbitrary units) for crushed rock from the Betts Cove ophiolite complex. (a) M–T heating and cooling curves obtained in air in a field of 1000 Oe. (b) Generalized stratigraphy of the Betts Cove ophiolites (modified from Upadhyay *et al.*, 1971).

puzzling to find that some of our specimens appear to have retained a significant part of their high-T_c component on cooling (Fig. 2(a), pillow curves), as in a powder well exposed to air one would expect all except relatively large-sized inclusions of iron to have become oxidized at high temperatures (Strangway *et al.*, 1970). No iron or iron-nickel, however, has so far been identified by us under the microscope, although a Mössbauer absorption spectrogram, obtained at the NASA Manned Spacecraft Center on a magnetic extract from one of the Betts Cove pillow samples, confirmed iron.

Dykes, gabbro and serpentinite

We also measured the temperature dependence of low-field (0·22 Oe RMS) volume susceptibility (K) of a few dyke and pillow whole-rock cylinders in air. The K–T curves were obtained with an a.c. susceptibility bridge (Christie & Symons, 1969) adapted for high-temperature measurements by Mr. R. Pätzold at Memorial University of Newfoundland. Room-temperature susceptibilities typically were 5×10^{-5} Gauss or less. *K–T* curves for a Betts Cove dyke specimen (Fig. 3) show, on heating, a high Curie point suggestive of iron, similar to that indicated in the *M–T* curves (Fig. 2(a)). *K*-values on the cooling curve of Fig. 3 above 400°C are not significantly different from zero and are compatible with iron having been highly oxidized. As the *K–T* curves for different specimens were all very different, an interpretation of curve trends in Fig. 3 (e.g. the steep rise of the cooling curve below 400°C) is premature.

Figure 2(a) (lower diagrams) shows *M–T* behaviour typical of the two gabbro and four serpentinite specimens we measured. The drop of the gabbro heating curve occurs near 700°C, i.e. well below T_c for iron, whereas that for serpentinite resembles a simple magnetization curve for magnetite, except for a 'tail' above 600°C. The irreversibility of the serpentinite curves is similar to that reported from Alpine serpentinites (Wagner, 1971).

Other measurements

M–T curves (not shown here) were obtained also for a few specimens from the Bay of Islands and Hare Bay ophiolites. These have irreversible trends similar to the Betts Cove *M–T* curves, and usually magnetite or titanomagnetite appears to be the dominant component. A few specimens, mostly pillow lava, display a high-T_c component suggesting iron, but further data are needed to confirm this.

The natural remanence (NRM), measured for a few specimens from the three ophiolite areas, was typically in the range 10^{-4} to 10^{-6} Gauss, which is relatively weak for such rock types. The NRM tended to be largest for the ultramafics. A few pillow and serpentinite cylinders were then stepwise thermally demagnetized in nulled fields. The results suggest that a small high-temperature remanence in the pillow specimens is due only to magnetite or titanomagnetite. In one or two serpentinites it may be carried by iron as well, but this conclusion needs checking.

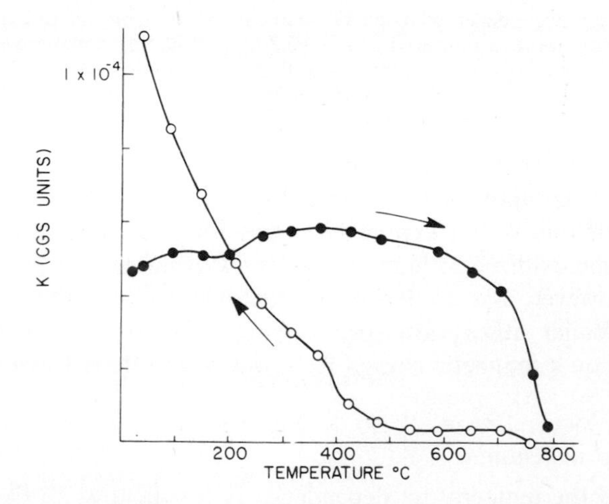

FIGURE 3 Temperature dependence of initial volume susceptibility, *K*, for a whole-rock dyke specimen from Betts Cove, measured in air in a field of 0·22 Oe RMS at 1·0 kHz.

Discussion

Results of the measurements described here are sufficient to indicate the presence of pure or nearly pure native iron in pillow lava and dyke rock from Betts Cove. Maximum amounts of iron may be a few tenths of one percent by weight. For the Betts Cove gabbro and ultramafics and for rock from the two other Newfoundland ophiolites sampled, thermomagnetic trends diagnostic of iron are either absent or inconclusive, and the main ferromagnetic constituent in these appears to be magnetite. Much more than this cannot be concluded at present.

Further investigations are needed to show whether these findings represent more than a very localized phenomenon. If they do, one will have to explain the existence of extreme reducing conditions at the time of genesis or emplacement of the ophiolites. Although native iron and native iron-nickel have been reported from only a few localities (Ramdohr, 1955), there is a possible clue in the fact, pointed out by Chamberlain *et al.* (1965), that virtually all the occurrences of known terrestrial origin occur in serpentinized ultramafic rocks. From a study of the Precambrian Muscox Intrusion, northwest Canada, which contains native metals including nearly pure iron, these authors argue that the serpentinization itself may set up the conditions favouring formation of native metals in serpentinized ultramafic rocks generally. Assuming sufficiently reducing conditions, they invoke a redox reaction forming serpentine, magnetite and hydrogen from olivine and water, followed by reduction of some magnetite to iron by the hydrogen. If our present initial findings are confirmed, the occurrence of iron in rock layers *above* the serpentinite in Newfoundland ophiolites would require a separate explanation, but there may be a clue in the strongly co-genetic type of origin generally attributed to such ophiolite bodies.

Acknowledgements

At the Physics Department, Memorial University of Newfoundland, special thanks are due to Wesley Drodge for much assistance with sampling, instrumentation and measurements. We are indebted also to Dr. Satyanaranaya Murthy and Peter Annan, Leo Kristjansson and Raymund Pätzold for extensive help. At the Geology Department, we thank Professor David Strong and Hansa Upadyhay for valuable discussions and for providing rock samples. We are pleased to acknowledge the generous co-operation of Professor David Strangway and of Wulf Gose and William Pearce at the NASA Manned Spacecraft Center, Houston, who obtained the Mössbauer spectrogram and are commencing a joint study with us. We are also grateful to Memorial University of Newfoundland for the award of a fellowship (to K.V.R.), and to the National Research Council of Canada, who supported this research through Grant A–1946.

References

Chamberlain, J. A., McLeod, C. R., Traill, R. J. and Lachance, G. R., 1965. *Can. J. Earth Sci.*, **2**, 188–215.
Christie, K. W. and Symons, D. T. A., 1969. *Geol. Surv. Can. Paper 69–14.*
Church, W. R. and Stevens, R. K., 1971. *J. Geophys. Res.*, **76**, 1460–6.
Deutsch, E. R., Kristjansson, L. G. and May, B. T., 1971. *Can. J. Earth Sci.*, **8**, 1542–52.
Dewey, J. F. and Bird, J. M., 1971. *J. Geophys. Res.*, **76**, 3179–206.

Helsley, C. E., 1970. *Proceedings, Apollo 11 Lunar Sci. Conf.*, **3**, 2213–9.
Larochelle, A. and Schwarz, E. J., 1970. *Proceedings, Apollo 11 Lunar Sci. Conf.*, **3**, 2305–8.
Moores, E. M. and Vine, F. J., 1971. *Phil. Trans. Roy. Soc., London*, **268**, 443–66.
Ramdohr, P., 1955. Die Erzmineralien und Ihre Verwachsungen. Akademie-Verlag, Berlin.
Strangway, D. W., Larson, E. E. and Pearce, G. W., 1970. *Proceedings, Apollo 11 Lunar Sci. Conf.*, **3**, 2435–51.
Upadhyay, H. D., Dewey, J. F. and Neale, E. R. W., 1971. *Proceedings, Geol. Ass. Can.*, **24**, 27–34.
Wagner, J. J., 1971. *Zeits. Geophys.*, **37**, 589–93.
Wilson, J. T., 1966. *Nature*, **211**, 676–81.

10

CONTINENTAL EVOLUTION

10.1

S. K. RUNCORN
School of Physics,
University of Newcastle upon Tyne,
Newcastle upon Tyne, England

Polar wandering and continental drift

The idea of polar wandering was early invoked in the discussion of the Permo-Carboniferous glaciation. Later the more radical idea of relative movement of the continents, or continental drift, was first established quantitatively from the non-coincidence of the polar wandering curves calculated from the remanent magnetic directions of rocks from different continents (e.g. Creer *et al.*, 1957).

Polar wandering requires that the earth's equatorial bulge slowly adjusts by the plasticity of the earth's mantle. Assuming this could happen, Gold suggested that excess masses on the crust, e.g. mountains, could cause polar wandering, but Runcorn (1957) suggested that motions in the mantle were a more likely agency. Since the evidence for continental drift has become accepted and convection currents in the mantle have been seriously considered as an explanation, these distinctions between the two phenomena have been blurred. In fact, it has become accepted by many that no meaningful distinction can be made between continental drift and polar wandering. However, it is clear that the palaeomagnetic evidence for the two is distinct in the sense that coincidence of polar wandering paths based on continents as positioned at any one time implies polar wandering without continental drift prior to that time. Non-coincidence of polar wandering paths demonstrates that movements of the continents both relative to the pole and to each other have occurred up to the time that the polar wandering curves coincide. Thereafter only very restricted drift is admissible. Creer (1968a, b) has argued that the palaeomagnetic data of the middle Palaeozoic provide evidence that polar wandering was occurring before Wegenerian continental drift, that is, before the dispersal of the continents which once formed part of Gondwanaland and Laurasia began.

While both require a mantle in which solid state creep can allow flow under small stress differences and finite strength is attributed only to the crust or lithosphere, it can be shown that the motions needed to produce the two phenomena are different mathematically and in their physical origin.

Motions in a spherical shell obeying the continuity relation can be divided into a toroidal velocity vector $\mathbf{v}_T$ and a poloidal one $\mathbf{v}_P$

$$\mathbf{v}_T = \mathbf{r} \wedge \nabla V$$

and

$$\mathbf{v}_P = \nabla \wedge \nabla \wedge \mathbf{r}W = \nabla\left(\frac{r\,\mathrm{d}W}{\mathrm{d}r}\right) - \mathbf{r}\nabla^2 W$$

where V and W are scalar functions and $\mathbf{r}$ is the radius vector.

The poloidal motions diverging in some places at the surface and converging in others are an appropriate mechanism by which the lithosphere is stressed and broken and explain the essential fact of geotectonics: why compressional features occur in certain places on the globe while at the same time tensional features develop at others. The radial components of this motion suggest that the poloidal currents are driven by buoyancy forces arising from differences in density with latitude and longitude. Thermal instability or convection has been shown by Runcorn (1969) to be both quantitatively reasonable to account for these density differences and geophysically a likely phenomenon.

It is important to note that no angular momentum **H** is associated with poloidal motions, because the driving forces are the radial gravitational acceleration.

Thus

$$\mathbf{H}_P = \int_V \mathbf{r} \wedge (\nabla \wedge \nabla \wedge \mathbf{r}W)\, \mathrm{d}v$$

$$= \int_V \nabla \wedge \mathbf{r}\left(r\frac{\mathrm{d}W}{\mathrm{d}r}\right) \mathrm{d}v$$

However, the toroidal velocity field does possess angular momentum $\mathbf{H}_T$ for

$$\mathbf{H}_T = \int_V \mathbf{r} \wedge (\mathbf{r} \wedge \nabla V)\, \mathrm{d}v$$

$$= \int_V \left(r\frac{\mathrm{d}V}{\mathrm{d}r}\right)\mathbf{r} - r^2 \nabla V\, \mathrm{d}v$$

$$= \int_V - r^2(\nabla_1 V)\, \mathrm{d}v$$

where ∇_1 is the transverse component of grad.

Consequently it is only possible for the lithosphere to obtain angular momentum about an axis in its equatorial plane if toroidal motions in the mantle are present and convection or poloidal motions do not contribute to this. As polar-wandering implies the presence of such angular momentum in the lithosphere, it appears that it is physically distinct from continental drift and requires a separate physical mechanism from thermal convection.

The palaeomagnetic data in the Canadian shield (see Donaldson *et al.*, 1973) of a polar-wandering path which has sharp changes in direction at 1000 m.y., 1800 m.y. and 2600 m.y., suggests that possibly these two phenomena are present in the geophysical record. The dates are the well-known concentrations of those determined radiometrically from the metamorphic and igneous rocks on the Canadian shield. Runcorn (1966) has suggested that these epochs, evidently, are times of exceptional tectonic activity, or occasions in the Pre-Cambrian when "continental drift" occurred. He suggested an explanation in terms of changes in the convection pattern which would apply varying forces to the plates and cause important lithospheric displacements favourable to extensive metamorphism and igneous activity. It is interesting to find from Irving's data between these epochs, as there was before the Wegenerian dispersal of the continents, a more or less steady motion of the pole. The cause of the necessary "toroidal" type motion is unknown but is a theoretically possible mantle motion.

The geophysical likelihood of polar-wandering is of some importance in the study of the palaeoclimatic record, for example in the interpretation of the Eo-Cambrian glaciation. As the data on the latter stands at present the more-or-less contemporaneous glacial deposits indicate a world-wide glaciation. While this is not of course impossible from a meteorological point of view, a fairly rapid polar movement towards the end of the Pre-Cambrian combined with more precise dating of the deposits may eventually prove them to have been in high geographical latitude.

References

Creer, K. M., Irving, E. and Runcorn, S. K., 1957. Geophysical interpretation of palaeomagnetic directions from Great Britain, *Phil. Trans. Roy. Soc.*, **A250**, 130–43.

Creer, K. M., 1968a. Arrangement of the continents during the Palaeozoic era, *Nature*, **219**, 41–4.

Creer, K. M., 1968b. Palaeozoic palaeomagnetism, *Nature*, **219**, 246–50.

Donaldson, J. A., McGlynn, J. C., Irving, E., and Perk, J. K., 1973. Drift of the Canadian Shield, Vol. I, 3–17.

Runcorn, S. K., 1957. Convection currents in the mantle and recent developments in geophysics, Giedenkbock F. A. Vening Meinesz, 271–7.

Runcorn, S. K., 1969. Convection in the Mantle, A.G.U. Monograph No. 13, 692–8.

Runcorn, S. K., 1966. Change in the moment of inertia of the Earth as a result of a growing core. *In:* The Earth–Moon System, pp. 82–92, Plenum Press.

10.2

P. JAKEŠ
Geological Survey, Hradebni 9,
Praha, Czechoslovakia

Geochemistry of Continental Growth

Introduction

Convincing arguments of geochemists and geophysicists over past few years have shown that the Earth's crust is strongly stratified in respect of both geophysical properties and chemical composition. The vertical stratification of physical parameters is well documented for the most of the Earth's crust; but possible changes of chemical composition that may account for the changes in physical constants are less well understood. Most of the recent models designed to explain the crustal density stratification (e.g. Conrad's discontinuity) do, however, invoke a simplified chemical division of crust into 'granitic' and 'basaltic' layers.

Harris (1957), Gast (1960), Masuda & Matsui (1963), Taylor (1964), Ringwood (1966) and others gave an empirical basis for such compositional stratification in their descriptions of the differing behaviour of the various chemical elements during fractional crystallization governed by melt/solid equilibria. Taylor (1964), for instance, presented a set of estimated 'crustal enrichment factors' of oxyphile elements as a function of ionic radius and charge, and indicated regions of depletion and/or enrichment of these elements in the Earth's crust. Lambert & Heier (1968) presented evidence for strong stratification of Th, U, K in the Australian shield, Fahrig & Eade (1968) for the stratification of K, Ti, U, Th, Na, Cr, and Ni in the Canadian shield, and Shaw (1968) found the evidence for the stratification of K/Rb in the crust. There is also strong lateral variation of chemical composition within the Earth's crust. Poldervaart (1955), for instance, have shown that the abundances of major elements differ between shield areas, Palaeozoic and Mesozoic mobile belts, and island arcs. For this reason modern estimates of crustal abundances of elements are based on the areal proportions of individual rock types (e.g. Ronov & Yaroshevsky, 1969). Such estimates are, however, clearly biased towards the upper crustal abundances, since the sampling is limited to these rocks. The lateral and vertical variations of element abundances indicate that, in the continental areas, they do not result from a single event or uniformly operating process in the geological past. Compelling geological evidence, and the data of trace element distribution and isotopic ratios, indicate that the Earth's solid crust, and its hydrosphere and atmosphere as well, result from a continual geological evolution.

This paper does not present further tables of the major and trace element abundances in the crust, which would relate to the proportions of the different rock types and crust-forming units, etc. Such estimates have been computed in great detail and firm constraints have been established on the total amounts of the various elements in the crust, e.g. radioactive elements Th, U, K or large cations of Rb type (Shaw, 1968; Hurley, 1968; Russell & Ozima, 1971). The subject of this paper is the way in which the hypothesis of sea-floor spreading and its consequences (e.g. subduction and thermal evolution of rocks laying above the zone of subduction) influence our views of major and trace element abundances in the bulk continental crust. A simple model of fractionation and evolution of a geochemically stratified and rigid continental crust is presented. This model is a variation of Ringwood's (1969) petrological, Taylor's (1968) and Shaw's (1968) geochemical models; in the same way this model requires large scale recycling and progressive geochemical development of pre-existing materials. The mechanism for such recycling is provided by the spreading of lithospheric plates, with subduction and thermal evolution of the deeper parts of crust. This model uses (or 'misuses'!) the ultramafic rocks and eclogites (depleted in the low melting point liquid) to return the elements 'unusable' for the continental crust back to the Earth's mantle. Their complementary materials, such as oceanic basalts, island arc volcanic rocks, Andean type volcanic rocks and/or plutonic rocks, move upwards as partial melts that are enriched in the 'incompatible' elements and are considered to be the major agents contributing to the continental growth.

Oceanic Crust

Isotopic and trace element evidence suggests that the oceanic crust is a direct, uncontaminated product of fractional melting of the upper mantle. Geological and geophysical evidence indicate a mechanical and thermal upwelling, with consequent mantle fractionation, in the areas of mid-oceanic ridges—a process by which new geochemically primitive oceanic crust is created.

The present model for the composition and structure of the oceanic crust is based on the geophysical evidence and postulates that the upper part of the oceanic crust is formed by a 0·3 km thick layer of sediments, underlain by a 1·5 km thick layer of oceanic basalts underlain in turn by a layer of gabbroic complexes and deeper still by peridotitic rocks. The element abundances and later process of formation of oceanic crust, i.e. the oceanic tholeiites, or their metamorphic equivalents (greenschist facies rocks of the oceanic bottom), must be discussed first because of their volumetric importance (i.e. their great preponderance over the other rocks of the oceanic areas) and petrologic significance.

Compared with other terrestrial rocks the oceanic tholeiites contain extremely small amounts of incompatible elements (e.g. K, Rb, Cs, Ba) and have primitive isotopic composition (Hart, 1971). Oceanic tholeiites, however, represent already fractionated material when compared with primitive mantle material or meteoritic abundances. For example, Ni and Cr abundances are low in the oceanic tholeiites compared to hypothetical mantle rocks. The pattern of rare earth elements in oceanic tholeiites is chondritic, i.e. unfractionated (Kay *et al.*, 1970), suggesting their origin from primitive materials, so that they have primitive or chondrite-like patterns of residual materials. The 'oceanic tholeiite pattern', with little relative enrichment of medium-sized REE, is complemented by the peridotite patterns which have slight depletion of medium-size REE. The olivine-rich nature of the residuum of the partial melt from which oceanic tholeiite was derived is indicated not only by the REE evidence, but also by the experimental evidence. Hence oceanic sub-alkaline basalts in the areas of mid-oceanic ridges are considered to originate by partial melting of upper mantle peridotitic-pyrolite material. In Green & Ringwood's (1967) version of partial melting in diapirically upwelling upper mantle material, 25% melting is required to produce oceanic tholeiites, leaving a barren residuum of olivine-pyroxene rocks of 'alpine peridotite' type. The oceanic tholeiites not only contain high amounts of the elements forming the crust but, together with the cover of oceanic sediments, are most likely to produce partial melts at the subduction zones when the oceanic crust is destroyed and new island arc crust is formed. The petrological evidence shows that the ultramafic rocks underlying the layers of oceanic tholeiites and gabbros are not likely to melt because of their residual nature and the depletion in fusible constituents that they have already suffered.

Island Arcs

The volcanic rocks of island arcs are extremely diverse and can be divided chemically into tholeiitic, calc-alkaline and shoshonitic (trachybasaltic) associations. These are mutually related in space and time by transitional types (Jakeš & White, 1972). The rocks of each association vary widely in SiO_2 content and the 'andesites' and 'dacites' of each association not only have features characteristic of that association but, like basalts, conform to the following regular pattern of variation in composition across the arcs: tholeiitic rocks on the oceanic or trench side, followed inland by calc-alkaline and shoshonitic rocks. There is also a variation in composition with time, i.e. with stratigraphic level in the island arcs. The earliest rocks in island arc sequences are tholeiitic (Baker, 1968; Jakeš & White, 1969; Gill, 1970) and include tholeiitic basalts, tholeiitic andesites and tholeiitic dacites. These have been named 'island arc tholeiites' (Jakeš & Gill, 1970). They have 'chondritic' rare earth element patterns, very low K, Rb, Ba and Sr contents (but higher than oceanic tholeiites), high Na/K, and in andesitic rocks, i.e. intermediate SiO_2 contents ($\simeq 60\%$), they have high Fe/Mg ratios. The island arc tholeiites have many features in common with the subalkaline tholeiites of oceanic areas but can be distinguished from them by using trace elements (Jakeš & Gill, 1970) and isotopic data (Hart, 1972). In more advanced stages of island arc evolution, calc-alkaline rocks are erupted together with the tholeiitic rocks and, in the latest stage, shoshonites appear (Joplin, 1968; Jakeš & White, 1969; Jolly, 1971).

The stratigraphical variations of the primary volcanic contribution to a growing island arc implies that the lower parts of island arcs are built of tholeiitic rocks, with low contents of Rb, Ba, K and Sr. Their eruption is probably either accompanied by emplacement of a suite of complementary ultramafic rocks at the base of crust, or involves separation from a residuum of ultramafic material left in the upper mantle. The residuum is similar in the mineralogy and element abundances to the 'alpine peridotites'. The higher stratigraphical layers of island arcs are composed of tholeiitic plus calc-alkaline and/or shoshonitic rocks with greater abundances of K, Rb, Sr, Ba and displaying fractionated REE patterns. Thus the whole island arc crust has a primary stratification with respect to K, Rb, Sr, REE, radioactive elements, SiO_2, FeO/MgO and K_2O/Na_2O, resulting from the eruptive sequence just outlined.

Models of the generation of magmas in island arcs has been the subject of numerous critical reviews. The observation of a strong lateral and vertical stratification in island arcs imposes severe constraints on any models explaining the origin of magmas in the island arcs since it must account for lateral as well as vertical variations and cannot consider only one group of rocks (e.g. calc-alkaline rocks). A satisfactory model must explain the positive correlation of K_2O content of the rocks with increasing depth of the volcanic hearth down the Benioff zone (Dickinson & Hatherton, 1967) and also the increase of K_2O content with increasing thickness of the continental crust (Condie & Potts, 1969).

Several kinds of source material contributing to the formation of island arcs must be considered. Large volumes of hydrated oceanic tholeiites, with greenschist facies mineralogy and high water-content may possibly participate in the generation of tholeiitic and calc-alkaline island arc magmas. The contribution of the upper mantle overlying the descending oceanic plate must also be considered, especially in the case of tholeiitic island arc association. Water, or water-rich silicate liquid, may be introduced along the Benioff zone into the 'dry parageneses of the mantle' facilitating melting and the upwelling of magma. Furthermore, the oceanic bottom sediments must be considered in any concept involving melting along the Benioff zone, since these carry substantial amounts of incompatible elements and water. Although the oceanic alkaline volcanic rocks occur on the ocean floors in only small volumes, their high content of large cations and REE is significant, even though the volumes of rock involved are small. Their role in the formation of island arc lavas seems, at present, to be underestimated. Armstrong & Cooper (1972) have argued, using isotope evidence, for the involvement of both oceanic sediments and alkaline rocks in the generation of island arc lavas.

Composition of the Island Arc Crust

Sugimura (1959) showed that about 70% of Quaternary volcanic rocks in Japan were erupted in a relatively narrow, 50 km, belt behind the volcanic front. In the next 50 km wide belt there is only about 25% as much Quaternary volcanic rocks. Petrologically the 50 km belts described by Sugimura (1961) correspond roughly to the belts with different parent basalt compositions (Kuno, 1966; Sugimura *et al.*, 1963). The first belt corresponds to the tholeiitic parent magma (or the tholeiitic association) and the next belt to high-alumina basaltic parent magma (the calc-alkaline association). Further inland from the trench lies the shoshonitic and/or alkaline belt. The age relationships of the rock associations in island arcs (Gill, 1970; Jakeš & White, 1972) suggest that island arcs can be divided into three categories according to the rock series present (i.e. their stage of evolution). (1) tholeiitic (young) island arcs (e.g. South Sandwich Island, the central Kuriles); (2) tholeiitic + calc-alkaline (older) arcs (e.g. New Hebrides, Aleutian Islands) and (3) tholeiitic + calc-alkaline + shoshonitic (oldest) arcs (e.g. Kamchatka, New Guinea). If we assume that a fully developed arc comprises of these stages, it appears from Sugimura's proportions that the tholeiitic association forms about 85%, the calc-alkaline association 12·5%, and the shoshonitic (or alkaline) association 2·5% of the completely developed arc system. If, now, we estimate the proportions of the different rock-types (classified by SiO_2 content) in each island arc rock-association, and take the propor-

TABLE 1 Typical major and trace element abundances in representative rocks of island arcs according to Jakeš & White, 1971

	Tholeiitic			Calc-alkaline			Shoshonitic	
	basalt	'andesite' (icelandite)	'dacite'	high-Al basalt	andesite	dacite	shoshonite	latite
SiO_2	51.57	57.40	79.2	50.59	59.64	66.80	53.74	59.27
TiO_2	0.80	1.25	0.23	1.05	0.76	0.23	1.05	0.56
AL_2O_3	15.91	15.60	11.1	16.29	17.38	18.24	15.84	15.90
Fe_2O_3	2.74	3.48	0.52	3.66	2.54	1.25	3.25	2.22
FeO	7.04	5.01	0.90	5.08	2.72	1.02	4.85	3.19
MnO	0.17	—	—	0.17	0.09	0.06	0.11	0.10
MgO	6.73	3.38	0.36	8.96	3.95	1.50	6.36	5.45
CaO	11.74	6.14	2.06	9.50	5.92	3.17	7.90	5.90
Na_2O	2.41	4.20	3.40	2.89	4.40	4.97	2.38	2.67
K_2O	0.44	0.43	1.58	1.07	2.04	1.92	2.57	2.68
P_2O_5	0.11	0.44	—	0.21	0.28	0.09	0.54	0.41
H_2O	0.45	—	—	0.81	1.08	0.26	1.09	1.44
At SiO_2 wt%	≅52%	≅58%	≅63%	≅52%	≅58%	≅63%	52%	59%
Rb	5.0	6.0	14	10	30	45	75	100
Ba	75	100	175	115	270	520	1000	850
Sr	200	220	90	330	385	460	700	850
K/Rb	1000	890	870	340	430	380	200	200
La	1.1	2.4	5.5	9.6	11.9	14	14	18
Ce	2.6	—	15	19	24	19	28	35
Yb	1.4	2.4	2.7	2.7	1.9	1.4	2.1	1.2
La/Yb	1.0	1.0	1.9	3.5	6.2	10	6.6	15
Th	0.5	0.31	1.6	1.1	2.2	1.7	2	2.8
U	0.3	0.34	0.85	0.2	0.7	0.6	1.0	1.3
Th/U	1.6	0.9	1.88	5.9	3.2	2.7	2.0	2.1
Ni	30	20	1	25	18	5	20	—
V	270	175	19	255	175	68	200	—
Cr	50	15	4	40	56	13	30	—
Zr	70	70	125	100	110	100	40	150
Hf	1.0	1.0	2.6	2.6	2.3	3.8	1.0	3.2

TABLE 2 Comparison of various estimates of the composition of continental crust with the calculated "island arc crust"

	1	2	3	4	5	6	7	8
SiO_2	49.34	58.78	56.15	59.50	58.40	66.06	66.00	68.40
TiO_2	1.49	0.84	1.01	0.70	1.10	0.54	0.60	0.40
Al_2O_3	17.04	15.58	15.63	17.20	15.60	16.08	15.30	14.70
Fe_2O_3	1.99	2.63	2.43		2.80	1.42	1.90	
				6.10				4.80
FeO	6.82	5.04	7.70		4.80	3.14	3.10	
MnO	0.19	0.11	0.18	0.15	0.20	0.08	0.10	0.08
MgO	7.19	4.57	5.24	3.42	4.30	2.22	2.40	2.20
CaO	11.72	8.02	7.76	7.03	7.20	3.44	3.70	2.90
Na_2O	2.73	3.39	3.00	3.68	3.10	3.95	3.20	2.70
K_2O	0.16	0.82	0.68	1.60	2.20	2.90	3.50	3.20
P_2O_5	0.16	0.22	0.22	—	0.30	0.16	0.20	—

1. Oceanic tholeiite, Engel *et al.* (1965).
2. Composition of developed island arc calculated from abundances in table 1 and rock proportions in Sugimura (1961).
3. Average Archaean volcanic rock from Baragar & Goodwin (1969).
4. Taylor's average andesite from island arcs (1968).
5. Composition of young folded belts from Poldervaart (1955).
6. Composition of Canadian shield from Fahrig & Eade (1968).
7. Composition of Ukrainian shield from Ronov & Yaroshevsky (1969).
8. Average granodiorite from New South Wales according to Kolbe & Taylor (1966).

TABLE 3 Trace element abundances in the island arcs and continental areas

ppm	1	2	3	4	5	6	7	8
Rb	12		31	135	140	120	120	
Ba	145	194	270	720	820	590	590	730
Sr	230	214	385	420	425	190	290	380
K/Rb					250			
La	3.8	2.9	11.9		25	41	44	
Yb	1.9	2.1	1.9				3.4	
La/Yb	2.0	1.4						
Th	0.8		2.2		13	16	11.0	10.3
U	0.4		0.69		4	4	3.5	2.45
Th/U	2.0		3.2		3.2	4	3.1	4.2
Ni	22	89	18	30	2	15	44	
V	195	274	175	215	90	65	95	
Cr	30	199	56	85	12	34	70	
Zr	80	160	110	180	300	195	160	

1. Trace elements in the "developed island arcs", Jakeš & White (1971).
2. Average Archaean volcanic rock from Baragar & Goodwin (1969), rare earth element data from White *et al.* (1971).
3. Average andesite, Taylor (1968).
4. Typical trace elements in Andean andesite, Siegers *et al.* (1969).
5. Typical trace elements in "intracontinental andesite", Jakeš & Gill (1970).
6. Average granodiorite from New South Wales, Kolbe & Taylor (1966).
7. Average igenous upper crustal rock, Wedepohl (1968).
8. Trace element abundances in the shield areas, Fahrig & Eade (1968) and Shaw (1968).

tions of each rock-association contributing to a complete island arc system, it is possible to calculate the bulk composition of the primary magmatic material derived from along the Benioff zone.

Andean Volcanism—Late Island Arcs

The concept of the compositional evolution of island arcs based on the lateral and stratigraphical variations and major and trace element abundances is naturally hypothetical. However, it provides a means of estimating the contributions from the upper mantle and from magmas formed along the Benioff zone (down-plunging oceanic plate) supplied to nourish continental growth. This concept does not consider the processes of magma formation within the island arc crust itself, although melting must be expected here also in view of the thermal evolution of the arcs (Minear & Töksoz, 1970). It has been noted earlier that in the island arcs, there occur andesites and rhyolites of calc-alkaline affinities, with trace element abundances, strontium isotope ratios (Ewart & Stipp, 1968) and lead isotopes (Armstrong & Cooper, 1972), suggesting that they originate within the lower part of the island arc crust, or are substantially contaminated by crustal materials or oceanic sediments. The content of incompatible elements is high, K/Rb ratios low and Sr^{87}/Sr^{86} ratios relatively high (0·7045). Ewart & Stipp (1968) and Armstrong & Cooper (1972) describe rocks from New Zealand of this character, while Siegers *et al.* (1969) and Armstrong & Cooper (1972) describe them from the continental Andean margin. Jakeš & White (1971, 1972) argued in favour of the presence of two calc-alkaline suites in the older island arcs and, using mineralogical evidence, they found differences in the composition of hornblendes from hornblende andesites of island arc areas and those of continental margins. They suggested that the amount of H_2O is responsible for these differences (and comparable features in older intracontinental orogenic chains) and that the content of H_2O is higher in the continental margin volcanics due to their derivation from lower crustal levels.

The trace element abundances in the Andean (continental margin) calc-alkaline rocks are very similar to those in large plutonic masses of tonalites, adamellites and of the other granitic rocks (e.g. Kolbe & Taylor, 1966). It may be that the plutonic masses have been derived from both metamorphosed igneous (volcanic) rocks of the lower crust and from earlier sedimentary (geosynclinal) rocks.

The fractionation trends of Andean calc-alkaline rocks in terms of Fe/Mg or SiO_2/K_2O and the trace element abundances (high Rb, Ba, low K/Rb), clearly differ from the calc-alkaline rocks of island arcs and, together with the petrological arguments (composition of hornblendes and experimental data), suggest that Andean rocks were in equilibrium with hornblende (Jakeš & White, 1972). If, however, the source materials included island arc tholeiites, i.e. relatively slightly fractionated rocks, then the presence of garnet in the residuum, together with hornblende, would be expected at the base of crust. It is likely that this requirement is satisfied since the high Al sub-alkaline tholeiites, which form the base of the early island arcs, might attain the metamorphic stage of garnet-bearing amphibolites. Such amphibolites may, on the partial melting, give rise to andesites and dacites of the Andean type with all their peculiarities—a model of wet, high-pressure, partial melting of basaltic composition (Green & Ringwood, 1967). There is no doubt that continental margin (Andean) volcanic rocks are underlain by extensive masses of plutonic rocks (Hamilton, 1969). It is believed that the process of Andean-type volcanism is accompanied by plutonic activity that probably ends with the major differentiation of element abundances from the upper mantle, via island arcs, into a continental crust. The trend of such compositional evolution is best recorded in the sedimentary rocks of the orogenic regions, greywackes. Even Pre-Cambrian greywackes (Fig Tree Group, South Africa) display similar patterns of compositional evolution (Condie *et al.*, 1970).

There is yet another process that strongly fractionates the elements within the crust This process was disclosed by Signinolfi (1970) and Shaw (1968) as a result of systematic study of changing element abundances during the transformation of selected rocks from amphibolite facies to granulite facies mineral assemblages, and has been called *continuous crustal evolution.* Heier (1965) Whitney (1969) and Sighinolfi (1970) noted fractionation of K/Rb, Eade *et al.* (1966) fractionation of K, while Lambert & Heier (1968) discussed Th, U and K. Some elements do not fractionate in this process, e.g. probably the REE (Green *et al.*, 1969). Hyndman & Hyndman (1968) presented geomagnetic and magnetotelluric measurements which demonstrated that the lower crust in areas of recent tectonic activity is probably water-rich (saturated), whereas in the stable shield areas the crust has become dehydrated through similar metamorphic processes. The fact that already stratified crust is further fractionated by weathering and the subsequent evolution of sedimentary (and/or metamorphic) rocks is beyond the scope of this paper (c.f. Wedepohl, 1968; Ronov & Midgasov, 1970; Fahrig & Eade, 1968).

Conclusions

The first fractionation of elements from the upper mantle to form continental crust takes place at the mid-oceanic ridges; oceanic crust generated there is carried to the continental margins or island arcs where it is destroyed. During subduction, the low melting fraction is removed as a liquid from the upper layers of the oceanic plate. Oceanic tholeiites, the cover of sediments and alkaline rocks are all thought to contribute to the melts formed along the Benioff zone. These melts, and melts formed from the mantle lying above the Benioff zone, give rise to island arcs. The island arc volcanism is laterally and vertically zoned and produces a primitive, but stratified, crust with tholeiitic rocks at the bottom and more alkaline rocks at the top. This stratification is accentuated by the process of Andean volcanism accompanied by plutonic activity. In this process the lower parts of an island arc crust of tholeiitic composition are partially melted to produce masses of tonalitic plutonic rocks and andesites and rhyolites with trace element abundances similar to those of continental areas. The last steps in fractionation to produce stratified crust take place during the process of the 'ripening' of crust. Drying of lower crust, during which rocks of wet amphibolite facies mineralogy are metamorphosed into granulitic facies, produces dry parageneses and probably 'metasomatic' solutions which finally enrich the upper parts of the crust in incompatible elements.

Acknowledgements

The author is greatly indebted to Drs D. H. Tarling and M. H. Battey for expressing the author's ideas in a more understandable and readable English form.

References

Armstrong, R. L. and Cooper, J. A., 1972. Lead isotopes in island arcs, *Earth and Planet Sci. Letts.*, in press.

Baker, P. E., 1968. Comparative volcanology and petrology of the Atlantic island arcs, *Bull. Volcan.*, **32,** 189.

Baragar, W. R. A. and Goodwin, A. H., 1969. Andesites and Archaean volcanism of the Canadian shield, *Proc. Andesite Conf., Oregon Dept. of Geol. Min. Ind., Bull.*, **65,** 121–42

Condie, K. C. and Potts, M. J., 1969. Calc-alkaline volcanism and the thickness of the early Precambrian crust in North America, *Canad. Journ. Earth Sci.*, **6,** 1179.

Condie, K. C., Macke, J. E. and Reimer, T. O., 1970. Petrology and geochemistry of early Precambrian graywackes from the Fig Tree Group, South Africa, *Bull. Geol. Soc. Amer.*, **81,** 2759.

Dickinson, W. R. and Hatherton, T., 1967. Andesitic volcanism and seismicity around the Pacific, *Science*, **157,** 801.

Eade, K. E., Fahrig, W. F. and Maxwell, J. A., 1966. Composition of crystalline shield rocks and fractionating effects of regional metamorphism, *Nature*, **211,** 1245.

Engel, C. G., Engel, A. E. and Havens, R. G., 1965. Chemical characteristics of oceanic basalts and the upper mantle, *Bull. Geol. Soc. Amer.*, **76,** 719.

Ewart, A. and Stipp, J. J., 1968. Petrogenesis of the volcanic rocks of the central North Island, New Zealand, as indicated by a study of Sr^{87}/Sr^{86} ratios and Sr, Rb, K, U, and Th abundances, *Geochim. Cosmochim. Acta*, **32,** 699.

Fahrig, W. F. and Eade, K. E., 1968. The chemical evolution of the Canadian Shield, *Canad. J. Earth Sci.*, **5,** 1247.

Gast, P. W., 1960. Limitations on the composition of the upper mantle, *Journ. geophys. Res.*, **65,** 1287.

Gill, J. B., 1970. Geochemistry of Viti Levu, Fiji, and its evolution as an island arc, *Contr. Miner. Petrol.*, **27,** 179.

Green, D. H. and Ringwood, A. E., 1967. The genesis of basaltic magmas, *Contr. Miner. Petrol.*, **15,** 103.

Green, T. H., Brunfelt, A. O. and Heier, K. S., 1969. Rare earth element distribution in anorthosites and associated high grade metamorphic rocks Lofoten–Vestraalen, Norway, *Earth and Planet Sci. Letts.*, **7,** 93.

Hamilton, W., 1969. The volcanic central Andes, a modern model for Cretaceous batholiths and tectonics of western North America, *Proc. Andesite Conf., Oregon Dept. Geol. Min. Ind. Bull.*, **65,** 175.

Harris, P. G., 1957. Zone refining and the origin of potassic basalts, *Geochim. Cosmochim. Acta*, **12,** 6119.

Hart, S. R., 1971. The geochemistry of basaltic rocks. Ann. Rep. Dir. Dept. of Terr. Magn. 1970–1971, p. 353. Carnegie Inst. Washington.

Hart, S. R., 1971. K, Rb, Cs, Sr and Ba contents and Sr isotope ratios of ocean floor basalts, *Phil. Trans. Roy. Soc. London A*, **268,** 573.

Hurley, P. M., 1968. Absolute abundances and distribution of Rb, K, and Sr in the Earth, *Geochim. Cosmochim. Acta*, **32,** 273.

Heier, K. S., 1965. Metamorphism and the chemical differentiation of the crust, *Geol. Fören. Stockh. Förhandl.*, **87,** 249.

Hyndman, R. D. and Hyndman, D. W., 1968. Water saturation and high electrical conductivity in the lower continental crust, *Earth and Planet Sci. Letts.*, **4,** 427.

Jakeš, P. and White, A. J. R., 1969. Structure of the Melanesian arcs and correlation with distribution of magma types, *Tectonophysics*, **8,** 223.

Jakeš, P. and Gill, J. B., 1970. Rare earth elements and the island arc tholeiitic series, *Earth and Planet Sci. Letts.*, **9,** 17.

Jakeš, P. and White, A. J. R., 1971. Major and trace element abundances in the volcanic rocks of orogenic areas, *Bull. Geol. Soc. Amer.*, **83,** 29.

Jakeš, P. and White, A. J. R., 1972. Hornblendes from calc-alkaline volcanic rocks of island arcs and continental margins, *Amer. Miner.*, **57,** 887.

Jolly, W. T., 1971. Potassium rich igneous rocks from Puerto Rico, *Bull. Geol. Soc. Amer.*, **82,** 399.
Joplin, G. A., 1968. The shoshonite association: A review, *J. Geol. Soc. Austr.*, **15,** 275.
Kay, R., Hubbard, N. J. and Gast, P. W., 1970. Chemical characteristics and origin of oceanic ridge volcanic rocks, *J. Geophys. Res.*, **75,** 1585.
Kolbe, P. and Taylor, S. R., 1966. Geochemical investigation of the granitic rocks of the Snowy Mountains area, New South Wales, *J. Geol. Soc. Austr.*, **13,** 1.
Kuno, H., 1966. Lateral variation of basalt magam type across continental margins and island arcs, *Bull. Volcan.*, **29,** 195.
Lambert, I. B. and Heier, K. S., 1968. Geochemical investigation of deep seated rocks in the Australian shield, *Lithos*, **1,** 30.
Masuda, A. and Matsui, Y., 1966. The difference in lanthanide abundance pattern between the crust and the chondrites and its possible meaning to the genesis of crust and mantle, *Geochim. Cosmochim. Acta*, **30,** 239.
Minear, J. W. and Töksoz, N. M., 1970. Thermal regime of a downgoing slab and new global tectonics, *J. Geophys. Res.*, **75,** 1397.
Poldervaart, A., 1955. The chemistry of the Earth's crust, *Geol. Soc. Amer. Spec. Pap.*, **62,** 119.
Ringwood, A. E., 1966. Composition and origin of the Earth. *In:* Advances in Earth Sciences, edited by P. M. Hurley, p. 287, MIT Press.
Ringwood, A. E., 1969. Composition and evolution of the upper mantle, *Amer. geophys. Un., Mono.*, **13,** 1.
Ronov, A. B. and Yaroshevsky, A. A., 1969. Chemical composition of the Earth's crust, *Amer. Geophys. Un., Mono.*, **13,** 37.
Ronov, A. B. and Midgasov, A. A., 1970. Evolution of the chemical composition of the rocks in the shield and sediment cover of the Russian and North American platforms, *Geokhimiya*, **173,** 403.
Russell, R. D. and Ozima, M., 1971. The potassium/rubidium ratio of the Earth, *Geochim. Cosmochim. Acta*, **35,** 679.
Shaw, D. M., 1968. A review of K-Rb fractionation trends by covariance analysis, *Geochim. Cosmochim. Acta*, **32,** 573.
Siegers, A., Pichler, H. and Zeil, W., 1969. Trace element abundances in the "Andesite" formation of Northern Chile, *Geochim. Cosmochim. Acta*, **33,** 882–8.
Sighinolfi, G. P., 1970. K-Rb ratio in high grade metamorphism. A confirmation of the hypothesis of a continual crustal evolution, *Contr. Miner. Petrol.*, **21,** 346.
Sugimura, A., 1961. Regional variation of the K_2O/Na_2O ratios of volcanic rocks in Japan and environs, *J. Geol. Soc. Jap.*, **67,** 292.
Sugimura, A., Matsuda, T., Chinzei, K. and Nakamura, K., 1963. Quantitative distribution of late Cenozoic volcanic materials in Japan, *Bull. Volcan.*, **26,** 125.
Taylor, S. R., 1964. The origin and growth of continents, *Tectonophysics*, **4,** 17.
Taylor, S. R., 1968. Geochemistry of andesites. *In:* Origin and Distribution of the Elements, edited by L. H. Ahrens, p. 559, Pergamon Press.
Wedepohl, K. H., 1968. Chemical fractionation in the sedimentary environment. *In:* Origin and Distribution of the Elements, edited by L. H. Ahrens, p. 999. Pergamon Press.
White, A. J. R., Jakeš, P. and Christie, D. M., 1971. Composition of greenstones and the hypothesis of sea-floor spreading in the Archaean, *Spec. Publ. Geol. Soc. Austral.*, **3,** 115.
Whitney, P. R., 1969. Variations of the K/Rb ratio in migmatitic paragneisses of the Northwest Adirondack. *Geochim. Cosmochim. Acta*, **33,** 1203.

10.3

R. W. R. RUTLAND

Department of Geology and Mineralogy, University of Adelaide, Adelaide, South Australia

Tectonic Evolution of the Continental Crust of Australia

Introduction

A principal proposition arising from the concepts of plate tectonics has been that ocean floor spreading and subduction provide the driving force for orogeny on plate margins. If this proposition is accepted it follows that similar processes are likely to be responsible for older orogenic belts, perhaps even in the Archaean (e.g. Dewey & Horsfield, 1970; Dewey & Bird, 1970; Dickinson, 1971; White *et al.*, 1971).

Geochronological studies, however, have led to the recognition of long-term cyclic phenomena in the earth's crust associated with changes in tectonic style (e.g. Gastil, 1960; Sutton, 1963, 1967, 1971). It is usually suggested that these changes can be related to a thinner continental crust (or better, a thinner lithosphere) in the Precambrian. This weakens the actualistic argument for the interpretation of Precambrian orogeny in terms of strict analogy with Phanerozoic collision and cordilleran orogenies and suggests that ocean floor spreading is likely to have had rather different expression in the Precambrian continental crust. Thus the different lithospheric conditions could probably account for the absence of blueschists as guides to Precambrian subduction zones; and, in general, some differences in the relations of volcanic, sedimentary and metamorphic facies in the Precambrian relative to the Phanerozoic should be expected. In short, the two fundamental Phanerozoic orogenic types, collision and cordilleran (Dewey & Bird, 1970), should be recognizable but in substantially modified form.

Demonstration of continental drift in the Precambrian would establish the validity of the interpretation in terms of modified plate tectonic models. The available palaeomagnetic evidence suggests large movements of individual continents relative to the poles (e.g. Spall, 1971) and some inconclusive geological evidence for drift and for collision type orogeny in the Precambrian has been produced (Gibb, 1971; Gibb & Walcott, 1971; Muehlberger, 1971). The geological evidence has also been interpreted by some as opposing drift (Hurley & Rand, 1969; Hurley, 1970; Engel, 1971). However, ocean floor spreading and subduction may still have occurred on active continental margins in the absence of drift and led to the development of cordilleran type orogenic belts. In addition, if plate motions are regarded as surface effects of the earth's major thermal cycles, there may be other effects not directly related to plate motions. Intra-continental, as distinct from inter-continental (collision), orogeny is of particular significance in this context (Shackleton, 1969) as also are crossing orogenic belts (e.g. de Swardt *et al.*, 1965; Vail *et al.*, 1968).

An important task of comparative tectonics is, therefore, the critical evaluation of Precambrian mobile belts in terms of the models of plate tectonics being developed for Phanerozoic belts. Models of Archaean tectonics have tended to be inductive models independent of plate tectonic concepts (e.g. Annhauser *et al.*, 1969) or else unmodified applications of Phanerozoic models (e.g. White *et al.*, 1971). It is suggested here that consideration of Proterozoic tectonics as an intermediate stage of tectonic evolution allows the application of a modified plate tectonic model to both the Proterozoic and the Archaean. Australia has perhaps the fullest record of Proterozoic cover rocks which allow the similarities and differences to both the Phanerozoic and the Archaean to be recognized.

Conclusions must necessarily be tentative, not only because field and analytical data are incomplete for the Precambrian belts but because the detailed features of Phanerozoic belts themselves are not fully understood in terms of plate tectonics. Moreover, it must be accepted that some of the assemblage of features which serve to define a Phanerozoic orogenic belt cannot be employed for Precambrian belts. Many geological and geophysical features of the latter may have been destroyed by subsequent mantle and crustal developments and by erosion.

It is argued that, in Australia, collision type orogenic belts are absent and that two kinds of belt can be recognized. The dominant type corresponds to the mio-

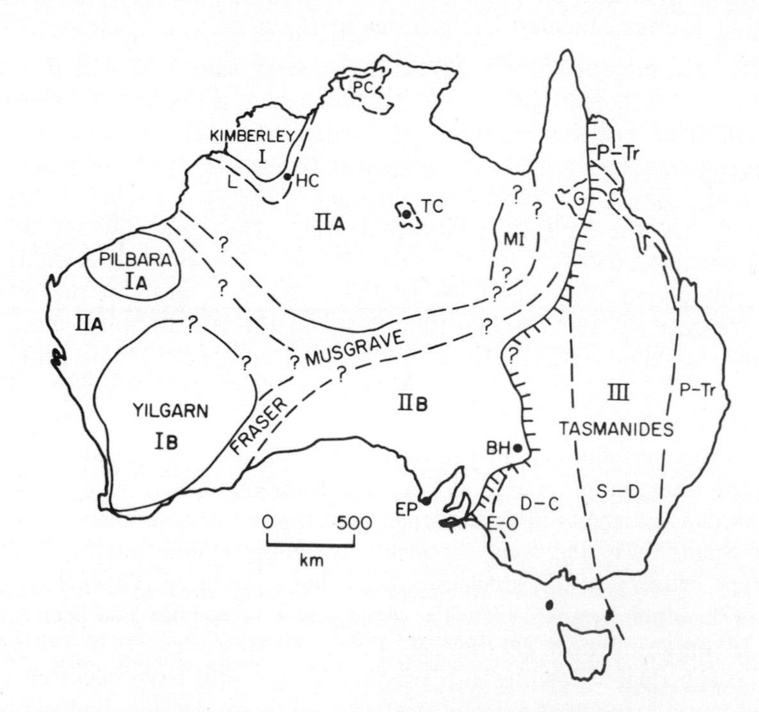

FIGURE 1 Major plutonic provinces of Australia: I, Archaean; II, Proterozoic; III, Phanerozoic. Abbreviations for Proterozoic orogenic domains: L, King Leopold; HC, Halls Creek; PC, Pine Creek; TC, Tennant Creek; MI, Mt. Isa; EP, Eyre Peninsula; BH, Broken Hill; G. Georgetown. For further explanation see text.

tectonic* (miogeosynclinal or external) belts of Phanerozoic cordilleran orogenies but there are substantial differences due to the greater mobility and lesser thickness of the lithosphere during the Precambrian. The other type has no close analogy in the Phanerozoic and is represented by the Fraser and Musgrave belts. It is attributed to high heat flow in narrow zones of lithosphere reactivation rather than to continental collision.

The available evidence also opposes notions of continental growth by marginal accretion and supports concepts of periodic cratonization in chelogenic cycles (Sutton, 1963, 1967; Muelberger, 1971).

Main Provinces of the Australian Crust

The Australian continent is particularly rich in platform cover sequences of both Phanerozoic and Pre-Phanerozoic age. The main Mesozoic platform cover was initiated in the Permian and covered roughly half the continent. The Adelaidean–Palaeozoic sequences (c. 1400–200 m.y.) cover much of the remainder of the continent while the Carpentarian sequences (c. 1800–1400 m.y.) occur mainly in N. Australia and the Nullaginian platform sequences (2200–1800 m.y.) are most limited in area in the north-west† (Figs. 2 and 4).

Consequently, platform basement forms a relatively small area of the continent and even then is often obscured by deep weathering or by thin alluvial and colluvial cover. Nevertheless the available evidence is sufficiently consistent to recognize several major divisions based on the occurrence of widespread episodes of plutonism in the basement rocks (Fig. 1).

(a) *Phanerozoic province*

The Tasmanides clearly form one major province in which plutonism is confined to linear belts (Tectonic map of Australia and New Guinea, 1972). The plutonism has migrated from west to east following the migration of stratotectonic zones. These zones are generally meridional and the most important plutonic belt (c. 400 m.y.) separates the essentially non-volcanic facies of the western Lachlan geosyncline from the volcanic facies of the eastern New England geosyncline (Packham, 1969). It is notable, however, that the sinuous early margin of the belt in S. Australia has a general

* The morphotectonic orogenic zone between the arc-trench complex and the craton has been called the external zone and generally coincides with the 'miogeosynclinal' zone in Phanerozoic orogenies. The terms internal and external were developed for apparently intracontinental (now collision) orogenies and are somewhat confusing when applied to Cordilleran orogenies where the terms Inner and Outer are also used in the island arc systems (see Fig. 7). The terms eutectonic (eu- ≡ true) and miotectonic (mio ≡ less) are therefore introduced here. The eutectonic zone is here defined as the tectonic zone developed from the arc-trench complex. The miotectonic zone retains the older continental basement on which it was developed and lies between the arc-trench complex and the craton. In the Precambrian the miotectonic zone does not necessarily show the same stratotectonic facies as in the Phanerozoic.

† It has been suggested (Dunn *et al.*, 1966) that the terms Adelaidean, Carpentarian and Nullaginian be accorded system status but following Trendall (1966) they are here regarded as limited in their use to the Australian continent. There is still considerable doubt as to the dating and correlation of the youngest groups in the Carpentarian (Roper and South Nicholson Groups) and the oldest formations of the Adelaidean (e.g. Compston & Arriens, 1968 p. 572).

north-easterly trend, while a belt of Carboniferous granites in Queensland shows a north-westerly trend. It is evident, therefore, that the plutonic belts are not always parallel to dominant strato-tectonic trends. Adelaidean and younger sediments equivalent in age to the sediments involved in the deformation and plutonism of the Tasman province, are preserved little disturbed on the adjacent Proterozoic province and demonstrate by their continuity that the present Australian continent west of the Tasmanides has been a single craton for at least the last 1050 m.y.

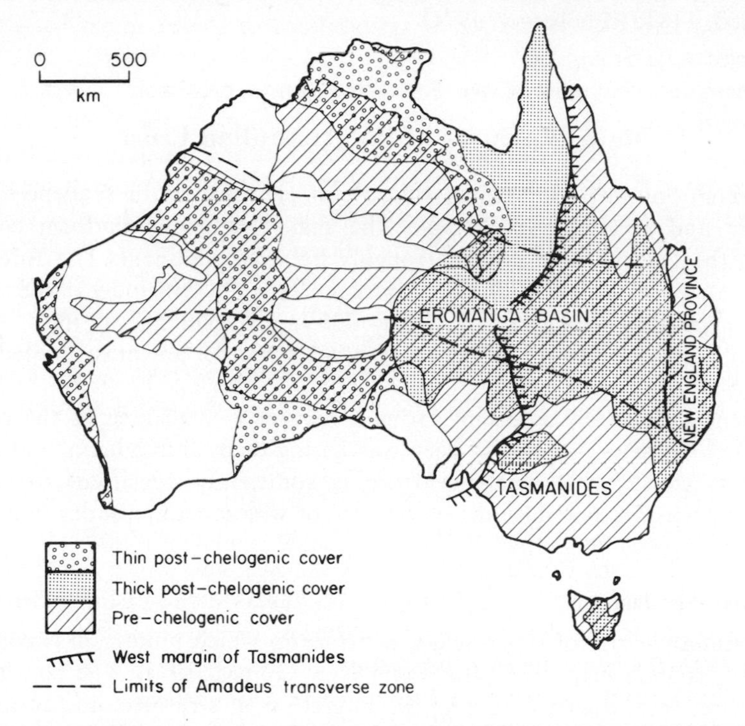

FIGURE 2 Present distribution of pre-chelogenic (Adelaidean and Palaeozoic) and post-chelogenic cover (Permian ad Mesozoic) of the Tasman chelogenic cycle. After Tectonic Map of Australia and New Guinea, 1972.

(b) *Proterozoic province*

The Musgrave and Fraser belts were produced between 1400 and 1050 m.y. ago and the former is unconformably overlain by basal quartzites of the Adelaidean. South and east of the Musgrave–Fraser system an older crystalline basement is exposed in the Gawler nucleus and other inliers, ranging in metamorphic and plutonic age from 1780–1650 m.y. and very little cover now supervenes between this basement and Adelaidean sediments (possibly as old as 1400 m.y. but mainly younger than 1050 m.y. (Compston *et al.*, 1966; Thomson, 1966)). North of the Musgrave belt, however, plutons dated at 1860–1730 m.y. intrude folded sedimentary sequences in areas as far apart as the King Leopold and Halls Creek mobile zones, the Pine Creek 'Geosyncline' and the Mt. Isa 'Geosyncline' (Fig. 1). These sequences themselves rest on an older basement (> 2000 m.y.) and are overlain unconformably by little deformed or metamorphosed members of the Carpentarian platform sequence which often has volcanics at the base similar in age to the Pre-Carpentarian plutons. The Carpentarian, in turn, is overlain by the Adelaidean.

Thus a major plutonic event, here called the Pre-Carpentarian plutonism (*cf.* the Hudsonian in Canada) can be recognized throughout the Proterozoic basement rocks of Australia west of the Tasmanides. This major plutonic event marks the cratonization of more than one-third of the continent. The later Musgrave–Fraser system now separates an infra-crustal province in the southeast from a supra-crustal province in the north and west. This suggests that the Gawler nucleus represents a deeper erosion level (where the Rb/Sr system became closed somewhat later) than the north and west provinces.

The widespread character of the Pre-Carpentarian plutonism contrasts with the linear expression of the various episodes of plutonism in the Tasmanides. There is, however, some evidence that gneissic basement south of the Musgrave belt and the granitic plutons in the metasediments north of the belt both decrease in age from west to east. Thus the plutons have been dated by Rb/Sr methods (m.y.) as follows: Halls Creek, 1855; Katherine-Darwin, 1830; McArthur Basin area, 1825; S. Nicholson Basin area, 1820–1760; Mt. Isa, 1800–1760; but the plutons may be still younger at about 1700 in the Gascogne Block (Leggo *et al.*, 1965). The southern gneissic area dates at about 1780 in Eyre peninsula and 1700–1650 at Broken Hill (all dates from Compston & Arriens, 1968). Gellatly (1971) gives a date of 1880 for the Halls Creek granites, and 1940 m.y. as compared with 1820 m.y. by Compson & Arriens for the Whitewater volcanics.

In addition it is notable that older plutonism (c. 1960 m.y.) is well developed in the Halls Creek belt on the west side of the Pre-Carpentarian province (and Gellatly (1971) has reported a granite date of 2100 m.y.) while younger plutonism (c. 1550 m.y.) is developed both in the supracrustal area on the east side of the Pre-Carpentarian province near Mt. Isa, and also in the infracrustal areas such as Broken Hill, around the Adelaide geosyncline. This younger plutonism probably post-dates suites 1–3 of the Carpentarian but antedates deposition of the Roper and South Nicholson Groups. It compares in age with the Elsonian Event in Canada.

Three cratonic areas escaped the Pre-Carpentarian plutonism—the Yilgarn, Pilbara and Kimberley blocks. Nullaginian platform sediments equivalent in age to those affected by the Pre-Carpentarian plutonism are preserved as the Mt. Bruce supergroup on the south side of the Pilbara block.

(c) *Archaean province*

The Kimberley Block is concealed by Carpentarian cover. It is inferred that this cover is underlain by Archaean rocks which were not involved in the Pre-Carpentarian plutonism because of the development of the Halls Creek and Leopold belts on its margins (*cf.* Gellatly, 1971). The other two blocks are composed of greenstone belts in granite gneiss terrain. In the Pilbara block, however, there is no well developed structural trend and the granitic plutonism is dated at about 3100 m.y., while in the Yilgarn block there is a north-northwesterly trend and post-greenstone plutonism is dated at about 2650 m.y. (Arriens, 1971). The two blocks are regarded as parts of one major province on the basis of their rock associations and structures.

Three major provinces can therefore be recognized in terms of their orogenic characters and associated plutonism.

I. *Archaean* Pilbara-Yilgarn Province
   (a) Pilbara Sub-province with 3100 m.y. plutonism.
   (b) Yilgarn Sub-province with 2650 m.y. plutonism.

II. *Proterozoic* Arunta-Gawler Province with Pre-Carpentarian plutonism (c. 1650–1800 m.y.)
   (a) Arunta supercrustal Sub-province* with older plutonic belt (c. 1960 m.y.) in west and younger belt (c. 1550 m.y.) in east.
   (b) Gawler infracrustal Sub-province.*

III. *Phanerozoic* Tasman Province
   (a) Lower Palaeozoic Lachlan Sub-province with main Silurian–Devonian (c. 400 m.y.) plutonism.
   (b) Upper Palaeozoic New England Sub-province with main Permian plutonism (c. 220 m.y.).

In addition the Musgrave Fraser belts characterized by granulite facies metamorphism (1300–1400 m.y.) and plutonism (1100–1200 m.y.) must be distinguished as a fourth province. These belts are notable for having older crust on both sides. They are also associated with a major episode of dyke emplacement (c. 1000–1100 m.y.) on north-northwest and east-northeast trends.

The boundaries between the major provinces are major discontinuities across which major changes of trend as well as plutonic age occur. Thus the boundary between Phanerozoic and Proterozoic provinces brings crust dominated by plutonism at 1700 to 1550 m.y. in the Broken Hill region against crust with local 500 m.y. and dominant 400 m.y. plutonism. That is to say, there is a plutonic discontinuity of more than 1000 m.y. Similarly, the Pilbara block with plutonism of 3100 m.y. lies against the Gascogne block with 1700 m.y. plutonism; and the Yilgarn block (2650 m.y. plutonism) lies against the Fraser Range (1330 m.y. metamorphism). No such large discontinuities occur within the four major provinces. The range of age of plutonism is about 250 m.y. in the Tasmanides (or < 500 m.y. if the island arc system is included) and 450 m.y. in the Proterozoic province. The range of plutonism later than the greenstone belts of the Yilgarn block is also about 500 m.y. (2700–2200) but this would be substantially reduced if the possibly unrelated plutonism later than the main east–west basic dyke suite (2400 ± 40 m.y.) were excluded. The age relations between the plutonic rocks and supracrustals of the Pilbara nucleus are still in some doubt (Arriens, 1971). But if the 3050 ± 180 m.y. granites of the Pilbara region do post-date the greenstone belts then both granites and greenstones must be allocated to an earlier volcano-plutonic cycle than is represented in the Yilgarn block. In this case the Pilbara block probably behaved as a craton while the Yilgarn block was mobile in the younger cycle. Reworked equivalents of the older Pilbara greenstones are probably represented by the granulite facies gneisses of the wheat belt in the Yilgarn province (Fig. 5 and Wilson, 1971; Arriens, 1971).

* The Arunta Block is the largest exposed portion of the Arunta Province which as here defined also includes the Gascogne, the King Leopold, Halls Creek, Granites-Tanami, Pine Creek, Tennant Creek and Mt. Isa orogenic domains. The Gawler Block is the largest exposed portion of the Gawler Province which as here defined (Fig. 1) also includes the Denison, Mt. Painter, Willyama and Wonaminta orogenic domains.

Phanerozoic–Proterozoic Tectonic Relationships

The two younger provinces can be related to two somewhat analogous chelogenic cycles. The younger cycle is here taken to begin with the Adelaidean sediments which have been considered to date back to about 1400 m.y. and which normally rest on older rocks in which the youngest plutonic activity was more than 1500 m.y. ago. In the region of the Musgrave mobile belt, mafic plutonism and dyke swarms were emplaced at about 1050 m.y. before the deposition of the quartzites which initiate the main development of Adelaidean cover.

The platform on which the Adelaidean cover was deposited was relatively mobile and the Amadeus Basin and Adelaide 'Geosyncline' can be regarded as aulacogenes or fault-bounded basins within the continental crust. In spite of local folding and unconformities Palaeozoic sedimentation on the craton tended to follow and accentuate the Adelaidean basin structures. The continental margin was east of the present Adelaide geosyncline and the presumed miogeoclinal development across that margin must be beneath the Murray Basin and the Palaeozoic rocks of the Lachlan geosyncline.

The first clear evidence that the continental margin had become active is provided by the 500 m.y. plutonism and metamorphism of the Kanmantoo belt in S. Australia. The major plutonic activity on the east margin of the Lachlan geosyncline is of Siluro–Devonian age and spread westwards to emplace post-tectonic granites across much of the Lachlan geosyncline in Devonian–Carboniferous time.

The main Siluro–Devonian plutonic belt probably corresponds to a volcano-plutonic arc complex close to the continental margin. The main part of the Lachlan geosyncline with post-tectonic granites is essentially non-volcanic, without lithic or feldspathic sediments and, although poor in carbonates, corresponds to the 'miogeosynclinal' zone of other Cordilleran belts. It is probably underlain by Pre-Adelaidean Precambrian crust as indicated by the inliers of Georgetown in the north and western Tasmania in the south (*cf.* Oversby, 1971).

Subsequent plutonic activity shows an eastward migration with Permian plutons on the east margin of the New England geosyncline and Cretaceous plutonism in New Zealand.

The Palaeozoic plutonism and orogenic activity led to the development of a new platform cover in the latest Carboniferous and Permian with a quite different pattern of sedimentation. This cover spread over the stabilized parts of the Tasmanides and on to the adjacent platform in the Jurassic and Lower Cretaceous. It can be described as a post-chelogenic platform cover in contrast to the pre-chelogenic Adelaidean and Palaeozoic cover which was involved in the Palaeozoic orogenic and plutonic activity (Fig. 2).

This reorganization of the strato-tectonic pattern on the stable continental crust corresponds with the shift of the mobile zones to the island arc regions of Australasia (e.g. to New Zealand and New Guinea); and more generally it corresponds with a general tectonic reorganization in the whole circum-Pacific region.

The general relations of the pre-chelogenic and post-chelogenic elements to Palaeozoic orogeny and plutonism in the Tasman chelogenic cycle are indicated diagrammatically in Fig. 3. Analogy with this cycle suggests that the previous chelogenic cycle should be taken to begin with the Nullaginian sediments (therefore analagous with the Adelaidean) younger than the 2400 m.y. phase of basic dyke emplacement in the Yilgarn block (see Fig. 4). Older sediments may also be involved in this pre-chelogenic cover since a pegmatite dated at 2700 m.y. intrudes sediments in the Halls Creek mobile zone. Thus over 700 m.y. may have elapsed before the pre-chelogenic cover became involved in the earliest plutonism (1960 m.y.) in the Halls Creek belt. This belt occurs on the margin of the Proterozoic province against the stable craton in the same way as the Kanmantoo plutons occur on the margin of the Tasman Province. The main plutonism at 1700–1800 m.y. then occurred across the whole Proterozoic province and corresponds to the main Palaeozoic plutonism of the Tasman Province. A new post-chelogenic platform cover, the Carpentarian (analogous to the Mesozoic of the Tasman cycle) was initiated after the main plutonism: on its eastern margin this cover became involved in orogeny and plutonism of the final stages of the cycle (e.g. 1550 m.y. at Mt. Isa) in the same way as the Mesozoic cover became involved in Cretaceous and Tertiary orogeny in the final stages of the Tasman cycle (Fig. 4).

The main and crucial distinction between the two chelogenic cycles is that the plutonism preceding the Carpentarian cover extended right across the Proterozoic province while the plutonism immediately preceding the Mesozoic cover is limited to the eastern margin of the Phanerozoic province, although earlier plutonism occurs right across the belt (Figs. 3 and 4).

Apart from this difference, each of these provinces has two major sedimentary tectonic elements related in similar ways to plutonic and orogenic activity (Fig. 3 and Table 1). The provinces are also of similar dimensions in time and space and orogenic trends in both are often closely parallel to the common roughly meridional boundary between the provinces. Orogenic trends are, however, much more variable in the Proterozoic province and its western margin is much less regular since it extends between the Archaean cratonic remnants.

The Nullaginian sediments (e.g. in the Pine Creek 'Geosyncline', Walpole *et al.*, 1968) are generally shallow water clastic and dolomitic rocks not associated with volcanics. Basic volcanics are prominent in the Carpentarian rocks of the Mt. Isa 'geosyncline' (Carter *et al.*, 1961) but even there they are interbedded with quartzites

TABLE 1 Timing in the Phanerozoic and Proterozoic chelogenic cycles

	Tasman chelogenic cycle (m.y.)	Intervals (m.y.)		Proterozoic chelogenic cycle (m.y.)
Earliest pre-chelogenic sedimentation	1400			2700
		350	300	
Basic dyke emplacement	1050			2400
		550	440	
Early plutonism against craton	500			1960
		100	110	
Main plutonism (followed by post-chelogenic sedimentation	200–400			1850–1650
		100	100	
Terminal plutonism (near inferred continental margin)	ca. 100			ca. 1550

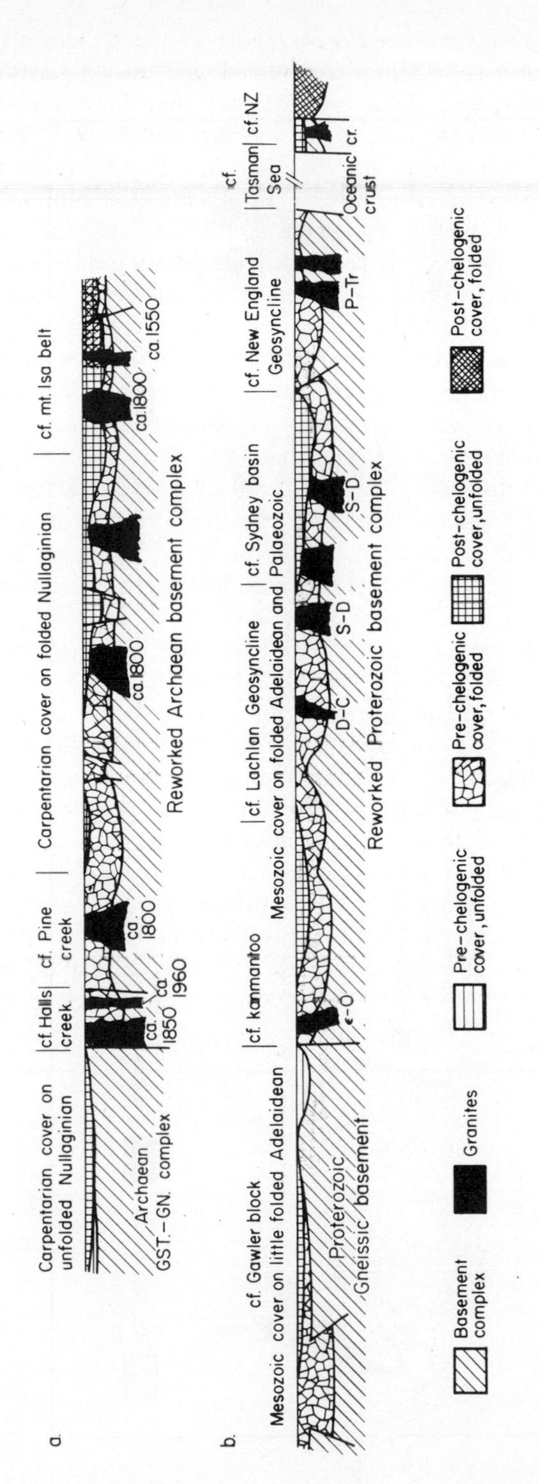

FIGURE 3 Diagrammatic sections to illustrate analogy between the early Proterozoic and Tasman chelogenic cycles.

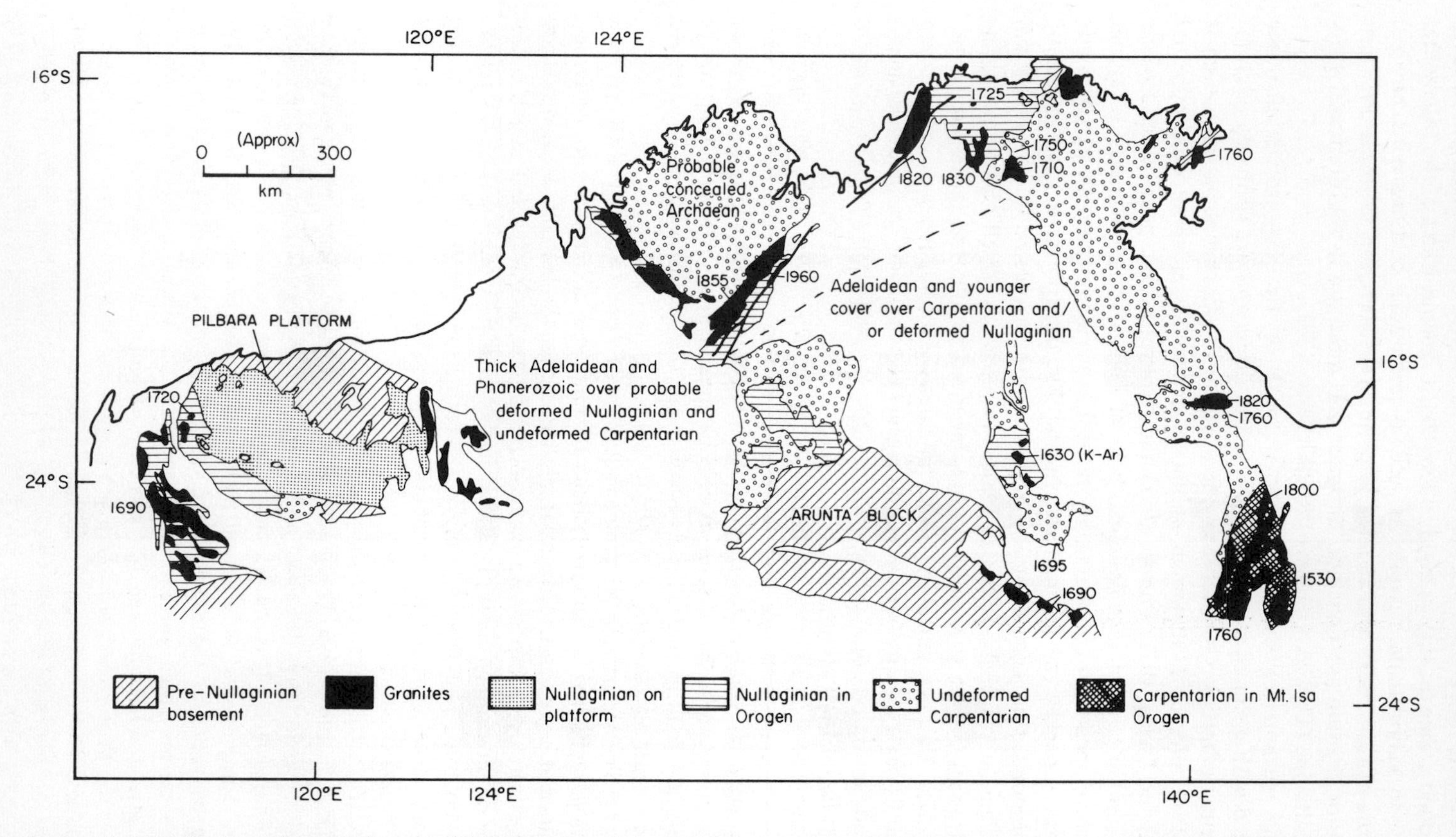

FIGURE 4 Distribution of pre-chelogenic (Nullaginian) and post-chelogenic cover of the early Proterozoic chelogenic cycle. Note that the Arunta Block contains deformed Nullaginian and little deformed Carpentarian in the northern part but these have not been differentiated from the older basement complex. Simplified after Tectonic Map of Australia and New Guinea, 1972.

and associated with thick dolomitic formations. In the Halls Creek belt pillow lavas have been found at one locality in highly altered basic igneous rocks of the Biscay Fm. associated with subordinate rhyolite as well as greywacke carbonaceous phyllite, dolomite, calc-silicate rocks and banded jaspilite. The overlying Olympia Fm. also contains minor rhyolite ash-flow tuff, dolomite and quartzite in the dominant greywacke and siltstone (Dow & Gemuts, 1969; Gellatly, 1971).

In general, it appears that the volcanic rocks of the Proterozoic province are associated with geanticlinal structures or major faults and are probably of continental tholeiitic affinities. Over most of the area the Pre-Carpentarian plutonism is known to have affected Nullaginian sediments deposited on an older continental crust (Fig. 3 and Dunn, 1971; Plumb, 1971) and the Proterozoic province as a whole is therefore similar to the miogeosynclinal belts of Phanerozoic cordilleran orogenies which also developed on a continental crust; and in particular analogy can be made with the western part of the Lachlan geosyncline. Metamorphism in the Proterozoic province supports this conclusion by indicating the expected high geothermal gradients. Thus in the Halls Creek belt local occurrences of sillimanite-almandine sub-facies transitional to low pressure hornblende-granulite sub-facies are found, while the Mt. Isa belt displays local andalusite-biotite-garnet-muscovite and cordierite-anthophyllite-garnet-mica assemblages. Over most of the province metamorphic grade in the Nullaginian rocks is low and the higher grade rocks can again be related to major fault zones.

Folds in the Pre-Carpentarian sediments are generally fairly open and upright, associated with steep cleavage and faults (e.g. Walpole and others, 1968. This miotectonic style is similar to that of the Lachlan geosyncline or of other Phanerozoic miotectonic belts such as the Eastern Andes or the S. Uplands of Britain. Isoclinal folding and listric thrusting are absent in contrast to external zones of the Cordilleran system in North America.

The Proterozoic of northern Australia thus shows a broad analogy with the miotectonic (miogeosynclinal or external) belts of Phanerozoic Cordilleran orogenies. The occurrence of large Pre-Carpentarian plutons across the belt, the local volcanism and low pressure–high temperature metamorphism can be interpreted as the consequence of a rather thinner lithosphere and higher heat flow than was characteristic of the Phanerozoic. The Pre-Carpentarian deformation which indicates flattening normal to the steep cleavage was probably responsible for substantial thickening of the continental crust.

This broadly analogous development of the Proterozoic and Phanerozoic provinces (Fig. 3 and Table 1) suggests that the Proterozoic province may have been related to ocean floor spreading and subduction in a similar way to Cordilleran belts in the Phanerozoic. This implies the existence during the Proterozoic of a volcano-plutonic arc complex and subduction zone east of the present Mt. Isa belt. The Etheridge geosyncline of the Georgetown inlier also contains rocks of 'miogeosynclinal' facies on an older gneissic basement, and the younger Croydon volcanics apparently give a minimum age of 1660 m.y. (Compston & Arriens, 1968). This suggests that the Proterozoic active margin must have been still further east beneath the Tasmanides. There is thus no evidence as to whether the active margin simply bordered on an ancestral Pacific Ocean or whether it may have been involved in collision orogeny with a continent which has subsequently drifted away again. The conclusion remains, however, that the whole Proterozoic province has developed on continental crust in a manner analogous to the miotectonic zones of Cordilleran orogenies. Collision orogeny is not responsible for individual belts such as the Halls Creek and King Leopold belts. It seems likely that similar conclusions apply to early Proterozoic mobile belts in the Canadian Shield such as the Coronation geosyncline (Hoffman *et al.*, 1970) and the Labrador geosyncline (Dimroth *et al.*, 1970; Dimroth, 1970) although the latter has been interpreted in terms of collision orogeny by Gibb & Walcott (1971). Both these belts occur adjacent to the Archaean cratons in a manner analogous to the Halls Creek and King Leopold belts, and their sedimentary, volcanic and metamorphic facies can probably be interpreted in a similar way. The Churchill Province of Canada with the Hudsonian orogeny can then be regarded as a reactivated miotectonic zone analogous to the Proterozoic Arunta-Gawler province of Australia.

The simplest hypothesis for the interpretation of the Proterozoic of Australia in terms of ocean floor spreading is illustrated in Fig. 8, which also indicates the differences due to the inferred thinner lithosphere. An important further indication of the thinner lithosphere in the Proterozoic is the lack of evidence for large vertical movements which characterize the Tertiary in the Phanerozoic chelogenic cycle. Such large vertical movements could probably not be sustained by the thinner and weaker Proterozoic lithosphere.

Proterozoic–Archaean Tectonic Relationships

No chelogenic cycle analogous to the Phanerozoic and Proterozoic cycles can be recognized in the Archaean super-province where earlier platform cover sediments are absent. As noted above, it seems probable that the Pilbara and Yilgarn blocks display two distinct volcano-plutonic cycles. The Pilbara is similar in age to most of the southern African greenstone terrains while the Yilgarn is similar to the Canadian Superior Province.*

Considerable emphasis has recently been placed on the distinctive tectonic style of the Archaean greenstone provinces (e.g. Annhauser *et al.*, 1969; Goodwin, 1968; Sutton, 1967) and on a rather abrupt change to the succeeding Proterozoic (e.g. Hoffman *et al.*, 1970; Goodwin, 1971).

Applications of the Phanerozoic island arc model to the Archaean generally imply that the greenstone belts were developed on oceanic crust and that any previous continental crust consisted of small nuclei which were swept together by ocean floor spreading (e.g. White *et al.*, 1971; Goodwin, 1968). The alternative view, which is preferred here, is that the greenstone belts were developed on a pre-existing continuous granitic crust with a high geothermal gradient (e.g. Annhauser *et al.*, 1969; Sutton, 1971). This is supported by the extensive development of quartzo-feldspathic sedi-

* It seems probable that approximately 500 m.y. cycles are characteristic of Archaean greenstone-granite gneiss tectonics and that broad global correlations can be made. The change from this style of tectonics to the Proterozoic style may appear to have taken place at different times in different areas according to which Archaean cycle is locally preserved. In general it appears that platform cover indicating the onset of cratonization began to develop on the older Archaean areas several hundred million years before the change of orogenic style which generally occurred after 2000 m.y. This difference broadly accounts for different views as to the position of the Archaean-Proterozoic boundary.

ments (Goodwin, 1968) as well as the occurrence of ancient high grade gneiss terrains (Windley & Bridgwater, 1971). Providing the lithosphere was thin (< 50 km) it is not necessary to postulate the existence of an unusually thin granitic crust although, as indicated below, the crust was probably substantially thinner prior to the main post-greenstone deformations. This model is apparently not inconsistent with the volcanic geochemistry which implies high partial melting at shallow depth (< 50 km) below a thin and weak lithosphere (Hart *et al.*, 1972). In Australia the volcanic rocks are mainly tholeiitic (White *et al.*, 1971; Hallberg, 1971) and McCall (1971) has suggested from the study of varioles that pillow lavas amongst them were deposited at depths of not more than 120 m. Deposition on oceanic crust therefore seems unlikely.

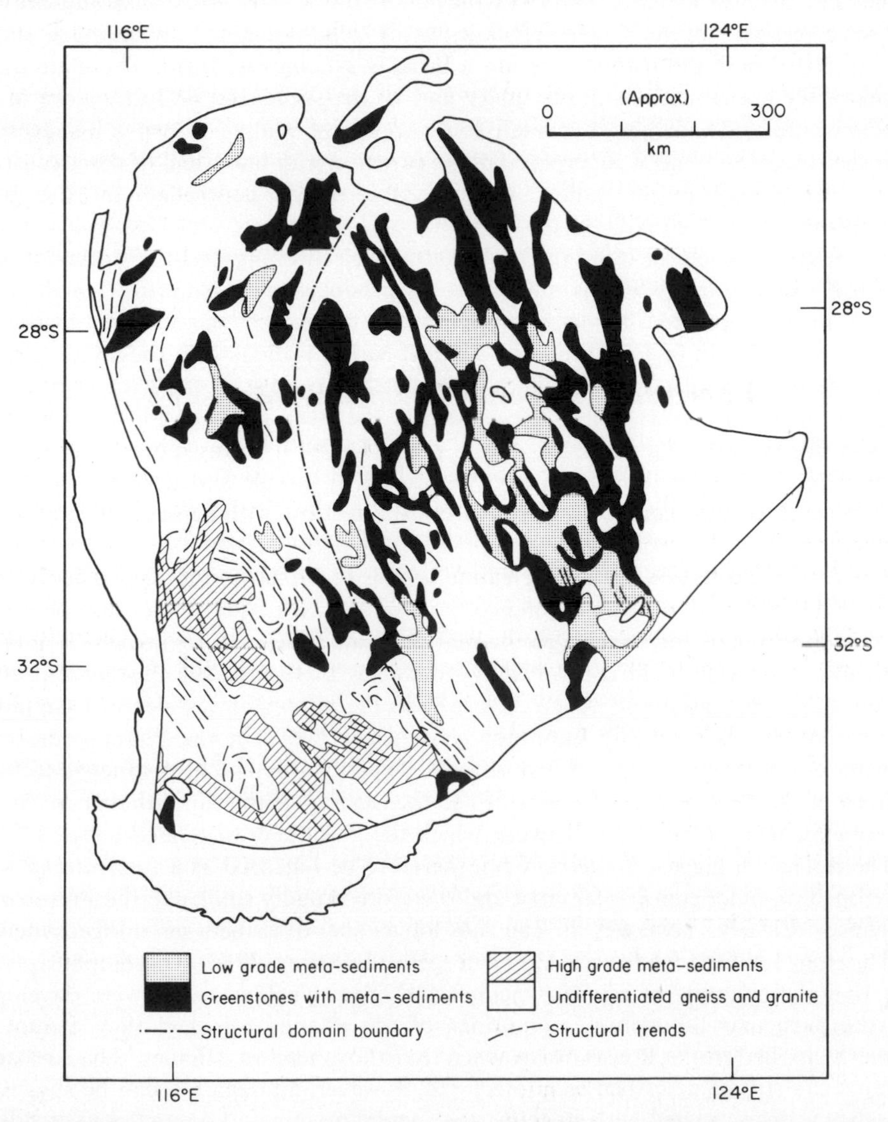

FIGURE 5 Tectonic pattern of the Yilgarn province. Simplified after Tectonic Map of Australia and New Guinea, 1972.

The greenstone belts of the Yilgarn block in Australia form a connected network and the geometry appears to demand that they formed a continuous area of deposition on pre-existing granitic crust. The older crust on which they were developed is apparently represented by the Pilbara block and by the older gneisses (2800–3100 m.y.) in the western part of the Yilgarn block (Fig. 5 and Arriens, 1971; Wilson, 1971). Recent seismic evidence shows that the surface and intermediate crustal layers thicken eastwards (Mathur S. P., Branson J. C. and Moss F. J., personal communication) and suggests that the denser and higher velocity granulite facies rocks in the west of the block extend eastwards under the greenstone-granite gneiss complex. Greenstone and meta-sedimentary belts in this western part of the Yilgarn block do not show well developed trends and this enhances the affinity with the Pilbara block. Fold axial traces and cleavage in the eastern part of the Yilgarn block generally trend north-northwest and as shown in Fig. 5 the main greenstone area has a broadly synclinorial form. The main meta-sedimentary accumulations in the upper part of the succession lie in the core of the synclinorium and indicate an overall south-southeast plunge. There is a suggestion that the structural trend is superposed on an earlier meridional trend of the greenstone belts which may reflect earlier stratotectonic trends controlled by the basic volcanism.

The regular structural trend in the Yilgarn block also suggests the operation of a uniform stress system which is unlikely to be produced by a pattern of small continental nuclei separated by subduction zones (for which there is no evidence). O'Driscoll (1971) in fact continues to support earlier notions of a rhegmatic pattern which he believes operated throughout the Australian crust in the Archaean as well as later times. However, if it is accepted that the steep cleavage in the greenstone belts is due to flattening (Glikson, 1971a, pp. 125–128) it seems more likely that the pattern in the Yilgarn block has been produced by simple shortening of a mobile lithosphere in an east-northeast–west-southwest direction, with consequent substantial thickening.

The deformation style of the greenstone belts is of similar miotectonic character to that in the Proterozoic. The difference between the two belts is ascribed here to a thinner lithosphere and higher geothermal gradient in the Archaean which allowed extensive basic volcanism controlled by fractures. In the thicker Proterozoic lithosphere basic volcanism still occurred locally in the miotectonic zones as exemplified by the Mt. Isa belt (or the Labrador geosyncline) while in the Phanerozoic miotectonic zones extensive basic volcanism was generally absent. The distinctive tectonic pattern of the Archaean can be ascribed to the loading of the thin lithosphere by the mafic volcanic accumulations between which the granitic crust tended to rise.

The Archaean greenstone terrain can therefore be regarded as a miotectonic zone developed on older continental crust and therefore broadly similar to the Proterozoic province. Clearly, however, it can no longer be described as miogeosynclinal and analogy has been made with Alpine orogeny (Glikson, 1971b). This emphasizes the fact that the concepts of miogeosynclinal and eugeosynclinal facies were developed for the particular lithospheric conditions of the Phanerozoic and they cannot be properly applied in the Precambrian when the lithosphere was thinner. The Archaean terrains can still be described as miotectonic, however, and again it can be suggested that they were associated with an active continental margin and ocean floor spreading. The active margin must have been outside the present Archaean areas and outside the

Proterozoic area known to be underlain by Archaean crust. It is not possible therefore to envisage lateral continental accretion. On the contrary, it is suggested that the crustal shortening and plutonism of the greenstone belts correspond to major episodes of ocean floor spreading in which the area of oceanic crust was extended and the area of continental crust reduced.

This miotectonic model for greenstone belt development is illustrated in relation to Proterozoic and Phanerozoic orogeny in Fig. 8. The model implies that the strong contrast between eutectonic ('eugeosynclinal' arc-trench complexes) and miotectonic ('miogeosynclinal') zones which characterizes Phanerozoic Cordilleran orogeny did not exist in the Archaean because of the thinner lithosphere. This explains the geochemical similarities between the Archaean miotectonic complexes and Phanerozoic eutectonic island arc complexes (Engel, 1971; White *et al.*, 1971; Goodwin, 1968, 1971).

Again the model does not appeal to collision orogeny to explain boundaries between major plutonic provinces in the Archaean. Thus the boundary between the Pilbara and Yilgarn provinces in late Archaean time is regarded as the boundary between craton (Pilbara) and miotectonic zone (Yilgarn). It is analogous to the boundary between the Kimberley craton and the Proterozoic miotectonic zone in Proterozoic time or the boundary between the Proterozoic craton and the Tasman miotectonic zone in Phanerozoic time. There therefore seem to be no grounds for postulating separation of the Pilbara and Yilgarn blocks (cf. Arriens, 1971) and, as indicated above, the western and northern parts of the Yilgarn block show some transitional features.

The Musgrave–Fraser Belts

The Musgrave and Fraser belts represent parts of a possibly continuous system of metamorphic and plutonic rocks dated between 1400 m.y. and 1050 m.y. which now separate parts of the Archaean and Proterozoic provinces described above. Tectonic trends in the latter provinces are somewhat variable but like those in the Tasman province they commonly lie within 30° of meridional trends. There is, however, a broad belt up to about 1000 km wide in which westerly or west-northwesterly trends are dominant (Fig. 6). The Archaean Pilbara province occurs within this belt but shows no very marked trends. However, the King Leopold, Granites–Tanami, Tennant Creek and Gascoyne domains all conform to these trends which must therefore have become established between about 2000 and 1800 m.y. ago. The westerly trends were renewed by the Musgrave (1400–1100 m.y.) Petermann Ranges (Late Precambrian) and Alice Springs (Mid-Palaeozoic) orogenies (Fig. 6). They are also reflected in the trends of basins within the belt and especially in the Canning Basin where a maximum of over 9000 m of sediment was accumulated. The inferred eastward continuation of the belt into the Tasman Province is marked by the widest and thickest development of Mesozoic post-chelogenic cover in the Eromanga Basin which causes a wide break in the exposure of the Lachlan Sub-province. The New England Sub-province lies east of this break and it is conceivable that it remained attached to the Australian continent when its northern and southern extensions were rafted away because it had an older continental basement with east–west rather than north–south structure.

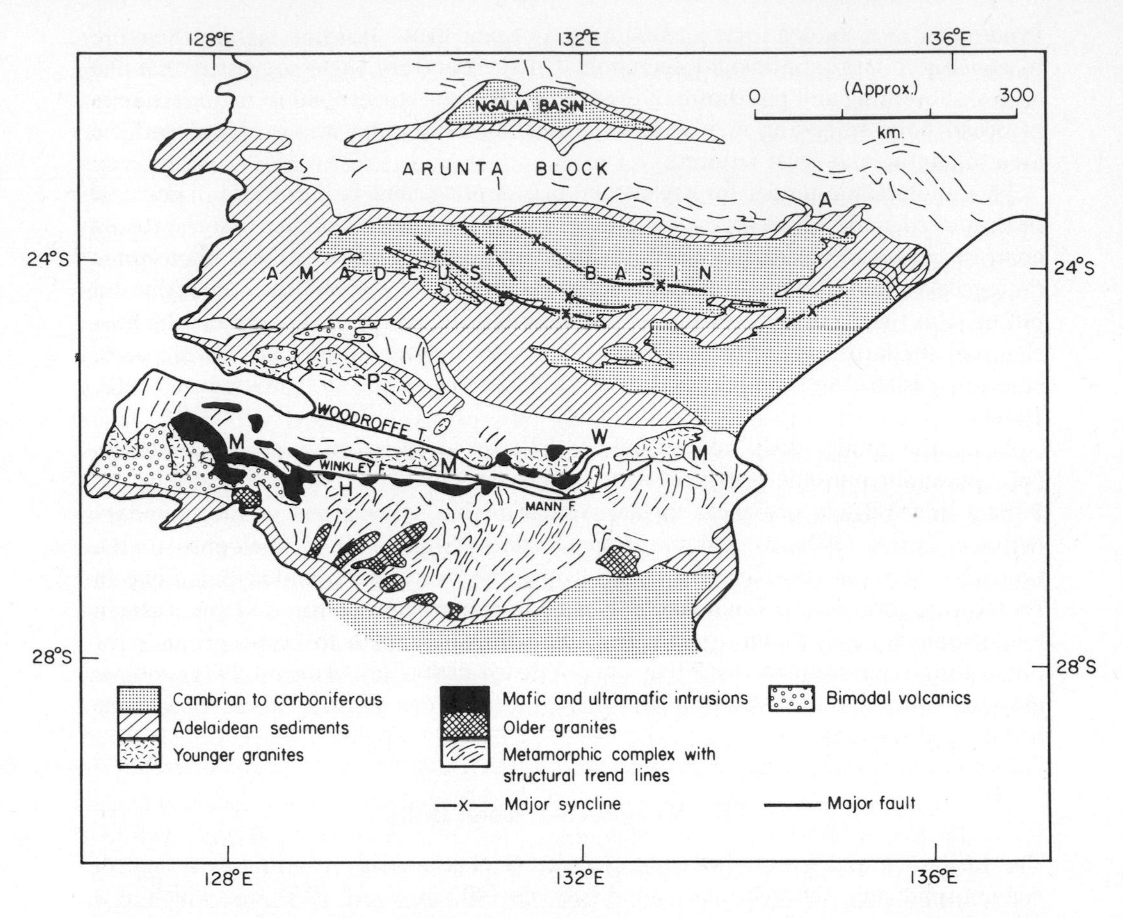

FIGURE 6 Tectonic elements of the Musgrave and Arunta Blocks and Amadeus Basin. M, Musgrave orogenic domain; P, Petermann Ranges nappe complex of Petermann Ranges 'orogeny'; A, Arltunga nappe of Alice Springs 'orogeny'. Simplified after Tectonic Map of Australia and New Guinea, 1972.

The Musgrave province must therefore be considered in the context of this broad belt (here named the Amadeus Transverse Zone) in which westerly or west-north-westerly trends have been important through most of Proterozoic and Phanerozoic time. In this context it seems likely that it should be regarded as an intra-continental orogenic belt rather than an inter-continental collision orogeny. It has already been noted that the Arunta and Gawler Provinces north and south respectively of the Musgrave belt show similar age patterns as would be expected if the Musgrave belt were due to a later superposed intra-continental orogeny but would be wholly unexpected from a collision orogeny.

The Musgrave belt is characterized by a relatively narrow zone of major faults, intermediate pressure granulite facies rocks and associated plutons (e.g. Wilson, 1969). The main metamorphism took place about 1330 m.y. ago (Arriens & Lambert, 1969). Some of the granulites represent reworked basement but others may not have been involved in any earlier orogenic cycle (Collerson *et al.*, 1972). In the Amata area the latter rocks consist largely of quartzo-feldspathic types (near

70% S_1O_2) which may be metavolcanic or more likely metasedimentary (they are interlayered with minor quartzites, calc-silicates and manganiferous units); but there are also subsidiary mafic granulites which are geochemically similar to high alumina and tholeiitic basalts (K. C. Collerson, personal communication). These rocks do not appear to represent a 'geosynclinal' development. The main penetrative deformation was synmetamorphic but deformation continued in mylonite thrust fault zones which carried the granulites northwards. The main fault movements were probably complete by the time of granite emplacement about 1120 m.y. ago since granite bodies lie across some fault lines but thrusting and folding continued locally so that it disrupted and tectonized parts of the Giles ultramafic igneous complex which was emplaced about 1050–1100 m.y. (Nesbitt *et al.*, 1970). It appears that the major faults controlled the emplacement of the Giles complex and although they show northward overthrusting at the surface they are probably steep at depth. The faults are slightly oblique to the belt of plutonism and it is possible that they have a strike-slip component.

The Giles complex consists of a group of layered intrusions of stratiform (e.g. Stillwater) type although they have been locally deformed and metamorphosed under granulite faces conditions. They are associated with a bi-modal basalt-rhyolite volcanic association and they are therefore interpreted as an intra-continental suite rather than as an ophiolite suite on an inter-continental collision suture zone.

The Musgrave 'orogeny' did not apparently result in exogeosynclinal development. Volcanics associated with the Giles complex are preserved at the western end of the belt suggesting that little uplift and erosion occurred prior to the deposition of the basal quartzites of the Adelaidean right across the belt.

The Musgrave orogenic belt is therefore here regarded as a narrow zone, controlled by major faults, of basement reactivation, in which high geothermal gradients have led to granulite facies metamorphism and plutonic activity. It is therefore one type of a broad class of orogenic belts which can be described as Intracontinental, in contrast to Cordilleran and Collision (inter-continental) orogenies.

The Fraser belt is broadly similar. There the Granulite facies rocks lie adjacent to a zone of reworked Archaean basement. The foliation throughout the belt is very steep and the rocks lack the strong lineation which is characteristic of the Musgrave belt.

The Petermann Ranges and Alice Springs 'orogenies' (e.g. Forman, 1970, 1971) can be regarded as later higher level orogenic effects related to the earlier Musgrave 'orogeny'. Both the later orogenies produced basement nappes which moved into the Amadeus basin. That is to say, the nappes were produced on the margins of basement uplifts.

The closest analogy of these structures elsewhere is probably provided by the thrusts on the margins of the Laramide basement uplifts of Central Wyoming and the Colorado Front Range (e.g. Eardley, 1963; Berg, 1963; Sales, 1968). These uplifts of the Rocky Mountains also represent an intra-continental orogenic belt and like the Musgrave belt, the Wyoming uplifts lie at a large angle to the trend of the Cordilleran orogenic belt with which they are associated.

It is suggested, therefore, that the Musgrave belt presents the deeper expression of a similar belt developed in the thinner Proterozoic lithosphere and also lying at a large angle to the continental margin (presumed meridional). The Wyoming uplifts are one expression of the Lewis and Clark transverse zone (Sales, 1968; King, 1969) and the Musgrave belt is one expression of the Amadeus transverse zone. Such transverse zones are probably loci of strike-slip movements in the continental crust and it seems likely that they were more extensive in the Proterozoic. Deformation high temperature metamorphism and plutonism along such zones may account for many apparently intra-continental orogenic belts, such as the Kapuskasing high in Canada or the Limpopo belt of Africa. It also seems likely that such zones may locally have become the sites of intra-continental geosynclines. The relationship of the King Leopold and Halls Creek belts suggests the presence of conjugate strike-slip systems.

The transverse zones are probably ultimately related to the pattern of ocean floor spreading and consequent plate movement. In the case of the Musgrave belt it can perhaps be inferred that the intra-continental orogenic activity represents a terminal phase of the Early Proterozoic chelogenic and spreading cycle in the same way as the Rocky Mountains represent a terminal phase of the Late Proterozoic–Phanerozoic cycle. The Early Proterozoic cycle therefore continued for 400 m.y. after the latest plutonism recorded by the Mt. Isa belt. The mafic dyke emplacement and plutonism of about 1050 m.y. can therefore be taken to end the Early Proterozoic chelogenic cycle and the basal quartzites of the main Adelaidean sequence can be taken as the beginning of the Late Proterozoic to Phanerozoic cycle.

Concluding Discussion

It has been shown that the tectonic phenomena of the Australian Precambrian differ

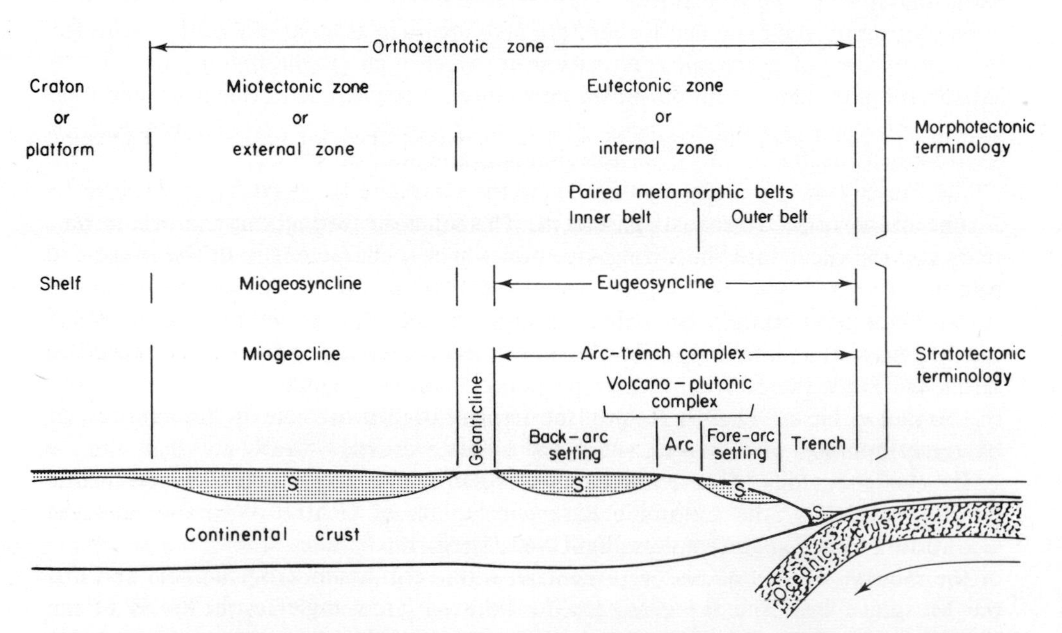

FIGURE 7 Terminology of Cordilleran orogenic belts. Supracrustal accumulations (s) are indicated diagrammatically and are normally earlier (Palaeozoic) in the miotectonic zone than in the eutectonic zone (Mesozoic). Note that the continental crust beneath the eutectonic zone may be formed by lateral accretion of earlier deposited supracrustal rocks. Exogeosynclinal deposits are normally developed across the margin between the miotectonic zone and the craton and mainly on the craton.

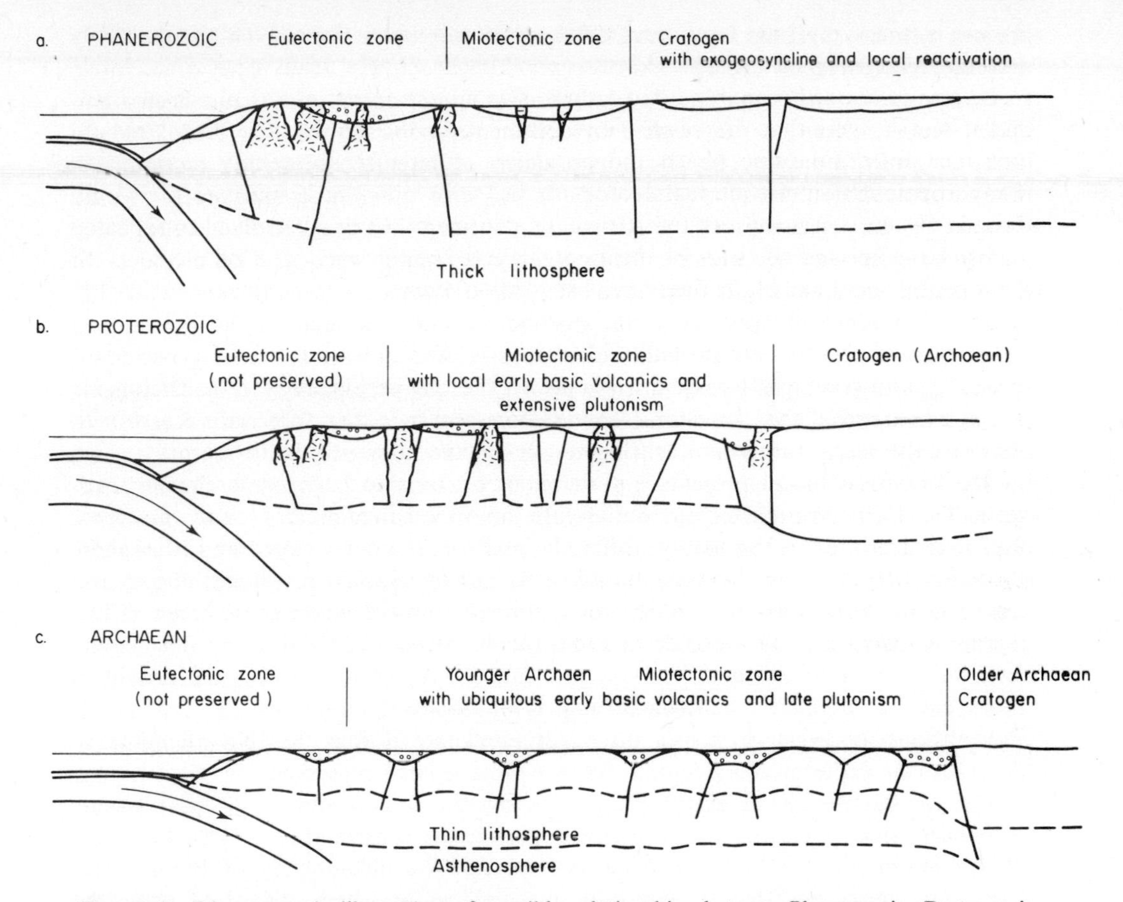

FIGURE 8 Diagrammatic illustration of possible relationships between Phanerozoic, Proterozoic and Archaean tectonics.

in important respects from those of the Phanerozoic. Sedimentary, metamorphic and volcanic facies show different associations which are related in different ways to plutonic activity. Nevertheless, it is suggested that the differences can be explained in terms of a progressively thinner lithosphere and higher geothermal gradient, moving back in Precambrian time; and that the orogenic activity can be related to ocean floor spreading.

The chelogenic cycles with their episodic plutonism probably relate to major cycles of ocean-floor spreading during which the oceans were periodically enlarged and the continents reduced in area by tangential compression. The volume of continental crust has probably not increased greatly but the volume of lithosphere has increased several-fold as the geothermal gradient has decreased.

No collision orogenies can be recognized within the Australian continent and it is possible that the continent has remained on the margin of an ancestral Pacific Ocean throughout geological time. Both the Proterozoic and Archaean orogenic domains can be interpreted as miotectonic zones on the continental side of active continental margins (eutectonic zones) and subduction zones. These active margins are not preserved and are supposed to be buried beneath the Tasmanides. The Musgrave–

Fraser belts are regarded as an intra-continental development at the end of the Early Proterozoic chelogenic cycle.

It follows from this analysis that lateral continental accretion has not been significant and has been far outweighed by continental contraction. (The New England province may, however, be the consequence of lateral continental accretion in Phanerozoic time.) An alternative hypothesis, with different consequences, would be that the eastern margin of the Australian continent has been involved in repeated collisions with and separations from other continental masses. The products of continental accretion might then have been rafted away.

The occurrence of apparently miogeoclinal successions such as the Adelaidean does suggest that the eastern continental margin was passive for a long period of time. This, however, can be explained in terms of a long period, early in the chelogenic cycle, when ocean floor spreading was inactive and it is therefore not necessary to invoke continental rifting and drifting to explain the inferred passive margin.

The preferred model presents a systematic relationship between Archaean Proterozoic and Phanerozoic tectonic activity. In the Archaean miotectonic and eutectonic (active-margin) zones probably differed little in character. As the lithosphere thickened, the contrast between miotectonic and eutectonic zones developed and pre-orogenic basic volcanic activity was excluded from the miotectonic zones. Thus, if only miotectonic zones are considered, the Proterozoic provides an intermediate term between Archaean and Phanerozoic; but the Archaean miotectonic provinces more closely resemble Phanerozoic eutectonic zones than do the Proterozoic. If Precambrian collision orogenies do occur elsewhere it may be that elements of Archaean or Proterozoic eutectonic zones will have been preserved.

The model provides a useful framework for the understanding of metallogenic provinces. Since eutectonic zones or active margins are preserved only in the Phanerozoic a major contrast is to be expected between the metallogeny of Phanerozoic eutectonic zones and that of Precambrian shield areas which consist of stabilized miotectonic zones. The Cu–Ni–Au associations of the Archaean province are related to the extensive basic volcanism and lesser acid igneous activity and can be regarded as largely primary mantle derivatives. The Cu–Au–U association of the Proterozoic super-province on the other hand can be regarded as the result of reworking of the underlying Archaean granitic crust. The major Pb–Zu deposits lie near the eastern margin of the Shield and were probably developed in sedimentary environments behind the eutectonic zone during Proterozoic time in a manner analogous to Phanerozoic deposits. The eutectonic zone of the Tasmanides is a complex metallogenic province where products of reworking of older crust, primary mantle derivatives and products of recycling by ocean floor spreading are probably associated.

Acknowledgements

This paper has developed from research carried out in collaboration with R. C. Oliver, R. W. Nesbitt and K. C. Collerson with the support of the Australian Research Grant Committee, and from the author's membership of the Tectonic Map committee of the Geological Society of Australia (Convenor E. S. Hills). Thanks are due to members of that committee for numerous fruitful discussions, and especially to B. Thomson, M. J. Rickard, F. Doutch and K. Plumb.

References

Annhauser, C. R., Mason, R., Viljoen, M. J. and Viljoen, R. P., 1969. A reappraisal of some aspects of Precambrian shield geology, *Bull. Geol. Soc. Amer.*, **80,** 2175–2200.

Arriens, P. A., 1971. The Archaean geochronology of Australia, *Spec. Publs. Geol. Soc. Aust.*, **3,** 11–23.

Arriens, P. A. and Lambert, I. B., 1969. On the age and strontium isotope geochemistry of granulite-facies rocks from the Fraser Range, Western Australia, and the Musgrave Ranges, Central Australia, *Spec. Publs. Geol. Soc. Aust.*, **2,** 377–88.

Berg, R. R., 1963. Laramide sediments along Wind River thrust, Wyoming. *In:* Backbone of the Americas, edited by O. E. Childs and B. W. Beebe. Amer. Assoc. Petroleum Geol., Mem. 2.

Carter, E. K., Brooks, J. H. and Walker, K. R., 1961. The Precambrian mineral belt of north-western Queensland, *Bur. Min. Resour. Aust. Bull.*, **51.**

Collerson, K. C., Oliver, R. L. and Rutland, R. W. R., 1972. An example of structural and metamorphic relationships in the Musgrave orogenic belt, central Australia, *J. Geol. Soc. Aust.*, **18,** 379–393.

Compston, W., Crawford, A. R. and Bofinger, V. M., 1966. A radiometric estimate of the duration of sedimentation in the Adelaide geosyncline, South Australia, *J. Geol. Soc. Aust.*, **13,** 229–76.

Compston, W. and Arriens, P. A., 1968. The Precambrian geochronology of Australia, *Can. J. Earth Sci.*, **5,** 561–83.

Dewey, J. F. and Bird, J. M., 1970. Mountain belts and the new global tectonics, *J. Geophys. Res.*, **75,** 2625–47.

Dewey, J. F. and Horsfield, B., 1970. Plate tectonics, orogeny and continental growth, *Nature*, **225,** 521–6.

Dickinson, W. R., 1971. Plate tectonic models of geosynclines, *Earth Planet Sci. Letts.*, **10,** 165–74.

Dimroth, E., 1970. Evolution of the Labrador Geosyncline, *Bull. Geol. Soc. Amer.*, **87,** 2717–42.

Dimroth, E., Baragar, W. R. A., Bergeron, R. and Jackson, G. D., 1970. The filling of the circum-Ungava geosyncline, *Geol. Surv. Can. Paper*, **70–40,** pp. 45–142.

Dow D. B. and Gemuts, I., 1969. Geology of the Kimberley region, Western Australia: The East Kimberley, *Geol. Surv. W. Aust. Bull.*, **120.**

Dunn, P. R., Plumb, K. A. and Roberts, H. G., 1966. A proposal for time-stratigraphic subdivision of the Australian Precambrian, *J. Geol. Soc. Aust.*, **13,** 593–608.

Dunn, P. R., 1971. Archaean of Northern Australia (Abstract only). *Spec. Publs. Geol. Soc. Aust.*, **3,** 152–3.

Eardley, A. J., 1963. Relation of uplifts to thrusts in Rocky Mountains. *In:* Backbone of the Americas, edited by O. E. Childs and B. W. Beebe. Amer. Assoc. Petroleum Geol. Mem. 2.

Engel, A. E. G., 1971. Global aspects of the Precambrian, *Geol. Soc. America Abstracts*, **3,** No. 7, 556–7.

Forman, D. J., 1970. *In:* Geology of the Amadeus basin, Central Australia, edited by A. T. Wells, *et al.*, *Bur. Min. Resour. Aust. Bull.*, **100,** 120–45.

Forman, D. J., 1971. The Arltanga nappe complex, Macdonnell Ranges, Northern Territory, Australia, *J. geol. Soc. Aust.*, **18,** 173–82.

Gastil, G., 1960. Continents and mobile belts in the light of mineral dating, Int. geol. Congr. 21st Sess. Pt. 9, pp. 162–9.

Gellatly D. C., 1971. Possible Archaean rocks of the Kimberley region, Western Australia, *Spec. Publs. Geol. Soc. Aust.*, **3,** 93–102.

Gibb, R. A., 1971. Origin of the great arc of eastern Hudson Bay: a Precambrian continental drift reconstruction, *Earth Planet Sci. Letts.*, **10,** 365–71.

Gibb, R. A. and Walcott, R. I., 1971. A Precambrian suture in the Canadian Shield, *Earth Planet Sci. Letts.*, **10,** 417–22.

Glikson A. Y., 1971a. Structure and metamorphism of the Kalgoorlie System, Southwest of Kalgoorlie, Western Australia, *Spec. Publs. Geol. Soc. Aust.*, **3,** 121–32.

Glikson, A. Y., 1971b. Archaean geosynclinal sedimentation near Kalgoorlie, Western Australia, *Spec. Publs. Geol. Soc. Aust.*, **3,** 443–60.

Goodwin, A. M., 1968, Evolution of the Canadian Shield, *Proc. Geol. Assoc. Canada,* **19,** 1–14.

Goodwin, A. M., 1971. Metallogenic patterns and evolution of the Canadian Shield, *Spec. Publs. Geol. Soc. Aust.*, **3,** 57–74.

Hallberg, J. A., 1971. Geochemistry of the Archaean basalt-dolerite association in the Coolgardie–Norseman area, Western Australia, *Spec. Publs. Geol. Soc. Aust.*, **3,** 151.

Hart, S. R., Brooks, C., Krogh, T. E., Davis, G. L. and Nava, D., 1972. Ancient and modern volcanic rocks; a trace element model, *Earth Planet Sci. Letts.*, **10,** 17–28.

Hoffman, P. F., Fraser, J. A. and McGlynn, J. C., 1970. The Coronation geosyncline of Aphebian age, District of MacKenzie, *Geol. Surv. Can. Paper*, **70–40,** pp. 201–12.

Hurley, P. M., 1970. Distribution of age provinces in Laurasia; *Earth Planet Sci. Letts.*, **8,** 189–96.

Hurley, P. M. and Rand, J. R., 1969. Pre-Drift Continental Nuclei, *Science*, **164,** 1229–42.

Jaeger, J. C., 1970. Heat flow and radioactivity in Australia. *Earth Planet Sci. Letts.*, **8,** 285–92.

King, P. B., 1969. The tectonics of North America—a discussion to accompany the tectonic map of North America Scale 1:5,000,000, U.S. Geol. Surv. Prof. Paper **628.**

Lambert, I. B., 1971. The composition and evolution of the deep continental crust, *Spec. Publs. Geol. Soc. Aust.*, **3,** 419–48.

Leggo, P. J., Compston, W. and Trendall, A. F., 1965. Radiometric ages of some Precambrian rocks from the Northwest Division of Western Australia, *J. Geol. Soc. Aust.*, **12,** 53–65.

McCall, G. J. H., 1971. Some ultrabasic and basic igneous rock occurrences in the Archaean of Western Australia, *Spec. Publs. Geol. Soc. Aust.*, **3,** 429–42.

Muelberger, W. R., 1971. Buried basement rocks of North America: their evidence for continental rifting and drifting, *Geol. Soc. Amer. Abstracts*, **3,** No. 7, 653–4.

Nesbitt, R. W., Goode, A. P. T., Moore A. C. and Hopwood, T. P., 1970. The Giles Complex Central Australia; a stratified sequence of mafic and ultamafic intrusions, *Spec. Publs. Geol. Soc. S. Africa*, **1,** 547–63.

O'Driscoll, E. S. T., 1971. Deformational concepts in relation to some ultamafic rocks in Western Australia, *Spec. Publs. Geol. Soc. Aust.*, **3,** 351–66.

Oversby, B., 1971. Palaeozoic plate tectonics in the southern Tasman geosyncline, *Nature Phys. Sci.*, **234,** 45–7 and p. 60.

Packham, G. H., 1969. *In:* The Geology of New South Wales, edited by G. H. Packham, *Geol. Soc. Aust.*, **16,** Part 1.

Plumb, K. A., 1971. The Archaean and Australian tectonics (Abstract only), *Spec. Publs. Geol. Soc. Aust.*, **3,** 385.

Sales, J. K., 1968. Crustal mechanics of Cordilleran foreland deformation: a regional and scale-model approach, *Bull. Amer. Ass. Petr. Geol.*, **52,** 2016–44.

Shackleton, R. M., 1969. Displacement within continents. *In:* Time and Place in Orogeny, *Geol. Soc. London Spec. Publ.*, **3,** 1–7.

Spall, H., 1971. Precambrian apparent polar-wandering; evidence from North America, *Earth Planet Sci. Letts.*, **10,** 273–280.

de Swardt, A. M. J., Garrard, P. and Simpson, J. G., 1965. Super-position of orogenic belts in part of central Africa with special reference to major zones of transcurrent dislocation. *Bull. Geol. Soc. Amer.*, **76,** 89–102.

Sutton, J., 1963. Long-term cycles in the evolution of the continents, *Nature*, **198,** 731–5.

Sutton, J., 1967. The extension of the geological record into the Precambrian, *Proc. Geol. Ass.*, **78,** 493–534.

Sutton, J., 1971. Some developments in the crust, *Spec. Publs. Geol. Soc. Aust.*, **3,** 1–10.

Thomson, B. P., 1966. The lower boundary of the Adelaide system and older basement relationships in South Australia, *J. Geol. Soc. Aust.*, **13,** 203–28.

Trendall, A. F., 1966. Towards rationalism in Precambrian stratigraphy, *Geol. Soc. Aust.*, **13,** 517–26.

Vail, J. R., Snelling, N. J. and Rex, D. C., 1968. Pre-Katangan geochronology of Zambia and adjacent parts of Central Africa. *Can. J. Earth Sci.*, **5**, 621–8.
White, A. J. R., Jakes, P. and Christie, D. K., 1971. Composition of greenstones and the hypothesis of sea-floor spreading in the Archaean, *Spec. Publs. Geol. Soc. Aust.*, **3**, 47–56.
Wilson, A. F., 1969. Granulite terrains and their tectonic setting and relationship to associated metamorphic rocks in Australia, *Spec. Publs. Geol. Soc. Aust.*, **2**, 243–58.
Wilson, A. F., 1971. Some geochemical aspects of the Sapphirine-bearing pyroxenites and related highly metamorphosed rocks from the Archaean ultamafic belt of South Quairading Western Australia, *Spec. Publs. Geol. Soc. Aust.*, **3**, 401–12.
Walpole, B. P., Crohn, P. W., Dunn, P. R. and Randal, M. A., 1968. Geology of the Katherine–Darwin region, Northern Territory, *Bur. Min. Resour. Aust. Bull.*, **82.**
White, D. A., 1965. The geology of the Georgetown/Clarke River area, Queensland, *Bur. Min. Resour. Aust. Bull.*, **71.**
Windley, B. F. and Bridgwater, D., 1971. The evolution of Archaean low and high-grade terrains, *Spec. Publs. Geol. Soc. Aust.*, **3**, 33–46.

10.4

K. BURKE
Erindale College,
University of Toronto, Mississauga,
Ontario, Canada

J. F. DEWEY
State University of New York at
Albany, New York 12222, USA

An Outline of Precambrian Plate Development

Introduction

Plate tectonic theory has provided geologists with a working hypothesis applicable to a great deal of geological data collected by thousands of geologists during the last 200 years. We here attempt a preliminary outline of a Plate Tectonic interpretation of Precambrian geological history based mainly on published results from Africa and North America. Severe limitations on the interpretation are imposed by the incompleteness of the Precambrian record. A great deal of the information which would be regarded as essential in interpreting the history of a Phanerozoic orogen is unavailable in the Precambrian either because it was never available (for example biostratigraphic information) or because it has been destroyed in the orogenic process (as for instance where continental collision has squeezed out rocks at a suture). A further problem is that some structural environments which are important in the Precambrian, for example reactivated areas, are inadequately known or have not been recognized in the Phanerozoic. These problems become greater going back in time until about $2{\cdot}7 \times 10^9$ years ago. We conclude that before that time the outer part of the earth was so mobile that no plate tectonic system operated.

Permobile Phase

Lithospheric plates appear to be driven by some kind of mantle convection produced by radiogenic heat (see, for example, Bott, 1971 p. 288). Although radiogenic heat generation was at a maximum in the early post-accretionary phase of the earth's history early tectonic processes apparently differed from plate tectonics because there was no rigid lithosphere. The moon's surface, which may represent an approximate analogue of the earth's surface at an early stage, contains basaltic products of two mobile episodes ($4{\cdot}6$ and $3{\cdot}1 \times 10^9$ years ago) perhaps related to convection (Strangway *et al.*, 1972) but there is no evidence indicating that a lunar plate system ever operated.

The oldest exposed parts of the earth ($\sim 3{\cdot}4 \pm 0{\cdot}3 \times 10^9$ years old) contain abundant rocks with compositions typical of continental crust. Although this material is older than the products of the younger mobile phase on the moon, crustal differentiation on the earth had already progressed farther than lunar differentiation ever reached. Structural environments represented by old terrestrial rocks differ greatly from those represented on the moon. For example although anorthosites occur on both planets those on the earth are associated with typical continental crustal rocks while those on

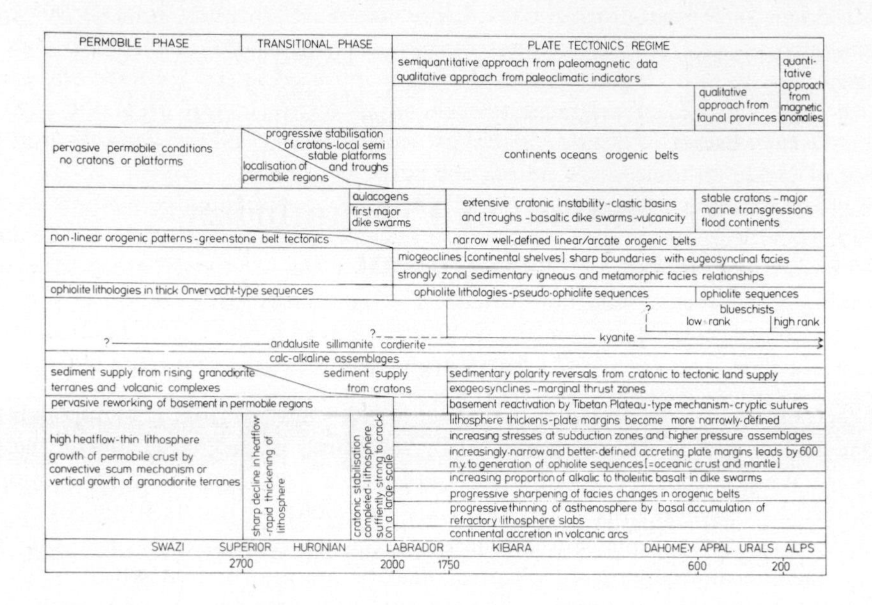

FIGURE 1 Diagrammatic representation summarizing the evolution of lithospheric structural systems from an early permobile phase, through a transition to the present Plate Tectonic regime.

the moon occur in a wholly basaltic association.

Terrestrial rocks formed prior to $2{\cdot}7 \times 10^9$ years ago (Fig. 1) include many with strikingly close compositional and structural resemblances to those of later times. Thus volcanic and sedimentary rocks of the Archaean greenstone belts present close analogies with those of modern island arcs (Goodwin, this volume). Although many small and medium scale tectonic processes were probably similar to those operating today (for example: calc-alkaline rocks are likely to have been produced by partial melting of down-going basaltic material) the global tectonic regime appears to have been characterized by great mobility (*cf.* Goodwin, 1971; Sutton, 1971) and an absence of rigid features. We call this earliest recorded phase the Permobile phase.

Characteristics of the Permobile phase are summarized in Fig. 1 where emphasis is laid on the absence of continents or large cratonic areas. Individual dominantly granodioritic areas rarely exceed 10^5 km^2 and these are bounded by and generally intrude curved greenstone belts consisting of tholeiitic and calc-alkaline volcanics and coarse, mainly volcaniclastic, sediments associated with ophiolite lithologies in thick Onverwacht type sequences (Viljoen & Viljoen, 1971). Greenstone belt outcrops are generally about 10 times as long as they are wide, in contrast with the rocks of comparable composition and grade locally preserved in the suture zones of later continental collision orogenies (which are about 100 times as long as they are wide). Outcrop areas in suture zones are generally much smaller than in greenstone belts and this we attribute to a contrast between the plate and permobile regimes. Under the plate regime thick, converging continent-laden plates squeeze out ocean floor material to leave narrow and cryptic sutures (see below). Under the permobile regime with thin lithosphere and no extensive rigid areas, relatively wide belts of greenstone lay between dominantly granodioritic areas. Large volume sedimentary units ($> 10^4$ km^3) are generally absent from the terranes formed in the permobile phase although an unknown volume of high grade metasediment is included in the granodioritic areas. Because large volumes of sediment accumulate either on top of or at the edge of continents, the absence of thick sedimentary sequences is an indirect indication of the absence of stable cratonic areas during the permobile phase.

The oldest rocks have as yet been the subject of few comprehensive isotopic studies but there is evidence of approximately simultaneous reactivation over extensive areas ($> 10^6$ km^2) some hundreds of millions of years after the reactivated greenstones and intervening granodiorites were first formed. The widespread preservation of very low grade assemblages in the greenstone belts through these events is a problematic feature. Kyanite is absent from high grade metamorphic assemblages among the oldest rocks indicating relatively low pressure environments.

In summary, we envisage the permobile phase as a regime of high heat generation and thin lithosphere during which large volumes of basaltic and sialic material differentiated from the mantle through rapid convective circulation, probably in two stages as Jakes (this volume) has suggested. The processes had perhaps been in operation for between 0·5 and $1{\cdot}0 \times 10^9$ years before the oldest rocks now preserved were formed.

Transitional Phase

Extensive dyke swarms in continental crust and the oldest intrusion of rift valley type (The Great Dyke of Rhodesia, Grantham, 1958; Burke & Whiteman, this volume) provide evidence of the existence of a rigid lithosphere before $2{\cdot}0 \times 10^9$ years ago. The spread in ages between the Great Dyke ($\sim 2{\cdot}6 \times 10^9$ years) and the youngest greenstone belt/granodiorite associations (those in the Birrimian aged about $1{\cdot}8 \times 10^9$ years ago) indicates the approximate duration of a transitional phase between the Permobile and Plate Tectonic regimes. During this phase rock associations and structures characteristic of the older regime became progressively less common. The transition, which we suggest reflects the development of a rigid lithosphere, may have resulted from a relatively rapid decline in radiogenic heat generation (see for example Lubimova, 1969, Fig. 1) and the accumulation, through the continuing operation of convective processes, of a substantial proportion of the presently existing sialic crust.

Early cratonic cover

Some rock associations appear to have been peculiar to the transitional phase. Perhaps the most remarkable of these are in the Witwatersrand Triad (Whiteside, 1970) which is the oldest thick extensive group of rocks deposited on continental crust. The Dominion Reef, Witwatersrand and Ventersdrop systems which make up the Triad, were deposited between 2·8 and $2{\cdot}3 \times 10^9$ years ago, crop out over 2×10^5 km^2 and reach a maximum thickness of 13 km (Fig. 2). They consist of slightly metamorphozed shallow water and fluviatile shales, sandstones and gold and uranium bearing conglomerates intercalated with more than 40% of calc-alkaline lavas. Calc-alkaline volcanics in the permobile phase were characteristic of the island arc type of association preserved in the greenstone belts, and in the later Plate Tectonic phase have been restricted to within a few hundred kilometres of the outcrop of Benioff

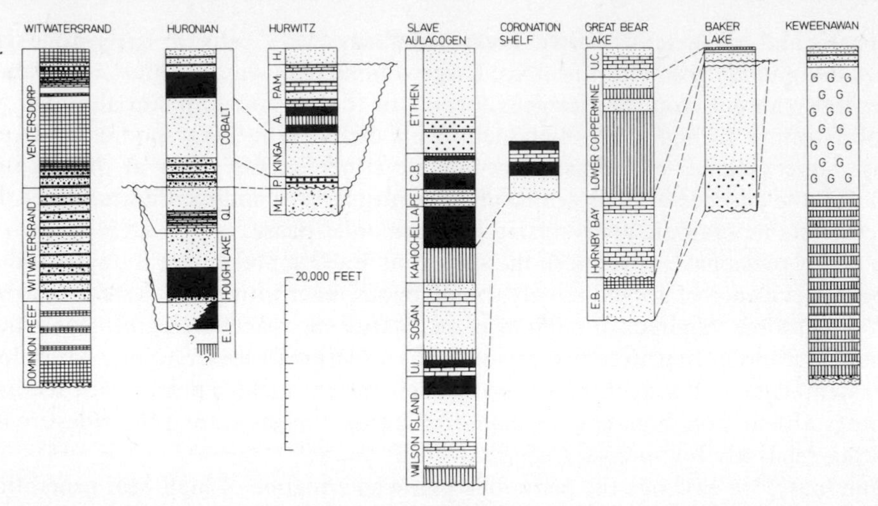

FIGURE 2 Sections representative of cratonic cover and aulacogen sedimentary fill developed mainly during the transitional phase. Small dots: sandstone, large dots: conglomerate, black: argillite-siltite, vertical lines: basalt, G: gabbro, white: rhyolite, checkered ornament: 'andesite', bricks: limestone.

zones. Their occurrence in the Triad is one of the unique features of the transition.

Thick sedimentary and volcanic sequences somewhat resembling the Witwatersrand Triad, but with less volcanics, mark the transitional phase in the Canadian shield (Fig. 2). Some of these (e.g., the Huronian and the Hurwitz) were laid down on continental crust but are on average thicker than the thickest sequences formed on continental crust in exogeosynclines during the Plate Tectonic regime. This may indicate that the earliest rigid continental lithosphere was generally thin.

Aulacogens (failed oceans?)

Other very thick, generally unfolded, sedimentary and basaltic sequences outcropping in narrow troughs which are locally radially disposed within cratonic areas have been termed aulacogens from the Russian word for graben (Hoffman, 1971). These we suggest may represent abortive rift valley-ocean systems which, after filling with basalt and sediment, have become part of the continental crust. Burke & Whiteman (this volume) have shown that numerous rift valley systems initiated within the last 200 m.y. have failed at various stages thus never reaching the complete ocean opening stage. However, Wilson (1968) pointed out that a cyclical process of ocean opening and closing dominates the Plate Tectonic regime and the history of the Benue trough (Burke *et al.*, 1971; Grant, 1971) indicates that even quite small oceans will close in response to changes in plate geometry. The persistence of tectonic features representing slightly opened oceans in the aulacogens thus presents a problem. We suggest that the reason such aulacogens as the Great Slave, the Goulburn and the Keewanan (Fig. 2) have not been destroyed by ocean closure is that they have been filled by accumulations of sediment and basalt. This material is less dense than ocean floor material and is not subductable. It has been preserved within the continents as effectively new continental crust. Although aulacogens are not restricted to the transitional phase they appear to have been particularly well developed during that time and are perhaps evidence of cratonic instability.

Plate Tectonic Phase

General

The theory of plate tectonics applies to the present structure of the lithosphere and can be extrapolated backward through time with some rigour as far as old rotation poles can be accurately located, which is roughly a maximum of 200 m.y. Before that time application of plate theory rests largely on the uniformitarian assumption that rocks and structures of the past were formed in plate environments like those in which similar rocks and structures occur today. Palaeomagnetic and climatic data provide some additional information. Analyses of mountain building in terms of plate tectonics have been made by many authors and it is unnecessary to review them here. Plate systems appear to have evolved since they first developed $2{\cdot}0 \times 10^9$ years ago. The oldest high rank blue schists and ophiolite sequences are no older than 250 and 600 m.y. old respectively and while this may be in part an accident of preservation, the place of the ophiolites in older orogens appears to have been taken by rather different assemblages ('pseudo-ophiolites') with ultramafics of a wider range of compositions and structures and without sheeted complexes. Plate margins were perhaps less sharply defined in the early phases of the regime when lithosphere thicknesses were generally less than those of today (Fig. 1).

Continental collision (Himalayan) orogenies dominate the later Precambrian because although there were island arc and cordilleran (Andean) orogenies the record of them has been overprinted during later continental collision.

Himalayan Orogeny

The Himalayas constitute the only active fully developed collision orogen of the present time and we distinguish five Himalayan tectonic zones analogues of which can be recognized in the Precambrian. From north to south these are: The Tibetan plateau; the Indus suture; the mobile core; the gliding nappe zone and the exogeosyncline or foreland basin of the Indo-Gangetic plain.

The Indus suture (Gansser, 1966) marks the place where the Tethyan ocean floor has been squeezed out by the collision of Indian and Asiatic continental lithosphere. The ophiolitic rocks marking the suture have been mainly squeezed upward and outward (some which have glided on thrusts are preserved far to the south of the suture on the tops of the Kiogas). Gansser (1964) has drawn a section showing the narrow nearly vertical ophiolites of the Indus suture wedging out downwards. If he had extended his section a few kilometres lower he might have shown the Indian continent directly against the Asian continent without any preserved rocks to indicate that an ocean had closed along the line of the suture. We suggest that such cryptic sutures occur in the Precambrian where they can be recognized by following them along strike to a place where rocks indicative of ocean closure are preserved (see for example Fig. 3).

The high Tibetan Plateau north of the suture we interpret as a product of continued convergence of the Asian and Indian plates after closure of the Tethyan ocean. The average height of the plateau is over 4 km above sea level and the depth of the Moho below it (Cummings & Schiller, 1971) is 60 to 80 km. The plateau is marked by numerous shallow (< 70 km) earthquake foci (Barazangi & Dorman, 1970; Wilson, 1972) aligned roughly parallel to the trend of the Himalaya and to the trend of reverse faults which bring up crystalline rocks through the shelf sediment succession that outcrops over much of the plateau (Gansser, 1964). The plateau is exceptional among modern orogenic features in being almost as wide (~ 600 km) as it is long (~ 1000 km). We suggest that deeply eroded equivalents of the Tibetan plateau may be represented in the mobile belts of the Precambrian. These belts, for example: the Grenville, the Pan-African, parts of the Churchill (Anhaeusser *et al.*, 1969: Pettijohn, 1970) are characterized by large areal extent rather than linear development and by reactivation of crustal material. That is rocks from them yield older whole rock Rb/Sr ages than their rock and mineral K/Ar ages.

We suggest that similar reactivation is going on now below Tibet. Since the last subductable ocean floor disappeared and the Indian and Asian continents collided, continued convergence has forced the Asian lithosphere to override itself. This is taking place mainly in the crust (hence the shallow earthquakes) and has led to an approximate doubling of the crustal thickness. If the Tibetan plateau is in isostatic equilibrium its elevation requires a crust about 70 km thick comparable to that indicated from seismic studies.

The overthrusting by which the crust below the Tibetan plateau appears to have been thickened has presumably thickened the upper crust in which potassium, uranium and thorium are concentrated so that the thermal gradient below the plateau is unlikely to be less than average and probably exceeds 20°C/km. If this is the case large volumes of old crust below a depth of about 40 km are likely to be partly molten and K/Ar clocks are now being reset over an extensive area below the plateau. S. Hedin (1916) reported Quaternary volcanoes in Tibet which might be a part of this igneous activity. We consider that the Tibetan plateau represents an analogue of the extensive reactivated areas of the Precambrian shields for which no uniformitarian analogue has previously been suggested. Partial melting of continental crust in an environment like the Tibetan plateau to produce granite is capable of leaving residual material including anorthosite (Green, 1969) and granulite at depth. Although these rocks are not confined to reactivated areas in the Precambrian there are major concentrations of granulite and anorthosite in such reactivated areas as the Mozambique and the Grenville.

Reactivated basement produced in a different way occurs in the mobile cores of cordilleran orogens but these zones are typically linear and only locally extend more than 200 km across the strike of the mountain belt.

The mobile core and gliding nappe (Dewey & Bird, 1970) zones of the Himalaya (Gansser, 1964) are closely analogous to similar zones in the Alps, Andes, Caledonides and other Phanerozoic mountain belts. It appears to be a consequence of erosion that these high level zones characterized by high T/P metamorphism and polyphase deformation which are the most studied parts of Phanerozoic mountain belts are relatively rarely well preserved in the Precambrian collision orogens. On the other hand the foreland basin or exogeosyncline which contains the classic flysch/molasse sequences generally overlying miogeoclinal wedges and which suffers relatively little during orogeny is in some places very well preserved (see Fig. 3) in the Precambrian.

Sutures around the Ungava and West African cratons

A comprehensive review of Precambrian collision orogenies is at present impossible. We draw attention here to two examples, one from Africa and one from North America.

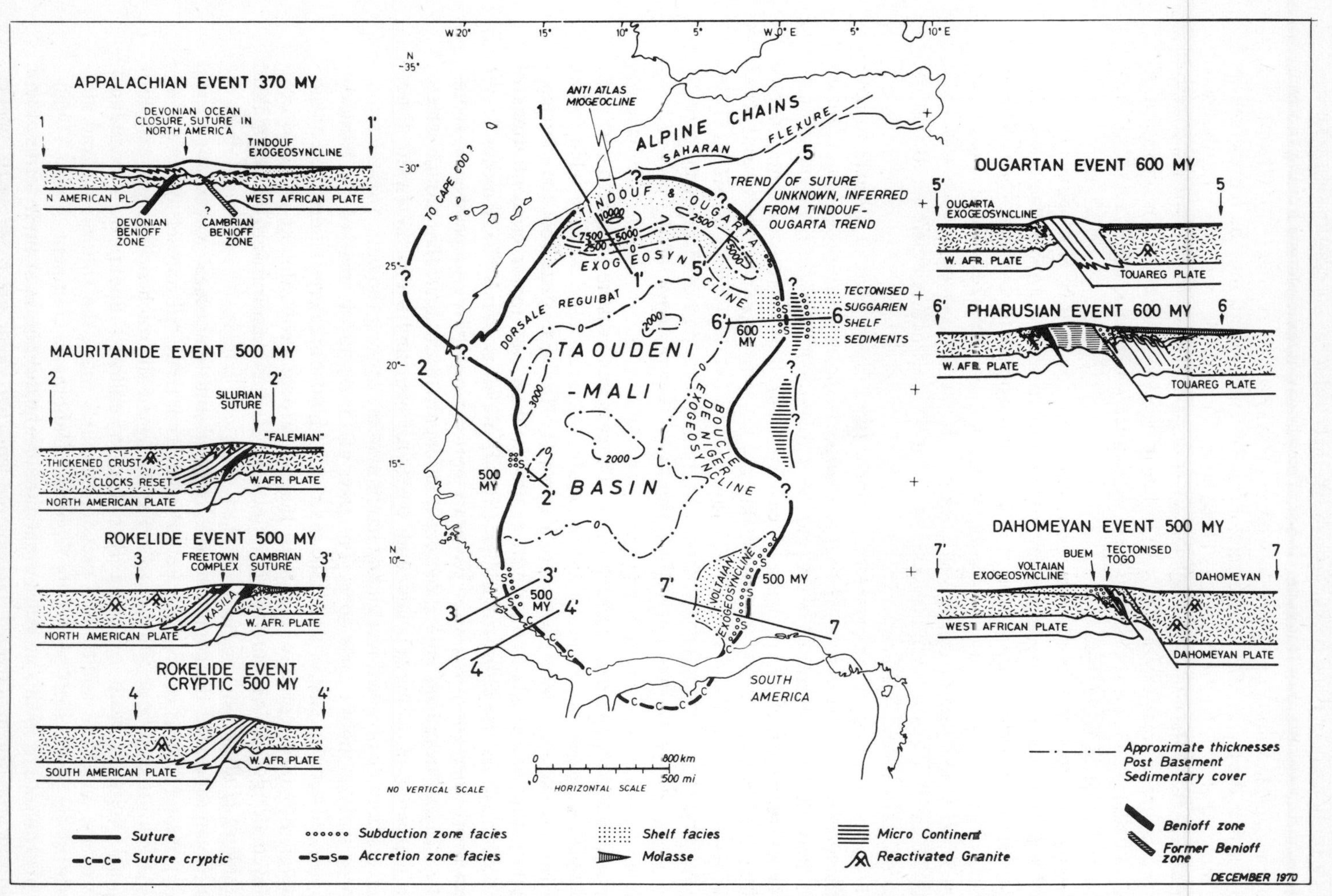

FIGURE 3 Pan-African collision orogenies around the West African craton. (From unpublished work by Burke and Whiteman). Exogeosynclinal or foreland basins are developed on the West African craton which is surrounded by reactivated basement interpreted as produced by continental thickening (Tibetan mechanism). The suture between these two groups of rocks shows 'eugeosynclinal' developments in the Pharusien, the Buem and the Rokellides but in some intervening areas it is cryptic and reactivated basement abuts against unreactivated basement.

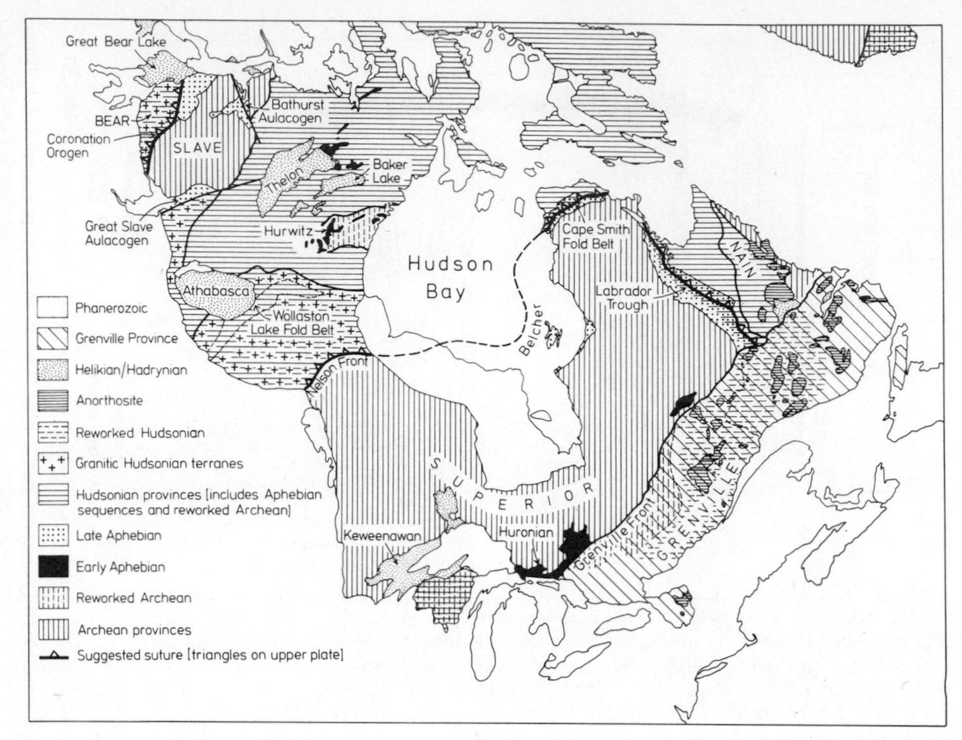

FIGURE 4 The Canadian shield showing the circum-Ungava suture.

Although separated in time by about $1{\cdot}2 \times 10^9$ years, both can be compared with the modern Tibetan system.

Wilson (1968) and later Gibb & Walcott (1971) have described the circum-Ungava suture in Canada as a product of Precambrian collision orogeny. A west to east section across the Labrador trough (Fig. 4) drawn at about 56° N passes from Superior province gneisses ($2{\cdot}7 \times 10^9$ years old) corresponding to peninsular India on the Himalayan model, onto Proterozoic sediments of the Labrador trough which were deposited in a miogeocline and overlying exogeosyncline. Eastward these sediments become involved in a gliding nappe system carrying tholeiitic volcanics and gabbros from a mobile core zone. The area of the suture is marked by pseudo-ophiolitic ultramafics east of which the reactivated Hudsonian crystalline rocks of the Churchill province represent a Tibetan plateau environment. Similar relations are displayed along a section from south to north across the Nelson front (Fig. 4) on the west side of Hudson Bay.

An ocean separated the continents of 'Birrimia' and 'Dahomea' which now form part of West Africa (Figs. 3, 5) in Late Precambrian times. Closure of this ocean in Pan-African times (about 500 m.y. ago) has preserved the miogeoclinal and exogeosynclinal Voltaian basin overlying Birrimian crystallines. Buem rocks including greywackes, andesites, spilites and pseudo-ophiolites mark the suture zone and are overthrust by Togo shelf sediments from the Dahomean continent. Grant (1969) provided an outline of this orogenic event although he did not discuss it in terms of an ocean closure. The reactivated Dahomeyan gneisses yield $2{\cdot}0 \times 10^9$ years whole

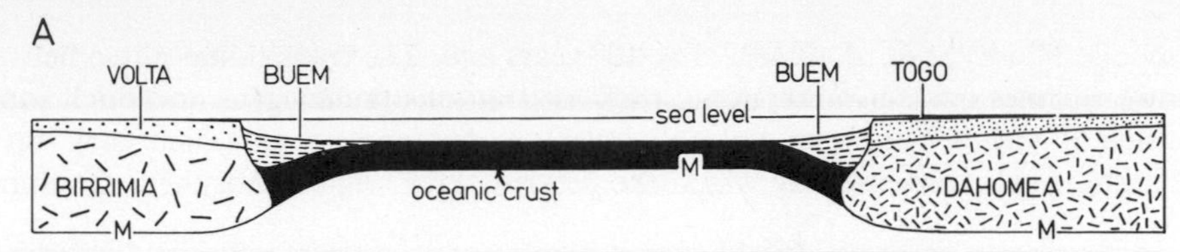

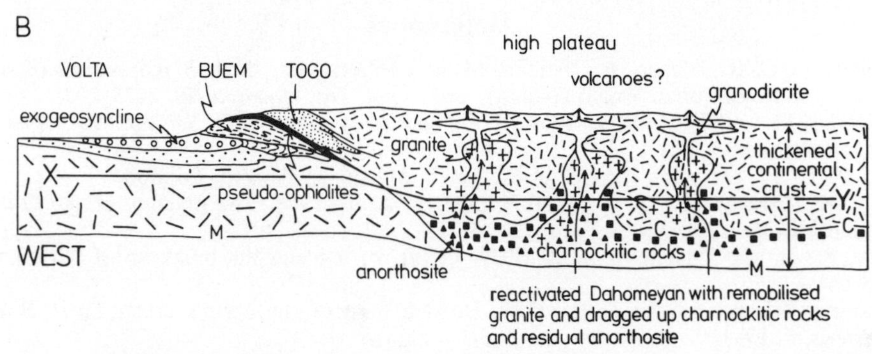

FIGURE 5 Collision orogeny at the Dahomey suture in Pan-African times. (A) The continents of Birrimia and Dahomea are envisaged as separated about 800 m.y. ago. Togo and Voltaian shelf sediments accumulated on the continents and the Buem represents deeper water accumulations. (B) After the continents collided, the Buem rocks came to occupy a suture between the Voltaian exogeosyncline and the Togo shelf rocks. The Dahomeyan was formed by thickening of continental crust which led to remobilization of 2.0×10^9 year old rocks; resetting of K/Ar ages; production of new granite and granodiorite at high levels and charnockite and anorthosite at depth. The section X-Y shows how erosion to a deep level would expose the site of the continental collision as a cryptic suture.

rock Rb/Sr ages and Pan-African (500 m.y.) mineral K/Ar ages. We interpret the Dahomeyan as an analogue of the Tibetan plateau.

The Dahomey suture is shown in relation to the rest of the margin of the West African craton in Fig. 3, from an unpublished study by Burke & Whiteman. Along strike from Dahomey, for example in Brazil and Liberia, there are parts of the suture where no rocks of 'eugeosynclinal' facies, like the Buem, outcrop. Reactivated continental crust is seen in contact with unreactivated crust. These areas are cryptic sutures and we suggest that they are common features of collision orogeny.

Conclusions

Only the most tentative conclusions are realistic at this state in the study of Precambrian plate tectonics. We suggest that there is evidence indicating that collision orogenies took place during the later Precambrian and that sutures (including cryptic sutures) and Tibetan type reactivated areas are important elements of these orogens. Precambrian exogeosynclines or foreland basins, because they were relatively undisturbed in orogeny provide a useful indicator of the orogenic process especially where (as for example in the $2{\cdot}0 \times 10^9$ year old Coronation orogen) they record a change in sedimentary polarity from craton to tectonic land derivation (Fig. 1).

There is evidence of secular development of the Plate Tectonic process through the later Precambrian and of the existence of a radically different tectonic regime (the

Permobile Phase) prior to about $2 \cdot 5 \times 10^9$ years ago. The transitional phase between the two regimes contains exceptional rock associations (aulacogens and thick supracontinental sedimentary and volcanic piles) understanding of which may reveal something of the processes by which the present thick, rigid lithosphere developed.

References

Anhaeusser, C. R., Mason, R., Viljoen, M. and Viljoen, R., 1969. A reappraisal of some aspects of Precambrian Shield Geology, *Bull. Geol. Soc. America*, **80,** 2175–200.

Barazangi, M. and Dorman, J., 1969. World seismicity maps, *Bull. Seismol. Soc. America*, **59,** pp. 369–380.

Bott, M. H. P., 1971. The Interior of the Earth, 316 pp. E. Arnold, London.

Burke, K., Dessauvagie, T. F. J. and Whiteman, A. J., 1971. The opening of the Gulf of Guinea, *Nature*, *Physical Science*, **233,** 51–5.

Burke, Kevin and Whiteman, A. J., 1972. Uplift, rifting and the break-up of Africa, this volume.

Cummings D. and Schiller G. I., 1971. Isopach map of the earth's crust, *Earth Science Reviews*, **7,** 97–110.

Dewey, J. F. and Bird, J. M., 1970. Mountain belts and the new global tectonics, J. *Geophys. Res.*, **75,** 2625–47.

Gansser, A., 1964. The Geology of the Himalayas, 289 pp. Interscience, New York.

Gansser, A., 1966. The Indian Ocean and the Himalayas, *Eclog. Geol. Helvetiae*, **59,** 831–48.

Gibb, R. A. and Walcott, R. I., 1971. A Precambrian suture in the Canadian Shield, *Earth Planet. Sci. Letts.*, **10,** 417–22.

Grant, N. K., 1969. The Late Pre-Cambrian to Early Palaeozoic Pan-African Orogeny in Ghana, Togo, Dahomey and Nigeria, *Bull. Geol. Soc. America*, **80,** 45–56.

Grant, N. K., 1971. South Atlantic, Benue Trough and Gulf of Guinea Cretaceous Triple Junction, *Bull. Geol. Soc. America*, **82,** 2295–8.

Grantham, D. R., 1958. Discussion on Great Dyke, *Proc. Geol. Soc. London* No. 1555, p. 7.

Green, T. H., 1969. High pressure experimental studies on the origin of anorthosite, *Can. J. Earth Sciences*, **6,** 427–40.

Goodwin, A. M., 1971. Metallogenic patterns and the evolution of the Canadian Shield. *In:* Symposium on Archaean Rocks, edited by J. E. Glover, *Geol. Soc. Australia*, *Spec. Publ.*, **3,** 157–74.

Goodwin, A. M., 1972. Plate tectonics and the evolution of the Precambrian crust, this volume.

Hedin, Sven, 1916. Southern Tibet, 9 vols. Lithographic Inst. Swedish Army, Stockholm.

Hoffman, P., 1971. Coronation Geosyncline, *Abstracts Geol. Soc. Meeting*, Washington, Nov. 1971.

Lubimova, E. A., 1969. Thermal history of the Earth, The Earth's Crust and Upper Mantle, *AGU Monograph*, **13,** 63–7.

Pettijohn, F. J., 1970. The Canadian Shield—A status report. *In:* Basins and Geosynclines of the Canadian Shield, *Geol. Surv. Canada Paper*. **70–40.**

Jakes, P., 1972. Model for geochemistry of continental growth, this volume.

Strangway, D. W., Pearce, G. W. and Gose, W., 1972. Lunar magnetic history, *EOS* (*Trans. Amer. Geophys. Union*), **53,** 431.

Sutton, J., 1967. The extension of the geological record into the Precambrian, *Proc. Geol. Ass.*, **78,** 493–534.

Sutton, J., 1971. Some developments in the crust. *In:* Symposium on Archaean rocks, edited by J. E. Glover, *Geol. Soc. Australia Spec. Publ.*, **3,** 1–10.

Viljoen, M. J. and Viljoen, R. P., 1971. The geological and geochemical evolution of the Onverwacht group. *In:* Symposium on Archaean rocks, edited by J. E. Glover, *Geol. Soc. Australia Spec. Publ.*, **3,** 133–50.

Whiteside, H. C. M., 1970. Volcanic rocks of the Witwatersrand Triad. *In:* African Magmatism and Tectonics, edited by T. N. Clifford and I. G. Gass, pp. 73–88. Oliver and Boyd, Edinburgh.
Wilson, J. T., 1968. Static or mobile earth. *In:* Gondwana Revisited, *Proc. American Phil. Soc.*, **112**, 309–20.
Wilson, J. T., 1968. *In:* Science, History and Hudson Bay, *Dept. Energy, Mines and Resources*, p. 1015. Ottawa.
Wilson, J. T., 1972. Mao's Almanac, *Saturday Review*, Feb. 19.

10.5

A. M. GOODWIN
Department of Geology,
University of Toronto,
Toronto, Ontario, Canada

Plate Tectonics and Evolution of Precambrian Crust

Introduction

According to the plate tectonic model, new crust is developed from a mantle source by way of (1) outward spreading oceanic ridges at accreting plate boundaries and (2) downward plunging subduction zones at consuming plate boundaries. New sialic material is added mainly to existing continents; crustal addition of notable calc-alkaline affinity is vividly executed at volcanic island arcs situated at continental-oceanic interfaces. Plate boundaries are distributed in recognizable global patterns (Morgan, 1968) which feature complementary oceanic ridges, with transform faults, and arcs. The arcs may be island arcs or continental arcs, depending on their setting. Assuming that plate tectonics has been operating in Mesozoic–Cenozoic time, the question arises whether it, or a variant mechanism, operated in the more distant geological past. Obviously the question becomes more difficult to resolve with successive geological eras and is most difficult in the Precambrian.

In lieu of direct rigorous analysis of plate motion, any evaluation of the plate tectonic model to evolution of Precambrian crust depends primarily upon geological analogy. With due allowance for possible secular changes in mantle-crust relations since early Precambrian time the current examination is directed to recognition of reasonable evolutionary analogies rather than exact analogies. Unequivocal answers are unlikely to be forthcoming at this early stage of examination.

In this vein two supracrustal elements of Precambrian crust; (1) Archean volcanic belts (c. 2700–3100 m.y.) and, (2) early Proterozoic banded iron formation (c.1900–2100 m.y.) have been examined for possible evidence of plate motion. The examination has led to recognition of certain petrogenetic and distribution patterns within Precambrian crust which are compatible with some form of plate motion. This does not preclude eventual application of some other model to explain them. At present, however, the plate tectonic model offers a satisfactory explanation.

Archean Volcanic Belts

Introduction

Archean volcanic-rich belts of the Canadian Shield and elsewhere present close analogies with modern island arcs and fold belts situated at consuming plate boundaries. The analogies are expressed in terms of igneous affinities, geochemical patterns, stratigraphic successions, parallelism of successively younger belts, and sedimentary associations.

The classical field studies of Cooke *et al.* (1931), Gunning & Ambrose (1940), M. E. Wilson (1941) and others contributed greatly to an understanding of the physical relations of Archean volcanic rocks and established their gross similarities to modern volcanic rocks. Recent studies have served to emphasize a close comparison with modern island arcs and fold belts.

Goodwin (1961, 1962) reported on volcanic relations in the Michipicoten area as representative of Superior province 'greenstone' belts. Michipicoten volcanics are distributed in mafic to felsic volcanic cycles of the andesite-rhyolite association. Chemical compositions of typical Michipicoten volcanic rocks were presented, their low K_2O content noted, and close similarity to volcanic rocks of the Cascade Province of the Pacific border illustrated by means of variation diagrams.

Wilson *et al.* (1965) reported on 261 new analyses from ten volcanic belts in Superior Province and compared these with various volcanic associations. The comparison shows that Archean volcanic rocks of all ten belts belong to the basalt-andesite-rhyolite association, typical of continental orogenic belts or island arc systems.

Several detailed stratigraphic and geochemical studies of Archean volcanic assemblages supported by hundreds of chemical analyses for major and minor elements have been published in recent years. Included are the Yellowknife area (Baragar, 1966), Birch–Uchi Lakes area (Goodwin, 1967), northwestern Noranda area (Baragar, 1968), Lake of the Woods–Manitou-Wabigoon area (Goodwin, 1970), Kirkland Lake area (Ridler, 1969), and the Rice Lake–Gem Lake greenstone belt in southeastern Manitoba (Weber, 1970). Review papers of Archean volcanic stratigraphy and geochemistry have been published by Goodwin (1968) and Baragar & Goodwin (1969).

These and other workers are contributing to a growing body of stratigraphic and geochemical data on Archean volcanic rocks of the Canadian Shield. Similar studies in other shields of the world including those of Anhaeusser *et al.* (1969) and Viljoen & Viljoen (1971) in Africa and Glikson (1971) in Western Australia and others are increasing our understanding of Archean volcanism. A salient result has been to strengthen the analogy with that of modern island arcs and fold belts at consuming plate boundaries.

Distribution and character

Archean volcanic belts are widely distributed across the Canadian Shield. The units range from isolated ribbons a few kilometers long to large irregular belts several hundred kilometers long. They are characteristically east-trending in Superior Province, north-trending in Slave province, and northeast to east-trending in Churchill province. The largest continuous belt, the Abitibi in northwest Quebec and northeast Ontario, is 650 km long and 150 km broad (Goodwin & Ridler, 1970).

Archean volcanic belts contain a wide variety of supracrustal and plutonic rocks. Volcanic rocks are the principal components. They comprise flows and pyroclastics of the basalt–andesite–dacite–rhyolite association typically arranged in mafic to felsic cycles or sequences (see Goodwin, 1968, p. 3, Fig. 4). The volcanics are intercalated, especially in upper stratigraphic parts, with volcanogenic sediments, mainly greywacke, argillite, tuff, conglomerate and cherty iron formation. Granitoid plutons or diapirs are common. Mafic to ultramafic intrusions in the form of discrete sills, dikes and small irregular plutons are widespread although in small proportions.

A variety of volcanic rocks is present in the belts. Flows and pyroclastics of tholeiitic to calc-alkaline chemical affinity predominate but some alkalic rocks are present at Kirkland Lake in the Abitibi belt. Basalt lava flows and associated mafic intrusions predominate in lower stratigraphic parts. Andesitic flows and pyroclastics are intercalated with basalt and, in many assemblages, increase proportionately upwards in the assemblage. Felsic rocks, mainly pyroclastics but locally massive lava flows as at Noranda, are generally present in upper stratigraphic parts. Many Archean assemblages display a single generalized mafic to felsic sequence; some have, in addition, a mafic capping; still others display two or more superimposed mafic to felsic cycles. The total stratigraphic thickness is in the order of 12,000 m but locally attains 20,000 m (Ridler, 1969).

Within Archean volcanic belts the felsic volcanic portion is commonly concentrated in clusters each representing an ancient volcanic centre. Remnants of structurally deformed individual volcanoes have been identified within the belts. Although mineralogically altered in large part Archean volcanic rocks have retained original textures and structures to a remarkable degree.

Volcanic Classes

The abundance of volcanic classes in three representative Archean volcanic belts in Superior Province are as follows (in percentages):

	Mafic		Felsic
Volcanic Belt	Basalt	Andesite	Dacite + Rhyolite
Birch–Uchi	57·8	29·2	13·0
Lake of the Woods–Wabigoon	62·3	25·8	11·9
Timmins–Noranda	54·9	37·1	8·0
Average Superior Province	58·3	30·7	11·0

The proportions of volcanic classes, in summary form, are basalt:andesite:felsics = 6:3:1. The proportions of Archean andesite and felsic volcanics are somewhat lower than in modern developed island arcs but correspond closely with those of the more primitive or tholeiitic modern island arcs (Jakeš & White, 1971 p. 225).

Chemical Compositions

The average chemical composition of the five main volcanic classes (basalt, andesite, dacite, rhyodacite and rhyolite) in a representative volcanic assemblage (Lake of the Woods–Wabigoon) in Superior Province have been recently published (Goodwin, 1972 Table 3). In general Archean volcanic classes compare closely with Cenozoic island arc tholeiitic and calc-alkaline volcanics (Jakeš & White, 1971 Table 2).

The average chemical composition of volcanic assemblages in three representative belts in Superior Province has been recently published (Goodwin, 1972 Table 4). The three Archean belts are remarkably similar in average chemical composition. They correspond closely to the computed composition of modern 'developed' island arcs (Jakeš & White, 1971 p. 227 Table 3).

Southern Superior Province Volcanic Belts

Southern Superior Province includes three well-defined, east trending Archean volcanic-rich belts named respectively from north to south, Uchi, Wabigoon (Keewatin) and Abitibi (see Goodwin, 1968 Fig. 9). Recent preliminary zircon U-Pb ages of volcanic rocks in the three belts (Krogh & Davis, 1972) reveals a simple pattern with ages of volcanic belts decreasing i.e. younging, to the south in a time interval of 200 m.y. (2950–2750 m.y.).

Thus Archean volcanic belts present close analogies to modern island arcs in terms of (1) prevailing mafic to felsic stratigraphic successions, (2) abundant calc-alkaline volcanism including much highly explosive felsic eruptives, (3) predominant subaqueous accumulation of volcanics and closely associated volcanogenic sediments, and (4) parallel arc-like arrangement of successively younger linear belts.

Abitibi Orogenic Belt

A characteristic feature of Cenozoic island arcs is a systematic lateral variation of volcanic suites; this is marked particularly by increasing alkalinity of components from ocean-side (younger) to continent-side (older) of the arc. Lateral variation is commonly attributed to increasing depth to source magma, a linear function of vertical distance to the continent-inclined subduction zone and at the same time horizontal distance from the ocean-side trench. (Dickinson & Hatherton, 1967.) According to the classical studies of Kuno (1966, 1968) three main volcanic series—tholeiitic, high alumina, and alkaline—are arranged in that order across an arc from ocean-side to continent-side.

Recently, the southwestern portion of the Abitibi orogenic belt in Superior Province of the Canadian Shield has been geochemically studied in detail (Goodwin, in preparation). Evaluation of the results demonstrates that the Abitibi belt contains all the principal volcanic series of modern island arcs. This is demonstrated by reference to six volcanic sections as illustrated in Fig. 1.

(*a*) Localities 1 and 2—Tholeiite: the two localities respectively 50 km south of Timmins in McArthur Twp. (Locality 1) and 80 km east of Timmins in Munro Twp. (Locality 2) are similar in containing peridotite lava flows within a tholeiitic volcanic assemblage. Chemical data is available for the McArthur Twp. tholeiites (Locality 1) and detailed petrographic data on the ultramafic flows in Munro Twp. (Locality 2). The two areas combined illustrate the Archean tholeiitic series.

Ultramafic lava flows exhibiting quench ('spinifex') texture indicative of rapid cooling are unusually fresh and well exposed in Munro Twp. (Pyke, Naldrett and Eckstrand, in preparation). Numerous superimposed ultramafic units ranging from ½ to 17 m thick with a strike length varying from a few metres up to nearly 200 m occur. At least 52 and probably 63 units are well exposed.

As described by Naldrett (in press) a typical unit consists of a thin basal layer, 10 cm thick, over which the grain size decreases downwards progressively; an overlying layer, 1·5 m thick, composed of 75% equant olivine crystals (interpreted as phenocrysts) set in a matrix of acicular clinopyroxene needles and chloritic material apparently pseudomorph after glass; a 1·5 m layer showing the rapid cooling (now called 'spinifex') texture which decreases in grain size upwards; and a very fine-grained, polygonally jointed brecciated layer that is interpreted as a flow top. Small rounded pillow-like bodies occur in some of the thicker units.

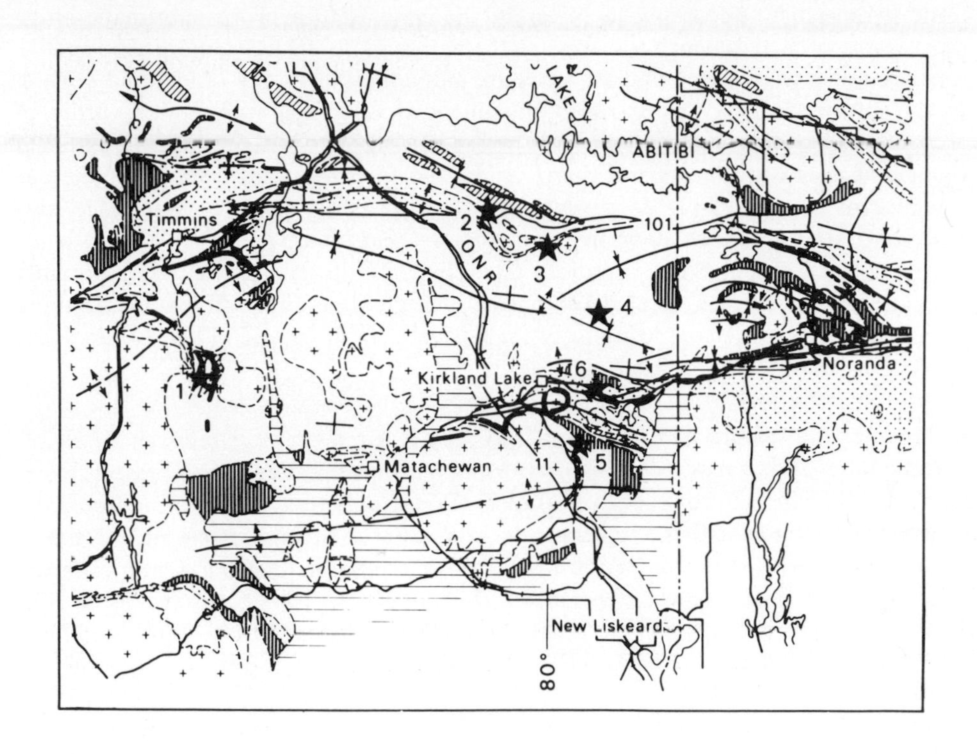

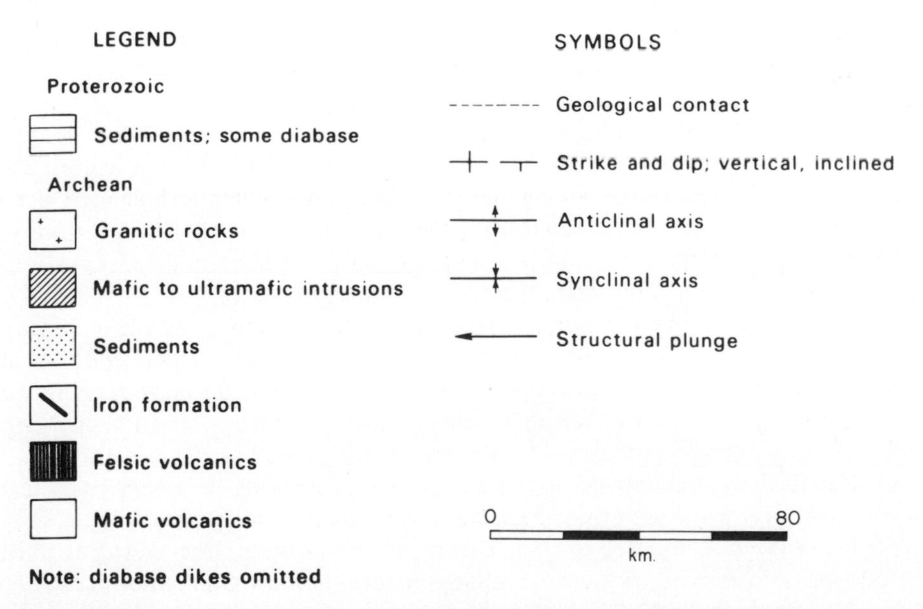

FIGURE 1 Geological map of Timmins-Kirkland Lake-Noranda region in northwestern Quebec and northeastern Ontario illustrating geological relationships in the southwestern portion of Abitibi orogenic belt. The following stratigraphic sections are located and numbered: 1. McArthur Lake, 2. Munro, 3. Garrison, 4. Magusi, 5. Catharine, 6. Timiskaming.

The ultramafic units occur in a sequence of pillowed basalts and andesites, and felsic pyroclastics. According to Naldrett, the concordance of the ultramafic units, their asymmetry, the quench textures and the nature of the surfaces interpreted as flow tops are taken as evidence that the units are individual flows. If so the liquid portion of the ultramafic lava was extruded subaqueously at temperatures in the order of 1350–1400°C. Their association with pillowed basalt together with absence of significant clastic sediments and cherty oxide iron formation has been interpreted as evidence of comparatively deep water environment. This type of assemblage may represent Archean ocean floor material, the Archean analogue of modern oceanic crust. A total alkali–SiO_2 variation plot (Fig. 2) illustrates the predominantly tholeiitic character with local development of high-aluminous affinity in the more felsic parts. The mean compositions of the normal lava flows and of peridotites in the McArthur Lake area are listed in Table 1.

AVERAGE CHEMICAL ANALYSES

1. McArthur Lake Section			McArthur Lake Section la Peridotites			3. Garrison Section		
Sample nos. A.166-184,191,196-202			Samples nos. A.185-190,192-195			Sample nos.A.389-395		
N = 27	mean	st.dev.	N = 10	mean	st.dev.	N = 7	mean	st.dev.
SiO_2	57.1%	6.7	SiO_2	46.3%	2.0	SiO_2	49.5%	2.2
Al_2O_3	16.1	1.7	Al_2O_3	4.8	3.2	Al_2O_3	15.1	0.7
Fe_2O_3	1.8	1.3	Fe_2O_3	4.5	1.9	Fe_2O_3	3.3	1.6
FeO	6.7	3.1	FeO	6.2	2.0	FeO	10.7	1.7
CaO	7.7	3.3	CaO	5.0	2.3	CaO	9.7	1.4
MgO	6.0	2.9	MgO	31.6	7.5	MgO	5.8	0.9
Na_2O	2.8	1.1	Na_2O	1.1	1.4	Na_2O	3.5	0.7
K_2O	0.95	1.0	K_2O	0.01	0.0	K_2O	0.48	0.3
TiO_2	0.74	0.2	TiO_2	0.35	0.08	TiO_2	1.84	0.5
MnO	0.16	0.05	MnO	0.18	0.01	MnO	0.22	0.03

4. Magusi Section			5. Catharine Basalts			6. Timiskaming		
Sample nos.A.409-443			Sample nos.A.954-959			Sample nos.A.903-913,922-930		
N = 35	mean	st.dev.	N = 6	mean	st.dev.	N = 21	mean	st.dev.
SiO_2	60.1%	10.2	SiO_2	47.4%	1.3	SiO_2	56.0%	4.8
Al_2O_3	17.9	1.6	Al_2O_3	14.8	0.9	Al_2O_3	16.7	3.3
Fe_2O_3	1.8	1.4	Fe_2O_3	5.2	1.7	Fe_2O_3	2.9	1.7
FeO	4.6	1.8	FeO	10.9	2.5	FeO	5.1	2.3
CaO	5.1	3.2	CaO	10.4	1.6	CaO	6.1	3.3
MgO	3.6	2.4	MgO	5.9	1.7	MgO	4.7	2.0
Na_2O	4.0	0.9	Na_2O	3.5	0.9	Na_2O	4.0	1.7
K_2O	1.2	0.9	K_2O	0.5	0.2	K_2O	3.6	3.0
TiO_2	0.7	0.3	TiO_2	1.2	0.1	TiO_2	0.8	0.3
MnO	0.12	0.05	MnO	0.3	0.08	MnO	0.18	0.07

TABLE 1 Average chemical analyses of Archean volcanic assemblages in the Timmins-Kirkland Lake-Noranda region of Abitibi orogenic belt. Sample numbers refer to "A" series analyses (Goodwin, in preparation). N = number of analyses; st.dev. = standard deviation. la Peridotites are included in the McArthur Lake section. Column numbers 1–6 correspond to location numbers in Fig. 1. For corresponding stratigraphic thicknesses refer to captions accompanying Figures 2–8.

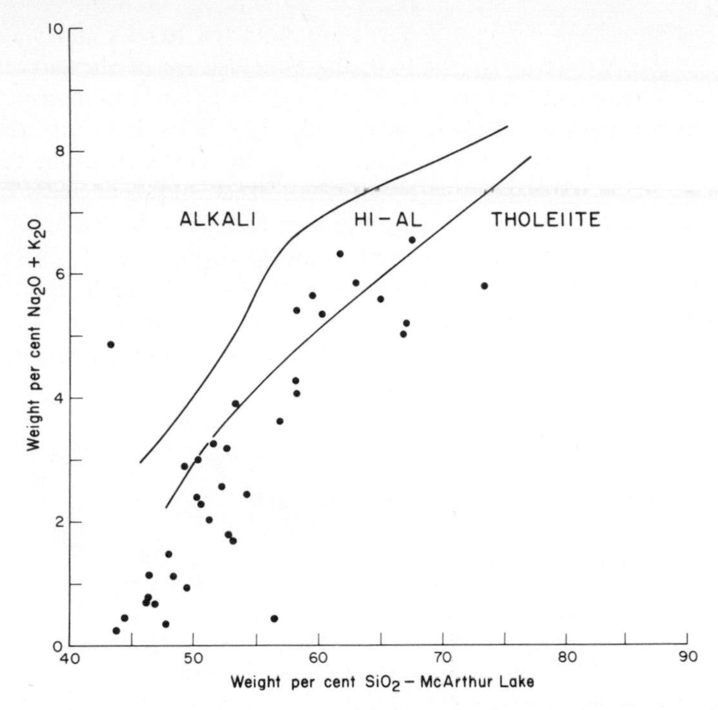

FIGURE 2 Total alkali—SiO_2 relations in the McArthur Lake section. The stratigraphic section represented by the samples is 40,000 feet thick. The predominant tholeiitic character of this suite is illustrated. For average chemical composition see Table 1, column 1.

(*b*) Locality 3—Garrison tholeiitic flows: 20,000 feet of superimposed mafic flows from the lower part of the Blake River Group along the north limb of an east-trending regional syncline 25 miles north of Kirkland Lake.

An MgO–Fe_2O_3 variation diagram (Fig. 3) for this 20,000 foot-thick section illustrates a moderately high iron enrichment trend (Thingmuli trend) which is characteristic of tholeiites. The corresponding mean composition is listed in Table 1, column 3.

(*c*) Locality 4—Magusi high-alumina flows: A thick succession of high-Al mafic flows is present in the upper stratigraphic part of the Blake River Group north of Kirkland Lake. The high-alumina character is well-illustrated in the alkali-SiO_2 plot (Fig. 4) as well as in two alkali-Al_2O_3 plots (Figs. 5a, b). The calc-alkaline character (Cascade or no-iron enrichment trend) of the suite is well-illustrated in Fig. 6. This trend is in sharp contrast to the tholeiitic (Thingmuli) trend in Garrison volcanics (Fig. 3). The mean composition of the Magusi section is listed in Table 1, column 4.

(*d*) Locality 5—Catharine alkali basalt: The Skead Group south of Kirkland Lake includes a substantial thickness of alkali mafic volcanics including basanite and alkali andesite (Ridler, 1969 p. 18–23).

The alkali-SiO_2 variation diagram (Fig. 7) illustrates the alkalic nature of the basalts. The mean composition is listed in Table 1, column 5.

(*e*) Locality 6—Timiskaming trachyte: Stratigraphically overlying the Skead and Blake River volcanic rocks is the Timiskaming Group which includes substantial thicknesses of trachyte, phonolite and trachybasalt (Ridler, 1969 p. 35).

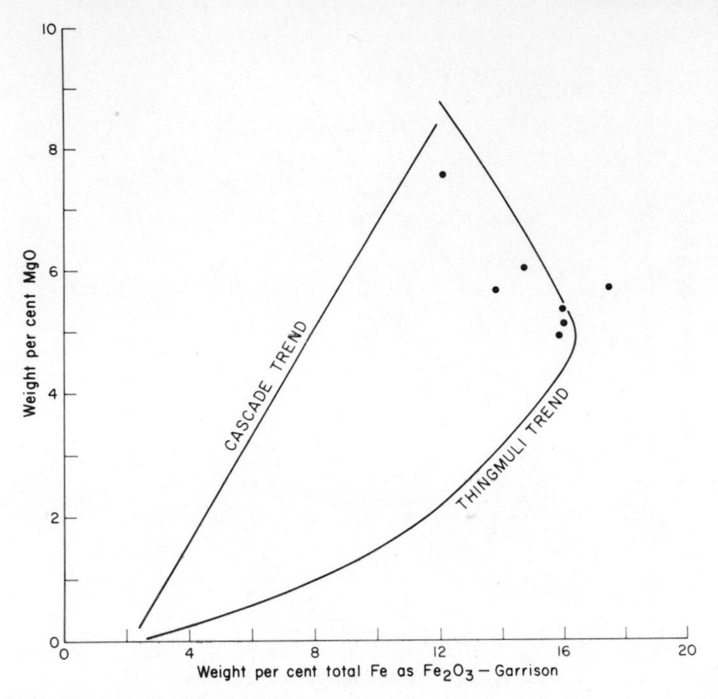

FIGURE 3 $MgO-Fe_2O_3$ variation diagram of Garrison tholeiites illustrating the moderate iron enrichment trend. This corresponds to the Thingmuli trend (Carmichael, 1964). The stratigraphic section represented by the samples is 6100 m. thick. For average chemical composition see Table 1, column 3.

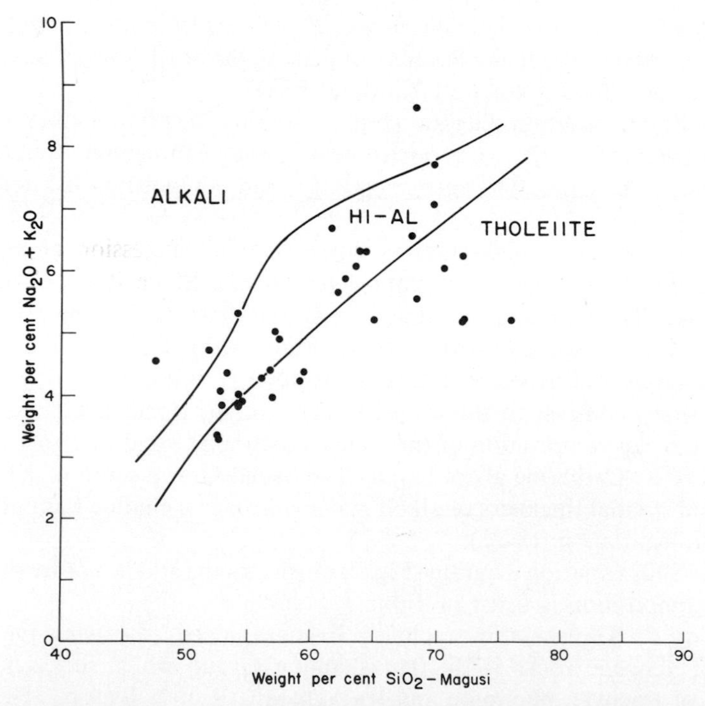

FIGURE 4 Total alkali—SiO_2 relations in Magusi volcanic assemblage. The stratigraphic section represented by the samples is 7925 m. thick. The predominant high-alumina character of this suite is illustrated. For average chemical composition see Table 1, column 4.

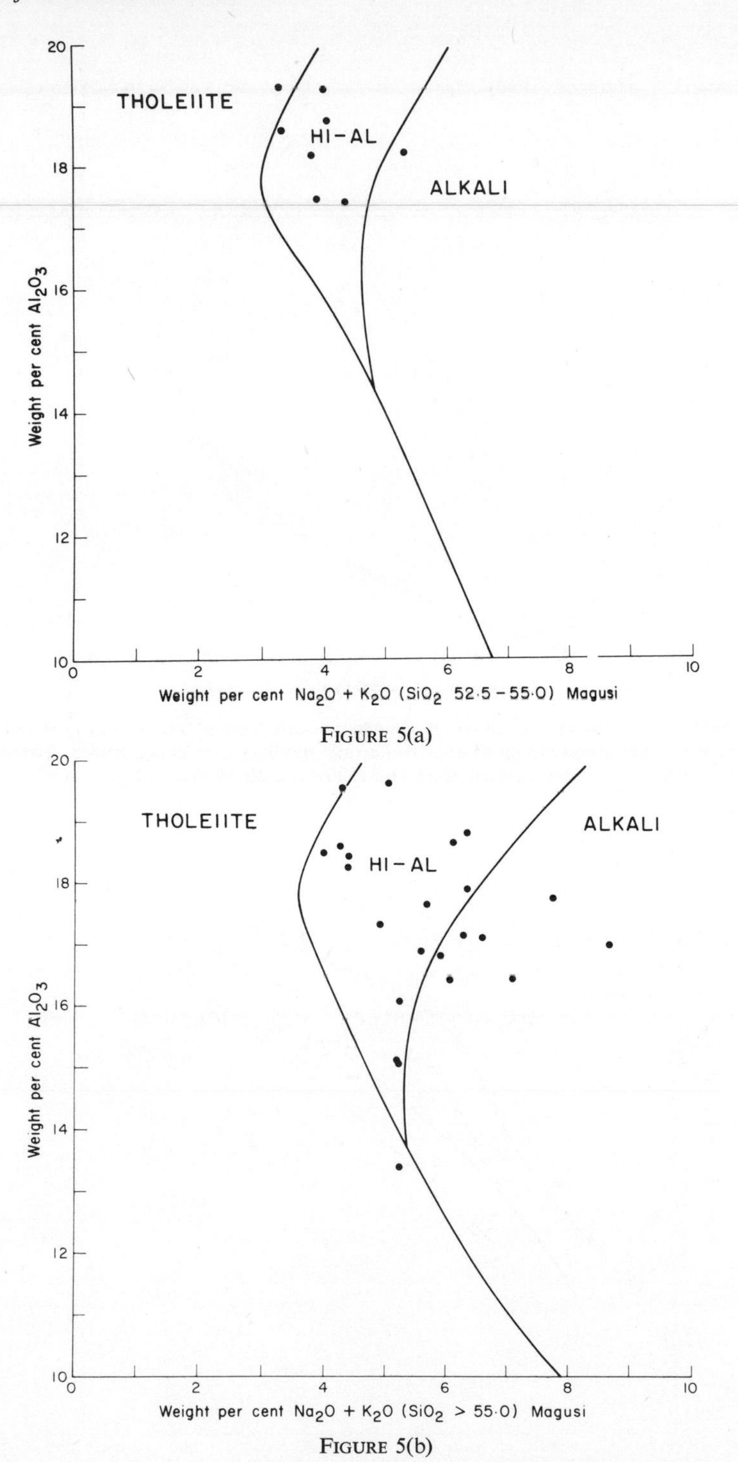

FIGURE 5 Al_2O_3—alkali—SiO_2 relations in Magusi volcanic assemblage (as in Fig. 4) illustrating high-alumina character of the volcanic rocks: (a) SiO_2 range 52.5–55 percent, (b) SiO_2 range greater than 55 percent.

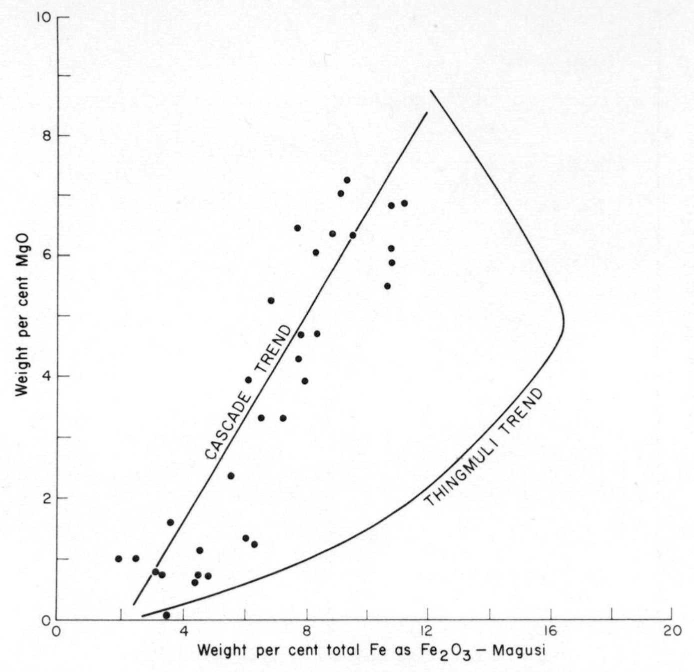

FIGURE 6 MgO–Fe_2O_3 variation diagram of Magusi volcanic assemblage (as in Fig. 4). The predominant no-iron enrichment trend characteristic of the calc-alkaline suite is illustrated. This corresponds to the Cascade trend (Carmichael, 1964).

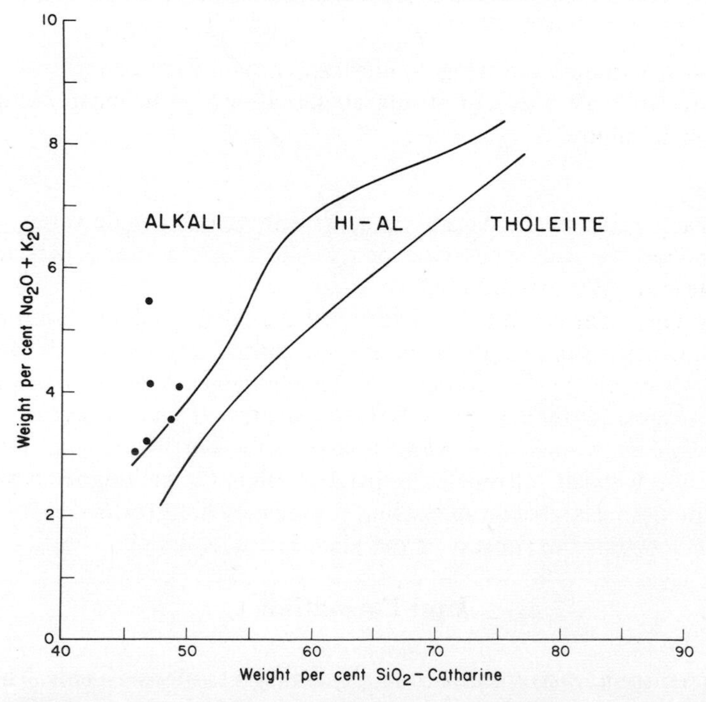

FIGURE 7 Total alkali—SiO_2 relations in Catharine basalt assemblage. The stratigraphic section represented by the samples is 2440 m. thick. It forms a part of the Skead Group of volcanic rocks. The predominant alkalic character of the basalts is illustrated. For average chemical composition see Table 1, column 5.

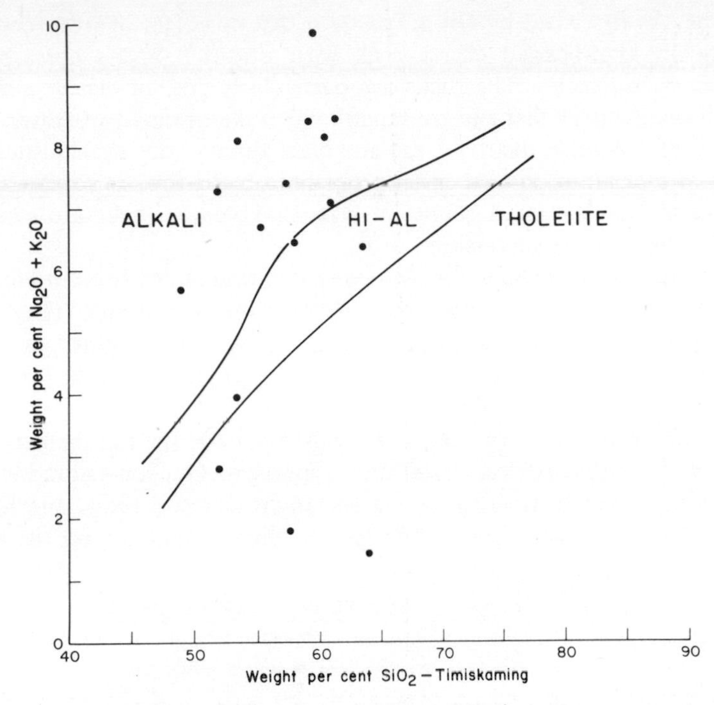

FIGURE 8 Total alkali—SiO_2 relations in Timiskaming volcanic assemblage. The stratigraphic section represented by the samples is 2000–3000 m. thick. The predominant alkalic character of the volcanic rocks is illustrated. For average chemical composition see Table 1, column 6.

An alkali-SiO_2 variation plot (Fig. 8) illustrates a significant concentration of mafic to intermediate volcanic rocks of strong alkalic affinity. The mean composition is listed in Table 1, column 6.

Conclusions

Archean volcanic rocks of the Abitibi orogenic belt include a wide variety of igneous types. The major volcanic types characteristic of modern island arcs are present including tholeiitic, high-Al and alkaline affinities.

Within this particular part of the Abitibi belt the alkali content (K_2O) of volcanic components increases from north (Munro Twp., locality 2) to south (Kirkland Lake, locality 6). This north-south variation may be analogous to Kuno's classical lateral variation of magma types across modern island arcs. If so, the ocean side of the Archean assemblage would have been situated to the north.

In conclusion, Archean volcanic assemblages contain the main igneous series which characterize modern island arcs. Accordingly, they may have formed in a similar type of environment within the context of the plate tectonic model.

Iron Formation

Introduction

Precambrian sedimentary iron formations are common to all continents. The principal type which is particularly common in early Proterozoic rocks, is classified as Lake

Superior type. As described by Gross (1970a p. 20) this class of iron formation is a characteristically thin-banded cherty rock with iron-rich layers corresponding to various sedimentary facies. The rocks are particularly free of clastic material. The textures and sedimentary features are remarkably alike in detail wherever examined. The sequence of dolomite, quartzite, red and black ferruginous shale, iron formation, black shale and argillite, in that order from bottom to top, is so common on all continents as to be considered almost invariable. Volcanic rocks are nearly always present somewhere in the succession.

Continuous belts of Lake Superior type iron formation typically extend for hundreds or thousands of miles along the margins of geosynclines and basins. They apparently formed in fairly shallow water on continental shelves or margins. As stressed by Gross, it is still uncertain whether the iron and silica contained in the iron formation was derived from a land or a marine source.

This type of cherty iron formation is the host rock for the rich hematite–geothite ore bodies of Australia, Africa, Brazil and Venezuela, Quebec–Labrador and Lake Superior region in North America, Orissa and Bihar states in India, and Krivoy Rog and Kursk areas in USSR (Gross, 1970a p. 20). They include by far the largest and

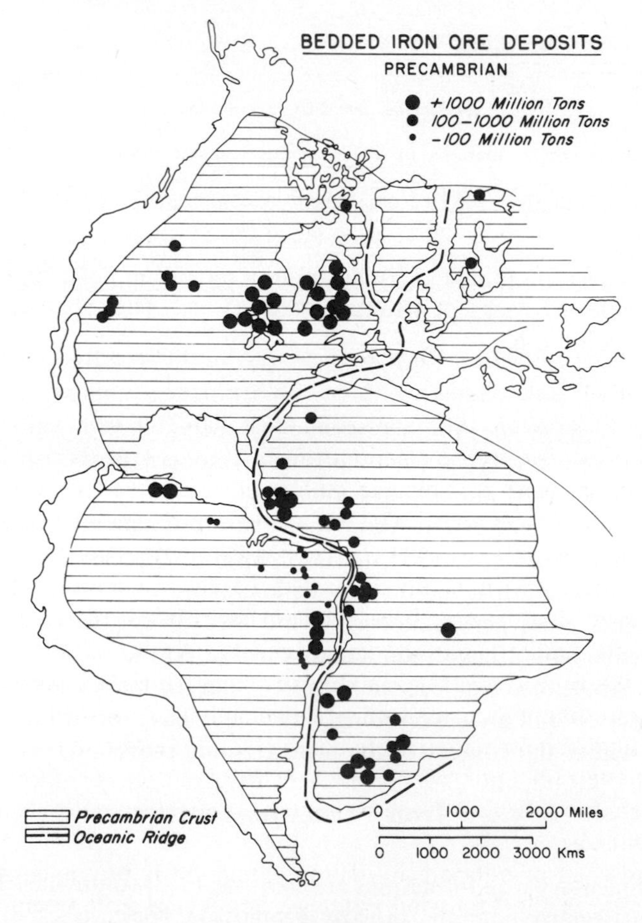

FIGURE 9(a)

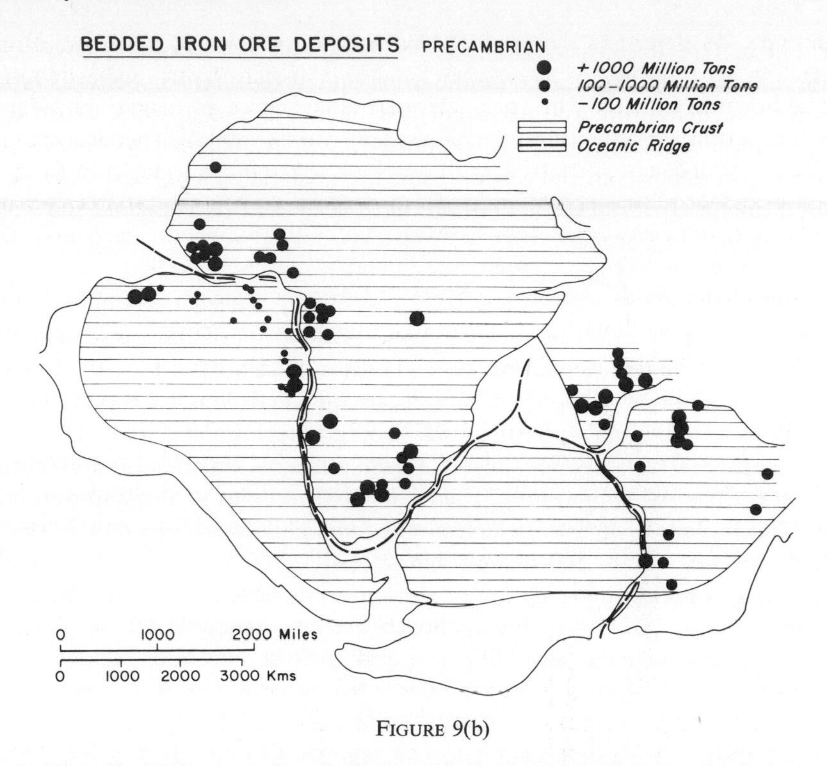

FIGURE 9(b)

FIGURE 9 Plot of bedded iron ore deposits in Precambrian iron formation: (*a*) Atlantic reconstruction (Bullard *et al.*, 1965), (*b*) Gondwanaland reconstruction (Veevers *et al.*, 1971). Iron deposits are plotted in three size ranges. All iron ore data is taken from United Nations Survey of World Iron Ore Resources (1970). The approximate distribution of Precambrian crust including those parts with Phanerozoic cover are illustrated. Oceanic ridges are Cenozoic in age.

economically most valuable iron formations in the world. As discussed below, the largest proportion of Lake Superior type iron formation appears to have been deposited in early Proterozoic time approximately 1900–2100 m.y. ago.

A second class of bedded iron formation, called Algoma type, is present in nearly all early Precambrian belts of volcanic and sedimentary rocks in many shields (Gross, 1970a p. 20). Characteristically, they are thin-banded or laminated with interbands of ferruginous grey or jasper chert, hematite and magnetite. Carbonate and sulphide facies are locally prominently developed. Usually a number of lenses of iron formation are present in a volcanic belt. They are intimately associated with various volcanic rocks and fine grained clastic sediments. As stressed by Gross, the associated rocks indicate a eugeosynclinal environment for their formation and a close relationship in time and space with volcanic activity.

Much Algoma type iron formation is present in older Precambrian rocks particularly Archean basement terrain upon which much of the younger, predominantly Lake Superior type iron formations have been deposited together with their associated supracrustal assemblages.

A possible third class of oolitic bedded iron formation is prominently developed in the Pretoria Series of the Transvaal system in South Africa. It resembles Minette type iron formation (Gross, 1970a p. 20–21) and has been so classified in the Survey of World Iron Ore Resources (U.N., 1970). Although not normally developed in Precambrian rocks, the arenaceous oolitic ironstone of the Pretoria Series is reported to constitute an extraordinarily persistent bed from 1·5 to 9 m thick which can be traced for hundreds of kilometers. It is reported by P. A. Wagner to contain 'one of the greatest iron deposits in the world, if not the greatest' (Haughton, 1969 p. 144). Normally Minette type iron formations are most abundant in Mesozoic and Tertiary rocks, particularly in Europe.

Precambrian Distribution

The distribution of Precambrian bedded iron ore deposits of the world taken from Survey of World Iron Ore Resources (U.N., 1970) is illustrated in Figs. 9a, b and 10c. All types of Precambrian bedded iron deposits are included. Massive Precambrian iron deposits (Bilbao, Magnitnaya, Kiruna and Taberg types) are excluded. The deposits are shown in three size categories: over 1,000 million tons, 100 million to 1000 million tons, and less than 100 million tons. The total listed reserves of Precambrian bedded iron deposits amounts to 469,409 million metric tons. This constitutes 60 % of the listed total world reserves of 782,500 million metric tons.

The greatest concentrations by far of Precambrian bedded iron ore deposits are present in western Australia, India, South Africa, western Africa, Brazil and Venezuela in South America, Lake Superior and Quebec–Labrador regions of North America, and Krivoy Rog and Kursk regions in western USSR. In each of these regions iron formations of different Precambrian ages may be present as for example in Australia, India, Africa, and Lake Superior regions. Commonly iron formations of different ages in a particular region are separated by major unconformities demonstrating that conditions favourable for iron deposition recurred over major intervals of Precambrian time.

Figure 9a illustrates the distribution of Precambrian bedded iron ore deposits in a pre-drift Atlantic reconstruction after Bullard *et al.* (1965). Particularly notable are the iron concentrations in southern Africa, Brazil and Venezuela, and Lake Superior and Quebec–Labrador regions in North America. Figure 9b illustrates corresponding distributions in India and Australia in a pre-drift Gondwanaland reconstruction after Veevers *et al.* (1971). In Fig. 10c the main reserves present in the Krivoy Rog and Kursk regions with lesser reserves in Siberia, China and North Korea are illustrated in a Laurasian reconstruction after Hurley & Rand (1969).

Early Proterozoic Distribution

In Figs. 10a, b, c only ore reserves in iron formations of established or interpreted early Proterozoic age (c. 1800–2100 m.y.) are shown.

Total ore reserves in iron formations of Lower Proterozoic age are calculated at approximately 372,807 million metric tons or 48 % of total listed world reserves. Immense tonnages as listed in the Survey of World Iron Ore Resources (tables accompanying regional appraisals) are present in western Australia, India, South Africa, western Africa, Brazil and Venezuela in South America, Lake Superior and Ungava regions of North America and Krivoy Rog and Kursk areas of USSR. Even though some iron formations may be incorrectly dated the fact is clear that early Proterozoic iron formations of the world contain vast iron ore reserves totalling approximately 50 % of the total listed world reserves.

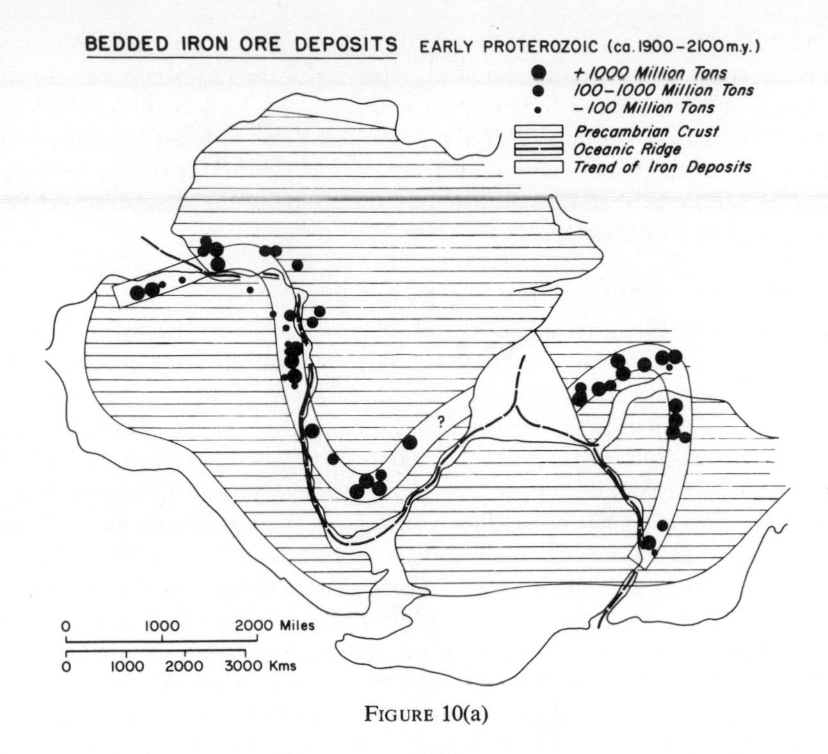

FIGURE 10(a)

In Western Australia, the Hamersley Group which contains the Brockman Iron Formation, includes the Woongarra Volcanics dated at 2000 ± 100 m.y. (Trendall, 1970 p. 5). In South Australia the Middleback Group of iron formation lies at the base of more than 9000 m of metasediments; granulites from the Gneiss Complex yield an age of 1780 ± 120 m.y. which sets a minimum depositional age for this iron formation (Trendall, 1970 p. 6).

In India, iron ores of the Singhbhum Group with an age of 2000–1700 m.y. are to be correlated with the Iron Ore Stage of Dunn and the Iron Ore Series of Krishnan. These iron formations would then correlate with the Australian iron formations of the Hamersley Group in Western Australia (Crawford, 1969 p. 383).

Principal iron ores in South Africa which include some of the largest concentrations of iron in the world are contained in the Dolomite and Pretoria Series and equivalents of the Transvaal System dated between 2300 m.y. and 1950 m.y. (Haughton, 1969). Hematite-magnetite deposits of similar age are present in Namibia (Southwest Africa) within the Kaokoveld of Pre-Damara age as indicated by the age of the Huabian tectonic episode estimated at approximately 1760 m.y. (Haughton, 1969 pp. 286, 479).

Large iron reserves are present in western Africa particularly in Angola, Congo, Gabon, Ivory Coast, Liberia, Sierra Leone and Guinea. In some localities, the exact age of the iron formation is in dispute. Various workers report either an Archean or a Lower Proterozoic age depending upon whether they consider the iron formation to be an integral part of the Archean basement or to belong to young Lower Proterozoic infolds in the basement complex. Thus the Kambui schists of Sierra Leone show whole rock Rb–Sr ages of about 2700 m.y. But the pelitic and iron-bearing metamor-

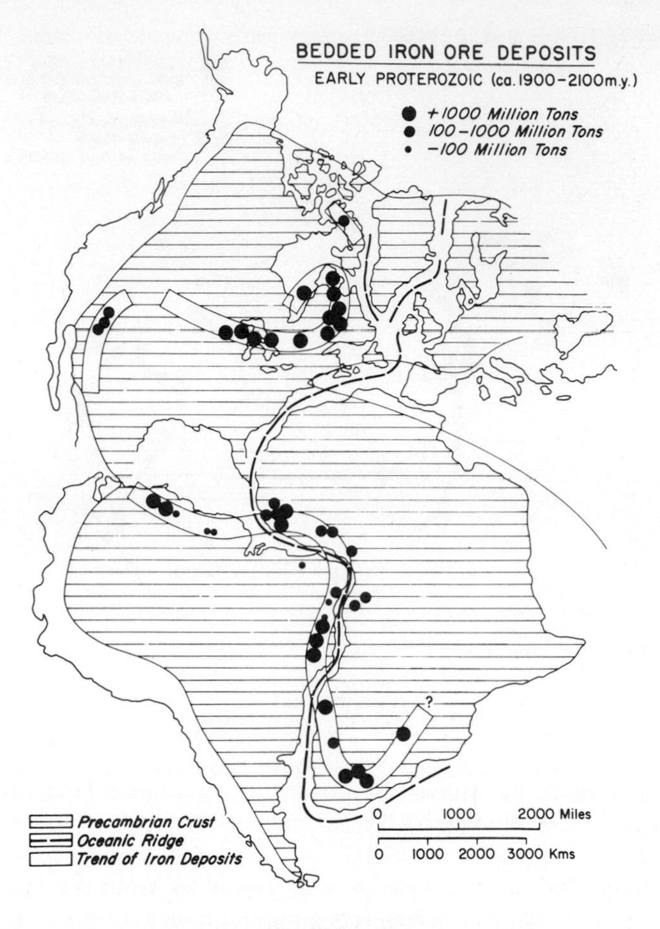

FIGURE 10(b)

phic rocks in the Marampa Formation of Sierra Leone and similar sections in the Nimba Range, Liberia, appear to be about 2200 m.y. old (Hurley *et al.*, 1971). Precise dating of the iron formation in adjoining countries is not yet available. Assessment of available literature indicates that a significant part, if not most of the itabirite, is of Lower Proterozoic age.

In South America, the main Precambrian iron formations contain immense reserves of iron ore particularly in Brazil and Venezuela. The Quadrilatero Ferrifero in Brazil contains one of the largest concentrations of iron ore in the world ranking in this respect with Western Australia. The Itabirite Group of the Minas Series has not been adequately dated. It is known to fall in the interval 2500–1350 m.y. (Dorr, 1969). It is commonly considered to be of Lower Proterozoic (i.e. Middle Precambrian) age. Exposures of itabirite are present northward to the Amazon River. Similar iron-bearing rock in Venezuela contain very large reserves of iron ore. There is controversy concerning the age of Venezuelan iron formation. Bucher (1952) places the iron-ore bearing formation above the older rocks of the Imataca Series. Lopez (1956) also considers the iron ferruginous quartzites to be part of a series which overlies the early

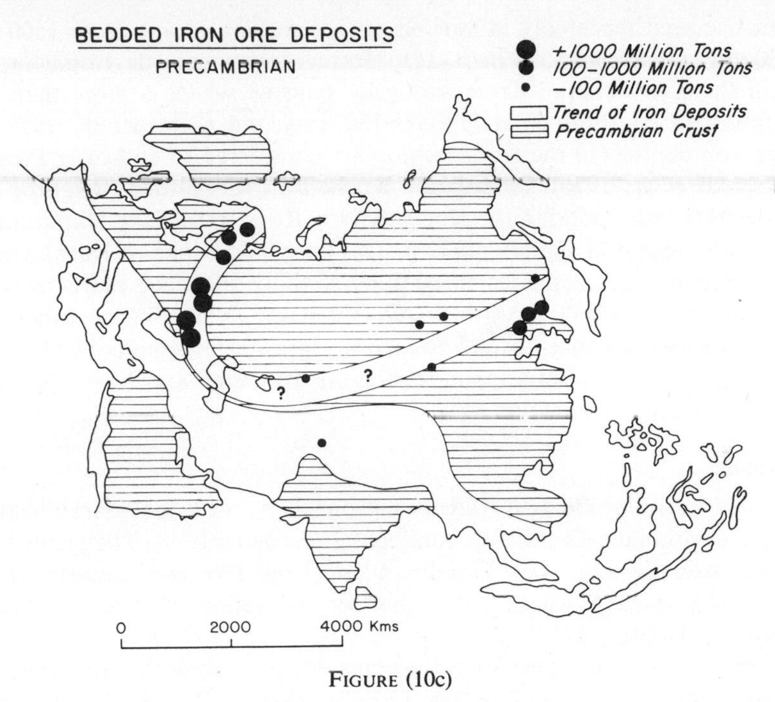

FIGURE (10c)

FIGURE 10 Plot of representative bedded iron ore deposits in early Proterozoic iron formations: (*a*) Gondwanaland reconstructions (Veevers *et al.*, 1971) (*b*) Atlantic reconstruction (Bullard *et al.*, 1965) (*c*) Laurasian reconstruction (Hurley, 1970). Approximate trends of early Proterozoic iron formations are delineated. The approximate distribution of Precambrian crust including those parts with Phanerozoic cover are illustrated. Oceanic ridges are Cenozoic in age. The distribution of iron deposits in southern Asia e.g. India, is not repeated in Fig. 10c.

Precambrian complex (Guyana System). Evidence supporting the younger age for the itabirite is presented by Morrison (in press). However, according to the Stratigraphic Lexicon of Venezuela (2nd ed., 1970), the ferruginous horizons are contained in the intensely metamorphosed and structurally complex sequence of Lower Precambrian; i.e. Archean age.

In North America, the main iron-bearing rocks in the Lake Superior and Quebec–Labrador regions have been comparatively well dated. Although the iron-bearing Animikian rocks of Lake Superior region are known to be younger than 2000 m.y. (Morey, 1970) there is still some uncertainty whether the minimum age is 1685 $\pm$ 24 m.y. (Faure & Kovach, 1969), 1750 $\pm$ 25 m.y. (Peterman, 1966) or 1900 $\pm$ 200 m.y. (Hurley *et al.*, 1961). A corresponding Lower Proterozoic age of 1900–2000 m.y. has been indicated by recent radiometric dating for the cirum-Ungava iron formations including Quebec–Labrador and Hudson Bay regions (Fryer, 1972). Similar ages have been suggested for the Mary's River hematite deposits of Baffin Island (Gross, pers. comm.).

In the eastern part of the Baltic Shield, the ferruginous quartzites of Kola Peninsula and Karelia are reported to be confined almost exclusively to the Lower Proterozoic age of 2600–2000 m.y. (Chernov, 1970) and to everywhere overlie the Archean basement (Goryainov, 1970). Ukrainian Shield data shows that cycles of iron-silica deposition occurred repeatedly in various stages ranging in age from 3500 m.y. to 1800–1700 m.y. (Semenenko, 1970 p. 16). However, the main development of iron deposits in the Krivorozhsko-Kremenchogsky syncline which is more than 200 km long, is found to be approximately 2000–1800 m.y. old (Semenenko, 1970 p. 11). Very large iron deposits in the Kursk region are considered to be Lower Proterozoic age (Plaksenko *et al.*, 1970). Indeed, the Kursk iron formations (Belgorod region) are considered to correspond to the Upper Krivoy Rog series of the Ukrainian Shield (Kalyaev, 1966 quoted in Zaitsev, 1970 p. 12). Iron formations of Kazakhstan have much in common with analogous formations of the world. The estimated absolute age of the formations is 2600–1900 m.y. (Novokhatsky, 1970). Data on other Siberian iron formations indicate that Lower Proterozoic ages are represented. Little information is available on Precambrian iron formations of China and North Korea other than their designated Proterozoic age.

Interpretation

The trends of Early Proterozoic iron formations as defined above are illustrated in Figs. 10a, b, c within pre-Cretaceous continental reconstructions. The reconstructions used for the Atlantic (Fig. 10a), Gondwanaland (Fig. 10b) and Laurasia (Fig. 10c) regions are as previously listed. The global reconstruction of Pangaea (Fig. 11) is after Dietz and Holden (1970).

Lower Proterozoic iron formations occupy distinct trends in the reconstructed land masses. Starting in southeastern Gondwanaland (Fig. 10a) the iron trend crosses Australia and India as shown. To the west of the gap in the Veever *et al.* reconstruction it loops around southern Africa and passes northward and westward alternately through western Africa and eastern South America with undetermined trend north of Venezuela. The North American iron trend (Fig. 10b) is uncertain in western United States, well-established in the Lake Superior and Ungava regions, including the Hudson Bay loop, and indefinite in Baffin Island. In northern Europe, USSR and Asia (Fig. 10c) the iron trend runs southerly through the Baltic Shield to the Kursk and Krivoy Rog regions. Scattered occurrences to the east culminating in those of North Korea–China suggest an easterly trend as illustrated.

The global iron trend of Lower Proterozoic iron formations is summarized in the Pangaela reconstruction (Fig. 11). Regardless of the precise age of specific iron formations, vast quantities of iron ore lie along this global trend. A salient feature of the global iron trend is that it lies mainly well within pre-drift continental reconstructions rather than towards the margins of Precambrian crust. The exceptions are in western United States and North Korea–China. Elsewhere the indicated iron trends are intracratonic. If we accept the shallow marine shelf environment (Gross, 1970b) and designated ages of these iron formations, then this particular environment extended intermittently along the proposed iron trend during early Proterozoic time.

Another salient feature of the global iron trend is its subparallelism to modern plate boundaries mainly those represented by oceanic ridges but including some fold belts. The degree of parallelism to oceanic ridges is moderately high in Gondwanaland including the loop in southern Africa (Fig. 10a); in the north Atlantic region the iron trend is subparallel to the mid-Atlantic ridge system; in USSR and Asia it is in part parallel to the Himalayan fold belt.

The relations suggest that some global plate pattern in early Proterozoic time

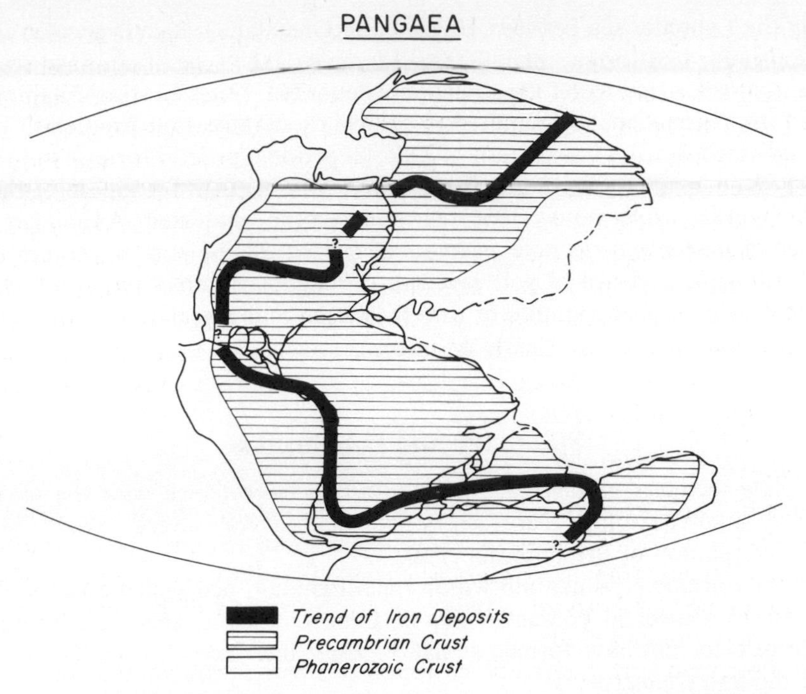

FIGURE 11 Global distribution of early Proterozoic iron formations and Precambrian crust shown on a reconstruction of Pangaea, the universal continent (Dietz & Holden, 1970). The zones of undetermined or questionable continuity are indicated by question marks. Approximately 50% of the total iron reserves of the world listed in the United Nations Survey of World Iron Ore Resources (1970) fall on or in the immediate vicinity of this global iron trend.

influenced the designated alignment of geosynclines and basins in which accumulated vast quantities of iron-silica sediments. Some unique feature of crustal evolution would be required to explain the scope and alignment of the deposits. The answers are, in all likelihood, to be found in first-order features of crustal evolution. In this regard it is noted that this time of massive iron-silica deposition shortly followed that all-important Archean-Proterozoic transition during which the main Precambrian shields of the world were stabilized. Unique consequences of this stabilization were (1) large-scale development of stable continental shelves, (2) global transition to predominantly oxidizing environment, and (3) accelerated biological evolution and activity. These interrelated developments may, in turn, have been a function of some significant stage in core-mantle-crust evolution. They may all have been causally related to development of the iron formations.

The indicated time-space restriction of major iron-silica deposition makes it unlikely that any single factor such as simple change in atmospheric environment is adequate explanation. Rather, it is suggested that explanation is to be sought within the context of crustal evolution in terms of (1) a global pattern of plate boundaries in early Proterozoic time; (2) introduction of large quantities of iron possibly from deep sources to the resulting basins and geosynclines and (3) deposition of iron formation in response to the prevailing oxidizing environment and accelerated organic activity. The exact nature of the early Proterozoic ocean environments is largely unknown.

Possibly the Labrador sea between Ungava and Greenland which is parallel in trend to the Labrador iron formations offers an example of the type of confined seaway in which early Proterozoic iron formations were deposited. Detailed stratigraphic studies of world iron formations are required to resolve these important problems.

The indicated distribution pattern of Lake Superior type iron formation is subject to test by paleomagnetic methods. Preliminary work on iron formation in the Lake Superior region by Symons (1966) has yielded an indicated Animikian (early Proterozoic) paleomagnetic pole position of 94° W, 28° N, site in western United States, with some evidence of pole reversals during deposition of the iron formation. Additional paleomagnetic studies of iron formation and associated rocks to test the proposed global pattern are clearly desirable.

Summary and Conclusions

1. The plate tectonics model, suitably modified in accordance with the great time interval involved, offers satisfactory explanation for several established and indicated features of Precambrian crust.
2. Archean volcanic belts contain within them the main petrogenetic suites and display the fundamental characteristics of Cenozoic island arcs. Accordingly, the Archean belts may have formed at sialic-oceanic interfaces in a similar manner to their modern counterparts.
3. The world-wide distribution pattern of Precambrian sedimentary iron formations in general and those of early Proterozoic age in particular indicate their deposition according to some global plate boundary pattern which is generally subparallel to that of modern oceanic ridges at accreting plate boundaries. The origin and pattern of the iron formations may be related to some type of plate motion that affected Proterozoic crust. If so, the apparent subparallelism in the distribution of early Proterozoic iron formations and modern plate boundaries supports Hurley's contentions of the stability of 'pre-drift' continental nuclei. This would impose some broad restrictions on the degree of pre-Cretaceous dispersion, disorientation and reassemblage of Precambrian crustal nuclei.
4. Additional widespread examination of the Precambrian record in the context of the plate tectonic model is warranted.

Acknowledgements

I have drawn freely from the United Nations Survey of World Iron Ore Resources 1970 as well as from unpublished papers presented at the International Symposium on the Geology and Genesis of Precambrian Iron/Manganese Formations and Ore Deposits, Kiev, 20–25 August 1970 which was organized jointly by Unesco and the International Association of Geochemistry and Cosmochemistry of the International Union of Geological Sciences in collaboration with the Ukrainian Academy of Sciences.

I have benefited greatly from discussion of iron formation problems with G. A. Gross, Geologic Survey of Canada. However, interpretative aspects including any shortcomings of this paper are my responsibility.

References

Anhaeusser, C. R., Mason, R., Viljoen, M. J and Viljoen, R. P., 1969. A reappraisal of some aspects of Precambrian shield geology, *Geol. Soc. America Bull.*, **80,** 2175–2200.

Baragar, W. R. A., 1966. Geochemistry of the Yellowknife volcanic rocks, *Can. J. Earth Sci.*, **3,** 9–30.

Baragar, W. R. A., 1968. Major element, geochemistry of the Noranda volcanic belt, Quebec, Ontario, *Can. J. Earth Sci.*, **5,** 773–90.

Baragar, W. R. A. and Goodwin, A. M., 1969. Andesites and Archaean volcanism of the Canadian Shield. *In:* Proceedings of the Andesite Conference, edited by A. R. McBirney. Bull. 65, Oregon Department of Geology and Mineral Industries, Portland.

Bucher, W. H., 1952. Geologic structure and orogenic history of Venezuela, *Geol. Soc. Amer., Memoir* 49.

Bullard, E. C., Everett, J. E. and Smith, A. G., 1965. The fit of the continents around the Atlantic, *Phil. Trans. Roy. Soc. London, Ser. A*, **258,** 41.

Carmichael, I. S. E., 1964. The petrology of Thingmuli, a Tertiary volcano in Eastern Iceland, *Jour. of Petrology*, **5,** 435–60.

Chernov, V. M., 1970. The ferruginous-siliceous formations of the eastern part of the Baltic Shield; Intl. Symposium on the geology and genesis of Precambrian Iron/Manganese Formations and Ore Deposits; Kiev.

Clifford, T. N., 1970. The structural framework of Africa. *In:* African magmatism and tectonics, edited by T. N. Clifford and I. G. Gass. Oliver and Boyd, Edinburgh.

Cooke, H. C., James, W. F. and Mawdsley, J. B., 1931. Geology and ore deposits of the Rouyn-Harricanaw region, Quebec, *Geol. Surv. Canada, Memoir 166.*

Crawford, A. R., 1969. Indian, Ceylon and Pakistan: new age data and comparisons with Australia, *Nature*, **223,** 380–4.

Dickinson, W. R. and Hatherton, T., 1967. Andesitic volcanism and seismicity around the Pacific, *Science*, **157,** 801–3.

Dietz, R. S. and Holden, J. C., 1970. Reconstruction of Pangaea: Break-up and dispersion of continents, Permian to Present, *J. Geophys. Res.*, **75,** 4939–4956.

Dorr, J. V. N., 1969. Physiographic, stratigraphic and structural development of the Quadrilatero Ferrifero, Minas Gerais, Brazil, *Geol. Surv. Professional Paper 641—A.*

Faure, G. and Kovach, J., 1969. The age of the Gunflint Iron Formation of the Animikie Series in Ontario, Canada, *Geol. Soc. America Bull.*, **80,** 1725–36.

Fryer, B. J., 1972. Age determinations in the Circum-Ungava geosyncline and the evolution of Precambrian banded iron-formations, *Can. J. Earth Sci.*, accepted for publication.

Glikson, A. Y., 1971. Structure and metamorphism of the Kalgoorlie System southwest pf Kalgoorlie, Western Australia, *Geol. Soc. Australia, Spec. Publ., No. 3*, 121–32.

Goodwin, A. M., 1961. Genetic aspects of Michipicoten iron formation, *Can. Inst. Min. Met. Trans.*, **LXIV,** 32–6.

Goodwin, A. M., 1962. Structure, stratigraphy and origin of iron formations, Michipicoten area, Algoma District, Ontario, Canada, *Bull. Geol. Soc. America*, **73,** 561–86.

Goodwin, A. M., 1967. Volcanic studies in the Birch-Uchi Lakes area, *Ont. Dept. Mines. M.P. No. 6*, 96 pp.

Goodwin, A. M., 1968. Evolution of the Canadian Shield; Presidential Address, The Geological Association of Canada, Proceedings, **19,** 1–4.

Goodwin, A. M., 1970. Archaean volcanic studies in the Lake of the Woods—Manitou Lake—Wabigoon region of western Ontario, Ont. Dept. Mines, Open File Report No. 5042.

Goodwin, A. M., 1972. Superior Province. *In:* Variations in Tectonic Style in Canada, *Geol. Assoc. Canada, Special Paper*, in press.

Goodwin, A. M., in preparation. Volcanic Studies in the Timmins–Kirkland Lake–Naranda region, *Geol. Surv. Canada.*

Goodwin, A. M. and Ridler, R. H., 1970. The Abitibi orogenic belt, *Geol. Surv. Canada Paper, 70–74*, 1–30.

Goryainov, P. M., 1970. Structural and stratigraphic disposition of the iron-formation of the Baltic Shield and some concepts of the Lower Precambrian geology; Intl. Symposium on the geology and genesis of Precambrian Iron/Manganese Formations and Ore Deposits; Kiev.

Gross, G. A., 1970a. Nature and occurrence of iron ore deposits. *In:* Survey of World Iron Ore Resources. United Nations, New York.

Gross, G. A., 1970b. Continental drift and the depositional environments of principal types of Precambrian iron-formation; in Abstracts, Intl. Symposium on the geology and genesis of Precambrian Iron/Manganese Formations and Ore Deposits: Kiev.

Gruss, Hans, 1970. Itabiritic iron ores of the Liberia and Guyana Shields; Intl. Symposium on the geology and genesis of Precambrian Iron/Manganese Formations and Or Deposits; Kiev, August, 1970.

Gunning, H. C. and Ambrose, J. W., 1940. Malartic Area, Quebec, *Geol. Surv. Canada, Memoir 222.*

Haughton, S. H., 1969. Geological History of Southern Africa, Geol. Soc. of South Africa, 535 pp.

Hurley, P. M., 1970, Distribution of age provinces in Laurasia, *Earth Planet. Sci. Letts.*, **8,** 189–96.

Hurley, P. M., Fairbairn, H. W., Pinson, W. H. and Hower, J., 1961. Unmetamorphosed minerals in the Gunflint Formation used to test the age of the Animikie, *J. Geol.*, **70,** 489–92.

Hurley, P. M., Leo, G. W., White, R. W. and Fairbairn, H. W., 1971. Liberian age province (about 2,700 m.y.) and adjacent provinces in Liberia and Sierra Leone, *Geol. Soc. America Bull.*, **82,** 3483–90.

Hurley, P. M. and Rand, J. R., 1969. Pre-drift continental muclei, *Science*, **164,** 1229–42.

Jakeš, P. and White, A. J. R., 1971. Composition of island arcs and continental growth, *Earth and Planet. Sci. Letts.*, **12,** 224–30.

Krogh, T. E. and Davis, G. L., 1972. Zircon U-Pb ages of Archaean metavolcanic rocks in the Canadian Shield; Carnegie Institution year book 70, Geophysical Lab., preprint.

Kuno, H., 1966. Lateral variation of basalt magma across continental margins and island arcs. *In:* Continental Margins and Island Arcs, *Geol. Surv. Canada, Paper 66–15*, 327–35.

Kuno, H., 1968. Differentiation of basalt magmas. *In:* Basalts: The Poldervaart treatise on rocks of basaltic composition, edited by H. H. Hess and A. Poldervaart. John Wiley and Sons, Inc., New York.

Lopez, V. M., 1956. Venezuelan Guiana. *In:* Handbook of South American Geology, edited by W. F. Jenks, *Geol. Soc. Amer., Memoir 65.*

Morey, G. B., 1970. Mesabi, Gunflint and Cuyuna Range, Minnesota; Intl. Symposium on the Geology and Genesis of Precambrian Iron/Manganese Formations and Ore Deposits; Kiev.

Morgan, W. J., 1968. Rises, trenches, great faults and crustal blocks, *J. Geophys. Res.*, **73,** 1959–82.

Morrison, R. P., in press. The Geological Structure of South America. Longman, Essex, England.

Naldrett, A. J., 1972. Archaean ultramafic rocks; Earth Physics Branch, Ottawa, Vol. 42, No. 3, 141–152.

Novokhatsky, I. P., 1970. Pre-Cambrian ferruginous-siliceous formations of Kazakhstan; Intl. Symposium on the geology and genesis of Precambrian Iron/Manganese Formations and Ore Deposits; Kiev.

Peterman, Z. E., 1966. Rb-Sr dating of Middle Precambrian metasedimentary rocks of Minnesota, *Geol. Soc. America Bull.*, **77,** 1031–44.

Plaksenko, N. A., Koval, I. K. and Shchegolev, I. N., 1970. Ferruginous cherty formation of the Precambrian of the territory of the Kursk Magnetic Anomaly; Intl. Symposium on the geology and genesis of Precambrian Iron/Manganese Formations and Ore Deposits; Kiev, August, 1970.

Pyke, D. R., Naldrett, A. J. and Eckstrand, O. R., in preparation. Ultramafic Lavas in Munro Township, Ontario.
Ridler, R. H., 1969. The relationship of mineralization to volcanic stratigraphy in the Kirkland Lake area, Ontario. Unpublished Ph.D. thesis, University of Wisconsin.
Semenenko, N. P., 1970. The iron-chert formations of the Ukranian Shield; Intl. Symposium on the geology and genesis of Precambrian Iron/Manganese Formations and Ore Deposits; Kiev.
Stratigraphic Lexicon of Venezuela, 2nd ed., 1970. Boletin de Geologia, Publ. Esp. No. 4. Ministerio de Minas E Hidrocarburos, Caracas, Venezuela.
Symons, D. T. A., 1966. A palaeomagnetic study of the Gunflint, Mesabi and Cuyuna iron ranges in the Lake Superior region, *Econ. Geol.*, **61,** 1336–61.
Trendall, A. F., 1970. The iron formations of the Precambrian Hammersley Group, Western Australia, *Geol. Survey of Western Australia, Bulletin* **119,** 366 pp.
Trendall, A. F., 1970. Time-distribution and type-distribution of Pre-Cambrian iron-formations in Australia; Intl. Symposium on the Geology and Genesis of Precambrian iron/manganese Formations and Ore Deposits; Kiev.
Viljoen, R. P. and Viljoen, M. J., 1971. The geological and geochemical evolution of the Onverwacht Volcanic Group of the Barberton Mountain Land, South Africa, *Geol. Soc. Australia, Spec. Publ. No. 3*, 133–50.
Veevers, J. J., 1970. Phanerozoic history of Western Australia related to continental drift, *J. Geol. Soc. Australia*, **18,** pt. 2, 87–96.
Veevers, J. J., Jones, J. G. and Talent, J. A., 1971. Indo-Australian stratigraphy and the configuration and dispersal of Gondwanaland, *Nature*, **229,** 383–8.
Weber, W., 1970. The evolution of the Rice Lake–Gem Lake greenstone belt. Programme and Abstracts, Ann. Meeting, Geol. Assoc. Canada.
Wilson, H. D. B., Andrews, G., Moxham, R. L. and Ramlal, K., 1965. Archaean volcanism in the Canadian Shield, *Can. J. Earth Sciences*, **2,** 161–75.
Wilson, M. E., 1941. Noranda district, Quebec, *Geol. Surv. Canada, Memoir 229*, 162 pp.
Zaitsev, Yu. S., 1970. Geology of the Precambrian cherty-iron formations of the Belgorod iron-ore region; Intl. Symposium of the Geology and Genesis of Precambrian iron/manganese Formations and Ore Deposits; Kiev.

10.6

J. SUTTON
Department of Geology,
Imperial College, London S.W.7.

Some Changes in Continental Structure Since Early Precambrian Time

1. Introduction

Over the last fifteen years as the ages of Precambrian events have become progressively clearer, the mass of fact accumulated through observation in the field of Precambrian rocks and structures has taken on a completely new significance. It has emerged that the entire mode of operation of geological processes has changed with time and that the earth is a planet which in many respects has behaved in a non-uniformitarian fashion. There appear to have been at least four different styles of global tectonics of which the movements now described as plate tectonics are the latest.

Holmes was probably the first person to recognize similarities in the varying histories of widely separated Precambrian regions. In 1960 Gastil took the matter further when he showed that mineral dates were not uniformly distributed through geological time and concluded that world-wide long-term fluctuations occurred during the earth's history. About the same time several Russian workers, notably Voitkevich (1958), Vinogradov & Tugarinov (1961), had evolved schemes for the division of geological time into long periods—megachrons—which they suggested marked stages in the earth's evolution, the longest of which lasted as much as 800–900 m.y.

Analysis of Precambrian shields had shown that they could be divided into structural provinces, many of which, though in widely separated shields, had developed over similar time spans, suggesting that some form of global evolution had occurred. In 1963 I suggested that the earth had been affected by a succession of shield-producing cycles each lasting between 750 and 1000 m.y. During such a cycle plutonic activity and crustal mobility lessened only to return to stabilized parts of the crust at the onset of the succeeding cycle. A series of such cycles, which I called chelogenic cycles (Sutton, 1963) are superposed on an even longer time change which through geological time has taken the earth from a more mobile to a more rigid condition. The evolution of the crust then appears to reflect two long-term developments which I discussed at an earlier NATO meeting in Newcastle (Sutton, 1970):

(*a*) A progress to increased rigidity through geological time, punctuated by

(*b*) Cyclic changes through which periodically, a new system of mobile belts formed and subsequently decayed to leave progressively large stable areas or belts as one cycle followed another (Sutton, 1963, 1967, 1970).

A new development in the last year has been the demonstration by McGregor, Moorbath and his colleagues (Black *et al.*, 1971) that granitic crust formed before $3{\cdot}8 \times 10^9$ years is preserved near Godthaab in Western Greenland. These granitic rocks have a long history and this suggests that some form of granitic crust was in existence even earlier in the earth's history. Accordingly, any satisfactory hypothesis concerning the sialic crust has to account for its original development within the first 15% of geological time, for its increasing rigidity as it aged and also for the periodical return of regional metamorphism, partial melting and intense deformation as mobile belts were formed in crust which had been rigid in some instances for many hundreds of millions of years.

Such a model implies early differentiation and a falling off in heat flow with time coupled with some mechanism which caused long-term fluctuations in the way the crust reacted to movements deeper within the earth. It is the nature of these latter changes which present the most puzzling problems.

The chronology of these changes is, however, becoming clear for the first time. Taken together the two types of change have left a record in the continental crust in which one can recognize four evolutionary stages:

1. Production of a granitic crust made up of many intrusions and containing no more than small isolated fragments of recognizable non-granitic rock.

2. Development of granitic crust overlain by numerous large belts of volcanic and sedimentary rock many kilometres thick and several tens or hundreds of kilometres in extent. These belts retain many primary features, though all were folded and metamorphosed. It would seem that crust at this stage could not support permanently a gently inclined layer of sedimentary or volcanic rock 7 km thick. The earliest granulites yet identified date from this stage.

3. Development of a granitic crust divided into linear mobile belts and stable blocks, the latter rigid enough to support gently inclined non-metamorphic successions. Extensive fracture systems running for many hundreds of kilometres, in part filled by basic intrusions, may cut such blocks.

4. Production of a granitic crust divided into mobile belts and stiff plates dilated by dyke swarms, rift valleys and new ocean floors. Crust able to support mountain chains high enough to produce molase deposits after uplift.

The four stages are diachronous.

Stage 1 Oldest rock yet identified $3{\cdot}8 \times 10^9$ years, possible range of this stage $4{\cdot}5$–$3{\cdot}4 \times 10^9$ years.

Stage 2 $3{\cdot}4$–$2{\cdot}7 \times 10^9$ years.

Stage 3 $2{\cdot}7$–$2{\cdot}0 \times 10^9$ years.

Stage 4 $2{\cdot}2 \times 10^9$ years–present day.

2. The Earliest Granitic Crust

It is probable that the granitic crust within the North Atlantic Craton, that is to say the region extending from Northwest Scotland to the Labrador Coast, contains much material over $3{\cdot}8 \times 10^9$ years in age, although rocks as old as this have only been proved to exist in a limited area in West Greenland. In many regions in East Greenland and Northwest Scotland a widespread Pre-Ketilidian or Scourian metamorphism occurred before $2{\cdot}9 \times 10^9$ years ago. In Scotland this metamorphism affected two assemblages of rocks. It altered a number of narrow belts of sedimentary and volcanic rock, in some cases associated with highly deformed anorthosites, gabbros and tonalites. This association of rocks appears to be younger than the assemblages of grey gneisses and granulites which make up the bulk of the Precambrian in question—the Lewisian of Northwest Scotland. The two associations of rock are entirely conformable but the grey gneisses which themselves contain a great variety of rock types, form a more variable assemblage and one which has been migmatized to a greater extent than the supracrustal–anorthosite–gabbro–tonalite belts. For this reason several investigators in the last few years (Sutton, 1967; Dearnley & Dunning, 1968, Coward *et al.*, 1969) have suggested that the grey gneisses were derived from a Pre-Scourian assemblage of rock which was already in a crystalline state at the time when the supracrustal rocks were deposited, an event which must have occurred more than $2{\cdot}9 \times 10^9$ years ago, the date of the earliest metamorphic event known to have affected the sediments. This old crystalline complex in Scotland could be as old as the granitic complex in West Greenland now known to be at least $3{\cdot}8 \times 10^9$ years old. If this assumption is correct several consequences follow. In the first place we would have to accept that the Scourian gneisses and granulites were derived from a Pre-Scourian complex which had been reworked from time to time over a period of time at least 1000 m.y. long. It should be possible, though it would be difficult, to establish the times when some of these episodes of deformation and recrystallization occurred, because it is probable that this history of repeated performation and metamorphism (from about 4 to $2{\cdot}9 \times 10^9$ years) was punctuated either by the accumulation of supracrustal rocks or by the intrusion of basic, ultrabasic or anorthositic rocks which could be used as time markers once they are identified. It is known, for example, from McGregor's work that a basic dyke swarm, the Ameralik Swarm, formed over 3×10^9 years ago in West Greenland; highly deformed anorthosite sheets are widespread throughout the North Atlantic Craton and would provide other time markers probably rather late in Pre-Scourian times. A variety of basic and ultrabasic intrusions is known in Northwest Scotland. These have been carefully mapped and if dateable could provide further aids in identifying different phases of Pre-Scourian activity.

Establishing such a chronology would take a long time but it is already obvious that the Pre-Scourian rocks have been intensely deformed on several occasions. All the members of the anorthosite–gabbro–tonalite metasedimentary and metavolcanic associations are now conformable with each other and although primary igneous banding can be identified it is in an intensely deformed condition. Windley (1970) has suggested that anorthosite masses in West Greenland originally several kilometres in thickness have been reduced by deformation to bands a few hundred metres wide. Both in Scotland and in Greenland rocks older than this association had been intensely deformed at some earlier period of time. Such deformation can be seen in Scotland and East Greenland but at present is best understood through McGregor's work near Godthaab. There McGregor was able to identify two granitic complexes, both older than $3{\cdot}1 \times 10^9$ years. Each complex is made up of a variety of granitic rocks, and each is intensely deformed. These complexes which closely resemble the oldest parts of several other Precambrian regions, for example the earliest gneisses recognized in Rhodesia, appear to have been built up from a succession of intrusions many of which were deformed. Some of these may have been large plutonic bodies but a plexus of minor veins and dykes anastomosing in a manner comparable with arrangements seen in many Phanerozoic migmatite complexes is a characteristic feature of the early granitic crust wherever exposed. The distinction is that in Phanerozoic times such migmatization was restricted to mobile belts whereas in the early Precambrian the heatflow appears to have been so high that migmatization formed periodically throughout the crust.

The older of the two early Precambrian granitic complexes recognized by McGregor contains some non-granitic rocks and because of its complexity is likely to prove to have begun its evolution well before $3{\cdot}8 \times 10^9$ years ago. Bearing in mind the highly differentiated state that many meteoritic bodies and the Moon had reached by $4{\cdot}4 \times 10^9$ years, it is possible that the Earth may have produced a granitic crust within the first one or two hundred million years of its existence. The presence of a basic dyke swarm over 3×10^9 years near Gobthaab is of considerable interest. As known at present it is not a major feature of the Earth, for its established length is no more than 100 km, but it provides for the first time evidence that crust over $3{\cdot}0 \times 10^9$ years in age was rigid enough to be fractured and dilated. Although evidence of dilation remains scanty there can be no doubt from the earliest records we have that the early granitic crust was undergoing deformation, for the oldest granitic complexes were highly deformed. There must therefore have been some mechanism capable of deforming the granitic crust in operation more than $3{\cdot}8 \times 10^9$ years ago.

3. The Second Stage

This stage was marked by the development of greenstone belts and granites and by the widespread development of extensive high-grade metamorphic and migmatitic complexes in the amphibolite and granulite facies. No flat-lying rocks survive. The whole crust was deformed, though in contrast to the first stage in the Earth's history many supracrustal successions survived as clearly identifiable entities. There is now a large literature on crust of this type; it has often been suggested that the greenstone belts formed during a period when the granitic crust was developing for the first time. I think this view is wrong and that a pre-existing granitic crust formed a basement below the greenstone belts. The arguments for each view are amply set out in the literature (for example, Anhaeusser *et al.*, 1969). Crust at this stage in the Earth's development demonstrates rather clearly one way in which granitic material accumulates high in the crust and works its way up through the granitic layer. Heier (1972) and Lambert (1971) among others have argued that the heat-producing elements must be unusually concentrated in the upper sialic crust and that at depth potassium, thorium and uranium are less abundant, for unless this was so a higher heat flow than that observed would result. One step in such a concentration appears to have occurred in Archaean times. Evidence supporting a model of sialic crust with a granite rich upper part has come from the field observations of high level hood granites developed above less granitic gneisses in the high dissected Archaean terrain of Swaziland made by Hunter (1970) and from the synthesis put forward by Windley & Bridgwater

(1971). They have set up a model of crust of that age comprising a lower part in the granulite facies passing into an upper layer of greenstone belts around which granitic rocks occur. What in effect appears to have taken place in an area such as the Archaean Rhodesian Craton is the reworking of a granulitic or tonalitic basement on which supracrustal rocks had locally accumulated as greenstone belts, to give a more granitic assemblage of younger syn-tectonic and post-tectonic granites which have risen through the underlying basement to deform and locally to penetrate the greenstone belts. The resulting characteristic pattern in which the regional strike is deflected around rising and widening granite domes has been repeatedly discussed since MacGregor pointed out in 1951 the world-wide development in Archaean times of assemblages of this type. In many parts of the world there is a regional variation in the nature of the exposed crust dating from this stage. For example, in the northern hemisphere while much of the Canadian Shield of Kenoran age is formed of greenstone belts and granitic rocks, in the eastern part of the Shield higher grade rocks become more abundant and further east still in Britain and Scandinavia the Archaean consists predominantly of granulite or amphibolite facies assemblages with narrow strips of supracrustal material resting on a highly deformed basement probably of very great age in its original form.

Although apparently no stable blocks which may have formed rigid masses during this stage of the development of the crust survive at the present day, there may well have been limited periods when parts of the crust were tectonically stable. The presence of the Ameralik dyke swarm (Macgregor, 1972) provides definite evidence that the crust could fracture over 3×10^9 years ago. Several authorities (Goodwin, 1971; Anhaeusser *et al.*, 1969) have suggested that extensive crustal fracturing may have occurred when the greenstone belts evolved, and that the volcanicity that played so important a part in the greenstone belts was guided by underlying fractures. While it is impossible to prove, there seems much to support this hypothesis. The composition of the volcanic rocks indicates that a large part of a typical greenstone belt has come from mantle material and must therefore have penetrated granitic crust. When one allows for the subsequent deformation largely connected with the emplacement of younger granites, greenstone belts exhibit a certain regularity in any one region. Goodwin (1971) has shown how the Superior Province of the Canadian Shield is divisible into a number of subparallel belts rich in volcanic material separated by regions of granodioritic composition. If one accepts that the latter are derived from an older basement it would appear to follow that the granitic crust broke up in a way which allowed lavas from the mantle to pour out along a series of subparallel zones, the longest of which extends for over 1500 km. In Rhodesia, Wilson (1972) among others, has drawn attention to three principal directions which greenstone belts follow and has suggested that this arrangement is controlled by deep fractures due to deformation of the Rhodesian Craton. In Western Australia, O'Driscoll (1971) has paid particular attention to the relationship between mafic and ultramafic intrusion and crustal deformation, and has indicated the importance of shears within the crust. What is not clear at the moment is whether such deformation formed part of a global tectonic system or whether we are observing the results of comparatively local independent movements in the Archaean. What is striking, however, is the abundance of mafic and ultramafic rock erupted in those times on to the sialic crust. Comparisons have been made with present day ocean ridges and island arcs; in my opinion the structural analogies are slight and I would suggest that in that period of time we had a tectonic situation wholly different from any that exists at the present day. The greenstone belts are smaller than present day mobile belts. Where studies of Archaean sediments have been made, evidence has been found of the transport of debris from the surrounding granitic crust into greenstone rich depositories (Goodwin, 1971). On the other hand, as far as I know we have no evidence that, when deformation of such deposits had ended, uplift occurred, such as characterizes more recent mountain chains, and that post-tectonic sediments were transported into an adjoining foreland. One distinctive feature of Archaean volcanicity was the enormous extent of mafic and ultramafic igneous activity within the sialic crust. Nothing that follows in later stages of earth history appears to be comparable.

A second distinctive feature of this stage of evolution of the crust, no doubt shared with the earlier stage discussed previously, is the extent to which the sialic crust was deformed. The deformation was not restricted to mobile belts but affected the whole of the crust if we are to judge from surviving examples.

4. The Third Stage

This is the earliest stage at which a distinction between stable blocks and mobile belts becomes apparent. It was a critical period in the history of the earth for during this stage changes occurred which eventually led to the present day system of moving plates and restricted mobile belts. The restriction of mobility did not occur everywhere at the same time, but it appears to be true that every shield area records a change which took place at some time between 2·9 and $2{\cdot}4 \times 10^9$ years and which could be expressed in the following fashion. Widespread plutonic activity, i.e. regional metamorphism, the formation of acid rocks and migmatization, ceased to affect the entire crust and was gradually restricted. In Western Australia, for example, Arriens (1971) has shown that a belt 600 by 900 km was abundantly affected in these ways until $2{\cdot}6 \times 10^9$ years ago, thereafter was only sporadically intruded by isolated acid masses. A precisely similar situation is seen in Scotland, where a widespread granulite and amphibolite facies metamorphism ended about $2{\cdot}9 \times 10^9$ years ago and was succeeded by localized metamorphism and by the intrusion of isolated pegmatite bodies which continued intermittently until about $2{\cdot}4 \times 10^9$ years ago. There is no systematic arrangement as far as can be seen in the way in which these later events affected the crust. It appears to be the case that we are looking at a slow transition from a highly mobile crust with abundant plutonium, to a more or less completely stabilized mass sporadically intruded at isolated spots. The transition can take as much as 400 m.y.; for instance in Scotland the latest pegmatites in the Scourian complex are nearly 500 m.y. younger than the widespread granulite facies metamorphism which occurred there $2{\cdot}9 \times 10^9$ years ago. Furthermore, although on a restricted scale the stabilization of the crust appears to have taken place sporadically so that no regional pattern can be discerned within an individual shield, on a global scale the period of time between 2·9 and 2×10^9 years ago certainly showed a world-wide change. At the end of that period a major mobile belt extended southwards from the vicinity of Murmansk. The Belomoride chain as it is named in Eastern Europe was marked by abundant igneous activity between 2·2 and $1{\cdot}9 \times 10^9$ years ago. Further south in West Africa and in eastern South America, as Hurley has shown, a belt of similar age—the Eburnean—formed in a period of particularly impressive igneous activity and crustal deformation. Here again the general trend was from north to south and it is possible that the African and European belts once formed parts of a single whole. In contrast, the Precambrian exposed in the northern hemisphere outside these regions appears to have been wholly stabilized by this time. Not only had plutonic activity ended in the Canadian Shield, in Greenland in Northwest Europe and in Siberia but widespread dolerite dyke swarms formed over this interval of time ($2{\cdot}2$–$1{\cdot}9 \times 10^9$ years) in Northwest Canada, Eastern Canada, Greenland, Scotland and Scandinavia. Similar changes occurred in the southern hemisphere. In Southern Africa, for example, the crust had been stabilized everywhere outside the Limpopo Belt. In the Transvaal dolerite dykes dating from about $2{\cdot}4 \times 10^9$ years are widespread, and in Rhodesia the Great Dyke formed at about $2{\cdot}55 \times 10^9$ years. These basic and ultramafic intrusions in the Transvaal and Rhodesia developed in stable masses north and south of the Limpopo Belt and were contemporary with syntectonic plutonism in that belt. In the same way, though on a much larger scale, the fracturing of the crust which permitted the intrusion of the widespread northern hemisphere dolerite dyke swarms between $2{\cdot}2$ and $1{\cdot}9 \times 10^9$ years ago must have taken place at the same time as intensive deformation and igneous activity occurred in the Belomoride and Eburnean belts of Eastern Europe and Western Africa. Here for the first time we can see some sort of analogy with the present day situation in which the opening of rigid continental blocks as new oceans form is directly connected with igneous activity and deformation within contemporary mobile belts. On the other hand, there are important distinctions between the present day situation and the tectonic regime as it appears to have existed in this early stage of crustal development—between $2{\cdot}9$ and 2×10^9 years ago. An example of such a distinction is provided by the sedimentary basins which accumulated on the Archaean stabilized crust. The best-known instance is the Witwatersrand basin which from about $2{\cdot}7 \times 10^9$ years onwards was filled with Witwatersrand and Ventersdorp sediments and volcanic rocks which overlie a granitic basement in the Transvaal. As Anhaeusser (1971) has shown the well-known domes and depressions below the Witwatersrand basin were active during the accumulation of these sediments and volcanic rocks. The vertical relief of the underlying basement was at least 4 km and possibly very much greater. In contrast, the uppermost deposits, the Pretoria Series (c. $2{\cdot}3$–2×10^9 years) show through their continuity and sedimentary facies, a greatly increased stabilization of the basement (Sutton, 1970). Although when compared to the earlier greenstone belts it is legitimate to view the Witwatersrand basin as flat-lying, in fact early deposits of the basin accumulated on a crust where positive and negative areas continued to move vertically as the basin fillings were deposited.

The Limpopo Belt which lies immediately north of the Transvaal contains a succession of sedimentary rocks—the Messina Group—whose precise age is not yet known but which formed sometime in the interval between $2{\cdot}8$ and 2×10^9 years ago. These sediments have not been found outside the mobile belt and if in fact they are restricted to this belt they must form one of the earliest examples of pretectonic deposits laid down within what was to become a belt of mobility flanked by more stable regions on either side.

Mason (1972) has drawn attention to the ancient nature of the Limpopo axis which he regards as having evolved along a very early line of weakness—thus within a short distance of each other we find the earliest surviving flat-lying sediments laid down in a more or less stabilized basin in the Witwatersrand and an early example of sedimentation heralding a mobile belt in the shape of the Messina Formation. Associated with the Messina sediments are highly deformed anorthosites; a type of rock not widespread on the Rhodesian Craton although they are known further south in the Transvaal. They are a feature of other parts of the crust formed about 3×10^9 years ago and as Windley and Bridgwater have pointed out (1971) occur in one belt extending from North America through Greenland across Northwest Europe and in another stretching from Southwest Africa through Madagascar into India. Here again we may have an instance of an early linear feature below the earth's crust which was capable of controlling the nature of igneous activity. It is of interest that Mason has deduced from his experience of the Limpopo Belt that the belt might overlie an older crustal feature. It may be therefore that the intrusion of anorthosites and the accumulation of sediments such as the Messina sediments provide an indication of the differentiation of the crust into relatively mobile regions marked by deep fractures flanked by more stabilized areas where sedimentation and igneous activity took other forms.

Even as the stabilization of regions of the crust proceeded which was to lead eventually to the very extensive stable blocks that existed about 2×10^9 years ago, new mobile belts were starting to form along a series of sedimentary basins larger than any which had been present on earth up to that time.

In North America the Huronian began to accumulate about $2{\cdot}3 \times 10^9$ years ago. In Greenland the Ketilidian overlies a basement $2{\cdot}3 \times 10^9$ years in age and is cut by dolerites 2×10^9 years old. In Scandinavia the Svecofennian deposits similarly overlie rocks $2{\cdot}3 \times 10^9$ years in age but must be older than a widespread regional metamorphism which occurred about $1{\cdot}9 \times 10^9$ years ago. In practically every Precambrian shield area a new system of elongated sedimentary belts had begun to evolve on newly stabilized crust between about $2{\cdot}3$ and 2×10^9 years ago. These belts ushered in the next stage of crustal evolution.

5. The Fourth Stage

The geosynclinal deposits such as the Huronian of Canada, the Ketilidian of Greenland and the Svecofennian of Northern Europe are the oldest rocks which can places be traced into equally old platform successions. A particularly clear example has recently been described by Hoffman (1972) from the Coronation Geosyncline in Northwest Canada.

A widespread system of metamorphic belts had developed over previously stabilized shield areas by about $1{\cdot}8 \times 10^9$ years. in some places general metamorphism continued until $1{\cdot}6 \times 10^9$ years and in certain areas such as the Southwest United States and Northwest Australia there was extensive granitic and volcanic activity rather later at about $1{\cdot}4 \times 10^9$ years. In general, however, by $1{\cdot}4 \times 10^9$ years extensive stabilization of the mobile belts active at $1{\cdot}8 \times 10^9$ years had occurred. In Greenland and Northern Europe this stabilization led to vulcanicity associated with faulting, to the accumulation of clastic deposits often restricted to graben as in the Gardar deposits of Southern Greenland and the Jotnian of Scandinavia. This led in turn to rather widespread intrusion of isolated alkaline bodies which in Greenland are between 1·2 and 1×10^9 years in age, and in Scandinavia are closely linked in time with the Jotnian deposits. Other evidence of fracturing within consolidated crust is provided by the extensive dyke swarms formed between 1·2 and 1×10^9 in Canada and in Southern Greenland. These are contemporary with the Grenville mobile belts of Canada and Scandinavia and Northern Greenland which itself are in part superimposed on Hudsonian and Svecofennid structures. By the time the Grenville mobile belt developed, a very large area in Canada and Greenland had been stabilized to the west of the belt and much of the Baltic Shield to the east. Whether or not rifting of these stabilized regions led to ocean floor formation is still uncertain, but it may have been at this time that the Pacific seas formed. In Late Precambrian times between about 900 and 700 m.y. ago a further network of thick geosynclinal deposits, in some parts reaching 15 km in thickness accumulated. These mark a new system of depressions ultimately leading to the mobile belts of the Phanerozoic, some parts of which are still active. In Britain the best example of such a Late Precambrian formation is the Torridonian; this formation, predominantly of clastic sediments with very subordinate volcanic material has been compared to the clastic wedge deposits formed on a stable aseismic continental margin. It may be that these Late Precambrian deposits formed in some such manner on the trailing edges of separating continental plates. On the other hand, I have suggested (Sutton, 1963, 1967) that perhaps they formed a network within a pre-existing extensive continent. The decisive question is whether or not the deposits lay at the contact of ocean floor and continent. This should be soluble, by palaeomagnetic work which should indicate whether the stable areas on either side of such belts were once parts of widely separated continents which approached one another as ocean floor disappeared, or whether they formed parts of a single continental mass on which an ensialic mobile belt evolved. At all events the Late Precambrian mobile belts were stabilized one by one through the Phanerozoic until the present day situation was reached in which we find two narrow mobile belts along the outer margins of two disrupted super-continental masses.

It is of considerable interest to try and establish whether an earlier phase of continental disruption took place between 2 and 1×10^9 years ago. There is now an extensive literature on the mobile belts of that period and in a large number of instances it has been discovered that these belts contain much reworked older material. In Scotland, for example, the Laxfordian belts (climax of metamorphism $1{\cdot}8 \times 10^9$ years) contain much reworked older Scourian rock. It is possible to trace the Scourian rocks into the younger belts and to observe their deformation. From the structures that are preserved there is nothing to contradict the hypothesis that the Laxfordian mobile belt formed within a continental mass on a basement that consisted mainly of pre-existing sialic rock, but here again the decisive evidence is likely to arise from palaeomagnetic measurements of neighbouring stable blocks. It is interesting to note that recently Escher & Watterson (1970) have found rather similar relationships in Western Greenland where the Nagssugtoqidian belt was superimposed on Pre-Ketilidian basement about $1{\cdot}8 \times 10^9$ years ago. There they were able to establish that old rocks cut by two dyke swarms were reworked as they entered the younger fold belt in such a way that a 100 km wide dyke swarm was reduced in the reworked region to a width of 20 km. In Scotland, Sutton & Watson (1962) had shown using a similar approach that dolerite dykes were thinned and the intervening rocks were compressed so that the distance between dykes was reduced to about one-third of its original magnitude as they and the older country rock were deformed in the Laxfordian mobile belt. Figures such as these give some indication of the great extent of the deformation the granitic crust has suffered. It seems likely that in the early stages of the earth's history the main manner in which the granitic crust was moved horizontally was not through the transport of rigid masses of continental crust as oceanic crust was formed and destroyed but by the intense deformation of sialic crust. With the passage of time a once continuous granitic cover formed early in the history of the earth was perhaps thickened and reduced by deformation to give the present day arrangement of continental masses isolated within a predominantly oceanic crust. Over the same period of time successive phases of crustal dilation became more and more marked with each chelogenic cycle as the restricted dolerite swarms 3×10^9 years old led ultimately to the present day arrangement of dyke swarms, rift valleys and ocean floors.

References

Anhaeusser, C. R., 1971. The geology and geochemistry of the Archaean granites and gneisses of the Johannesburg-Pretoria dome. Economic Geology Research Unit, University of Witwatersrand, Information Circular No. 62.

Anhaeusser, C. R., Mason, R. Viljoen, M. J. and Viljoen, R. P. 1969. A reappraisal of some aspects of Precambrian shield geology, *Geol. Soc. Am. Bull.*, **80,** 2175–200.

Arriens, P. A., 1971. The Archaean geochronology of Australia, *Geol. Soc. Aust. Spec. Publ. 3*, 11–24.

Black, L. P., Gale, N. H., Moorbath, S., Pankhurst, R. J. and McGregor, V. R., 1971. Isotopic dating of very early Precambrian amphibolite facies gneisses from the Godthaab district, West Greenland, *Earth and Planet. Sci. Letts.*, **12,** 245–59.

Coward, M. P., Francis, P. W., Graham, R. H., Myrs, J. S. and Watson, J., 1969. Remnants of an early metasedimentary assemblage in the Lewisian complex of the Outer Hebrides, *Proc. Geol. Assoc.*, **80,** 387–408.

Dearnley, R. and Dunning, F. W., 1968. Metamorphosed and deformed pegmatites and basic dykes in the Lewisian complex of the Outer Hebrides and their geological significance, *Quart. J. geol. Soc. London*, **123** (for 1967), 353–78.

Escher, A., Escher, J. and Watterson, J., 1970. The Nagssugtoqidian boundary and the deformation of the Kângamiut dyke swarm in the Søndre Stromfjord area, *Grønlands geol. Unders. Rapport*, **28,** 21–3.

Gastil, G., 1960. Continents and mobile belts in the light of mineral dating, *Int. Geol. Congr. 21st Session*, **9,** 162–9.

Goodwin, A. M., 1971. Metallogenic patterns and evolution of the Canadian Shield. *Geol. Soc. Aust. Sp. Publ. 3*, 157–75.

Heier, K. S., (in press). Geochemistry of granulite facies rocks and problems of their origin, *Phil. Trans. Roy. Soc. Lond.* A. 273, 429–442.

Hoffman, P., (1973). The Coronation geosyncline: lower Proterozoic analogue of the Cordilleran geosyncline in the northwestern Canadian shield, *Phil. Trans. Roy. Soc. Lond.*

Hunter, D. R., 1970. The ancient gneiss complex in Swaziland, *Trans. Geol. Soc. S. Africa*, **73,** 107.

Lambert, I. B., 1971. The composition and evolution of the deep continental crust, *Geol. Soc. Aust. Sp. Publ. 3*, 419–28.

MacGregor, A. M., 1951. Some milestones in the Pre-Cambrian of Southern Rhodesia, *Proc. Geol. Soc. S. Africa*, **54,** 27–71.

McGregor, V. R., (1973). The early Precambrian gneisses of the Godthaab district, West Greenland, *Phil. Trans. Roy. Soc. Lond.* A. 273, 343–358.

Mason, R., (1973). The Limpopo mobile belt—Southern Africa, *Phil. Trans. Roy. Soc. Lond.* A. 273, 463–486.

O'Driscoll, E. S. T., 1971. Deformational concepts in relation to some ultramafic rocks in Western Australia, *Geol. Soc. Aust. Spec. Publ. 3*, 351–66.
Sutton, J., 1963. Long-term cycles in the evolution of the continents, *Nature, Lond.*, **198,** 731–5.
Sutton, J., 1967. The extension of the geological record into the Pre-Cambrian, *Proc. Geol. Assoc.*, **78,** 493–534.
Sutton, J., 1970. Migration of high temprature zones in the crust. *In:* Palaeogeophysics, edited by Runcorn, pp. 365–75. Academic Press,
Sutton, J. and Watson, J., 1962. Further observations on the margin of the Laxfordian complex of the Lewisian near Loch Laxford, Sutherland, *Trans. Roy. Soc. Edinb.*, **56,** 89–10
Vinogradov, A. P. and Tugarinov, A. I., 1961. Age of Pre-Cambrian rocks of the Ukrainian and Baltic shields, *Ann. N.Y. Acad. Sci.*, **91,** 500–13.
Voitkevich, G. H., 1958. *Priroda*, 77.
Wilson, J. F., 1973. The Rhodesian Archaean craton—an essay in cratonic evolution, *Phil. Trans. Roy. Soc. Lond.* A. 273, 389–412.
Windley, B. F., 1970. Anorthosites in the early crust of the earth and on the moon, *Nature, Lond.*, **226,** 333–5.
Windley, B. F. and Bridgwater, D., 1971. The evolution of Archaean low and high-grade terrains, *Geol. Soc. Aust. Spec. Publ. 3*, 33–46.

10.7

P. M. HURLEY
M.I.T. Cambridge, Mass., USA

On the Origin of 450 ± 200 m.y. Orogenic Belts*

Introduction

The question of the origin of the 450 ± 200 m.y. orogenic belts (Pan-African type) has been reopened to accommodate the concepts of modern plate tectonic theory. However, there still remains a strong division of opinion between two extreme possibilities. Were they formed mostly *in situ* by the upwelling of hot materials from the mantle with subordinate subduction effects and little or no oceanic seaway or motion of the adjacent cratons? Or, was their origin similar to the perimeter of the Pacific Ocean involving major sea-floor spreading, subduction belts and the emplacement of new materials by partial fusion of the down-going slab? The viewpoints are diametrically opposed to each other, and the discussions carry with them many of the major problems that are currently being debated on the evolution of the Earth's crust.

Hypothesis 1, *in situ* Orogeny

The 450 ± 200 m.y. belts are long and sinuous, and if defined broadly enough (broader than Kennedy's original definition of the Pan-African, 1964) will be found to extend in a complex network through parts of almost all major regions of the present continental crust. The term '450 ± 200' is applied here simply to avoid the confusion that arises when one attempts to think only in terms of discrete sub-units such as Hercynian, Caledonian, Acadian, Appalachian, Avalonian, Damaran, Caririan, or other locally traceable parts of this Earth-circling orogenic system. The unifying characteristics which fuse this complex into a single definable entity in space and time are: (1) that it includes a period of geological activity within a specific time range, and (2) that if one starts at any point on one of these belts he can trace it without a break to all other belts by a continuity of foliated rocks, thermal effects, intrusive igneous materials or substantial structural or tectonic features. This, of course, does not include breaks resulting from subsequent drift motions.

Within these two principal characteristics of a limited span of time, and great spacial extension, there are many other characteristics common to these orogenic systems which may throw some light on their origin. There appear to be discrete periods of activity (igneous and metamorphic) which are demonstrated by careful geochronology to be isochronous within narrow limits, and any one part of a 450 ± 200 belt may have one or more of these discrete pulses that can be isolated by detailed

* M.I.T. Age Studies No. 106.

geological work. In addition, there generally is a peppering of post-kinematic granitic emplacements which range in age 100 m.y. or more younger than the principal thermotectonic activity within any region. The belts are marked by a predominance of granitoid rocks, including paragneisses, granite gneisses and plutonic rocks that appear to have an average composition approximately in the granodiorite range (exclusive of the post-kinematic group), and a definite scarcity of volcanic rocks relative to what might be expected in an island-arc system. Ophiolite suites are more abundant than in earlier Precambrian age provinces, but much less abundant than in the young Alpine and Pacific orogenic belts.

Sr^{87}/Sr^{86} ratios of the granitoid rocks at time of emplacement are too low to permit any more than a minor contribution of detrital materials from adjacent cratons, particularly in those regions where the adjacent cratons are over 2000 m.y. in age. For example, Fullagar (1971) recently demonstrated the existence of 25 plutons in the Piedmont of the southern Appalachians having an age of 600–300 m.y. and initial Sr^{87}/Sr^{86} ratios of 0·702–0·705. These statements apply even to many of the metasedimentary rocks in the central parts of the belts, suggesting some shedding of sedimentary primitive materials within the orogenic process itself.

It should be noted that Armstrong's model (1968), calling for a fixed volume of sial at the surface since early Earth history to accommodate the so-called freeboard problem of oceans maintaining a constant level with respect to the surface of ancient continental plates, is directly opposed to the first hypothesis and can only apply to the case of active subduction as the only mechanism. Armstrong's model states that all crustal materials showing a Rb–Sr whole-rock age less than Katarchaen must have been through a mantle purification in order to lose radiogenic Sr^{87}, and at the same time the steady state load of sial in the mantle cannot be too large or the fluctuations would destroy the constant freeboard effect. This means that granitoid rocks, wherever they occur, must have been carried down into the mantle and up again by subduction, and mostly at approximately the same location. In opposition to this, the upwelling hypothesis involves a direct differentiation of mantle materials into silica-saturated derivatives with primitive Sr^{87}/Sr^{86} ratios, and a growing total volume of sial at the surface which implies a continuing differentiation of the Earth.

In his classic report describing the Mozambique belt in Africa, Arthur Holmes (1951) attributed the great regions of N–S trending gneisses, migmatites and granites to a late orogenic cycle superimposed on pre-existing continental basement rocks. Within the belt he found a succession of orogenic cycles, and envisaged older rock sequences entering into and affected by the younger activity. In other similar belts in Africa this phenomenon has been described many times since then: French geologists have used the word 'rejeunissement'; others have referred to the process as a reworking or reactivation. A recent comprehensive review of the phenomenon has been given by Wynne-Edwards and Hasan (1970). African geologists have noted for many years such features as Archean granulites, cross-cutting features which replace rather than displace ancient structural elements, and measurably older relicts or crustal blocks, all of which suggest the presence of a former basement of much greater antiquity. These kinds of observations have led to the concept of 'cratonization' in Africa, where the nuclei of Archean and Proterozoic stable crust remained in place while the belts of Pan-African activity were in the process of generating new stable additions to the continent (Clifford, 1968).

Hypothesis 2, Pacific-Type Orogeny

The opposing hypothesis states that all orogenic belts are formed as a surface expression of down-going lithospheric slabs, and both the orogenic tectonism and the new crustal materials result from this process. It is possible to find all of the characteristics of the 450 ± 200 m.y. belts in the later Pacific-type orogenies. One can produce extremely long and sinuous belts of orogenic activity that encircle the earth and cover a time span of at least 200 m.y. by observing the present distribution of the group of Mesozoic and Cenozoic orogenies. These orogenies tend to coincide spacially and as an assemblage of orogenic belts they could resemble the 450 ± 200 m.y. assemblage. It is noteworthy that the Mesozoic–Cenozoic system is not randomly distributed, but follows in a general way an orthogonal pair of great circles, and also tends to mark the border between a continental hemisphere and an oceanic hemisphere. Parts of the broad belts are quite complex, as in southeast and northeast Asia where the orogenic sub-units are branched and split leaving older blocks of continental crust surrounded by the younger belts of tectonism or intrusion.

This second hypothesis allows for the generation of new magma with primitive Sr^{87}/Sr^{86} ratios by the partial fusion of down-going oceanic basalt in island-arc or Andean type structures, or by the collision or override of continental masses which thicken the crust and cause remelting of sialic materials at depth. In this hypothesis it is possible to have a continuously growing crust (Dickinson & Luth, 1971) by a two-stage process involving the oceanic lithosphere as an intermediate step; or a fixed, early-generated, total volume of sialic crust (Armstrong, 1968) in which the recycling of the crustal materials through subduction into the mantle will remove radiogenic strontium while returning the remainder of the sialic material to the surface.

The high thermal gradients postulated for the 450 ± 200 m.y. belts (Zwart, 1969) are present today, strongly correlated with zones of high seismic activity. They seem to occur above the central parts of Benioff zones, and may be due either to the rise of magmatic materials, or to a roller cell in the hanging wall of an island-arc subduction zone, or to some other process not yet clearly demonstrated, despite the fact that a down-going slab would be thought to bring about a region of low heat flow. Plutonic activity commonly occurs above the Benioff zone in its middle reaches with the result that the older continental craton is invaded by numerous anorogenic granites without accompanying tectonism. These granites are found to have low Sr^{87}/Sr^{86} ratios compared to their host rocks, and could be comparable to the post-kinematic granites of the 450 ± 200 m.y. belts. Korea is an excellent example of extensive invasions of post-Palaeozoic granitic masses occurring in rocks of much greater age. In the Mesozoic–Cenozoic orogenic assemblage we can find relicts of ancient crust, remobilized and refoliated basement rocks, 'micro-continents' or large angular blocks of continental crust within the region of activity. Although ophiolites are common in the younger orogenic belts, they are much more plentiful in island-arc type than in Andean type representatives.

The length of time involved in the complex Mesozoic–Cenozoic system can also be considerably greater than the time involved in a single sea-floor opening such as that of the Atlantic, and can match the time span seen in the 450 ± 200 m.y. belts. This can be accounted for by the moving of continental plates both into a new configuration by sea-floor motions, and subsequent movement in an outward direction against the sea-floor as the mass splits and breaks apart in another region. The edge would then suffer the activity during both the opening stage of the sea-floor moving it into position and the closing stage where the continent was moving against the sea-floor.

Discussion

If one extends the present day analogy into the past and states that the most continuous (in time and space) sections of the 450 ± 200 m.y. orogenic system marked the boundary between continental and oceanic hemispheres as in the circum-Pacific belt today, it would be clear that the present continents would have been split apart by oceanic distances of the magnitude of the Earth's diameter. Thus, for example, if the belt through east-central Australia, the Trans-Antarctic mountains of Antarctica, the Cape fold belt of South Africa, the Paraiba and Cariri of Brazil, the Pan-African of coastal West Africa, and the Appalachian and Caledonian of the North Atlantic occurred along a great circle (and they almost do so in a Pangaea reconstruction) and marked the dividing line between continental and oceanic hemispheres, it would mean that the continental regions of North and South America west of this line would have drifted across the oceanic hemisphere to reach their position on the boundary. Palaeomagnetic evidence and matching of earlier geologic features weakly suggest that this is not the case in the pre-drift separation of the Canadian and Baltic Shields.

A corollary to these inferences is that the distribution of land masses prior to 650 m.y. would bear little resemblance to the present continents, except for areas not transected by 450 ± 200 m.y. belts. If true, this would mean that there would be little chance of finding two ancient cratons with matching geological features on opposite sides of a younger thermotectonic belt. This is examined in the following paragraphs.

In opposition to the use of Hypothesis 2 is a matching of the earlier Precambrian features in the Guyana and West African Shields despite the Pan-African belt between them. Hurley (1972) has attempted to establish the fixed positions of the West African and Guyana Shield areas relative to each other since 3000 m.y. ago as an argument in favor of *in situ* development of the Pan-African belt without major motions of the adjacent ancient nuclei. He raises the question of how a block of ancient crust such as the West African Shield can respond to a succession of near-encirclements by oceanic seaways and bordering subduction zones which continue to be active for a period of at least 200 m.y., and finally find its former position and orientation opposite the Guyana Shield.

In both regions (Guyana and West Africa) an ancient basement 2700 m.y. or older is overlain unconformably by geosynclinical sediments which were folded, metamorphosed, and intruded by granitic rocks in the age range 1800–2000 m.y. (Figs. 1, 2). The stratified rocks have similar sequences with a lower volcanic-pelitic section covered by coarser clastic sediments in the upper section, and a ubiquitous horizon of Mn-bearing sediments occurring as gondites and manganiferous phyllites at the top of the lower section (McConnell & Williams, 1969). This horizon is found in the upper Birrimian of Ghana and Ivory Coast, the upper Paramaca of French Guiana, the upper Barama of Guyana, the Carichapo of Venezuela and the Serro

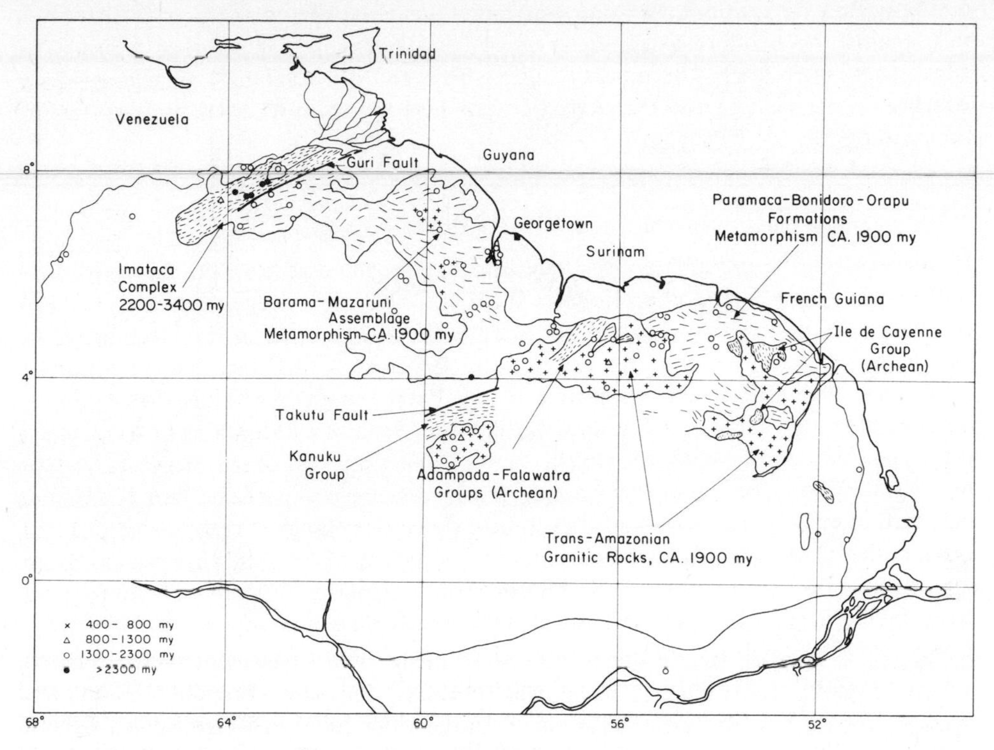

FIGURE 1

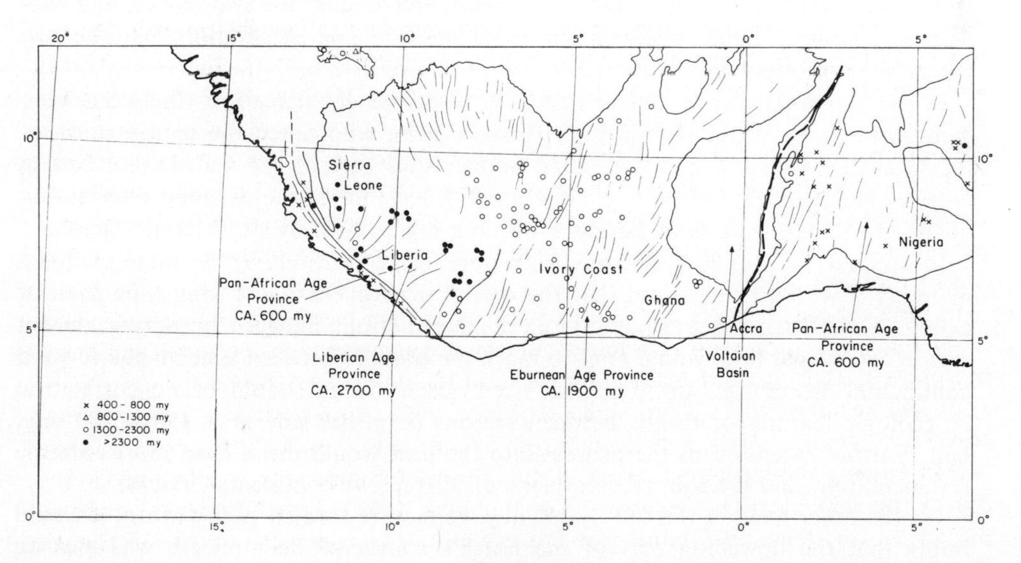

FIGURE 2

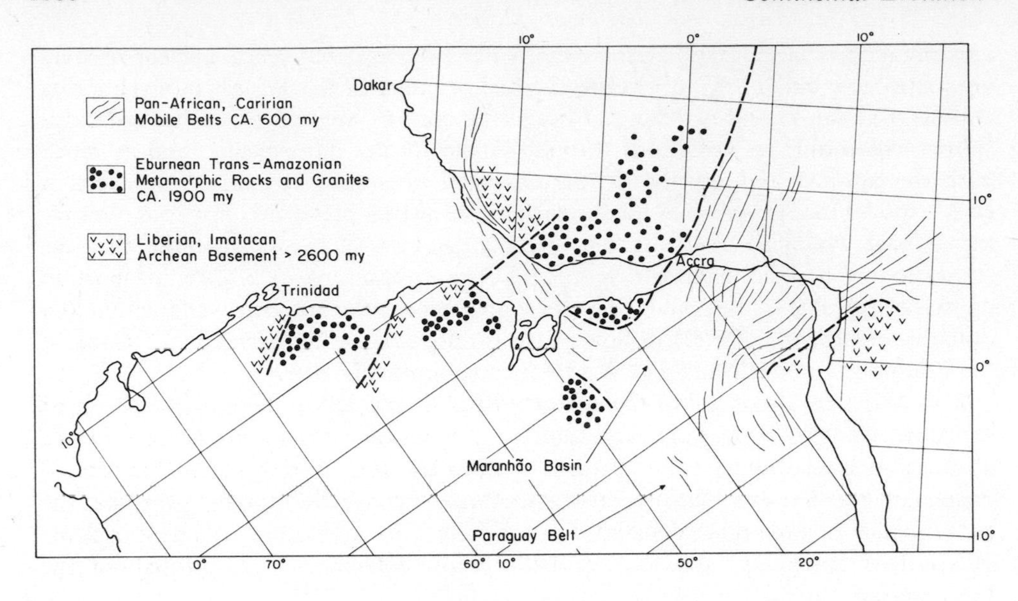

FIGURE 3

do Navio of Amapa, Brazil. Granulites occur in the oldest basement rocks in Sierra Leone, Liberia, Ivory Coast, Venezuela, Guyana, Surinam, French Guiana, and Amapa, Brazil. Iron formation also occurs in the older rocks in Sierra Leone, Liberia and Venezuela. When South America is rotated about 40° counterclockwise to fit into pre-drift reconstruction, the general structural trends and age provinces line up reasonably well (Fig. 3). It seems unlikely that the single and apparently continuous ancient block of Guyana–West Africa should have developed and existed from about 3000 m.y. to some time less than about 1800 m.y. and then have been split and moved apart with oceans forming between and around the two halves, and subsequently come together again in the same position and orientation that the two halves were in originally.

In the North Atlantic a similar matching of earlier geological provinces has been demonstrated by Wynne-Edwards & Hasan (1970), who noted the continuation of the Grenville Province and associated anorthosites east of the Caledonian belt in Norway and Sweden. Others (e.g. Hurley, 1970) have noted the coherent distribution of the most ancient parts of Laurasia within a region split by the Atlantic Ocean.

In summary, if the 450 $\pm$ 200 m.y. belts were formed entirely by the plate tectonics of today it would be expected that the spatial extensiveness and long time span of activity would demand extremely large relative motions between bordering ancient cratons. Principal belts would tend to mark the borders between a hemispheric-sized continental assemblage and a similar sized ocean. There should be no correlation of geologic features or trends between cratons on either side of a 450 $\pm$ 200 m.y. belt. Further extension of the process into the past would mean even more extreme fragmentation and random reassemblage of all present continental blocks.

On the other hand, if the 450 $\pm$ 200 m.y. belts were formed partly *in situ* it would imply that the upwelling part of the Earth's convected heat release was also an important process of orogeny, that new sial is added by direct differentiation in uprising diapirs of mantle material (with water present), that relics, ancient residual granulites, low initial Sr^{87}/Sr^{86} ratios, thermal resetting of Rb–Sr ages in neighboring cratons, refoliation and matching of bordering cratons were part of a hot rise-ridge system which only in part broke through to the surface dynamically, and in which both the upward and downward motions were essentially vertical. Presumably in such a model the resistance of the bordering continental plates to a horizontal movement apart would so constrain the thermal columns of mantle material that the separation of a sialic magmatic suite could occur, resulting in a lower proportion of mafic volcanic rocks, and only local evidences of turnover, subduction and ophiolites where the convective flow broke through to the surface. The matching of the Precambrian of Guyana and West Africa support this view.

It is clear that almost all of the characteristics of the 450 ± 200 m.y. belts can be explained by both hypotheses, and that the conclusion must await secure evidence from a matching of pre-existing features between cratons, or from acceptable palaeomagnetic data. The fact that the Earth loses heat by convection today means that the process must be even more demanded in the past. This is a strong argument in favor of sea-floor spreading, and thus probably continental break-up, throughout the Precambrian.

References

Armstrong, R. L., 1968. A model for the evaluation of Sr and Pb isotopes in a dynamic earth, *Rev. Geophys.*, **6,** 175.

Clifford, T. N., 1968. Radiometric dating and the pre-Silurian geology of Africa. *In:* Radiometric Dating for Geologists, edited by E. E. Hamilton and R. M. Farquhar, p. 299. Interscience.

Dickinson, W. R. and Luth, W. C., 1971. A model for plate tectonic evolution of mantle layers, *Science*, **174,** 400.

Fullagar, P. D., 1971. Age and origin of plutonic intrusions in the Piedmont of the Southeastern Appalachians, *Bull. Geol. Soc. Amer.*, **82,** 2845.

Holmes, A., 1948. The sequence of Precambrian orogenic belts in south and central Africa. Intern. Geol. Cong. 18th, London, p. 254.

Hurley, P. M., 1970. Distribution of age provinces in Laurasia, *Earth Planet. Sci. Letts.*, **8,** 189.

Hurley, P. M., 1972. Can the subduction process of mountain building be extended to Pan-African and similar orogenic belts? *Earth Planet. Sci. Letts.*, **15,** 305.

Kennedy, W. Q. The structural differentiation of Africa in the Pan-African (±500 m.y.) tectonic episode. Res. Inst. African Geol., Univ. Leeds 8th Ann. Rept. (1962–63), p. 48.

McConnell, R. B. and Williams, E., 1969. Distribution and provisional correlation of the Precambrian of the Guyana Shield. Paper presented at 8th Guyana Geol. Conf., Georgetown, Guyana.

Wynne-Edwards, H. R. and Hasan, Z., 1970. Intersecting orogenic belts across the North Atlantic, *Amer. J. Sci.*, **268,** 289.

Zwart, H. J., 1969. Metamorphic facies series in the European orogenic belts and their bearing on the causes of orogeny, *Geol. Assoc. Can. Spec. Paper No.* 5, p. 7.

10.8

R. M. SHACKLETON
Department of Earth Sciences,
University of Leeds, Leeds, England

Correlation of Structures Across Precambrian Orogenic Belts in Africa

Plate tectonic theory has led to the recognition of two types of mountain belts. The first, developed over a subduction zone where an oceanic plate is being consumed, may be an island arc if the non-consuming plate is oceanic, or Cordilleran if it is continental. The second is the result of collision either between two continental plates or between a continent and an island arc. The collision type is structurally complex but usually shows a dominant sense of thrusting on to the consumed plate. A characteristic feature is the presence of a suture zone, where two originally distant plates have been welded together. This suture is often marked by ophiolites, abyssal sediments and slices of suboceanic crust and upper mantle which represent the remnants of the consumed oceanic plate. Orogenic belts within continents, with older continental plates on both sides of them, such as those in Central Africa considered here, must, if they are the result of plate tectonics, be of the collision type. The plates on either side should have been originally far apart and separated by oceanic crust now represented only as fragments along a suture.

Few Precambrian orogenic belts can be clearly identified as either Cordilleran/island arc or collision type. The Coronation geosyncline, more than 1750 m.y. old, in northwestern Canada, is a rare example of a very old Cordilleran-type belt (Hoffman *et al.*, 1970). The chemistry of the volcanic rocks in the Archaean belts is similar to that of volcanic rocks on the trench side of island arcs (Hart *et al.*, 1970) and so these belts have been interpreted as island arcs, despite the absence of the diagnostic asymmetric metamorphism, magmatism and structure. The great wrench faults known in many Precambrian terrains may be transform faults, as all currently active major wrench faults appear to be. But even if all these various phenomena are accepted as the results of motions of plates relative to one another, there remain many others which seem to imply some other process which involves little or no relative motion of crustal plates or crustal shortening between them and no former oceanic plates which have been consumed.

The relative motions of crustal plates in a continental domain may be evaluated from palaeomagnetic, structural or tectonic evidence. I shall examine some of this evidence as it relates to motions of plates in the African continent during Precambrian times.

Palaeomagnetic evidence from Africa demonstrates extensive and complex motions relative to the magnetic dipole axis (McElhinny *et al.*, 1968) but motions of one plate relative to another are not implied by the data so far available. Precambrian polar-wandering curves for the Kaapvaal–Rhodesia and Tanzania Archaean cratons fit as though there has been no significant movement between them since 930 m.y. ago (Briden *et al.*, 1971), but as the curves are consecutive rather than overlapping, more data are needed before any firm conclusion can be reached.

The structural method depends on integration of measurements of finite strain (Ramsay, 1969) from many outcrops across the whole width of an orogenic belt. Even if enough suitable strain markers, such as pebbles or randomly orientated veins or dykes, can be found, the strains are usually so heterogeneous that an immense amount of work is needed before reliable conclusions can be drawn (estimates of the crustal shortening across the Alps have varied from 1500 km to zero), and at present no results are available across a whole orogenic belt anywhere in Africa. Some data from the southern part of the Zambesi belt (Talbot, 1967) indicate significant north–south crustal shortening but this cannot be extrapolated to the whole width of the belt.

One tectonic method is to measure the offsets of lineaments which are crossed obliquely by a later orogenic belt and can be recognized in the cratons on either side. Such lineaments may be major faults, the edges of earlier orogenic belts, dyke swarms or an individual dyke, geochemically or isotopically defined lines, or any other lineament. Such lineaments may be straight or curved but if curved the curvature must be regular. This method can be applied to the Zambesi and Kibaran belts in Central Africa. The lineaments recognized on either side of the Zambesi belt and within the adjacent part of the Mozambique belt are shown in Fig. 1. Across the Zambesi belt the most significant lineament is the southeast front of the Irumide belt. South of the Zambesi belt, in Rhodesia, this front must be between the Lomagundi fold belt, where a K/Ar age of 1655 m.y. in phyllites (Stagman, 1962) shows the absence of Irumide imprint and the Kamativi area (26°45′E, 18°30′ S), where Irumide ages (but also older ones) are found. North of the Zambesi belt, in southeast Zambia, this front appears to be defined by a difference between Irumide and older trends (Fig. 1). Farther northeast, the Irumide belt becomes narrower. The metasediments in the Mafingi Range (30°20′ E, 10° S) which show Irumide folding and metamorphism seem to be near the end of this Irumide belt. A second lineament is the eastern front of the Lomagundi fold belt. This is well-defined in Rhodesia and it appears to be recognizable as the eastern limit of a similar-trending system of folds in southeast Zambia. This belt is probably the Tumbide belt (Ackermann & Forster, 1960; Forster, 1965). A third lineament may be provided by the Great Dyke of Rhodesia, if the basic rocks in line with it north of the Zambesi are its northward continuation, as has been suggested (de Swardt *et al.*, 1965). A fourth lineament is provided by the western limit, in Mozambique, Malawi and Zambia, of granulite facies rocks which appear to connect the Limpopo and Ubendian belts, both of which are distinguished from the cratons on either side of them by the presence of granulite-facies rocks. The relations of these several lineaments to the Zambesi and Mozambique belts show that the deformation in these belts cannot have involved large relative movements of the cratons on either side, since it is difficult to believe that the cratons moved apart to form oceanic crust and then back to their previous positions; nor is there any sign of a suture.

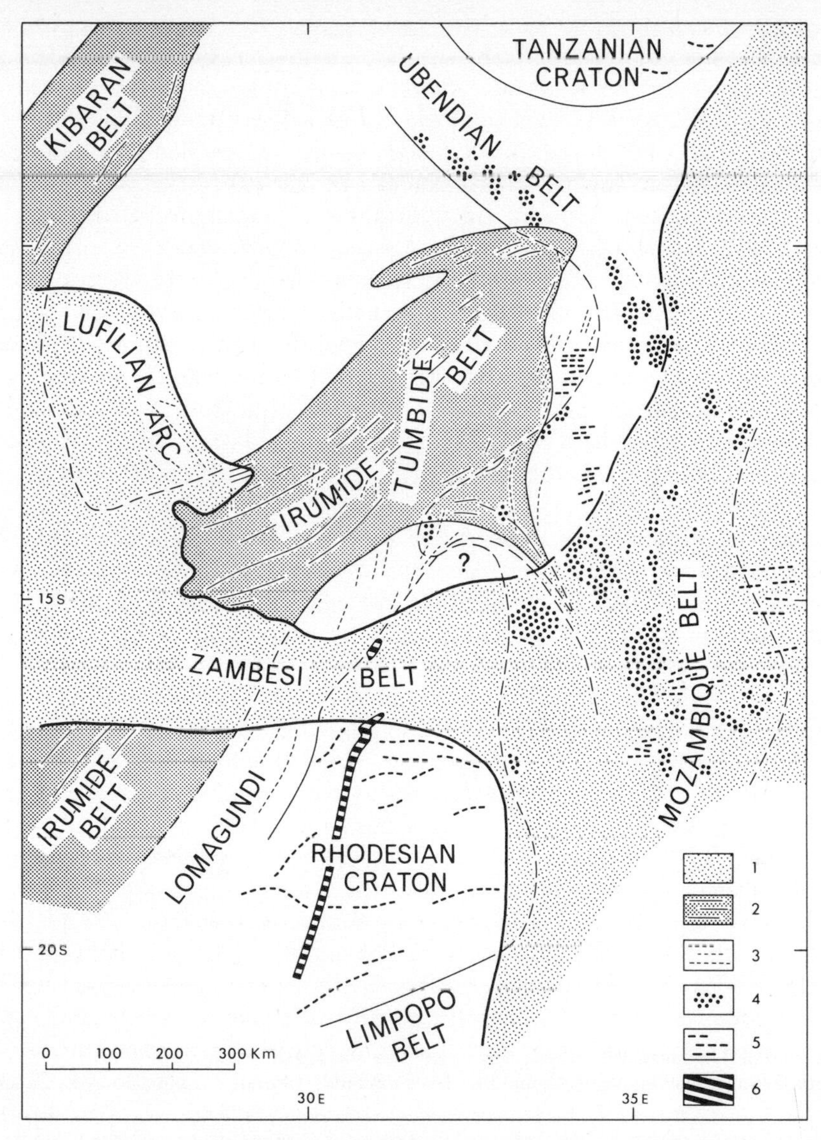

FIGURE 1 Interpretation of the tectonics of Central Africa. 1. Pan-African (Mozambiquian, Zambesi and Lufilian) fold belts ($\pm$500 m.y.). 2. Irumide and Kibaran fold belts ($\pm$1000 m.y.) with fold trends. 3. Lomagundi and Tumbide trends ($\pm$1700 m.y.). 4. Granulite-facies rocks in Ubendian and Mozambique belts and supposed connection with Limpopo belt. 5. Archaean trends (>2550 m.y.) in Rhodesian and Tanzanian cratons, and possible Archaean trends within younger fold belts. 6. Great Dyke of Rhodesia and its possible continuation on north side of Zambesi belt.

Farther north, in Central Africa, there is another oblique crossing where the Kibaran fold belt crosses the Ubendian–Ruzizi fold belt (Fig. 2). Again the older belt can be traced continuously into and through the younger one, from which it emerges without any appreciable offset. The margins could no doubt be fixed with more precision but from what is already known it may be concluded that here too a major fold belt with many extensive granite plutons, was the result of processes which

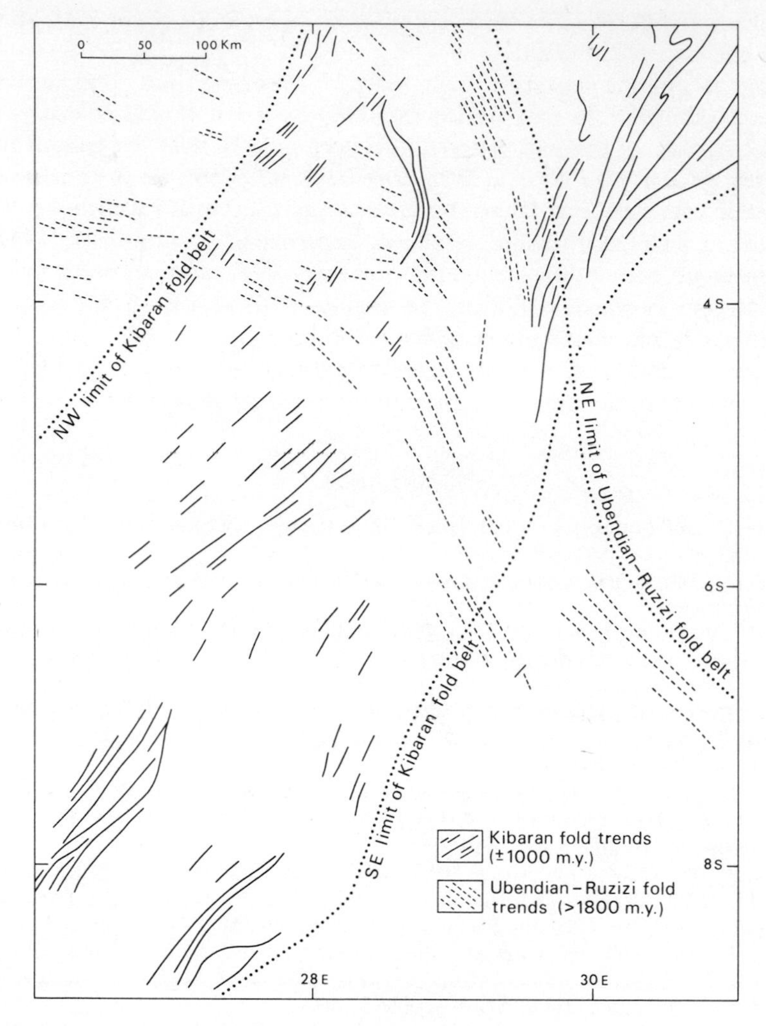

FIGURE 2 The Ubendian-Ruzizi fold belt crossed by the Kibaran fold belt.
–based on Esquisse tectonique du Congo Belge et du Ruanda-Urundi, 1:3,000,000 by L. Cahen 1952.

involved little or no movement of the plates on either side relative to one another.

A different, less precise tectonic method depends upon the distribution of distinctive groups of rocks. Those which have been used are anorthosites (Herz, 1969), kimberlites and the diamonds derived from them (Knopf, 1968), granulite facies rocks and banded ironstones. These rocks show simple distribution patterns if the continental plates are arranged as they supposedly were before the break-up of Gondwanaland and Laurasia. This arrangement would be difficult to explain if, during the Precambrian, they were distributed in a number of independently moving plates.

I have suggested elsewhere (Shackleton, 1969) that several independent arguments lead to the conclusion that many orogenic belts overlay upwelling currents in the mantle. It is supposed that the crust over such an upwelling current does not always break apart to form new oceanic crust. If it does not, the small proportion of acid magma which is differentiated from melts in the upwelling current may accumulate at or near the base of the crust.

Assuming an upward velocity of 5 cm/year, 10% melting (Bott, 1965) and 0·5% of sialic differentiate from the melt (in Iceland the proportion of acid volcanics is about 10%), the thickness of the acid material, if it accumulated near the base of the crust, would reach 1·0 km after 40 m.y. It is proposed that when the thickness reaches a limiting value, perhaps 2 or 3 km, the gravitational instability becomes sufficient to cause overturn and deformation, as shown experimentally (Ramberg, 1967). Thus this process might account for the intermittent or cyclical nature of orogenic deformation and plutonism which is difficult to explain if these are attributed to relative motions of plates and melting from a Benioff Zone.

It is concluded that many ancient orogenic belts may be the result of processes above upwelling currents in the upper mantle rather than plate movements.

References

Ackermann, E. and Forster, A., 1960. Inter. Geol. Congr. XXI Sess., Norden-Copenhagen, Part 18, 182–92.
Bott, M. H. P., 1965. Upper Mantle Symposium, New Dehli, 1964 of the Intern. Un. of Geol. Sci., Copenhagen.
Briden, J. C., Piper, J. D. A., Henthorn, D. I. and Rex, D. C., 1971. 15th Ann. Rept. Res. Inst. Af. Geol., Univ. Leeds, pp. 46–50.
Forster, A., 1965. *Geotekt. Forsch. Heft*, **20**, 1–115.
Hart, S. R., Brooks, C., Krogh, T. E., Davis, G. L. and Nava, D., 1970. *Earth Planet. Sci. Letts.*, **10**, 17–28.
Herz, N., 1969. *Science*, **164**, 944–6.
Hoffman, P. F., Fraser, J. A. and McGlynn, J. C., 1970. *Geol. Surv. Can.* 70–40, 201–12.
Knopf, D., 1968. Les kimberlites et les roches apparentées de Côte d'Ivoire. Thèse, D. ès Sc., Fac Sci. Univ. Lausanne, Sodeni-Abidjan.
McElhinny, M. W., Briden, J. C., Jones, D. L. and Brock, A., 1968. *Rev. Geophys.*, **6**, 207–38.
Ramberg, H., 1967. Gravity, Deformation of the Earth's Crust. Academic Press, London.
Ramsay, J. G., 1969. *In:* Time and Place in Orogeny, *Geol. Soc. Lond. Spec. Publ.* 3, 43–79.
Shackleton, R. M., 1969. *In:* Time and Place in Orogeny, *Geol. Soc. Lond. Spec. Publ. 3*, 1–7.
Stagman, J. G., 1962. *Bull. Geol. Surv. S. Rhod.*, **55.**
Swardt, A. M. J. de., Garrard, P. and Simpson, J. G., 1965. *Geol. Soc. Am. Bull.* **76,** 89–102.
Talbot, C. J., 1967. Rock Deformation at the eastern end of the Zambesi Orogenic Belt, Rhodesia (2 vols). Ph.D. Thesis, Dept. of Earth Sci., Univ. Leeds.

11

EPILOGUE

11.1

R. A. WELLS

Space Sciences Laboratory,
University of California,
Berkeley, California 94720, *USA*

Martian, Lunar, and Terrestrial Crusts: A 3-Dimensional Exercise in Comparative Geophysics

---

Five years ago Runcorn (1967) wrote that '. . . the possibility of convection makes a careful study of planetary surfaces for indications (however different from those of the Earth's surface) of internal activity, volcanic or tectonic, even of the most feeble kind, of great scientific significance.'

Although covering only about 10% of the Martian surface, the photographs returned by Mariners 4, 6, and 7 revealed little in the way of internal crustal activity. Some tectonic evidence in the form of parallel lineaments and polygonal crater walls was all that could be gleaned from these photographs pertaining to internal mechanisms at work on Mars.

However, this situation changed dramatically even with the first few photographs of the surface taken by Mariner 9 after the planet-wide dust storm cleared in mid-January 1972. These photographs revealed indisputable calderas with all of the terrestrial morphology present: rising shields surmounted by lava channels, the primary caldera with smaller subsidiary craters in their floors, stepped benches on the inside of crater walls, and both arcuate and radial fault patterns surrounding each province. Later photos showed widespread lava flows, sinuous lunar-like rilles, large-scale fractures, and massive faulting and slumping. In one case in particular, a rift zone was discovered near the Martian equator extending 5000 km in a nearly east–west direction with widths approaching 100 km. The depth of the main fracture varied from 2 to 7 km along this length. 5000 km on Mars is approximately 80° of longitude, nearly a quarter of the distance around the circumference. Thus, the feature is even more imposing than the East African Rift System.

In a large proportion of photographs, many features were clearly controlled by a tectonic grid pattern indicating orthogonal sub-surface prestressing prior to the eventual surface expression. (See McCauley *et al.* (1972) for photo geology.)

The major geological feature on Mars can be summarized with one word—volcanism. Although this revelation does not necessarily imply that convective processes have operated in the Martian mantle, it nevertheless points to far more crustal activity than had been supposed.

Prior to these later discoveries, the author (Wells, 1969a, b, 1971) put forward the hypothesis of continental drift and sea-floor spreading to account for *presumed* continental-type blocks and basin structures on Mars. This hypothesis gained support

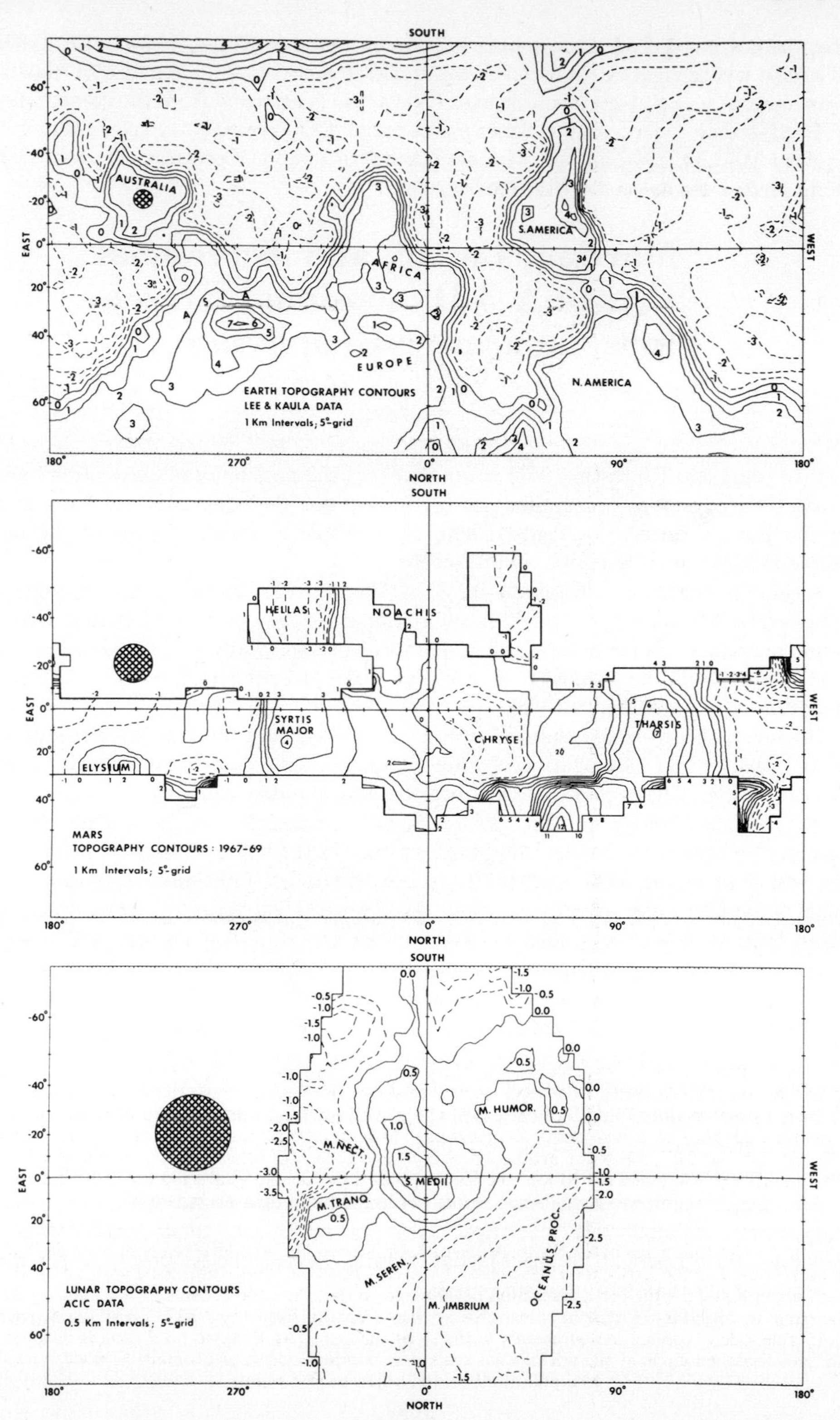
SOUTH
NORTH
EAST
WEST
AUSTRALIA
AFRICA
EUROPE
S. AMERICA
N. AMERICA
EARTH TOPOGRAPHY CONTOURS
LEE & KAULA DATA
1 Km Intervals; 5°-grid
HELLAS
NOACHIS
SYRTIS MAJOR
ELYSIUM
CHRYSE
THARSIS
MARS
TOPOGRAPHY CONTOURS: 1967–69
1 Km Intervals; 5°-grid
M. HUMOR.
M. NECT.
S. MEDII
M. TRANQ.
M. SEREN.
M. IMBRIUM
OCEANUS PROC.
LUNAR TOPOGRAPHY CONTOURS
ACIC DATA
0.5 Km Intervals; 5°-grid
180°
270°
0°
90°
180°

FIGURE 1

when actual blocks and basins were eventually mapped (Wells, 1969c, 1971, 1972) on a scale whose areal extent matched the blocks and basins of the Earth. The Martian features were revealed in a contour map (Wells, 1972) prepared from the combination of Earth-based radar ranging data, Earth-based CO_2 pressure measurements over resolved Martian areas, and the Mariner 6 and 7 IRS and UV pressure determinations along limited tracks on the Martian surface.

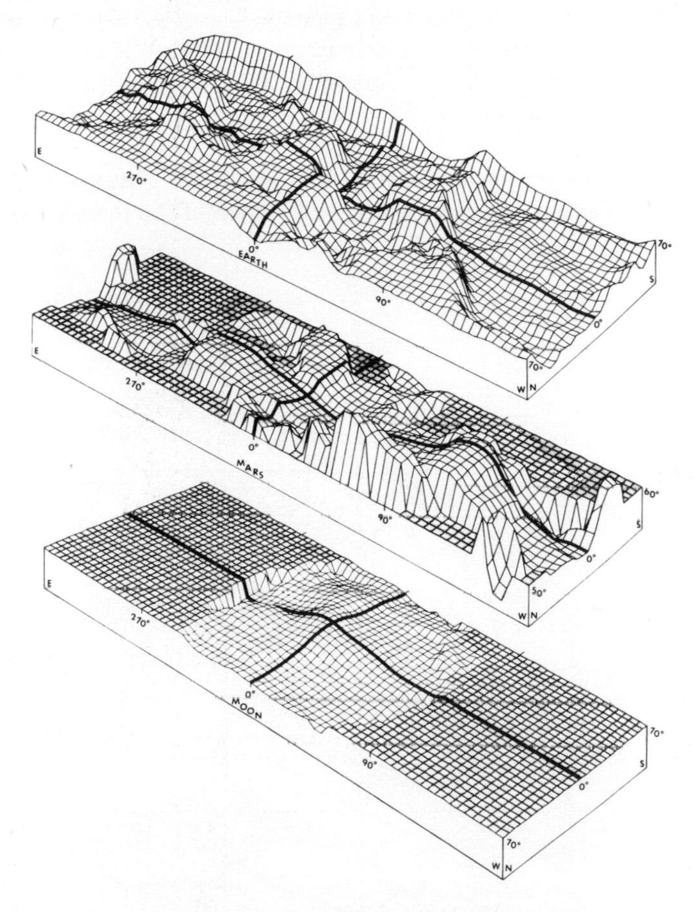

FIGURE 2 Perspective views of the Earth, Mars, and the Moon drawn by computer from the same 5° grid arrays used to construct the contour maps shown in Fig. 1. Views are looking towards the SE from a point inclined 20° to the horizontal. The vertical scale has been exaggerated twice and is the same for each body to permit direct comparisons. However, the blocks seen on Mars at 180°, 20° S; 150°, 45° N; and 245°, 40° N are artifacts produced by the lack of reliable data at the map edges in these areas. For Mars and the Moon, the flat, stippled areas denote unobserved regions and are set at the zero level; hence, slopes bordering their edges are fictitious.

FIGURE 1 Contour maps of the Earth, Mars, and the Moon. The zero level refers to the mean radius of each body. For the Earth, the 1° grid data of Lee & Kaula (1967) were processed onto a 5° grid and each data point smoothed to a resolution of 1000 km so that the scale of the contour map would be the same as for Mars. The lunar contour map was prepared from the A.C.I.C. (Meyer & Ruffin, 1965) selenodetic control measurements in the same manner. The 1000 km resolution is denoted by the cross-hatched circle at the left of each map. The computer contour program is based on linear interpolation between adjacent grid points.

The data from this contour map have been plotted by a computer in the form of a three-dimensional perspective view. Similarly, the total Earth topography (Lee & Kaula, 1967) and the selenodetic altitude measurements of the lunar front-face (Meyer & Ruffin, 1965) were processed by the same equations to the same scale used for the preparation of the Martian contour map.

Figure 1 compares the contour maps of the Earth, Mars, and the Moon prepared in this fashion. South is placed at the top to conform with the Astronomical convention of depicting maps of the Moon and planets. The cross-hatched circle on the left of each contour map represents the resolution (1000 km) of each data point in the 5° grid of values contoured by the computer. This resolution is set by the best available for the Martian contour map. The zero level in each case refers to the mean radius of the body concerned.

Figures 2 and 3 show the comparison of the large-scale features in perspective for the three bodies. Vertical scales are the same for each of the three views to permit

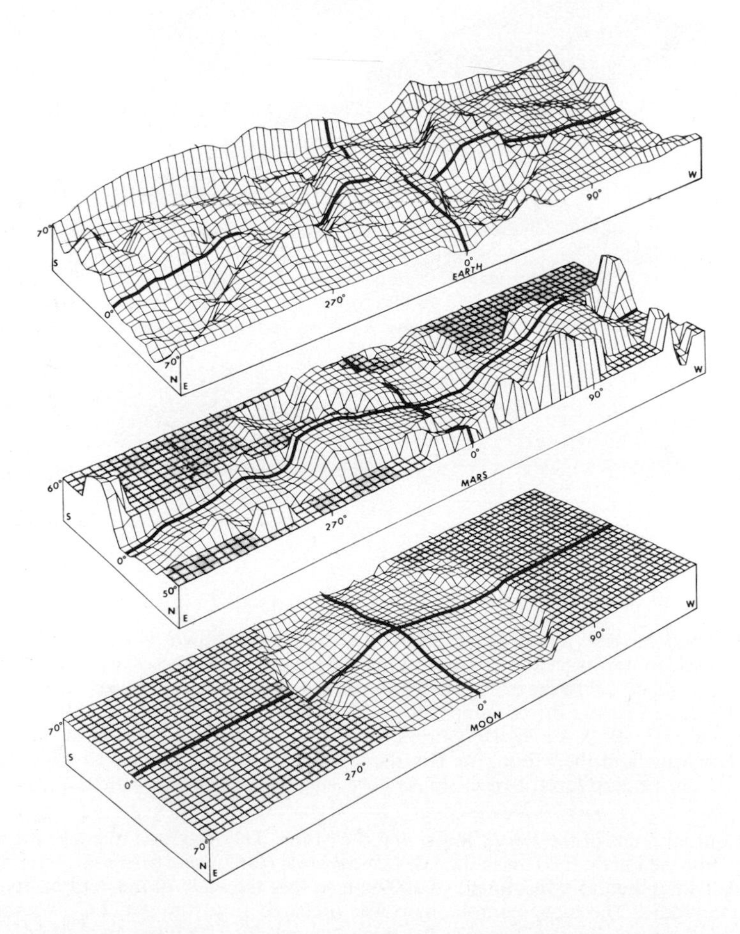

FIGURE 3 Perspective views of Earth, Mars, and the Moon rotated 90° with respect to Fig. 2. The views are looking towards the SW from an elevation angle of 20°.

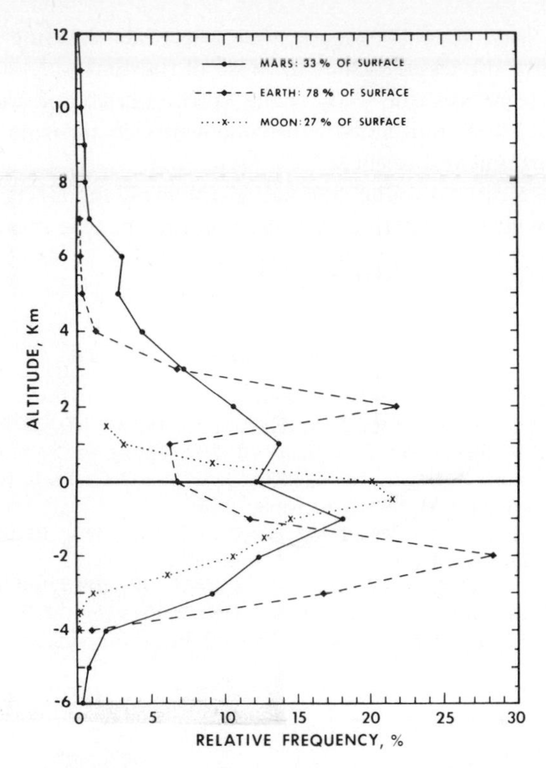

FIGURE 4 A comparison of the hypsometric curves of the Earth, Mars, and the Moon prepared by tracing the contours of Fig. 1 on an equal area projection. The abscissa denotes the areal frequency and the ordinate the altitude of the surface. The fractional areas mapped for each body are listed with the symbol definitions in the inset. Zero altitude refers to the mean radius of the body concerned.

a direct comparison of the sizes of the features seen. Grid intervals are 5° in both latitude and longitude for each case.

It is evident that the blocks and basins on Mars are more nearly comparable with those on the Earth than with those on the lunar crust which reveals only gradual undulations. The range of heights for the Earth is 11 km; for Mars 18 km; and for the Moon 5 km.

Hypsometric analyses of the three contour maps drawn on an equal area projection are given in Fig. 4. An interpretation of the nature of the peaks in areal frequency as a function of altitude in terms of isostatic mechanisms cannot be made without supplementary information regarding the density, temperature, and compositional distributions within the bodies. However, a bi-modal distribution does imply a change in slope from the highest features to the low points of basins; whereas a unimodal distribution signifies a constant slope from the tops of the highest areas to the bottoms of the lowest regions. Thus, Fig. 4 also implies that the Martian crust is, at least geometrically, more like the Earth than the Moon.

The presence of large-scale blocks and basins on Mars, and the recently detected widespread volcanism, extensive faulting and rifting of the Martian crust do not, of course, guarantee a 'continental drift and sea-floor spreading' theory for Mars, which may be even more difficult to firmly establish than it has been in the terrestrial case; but they at least strongly suggest that Mars has at some time in its history undergone formative processes similar to those of the Earth.

Within the next few years, the study of the Martian crust and the development of a Martian geophysics will undoubtedly provide valuable information of interest to the terrestrial environmental sciences.

A more complete discussion of Martian geophysics, including current spacecraft results integrated with past Earth-based observations, can be found in Wells (1973).

Supported by NASA grant NGR 05–003–431.

References

Lee, W. and Kaula, W. M., 1967. *J. Geophys. Res.*, **72,** 753.

McCauley, J. F., Carr, M. H., Cutts, J. A., Hartmann, W. K., Masursky, H., Milton, D. J., Sharp, R. P., and Wilhelms, D. E., *Icarus*, **17,** 289 (1972).

Meyer, D. L. and Ruffin, B. W. *Icarus*, 4, 513 (1965); A.C.I.C. Tech. paper No. 15, March (1965); A.C.I.C. Report RM–698–1, October (1966).

Runcorn, S. K., 1967. *In:* Mantles of the Earth and Terrestrial Planets, edited by S. K. Runcorn, p. 513. Interscience (Wiley), London.

Wells, R. A., *Geophys. J.R.A.S.*, **17,** 209 (1969a); *Ibid.*, **18,** 109 (1969b); *Science*, **166,** 862 (1969c); *Phys. Earth Planet. Interiors*, **4,** 273 (1971); *Geophys. J.R.A.S.*, **27,** 101 (1972); *Morphology of the Planet Mars*, D. Reidel, Dordrecht, in preparation (1973).

11.2

R. S. DIETZ and J. C. HOLDEN

National Oceanic and Atmospheric Administration, Atlantic Oceanographic and Meteorological Laboratories,
901 *South Miami Ave., Miami, Florida* 33130

Continents Adrift: New Orthodoxy or Persuasive Joker?

The modern rebirth of interest in continental drift commenced, we believe, in the mid-1950s when it was shown by a certain British geophysicist (who will remain nameless) from studies of rock magnetism that the polar-wander curves of North America and Europe, congruent in earlier times, diverged by about 30° in the Cretaceous and Cenozoic. This was widely, or perhaps wildly, hailed as proving the opening of the North Atlantic by continental drift.

This enthusiasm proved premature, because geologists rather carelessly handle their rocks, as Sir Harold Jeffries (1970) recently pointed out in his book, *The Earth.* He writes:

> 'When I last did a magnetic experiment (about 1909), we were warned against careless handling of permanent magnets, and the magnetism was liable to change without much carelessness. In studying the magnetism of rock the specimen has to be broken off with a geological hammer . . . It is supposed that, in the process, its magnetism does not change of any important extent, and, though I have often asked how this comes to be the case; I have never received any answer.'

For this, and other reasons, Jeffries dismisses Wegener's drift as being 'quantitatively insufficient and qualitatively inapplicable. It is an explanation which explains nothing which we wish to explain.' It would seem, therefore, that rocks collected for paleomagnetic measurements by hammers (and how else does a geologist collect his rocks) would have their magnetic memory scrambled. This being obviously so, rock magnetic studies must be regarded as worthless. The first law of geology, we suppose, is: The rocks remember, while liquids and gases forget. But it can hardly apply to rocks which have been hit on the head.

On the authority of Beloussov (1962) we have it that, 'the hypotheses suggesting horizontal drift of the continents, among them the hypothesis of Wegener, which was once famous, must be regarded as fantastic and having nothing to do with science . . . It is a source of profound amazement that such a hypothesis—based as it is on an overtly formalistic approach to the major problems and on a total and consistent disregard of the basic geotectonic data and, as already stated, explaining nothing of what must be explained in the first place—was not only seriously discussed in scientific literature but achieved considerable success and attracted some of the leading authorities into the ranks of its adherents. These men were apparently hypnotized by the boldness of Wegener's ideas and by his brilliant style of writing.'

This was a traumatic revelation to the zealots of continental drift and they fell into disarray. Like all true scientists, the drifters would 'rather be right than President,' but obviously they were neither.

The Animals Remember

All was not lost, however, for new evidence came in from an entirely unexpected quarter—the animal kingdom. Animals have some remarkably developed instincts, which sometimes recapitulate their evolutionary history. Witness the so-called loud bats with their FM sonar chirps by which they can search out and classify a moving target—usually moths, their favourite meal. But it remained for a bird, by its remarkable migratory path, to first demonstrate that the New World really has drifted away from the Old World.

This doughty bird, the sooty hoodwink, *Puffinus oceanicus*, winters in the Atlantic sector of Antarctica; then each spring it heads north, determined to nest in far away Spitzbergen. As if flying to this remote island is not a sufficient demonstration of fortitude, this bird chooses a zigzag path. First it touches down in Southwest Africa where, because of its confused and dazed habit of stumbling about (apparently searching for fresh bearings), it is locally termed the random walkabout. Then this bird executes further zigs and zags across the ocean as it threads its way north. On April 1st, the frayed remnants of the flock touch down on the British Isles at Lands End. (Remarkable as it is, their navigation sometimes goes awry. An errant flock was seen in 1967 far off course in the spaghetti fields of the Po valley.) Toward the end of April the sooty hoodwink finally reaches its destination, Spitzbergen.

FIGURE 1
. . . bat sonar

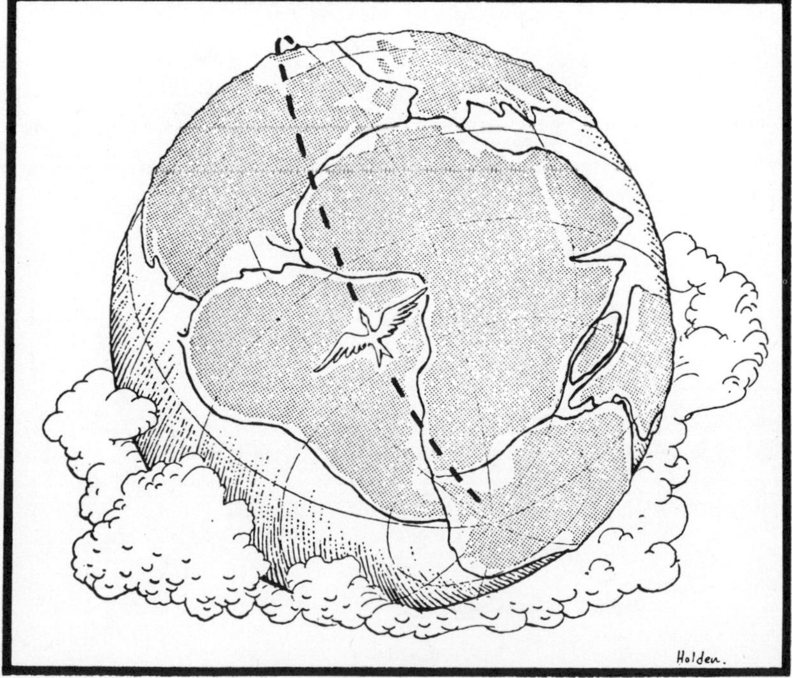

FIGURE 2 . . . a doughty bird

This curious migration path finds ready explanation once we recognize that continental drift has occurred. If we close the Atlantic Ocean, we find that the zigzag path becomes a straight line, a great circle route, or the shortest distance between Antarctica and Spitzbergen.

Another animal, albeit extinct, fills in still another facet of the continental drift puzzle. We refer here to *Glossopstompodon loathifoliata*, whose bleached remains are found in the Permian red beds of the Sahara. Critics of drift have argued that there never could have been a universal continent of Pangaea in the late Paleozoic, because, if this were true, the Glossopoteris flora on Gondwana would certainly have invaded the northern continents of Laurasia. These critics, however, failed to reckon with the tempestuous temper and unremitting phobia of *Glossopstompodon* for glossopterids. By setting up a rampaging patrol along the equator and trampling any young glossopterids sprouts, this reptile established a successful barrier against the northward migration of this flora. (We are reminded here in passing of the acceptance speech by a young English geophysicist when receiving a medal for his contribution to plate tectonics. If I recall correctly, he said with characteristic British modesty, 'If I have seen farther than others, it is because I have stomped on heads of giants.')

A somewhat similar explanation applies to the curious swimming behavior of the deep sea squid, *Architeuthis solenoides*. Nearly all squid swim backwards through life, apparently preferring not to look where they are going, but to see where they have been. By swimming forward, *A. solenoides* is exceptional, but apparently he has not always swum in this manner. Experiments show that he may be programmed to swim either forward or backward within an aquarium surrounded by a coil simply by

FIGURE 3
. . . *Glossostompodon loathifoliata*

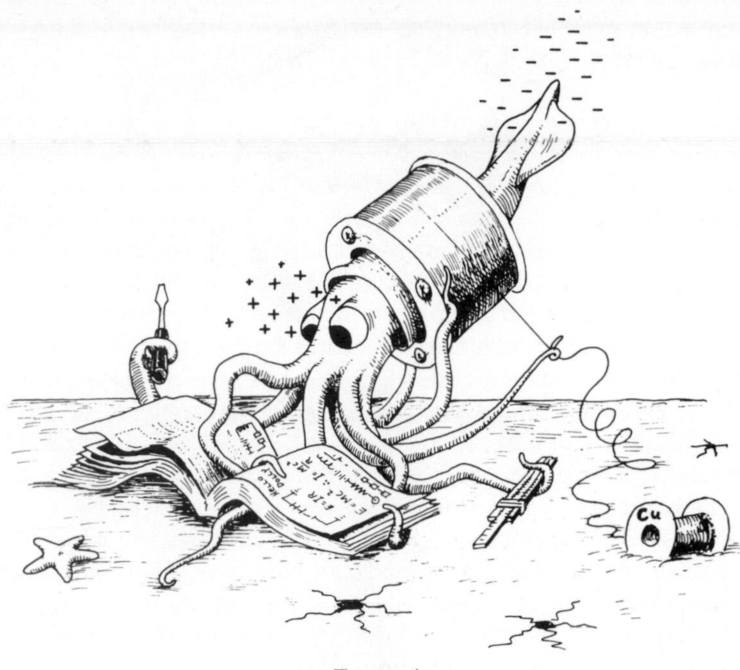

FIGURE 4
. . . *Architeuthis solenoides*

reversing the magnetic field. It would seem, then, that *A. solenoides* became a forward swimmer only 700,000 years ago when the earth's field switched from reversed to normal at the Matsuyama–Brunhes boundary.

We have learned of late from the paleomagicians that, although 'east is east and west is west, and never the twain shall meet,' this adage does not apply to north and south. Every so often, and quicker than you can say Willem Jean Marie van Waterschoot van der Gracht, north may become south and, we hope, vice versa. We have learned this from the 'fossil compasses' frozen in basalts and other rocks. Actually, animals which are senstive to the magnetic field of force have known this all along. It has been known for many years that, upon molting, crabs place a grain of sand in their inner ear, which then becomes a sensor for geotropism or, more simply, balance. Some crabs carelessly choose a grain of magnetite—a ferromagnetic mineral. Then, if a strong magnet is above their aquarium, they will forever crawl along the roof of their home. This much is evident by direct experimentation. But why do crabs crawl sideways? This mystery is solved when we recall the frequent flips in the earth dipole field. The crab becomes confused as to whether he should walk forwards or backwards, so, adapting to compromise, he instead walks sideways.

Yet another animal may be cited as proving continents drift. This is the common European eel, *Anguila*. After a few years in the streams of Europe, the eel heads for the Sargasso Sea on the far side of the mid-ocean spreading rift to spawn and then die. Then the newborn eel, the *Leptocephalus* stage, swims back to Europe—this instinct inherited from its parents. But by this time Europe is not where it was supposed to be,

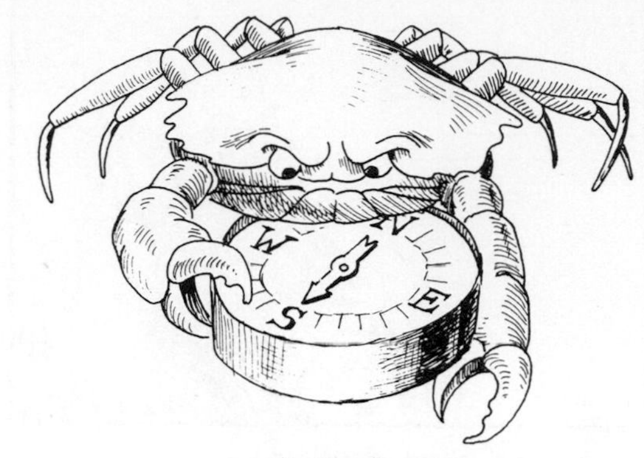

FIGURE 5
. . . and never the twain shall meet (?)

FIGURE 6
. . . one more last step

as this continent has drifted several centimeters eastward. No one knows for sure just what crosses the mind of a young eel in the face of this predicament. All we know for sure is that all eels go through life with a quizzical look. A chinese proverb runs that, in a journey of a thousand miles, the last step is no more important than any other. But suppose that every year someone adds *one more* last step. For eels, it would seem that one more step beyond the last step is needed for survival of the species.

A final animal which adds credence to drift is none other than the famous Loch Ness monster. Photos and descriptions of this elusive monster show its long neck and small, reptilian head, so that it is a swimming 'dinosaur' or pleisiosaur, a living fossil from the Jurassic. It also has been noted by boats plying the loch that Nessie yields to the right, thus obeying Napoleonic (or North American) rather than Caesarian (or British) rule-of-the-road. Clearly, then, Nessie is a beast of the New World now stranded in the Old World by the break away of Europe from North America in the Cretaceous. There seems now to be a new urgency for somehow corralling Nessie and, while treating her with the tender loving care that befits an endangered species, making such measurements as would permit identification as to species—for example, counting her teeth.

FIGURE 7
. . over the loch from Urquhart castle

Historical Beginnings

Before proceeding further, we would like to pay homage to the originators of continental drift. There seems little doubt that the concept is originally ascribable to Frank Bacon (Blackett, *et al.* 1965), who, in his *Novum Organum* of 1620, wrote: 'The very configuration of the world itself in its greater part presents Conformable Instances which are not to be neglected. Take, for example, Africa and the region of Peru with the continent stretching to the Straits of Magellan, in each of which tracts there are similar isthmuses and similar promontories, which can hardly be by accident.' Bacon undoubtedly thought that, given this hint, the reader would have sufficient intelligence to see that the South Atlantic Ocean can be closed in a continental drift reconstruction. This, of course, is simply a matter of flipping South America upside down, north for south, and then sliding this continent eastward such that the bulge of Peru fits beneath the bulge of Africa. (We should mention in passing that there is no truth to the rumor that Bacon wrote the Shakespearean plays. For that matter, Bill Shakespeare didn't either. They were written by another man of the same name—and that should be an object lesson in general semantics.)

Let us also set the record straight as to the first symposium on continental drift, because A. Meyerhoff (1972) accorded that niche in history to the American Association of Petroleum Geologists 1926 symposium on continental drift organized by W. J. M. van Waterschoot van der Gracht. This is incorrect. The distinction clearly belongs to Samuel Pepys, FRS.

Pepys' diary for 23 May 1661 reads:

> 'To the Rhenish wine-house in Crooked Lane, and there Mr. Jonas Moore, to us, and there he did by discourse make us fully believe that *England and France were once the same continent* (italics added), by very good arguments, and spoke very many things not so much to prove the Scripture false, as that the time therein is not well computed nor understood. In my black silk suit (the first day I have put it on this year) to my Lord Mayor's by coach, with a great deal of honourable company, and great entertainment. At table I had very good discourse with Mr. Ashmore, wherein he did assure me that frogs and many insects do often fall from the sky, ready formed.'

From the above, it is clear that Pepys accepted the view that England had drifted away from Europe. A symposium in the purest sense of that word, and, as its Greek roots reveal, is a wine-drinking party. To Samuel Pepys, Cheers!, or Drink Hail!—for convening the first symposium on continental drift.

Humpty Dumpty had a Great Fall

Piecing together all of the continents has been described as a jigsaw puzzle. This is hardly correct, because solving a jigsaw puzzle is child's play. In contrast, piecing the continents together would seem to be a Humpty Dumpty problem, for, as you will recall, all the king's horses and all the king's men could not put Humpty Dumpty together again. Similarly, all the world's eggheads have been unable to reconstruct Pangaea.

The congruency of the margins of Africa and South America has always been the inspiration for drift. Alfred Wegener (1922) opened his classical book on drift with: 'He who examines the opposite coasts of the South Atlantic Ocean must be somewhat struck by the similarity of the shapes of the coast line of Brazil and Africa . . . . This phenomenon was the starting point of a new conception . . . called displacement of continents.' Wegener apparently did not realize that many decades ago A. Snider (1859) had already quantitized this fit with nice precision. This is presented in his remarkable book—'The Creation and its Mysteries Revealed: A Work which Clearly Explains Everything Including the Origin of the Primitive Inhabitants of America, etc., etc.' Snider illustrated by lithograph the fit of the New World against the Old World (see cut). It will be noted that, in his closing of the Atlantic Ocean, the match is

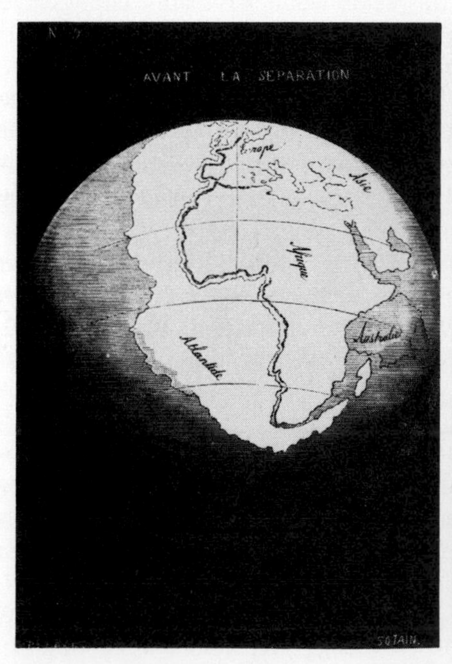

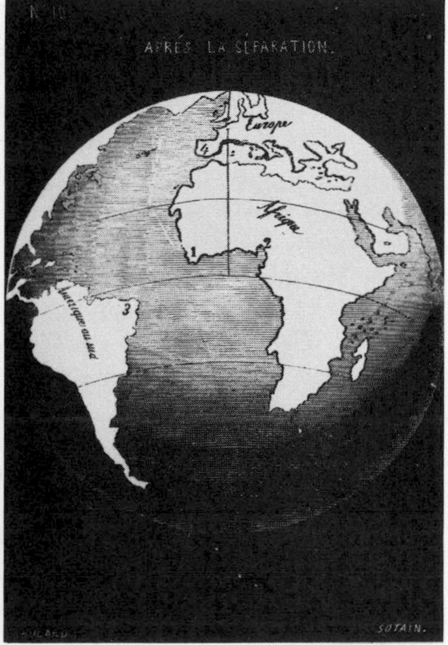

FIGURE 8
. . . Snider's fit of 1858 BC

perfect with neither overlaps nor underlaps. This fit is far superior to those attained by recent workers—for example, the Bullard *et al.* fit (1965). And Snider's fit was obtained in 1859 BC—yes, BC, Before Computers.

Subsequently, Carey (1958) has redone the Snider fit with cartographic precision, juxtaposing the 1000 fm isobaths on a common stereographic projection. This shows that the knee of South America fits snugly into the groin of Africa, that broad swales match broad re-entrant, bumps fit into bays, and even bumps on the bumps fit into bights in the bays. Of Carey's fit, Chester Longwell (1958) remarked, 'If the fit between South America and Africa (is not a gigantic rift), surely it is a device of Satan for our frustration.' Unlike the zealous drifters, Longwell at least has given us a choice. In view of this dilemma, we suppose we must resort to higher authority. In this respect, we can hardly do better than the eminent Sir Harold Jeffries (1970), who writes: 'On a moment's examination, the alleged fit of South America into the angle of Africa is seen to be really a misfit by about 12°.' It would seem that this supposed fit is no more than a persuasive joker.

Perhaps no one has shown better than Meyerhoff & Meyerhoff (1972) the folly of attempting to shoehorn continents together. In a diagram showing a family of squiggles said to represent the outline of Japan he has shown that Japan may be fitted almost anywhere in the world. Another masterwork of meyerhoffiana reveals that the eastern margin of North America, when turned upside down, fits nicely against the eastern coast of Australia. There has been much argumentation over the years about the position of Madagascar when the Indian Ocean is closed—whether this microcontinent fits against Tanzania or against Mozambique. In point of fact, both positions are morphologically poor. A proper fit is achieved only when Madagascar is leapfrogged over Asia and inserted into the Caspian Sea.

Wheels Within Wheels Within Wheels

It has been widely proposed that the continents can be piggybacked about by convection of the mantle. With convection one can do almost anything as the entire process is wonderfully amenable to mathematic manipulation. And there are many modes of convections—toroids, plumes with thunder-heads, helixes, etc.—all of which can be readily explained by arm waving which conjures up explicit models. Furthermore, simple but ingenious experiments can be performed at which the British scientists excel. The result has been a new third school of experimentalists (the earlier schools being the 'baling wire and sealing wax school' and the 'negative experiment school') which may be termed the 'kitchen experiment school'. But in all fairness it should be pointed out that this last-named school was anticipated in America by those early workers who successfully modeled lunar craters by dropping marbles into porridge. And we must mention here the important contribution by an early selenologist on the American frontier who discovered nearly a century ago that, regardless of the obliquity of the angle at which he shot a buffalo, the hole, like a lunar crater, was always round.

In defense of convection, a dimensionless formula is presented (see cut) for oboe and flute. Ideas that are too bizarre to record in writing may yet be sung with perfect propriety. The equation has no particular relevance to the present discussion, but it does add a touch of elegance. In this particular equation, it will be noted that all of

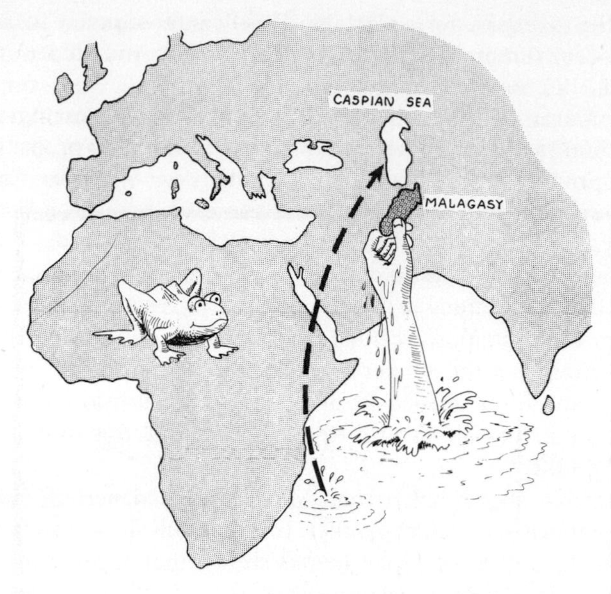

FIGURE 9
. . . a proper fit

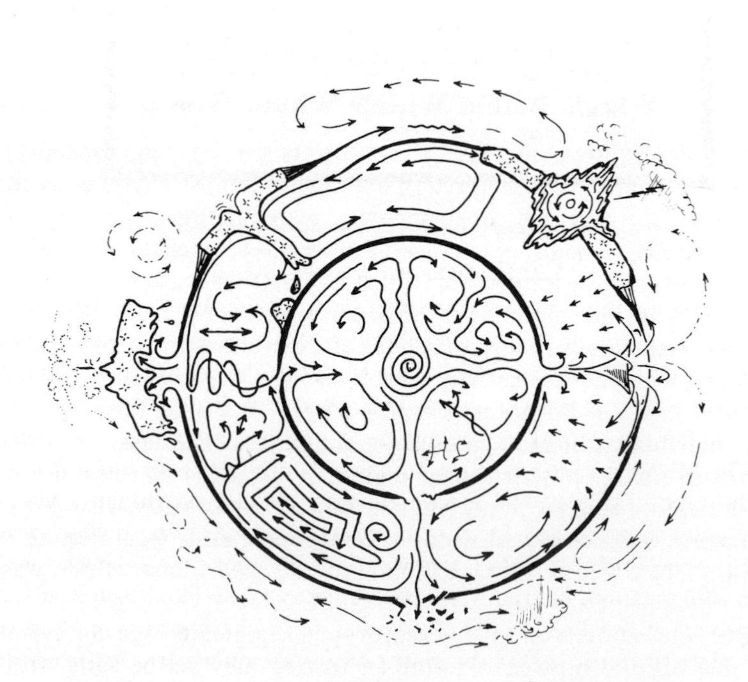

FIGURE 10
. . . wheels within wheels within wheels

the terms are either negligible or trivial; hence the entire equation is inconsequential. If we are ever to deal successfully mathematically with convection in the mantle, we must first establish some simplifying boundary conditions. For example, the rings around Saturn remained beyond understanding until a certain German mathematician cracked the problem by assuming just two simplifying conditions, the first being that the rings are square and the second being that the rings are at an infinite distance from Saturn.

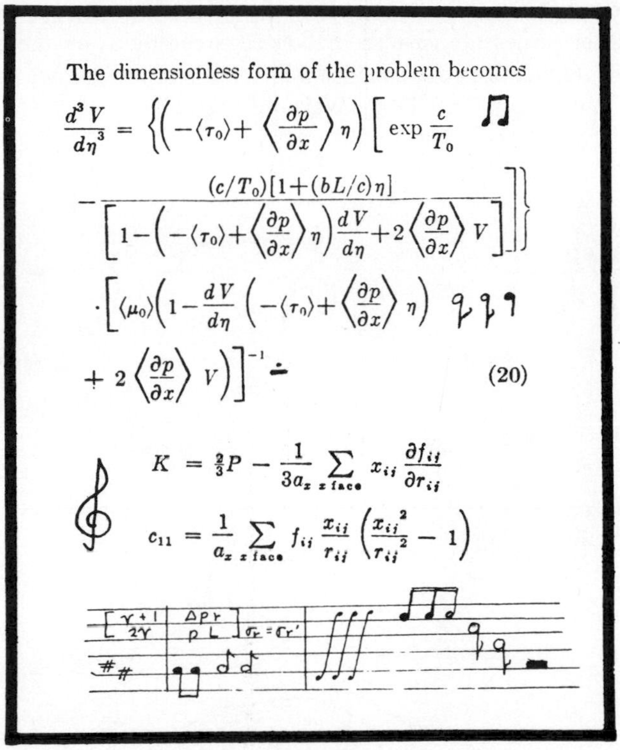

The dimensionless form of the problem becomes

$$\frac{d^3 V}{d\eta^3} = \left\{\left(-\langle\tau_0\rangle + \left\langle\frac{\partial p}{\partial x}\right\rangle\eta\right)\left[\exp\frac{c}{T_0}\right.\right.$$

$$\left.\left.-\frac{(c/T_0)[1+(bL/c)\eta]}{\left[1-\left(-\langle\tau_0\rangle+\left\langle\frac{\partial p}{\partial x}\right\rangle\eta\right)\frac{dV}{d\eta}+2\left\langle\frac{\partial p}{\partial x}\right\rangle V\right]}\right]\right\}$$

$$\cdot\left[\langle\mu_0\rangle\left(1-\frac{dV}{d\eta}\left(-\langle\tau_0\rangle+\left\langle\frac{\partial p}{\partial x}\right\rangle\eta\right)\right.\right.$$

$$\left.\left.+\,2\left\langle\frac{\partial p}{\partial x}\right\rangle V\right)\right]^{-1} \doteq \qquad (20)$$

$$K = \tfrac{2}{3}P - \frac{1}{3a_z}\sum_{z\ \mathrm{face}} x_{ij}\frac{\partial f_{ij}}{\partial r_{ij}}$$

$$c_{11} = \frac{1}{a_z}\sum_{z\ \mathrm{face}} f_{ij}\frac{x_{ij}}{r_{ij}}\left(\frac{x_{ij}^2}{r_{ij}^2}-1\right)$$

$$\left[\frac{\gamma+1}{2\gamma}\ \middle|\ \frac{\Delta p\gamma}{pL}\right]_{\sigma_r=\sigma_r'}$$

FIGURE 11
. . . dimensionless formula for oboe and flute

Continents on the Move

We are told that the earth's crust is a mosaic of six large, 100 km thick plates in relative motion and that the continents are embedded in these plates as passive passengers. The plates never impinge, for they subduct instead, but the continents do, for they are composed of sial which is not geo-degradable. And it is these continental collisions which cause orogenies—linear mountain belts thrown up along the margins of continents. The Miocene collision of the Indian subcontinent with the soft underbelly of Asia is the type example. The three-mile-high Himalayan rampart overlooking the Gangetic plain resulted and India received a bashed-up nose in the process. We may liken this to the English bulldog, renowned for his remarkable tenacity, and some say, but I am sure it cannot be true, stupidity. Anyway, the comparison holds, because the abrupt profile of the English bulldog comes from chasing after parked cars.

A tenet of plate tectonics is that the ocean basins are ephemeral features—they are either opening or closing, but the continents exist forever—suturing up, unsuturing and generally being shunted about. The Pacific plate is moving fast enough to entirely circumnavigate the globe during the Phanerozoic alone. Africa collided with North America in the Devonian 400 m.y. ago, triggering the abrupt Acadian orogeny. This collision must have stored elastic stresses, for 180 m.y. ago Africa took off again in the opposite direction, breaking apart generally along the earlier suture. In view of this historical scenario, we can expect Africa's return at any time, as the Atlantic Ocean cannot grow forever larger and larger. Surely a new plate boundary will form along the American shore and, first of all, a massive slab of the upper mantle will be obducted or thrust across the eastern seaboard, sweeping away the cities like some giant glacier. The implications are catastrophic; time is of the essence and we need now to establish a Continental Drift Early Warning System.

FIGURE 12
. . . Continents adrift

A New Twist

It is now well known that the Sea of Tethys, a vast arm off the former universal ocean of Panthalassa and lying between Gondwana and Laurasia, was the site of a subduction zone or trench along which there was also considerable transform slippage which has been termed the *Tethyan twist* (van Hilton, 1964). This slippage apparently was sinistral in the early opening phases of the North Atlantic Ocean, but became dextral during the Cenozoic, carrying Spain back from the open Atlantic to its present point of contact against Africa.

The Tethyan twist was one of the major scenarios of drift, but perhaps an even more important one, undiscovered until now, was the sinistral *Equatorial twist*. The proof that this twist has occurred lies in the remarkable symmetries achieved. The transformation may be performed using a suitable globe (e.g., the National Geographic 16″ Physiographic Globe) by slipping the northern hemisphere of the earth along the equator dextrally for precisely one quadrant of earth (90°). The result is an entirely new distribution pattern of land and oceans, but one which is entirely realistic. Indonesia is found to be exactly on register with the northern half of Africa, producing the new continent of Afrodesia. The mid-ocean rift of the North Atlantic lines up precisely with that of the southern Indian Ocean, producing an Indo-Atlantic rift. The southern half of Africa falls on register with the northern half of South America, producing an Amerafrica. Farther west, the North Pacific ridge can be connected to the South Atlantic ridge by imposing only a short segment of ridge-ridge transform fault, forming the Paclantic ridge. Proceeding even farther west, one finds the southern portion of South America standing alone in the central Pacific. The skeptic may pounce upon this as an oddity which vitiates entirely our Equatorial twist. But the reverse is true, for this is the lost continent of Mu whose former existence was adduced many years ago by James Churchward. This gentleman found the history of Mu written on secret tablets of stone amongst the archeological ruins of Mexico. He erred in just one respect: Mu did not sink beneath the waves of the Pacific, but instead it suddenly was wrenched eastward to become the southern portion of South America. It seems that some eons ago an errant asteroid, a cosmic cannonball, zeroed in on the earth, and . . . Bang!

The equatorial twist provides no answers, but it poses many questions.

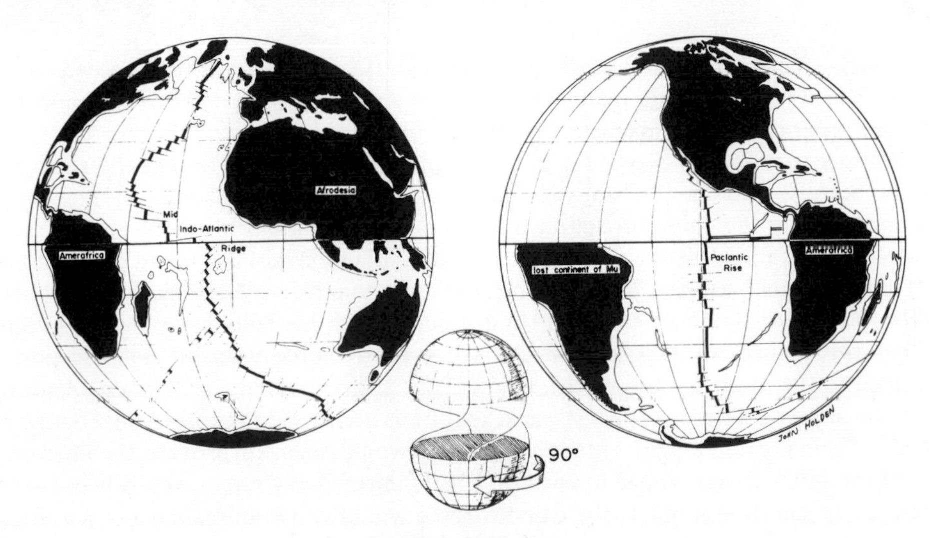

FIGURE 13
. . . the Equatorial Twist

Expansive Thoughts

There is a school of natural philosophy, prominent in the topsy-turvy world of Down Under, that our planet was once much smaller than it is now—and that it expanded to its present size only in the past 200 m.y. A common version of this concept is that the continents were all once joined and even the Pacific was closed, so that the world was without ocean basins. Continents, to the 1000 fm isobath, cover 40% of the earth, so that this earth would have only two-fifths the area of the modern world (510×10^6 km^2). The 'drifting' of continents could then be explained in a manner somewhat similar to the dispersion of points on the surface of an expanding

FIGURE 14
. . . small world of the Triassic

balloon—or like the stars in our expanding universe. However, only the oceans would grow while the continents maintained their original dimensions. The continents then would separate by displacement rather than drift.

Animals once again provide a key and appear to demonstrate that the earth may have been smaller in the Triassic than now. Assuming no change in Big G and accomplishing this miniaturization by phase change, the world of the early Triassic would have had a density of 22 and animals roaming the surface would have weighed $2\frac{1}{2}$ times as much as now. But this is not too unreasonable. Witness the squatty robust morphology of the fossil tetrapod, *Lystrosaurus*, with his belly scraping the ground. Does this stance not suggest that this sheep-sized reptile weighed not 100 pounds (scaled to the modern-sized earth), but actually 250 pounds in the *Triassic*? And why do you suppose all of those deep tracks and trails are found impressed on Triassic red beds. Their greater weight on a smaller earth would seem to provide the answer.

Many other things would happen on a 40% earth. Lazy rivers would become torrents. The nearly tripling of the Coriolis force would cause animals unconsciously to roam in circles of ever increasing concentricity, finally chasing their own tails. The high gravity field probably explains why plants of the Triassic such as the horsetail (Equisetum) were stiffened with stems high in silica. Conservation of the moment of inertia would increase the rate of the earth's spin, so that the Triassic day would have been a mere ten hours long. This may expain why the modern descendants of the Triassic reptiles remain today as the sleepiest of all God's creatures.

Admittedly, for all its advantages there are some aspects of the expanding earth concept which need a bit of tidying up. For example, what to do with all that water if there were no ocean basins to contain that stuff—1300×10^6 km^2. One solution would be to assume that it was all locked up as fresh water in enormous ice caps on the continents and, of couse, there was a great Permo-Carboniferous glaciation. But then what about all that residual salt. . . .

'I have discovered the length of the sea serpent, the price of the priceless and the square of the hippopotamus . . . and how many birds you can catch with the salt in the ocean—187,796,132 if it would interest you to know,' said the Royal Mathematician. 'There aren't that many birds,' said the King. 'I didn't say there were,' said the Royal Mathematician.
(Thurber/*Many Moons*)

Considering sodium chloride alone, the only type useful for taming birds, at 85% of total ocean salts we derive a figure of 43×10^{25} tons. Rounding the number of birds to be tamed to 188×10^{6} provides us with 230×10^{6} tons per bird. With due regard for the Royal Mathematician's wisdom, this does seem a bit much to pour on the tail of each bird.

Carey (1970) would configure the world into eight primitive (Paleozoic) polygons—one for each continent plus an Eo-Pacific polygon, all of which would have moved away from each other over the past 200 m.y., causing a 75% increase (220×10^{6} km^2) in the earth's surface area. This is a story of separating polygons, beginning with a roughly 60% earth, motivated by six convecting mantle toroids plus a jet stream. The old moribund passivity is replaced by a kaleidoscopic crust with gross churnings of the earth's interior. This may seem bizarre, but, as Carey points out, so was Wegener's drift a few years ago when 'the American bandwagon chanted *Ein Marchen*, a pipe dream, and beautiful fantasy.'

Carey adheres to the Egyed principle that the continents have dried off as the ocean was slowly withdrawn into the growing ocean basins. The ultimate cause, according to Egyed, was an expanding atom. The collapsed atoms of the inner core, he supposed, eventually evolved into outer-core substance with constrained electron shells and then into normal mantle rock. In other words, the core evolved into the mantle rather than the core being differentiated from the mantle as is usually supposed.

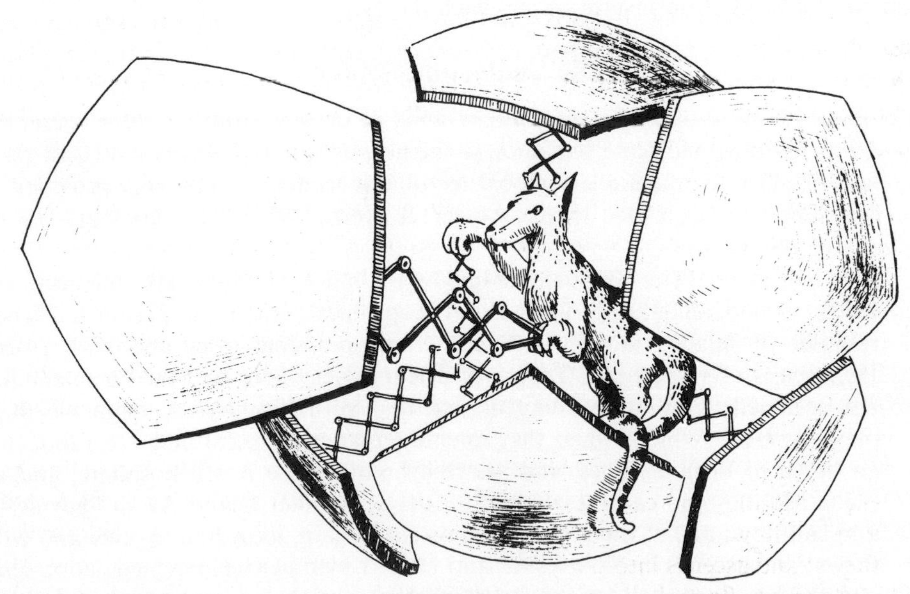

FIGURE 15
. . . Hungarian atom and the Tasmanian tesselated earth

Possibly Egyed of Budapest has provided ultimate key with which to unsuture Carey's zipped-up, tesselated earth of the Paleozoic. A Hungarian expanding atom would force apart the Tasmanian polygonal shells, causing the earth to grow and grow . . . .

In short, an expanding earth remains an exhilarating concept worthy of philosophic contemplation. It seems rather unkind that some unimaginative persons call it 'balloonus balonus.'

Drifters are Alive, . . . and Well

We drifters are now being pushed around, even as we push the continents around. There is a need to be more assertive—and to make some more flat-out bold statements. We need a return to the faith of Galileo Galilei, who, when forced under duress to recant his theory that the Earth rotates on its axis and revolves around the Sun, muttered aloud as he left the Inquisition, 'Eppur si muove' ('but still it moves'). Earlier, in his Sidereus Nuntius of 1610, Galileo wrote, 'The Moon is not perfectly smooth . . . but, on the contrary, it is full of inequities, uneven, full of hollows and protruberances just like the Earth itself.' The philosophers scoffed at this bizarre notion, for it was at variance with the writings of Aristotle. And even if the Moon did have irregularities, it must still be covered with a thick, smooth crystalline layer. Otherwise, how would the Moon so clearly reflect the map of the Earth on its face, with the dark region being our oceans and the light regions being our continents? Galileo retorted with an answer precisely suited to the merit of this argument. 'Let them be careful, for, if they provoke me too far, I will erect, on their crystalline shell, invisible crystalline mountains ten times as high as any I have yet described.'

So, let those fulminous forecasters of fixity take warning. If we devious disciples of drift are not permitted to move our continents a paltry few centimeters each year, then we will drift them several meters each day!

The Geotectonics Creed

In closing, it would seem that continental drift, as derived from sea floor spreading, transform faulting, and plate tectonics, is the new orthodoxy. As statement of faith, the oath of office for the modern global tectonicist seems in order as is provided by the new Geotectonics Creed (Scharnberger & Kern, 1972), with apologies to the Council of Nicea.

> 'I believe in Plate Tectonics Almighty, Unifier of the Earth Sciences, and explanation of all things geological and geophysical; and in our Xavier LaPichon, revealer of relative motion, deduced from spreading rates about all ridges; Hypothesis of Hypothesis, Theory of Theory, Very Fact of Very Fact; deduced not assumed; Continents being of one unit with the Oceans, from which all plates spread; Which, when they encounter another plate and are subducted, go down in Benioff Zones, and are resorbed into the Aesthenosphere, and are made Mantle; and cause earthquakes foci also under Island Arcs; They soften and can flow; and at the Ridges magma rises again according to Vine and Matthews; and ascends into the Crust, and maketh symmetrical magnetic anomalies; and the sea floor shall spread again, with continents, to make both mountains and faults, Whose evolution shall have no end.

And I believe in Continental Drift, the Controller of the evolution of Life, Which proceedeth from Plate Tectonics and Sea-Floor Spreading; Which with Plate Tectonics and Sea-Floor Spreading together is worshipped and glorified; Which was spake of by Wegener; And I believe in one Seismic and Volcanistic pattern; I acknowledge one Cause for the deformation of rocks; And I patiently look for the eruption of new Ridges and the subduction of the Plates to come. Amen.'

References

Beloussov, V. V., 1962. Basic Problems in Geotectonics. McGraw-Hill, London, 465 pp.

Blackett, P. M. S., Bullard, E. and Runcorn, S. K., 1965. A Symposium on Continental Drift. The Royal Society, London, 323 pp.

Bullard, E. C., Everett, J. and Smith, A. G., 1965. The fit of the continents around the Atlantic. *In:* Symposium on Continental Drift, *Phil. Soc.* 1066, *Roy. Soc.*, **258A,** 41–51.

Carey, S. W., 1958. Tectonic approach to continental drift. *In:* Continental Drift, A Symposium, pp. 177–358. Univ. Tasmania Press.

Carey, S. W., 1970. Australian, New Guinea and Melanesia in the current revolution in concept of the evolution of the earth, *Search,* **1,** No. 5, 178–89.

Jeffreys, H., 1970. The Earth: 5th edition. Cambridge Univ. Press, 525 pp.

Longwell, C. R., 1958. My estimate of the continental drift concept. *In:* Continental Drift, A Symposium, pp. 1–12. Univ. Tasmania Press.

Meyerhoff, A., 1972. Continental drift, *Geotimes,* **17,** No. 4, 34–6 (rev. of book by D. and M. Tarling).

Meyerhoff, A. A. and Meyerhoff, H. A., 1972. The new global tectonics: major inconsistencies, *Amer. Assoc. Petrol. Geol.,* **56,** 269–336.

Scharnberger, R. and Kern, E., 1972. Geotectonics creed, *Geotimes,* **17,** No. 1, 9–10.

Snider, A., 1859. La Creation et ses Mysteres devoiles. A. Franck and E. Dentu, Paris, 487 pp.

van Hilton, D., 1964. Evaluation of some geotectonic hypotheses by palaeomagnetism, *Tectonophysics,* **1,** 3–71.

Wegener, A., 1924. Origin of Continents and Oceans (English trans. 3rd edition by J. Skerl), 212 pp. Dutton and Co.

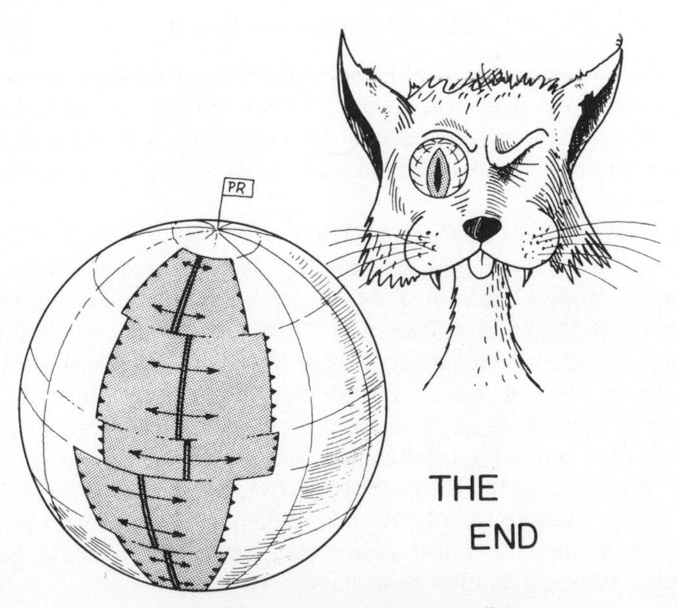

. . . hypotheses, like cats, have nine lives

Taxonomic Index

C

D

Author Index

B

D

E

F

G

H

K

L

S

T

U

V

Y

Z

Subject Index

B

C

D

E

F

I

J

K

L

M

N

S

T

X

Y

Z